THE ROUTLEDGE HANDBOOK OF INTERNATIONAL LAW AND ANTHROPOCENTRISM

This handbook explores, contextualises and critiques the relationship between anthropocentrism – the idea that human beings are socially and politically at the centre of the cosmos – and international law.

While the critical study of anthropocentrism has been under way for several years, it has either focused on specific subfields of international law or emanated from two distinctive strands inspired by the animal rights movement and deep ecology. This handbook offers a broader study of anthropocentrism in international law as a global legal system and academic field. It assesses the extent to which current international law is anthropocentric, contextualises that claim in relation to broader critical theories of anthropocentrism, and explores alternative ways for international law to organise relations between humans and other living and non-living entities.

This book will interest international lawyers, environmental lawyers, legal theorists, social theorists, and those concerned with the philosophy and ethics of ecology and the non-human realms.

Vincent Chapaux is the Research Manager of the Maison des Sciences Humaines of the Université Libre de Bruxelles, Belgium.

Frédéric Mégret is Full Professor and Dawson Scholar, as well as the co-Director of the Centre for Human Rights and Legal Pluralism at the Faculty of Law, McGill University, Canada.

Usha Natarajan is Edward W Said Fellow at Columbia University, USA and International Schulich Law Visiting Scholar at Dalhousie University, Canada.

THE ROUTLEDGE HANDBOOK OF INTERNATIONAL LAW AND ANTHROPOCENTRISM

Edited by Vincent Chapaux, Frédéric Mégret and Usha Natarajan

Designed cover image: André Giroux
Forest Interior with a Painter, Civita Castellana, 1825/1830
Gift of Mrs. John Jay Ide in memory of Mr. and Mrs. William Henry
Donner / Courtesy National Gallery of Art, Washington

First published 2023
by Routledge
4 Park Square, Milton Park, Abingdon, Oxon OX14 4RN

and by Routledge
605 Third Avenue, New York, NY 10158

Routledge is an imprint of the Taylor & Francis Group, an informa business

a Glasshouse book

© 2023 selection and editorial matter, Vincent Chapaux, Frédéric Mégret
and Usha Natarajan; individual chapters, the contributors

The right of Vincent Chapaux, Frédéric Mégret and Usha Natarajan to be
identified as the authors of the editorial material, and of the authors for
their individual chapters, has been asserted in accordance with sections 77
and 78 of the Copyright, Designs and Patents Act 1988.

With the exception of Chapter 6, no part of this book may be reprinted
or reproduced or utilised in any form or by any electronic, mechanical, or
other means, now known or hereafter invented, including photocopying
and recording, or in any information storage or retrieval system, without
permission in writing from the publishers.

Chapter 6 of this book is available for free in PDF format as Open Access
from the individual product page at www.taylorfrancis.com. It has
been made available under a Creative Commons Attribution-Non
Commercial-No Derivatives 4.0 license. Funded by University of
Gothenburg and Lund University.

Trademark notice: Product or corporate names may be trademarks or
registered trademarks, and are used only for identification and explanation
without intent to infringe.

British Library Cataloguing-in-Publication Data
A catalogue record for this book is available from the British Library

ISBN: 978-0-367-85822-3 (hbk)
ISBN: 978-1-032-50858-0 (pbk)
ISBN: 978-1-003-20112-0 (ebk)

DOI: 10.4324/9781003201120

Typeset in Bembo
by Deanta Global Publishing Services, Chennai, India

CONTENTS

List of Contributors   *vii*

Introduction: The Complexities of Anthropocentrism in International Law   1
Vincent Chapaux, Frédéric Mégret and Usha Natarajan

SECTION 1
Unveiling the Anthropocentrism of International Law   11

1 'One Vast Gasoline Station for Human Exploitation': Sovereignty as
Anthropocentric Extraction   13
Mario Prost

2 The Anthropocentrism of Human Rights   35
Frédéric Mégret

3 International Trade Law and the Commodification of the Living   61
Charlotte E. Blattner

4 Anthropocentrism and International Environmental Law   84
Vito De Lucia

5 The Law of the Sea's Fluid Anthropocentrism   102
Godwin E. K. Dzah

6 Ordering Human–Other Relationships: International Humanitarian Law
and Ecologies of Armed Conflicts in the Anthropocene   122
Matilda Arvidsson and Britta Sjöstedt

Contents

SECTION 2

Conceptualising the Anthropocentrism of International Law **143**

7 Anthropocentrism and Critical Approaches to International Law 145
Hélène Mayrand and Valérie Chevrier-Marineau

8 International Law, Legal Anthropocentrism, and Facing the Planetary 165
Anna Grear

9 Towards an Ecofeminist Critique of International Law? 183
Karen Morrow

10 Indigenous Knowledge and International (Anthropocentric) Law: The
Politics of Thinking from (and for) Another World 198
Roger Merino

11 Earth Jurisprudence: Anthropocentrism and Neoliberal Rationality 215
Peter Burdon and Samuel Alexander

12 Global Animal Law, Pain, and Death: An International Law for the
Dominion 231
Alejandro Lorite Escorihuela

SECTION 3

Imagining a Non-Anthropocentric International Law **261**

13 What Would a Post-Anthropocentric Legal System Look Like? 263
Ugo Mattei and Michael W. Monterossi

14 A Non-Anthropocentric Indigenous Research Methodology: The
Anishinabe Waterdrum, Residential Schools, and Settler Colonialism 279
Valarie G. Waboose

15 Non-Human Animals as Epistemic Subjects of International Law 295
Vincent Chapaux

16 Grounding Ecocide, Humanity, and International Law 307
Tim Lindgren

17 Formless Infinite: Law beyond the Anthropocene and the Earth System 323
Elena Cirkovic

Index *339*

CONTRIBUTORS

Samuel Alexander is currently Director of the Simplicity Institute, Melbourne, Australia. Prior to this he spent ten years as a lecturer and researcher at the University of Melbourne, teaching a course called 'Consumerism and the Growth Economy: Critical Interdisciplinary Perspectives' as part of the Master of Environment. He has also been a Research Fellow with the Melbourne Sustainable Society Institute. Alexander's interdisciplinary research focuses on degrowth, permaculture, voluntary simplicity, 'grassroots' theories of transition, post-capitalism, and the relationship between culture and political economy. His current research is exploring the aesthetics of degrowth and energy descent futures. He is author of 20 books, including *Degrowth in the Suburbs: A Radical Urban Imaginary* (2019, co-authored with Brendan Gleeson), *Carbon Civilisation and the Energy Descent Future* (2018, co-authored with Josh Floyd), and *Art Against Empire: Toward an Aesthetics of Degrowth* (2017).

Matilda Arvidsson is an Associate Professor of international law and a senior lecturer in law and theory at the Department of Law, the University of Gothenburg, Sweden. Her scholarship engages with feminist posthuman theory and international law, AI & law, as well as legal history and philosophy.

Charlotte E. Blattner is a Senior Lecturer and Researcher at the Faculty of Law, Institute of Public Law at the University of Berne, Switzerland, specialising in climate law, animal law, and environmental law. Dr. Blattner holds a doctoral degree in international law and animal law from the University of Basel, Switzerland, and an LLM from Harvard Law School, USA. Alongside numerous articles, she is the author of *Protecting Animals Within and Across Borders: Extraterritorial Jurisdiction and the Challenges of Globalization*, published in 2019. She is a co-editor, with Kendra Coulter and Will Kymlicka, of the volume *Animal Labour: A New Frontier of Interspecies Justice?* also published in 2019. For her outstanding work in animal law, she received the renowned Marie Heim-Vögtlin Prize 2020 by the Swiss National Science Foundation.

Peter Burdon is Associate Professor at the Adelaide Law School, Australia, and Deputy Dean (Learning and Teaching) for the Faculty of Arts, Business, Law and Economics. Peter is an expert in environmental theory and has written and edited books on Earth Jurisprudence and Earth Democracy. In 2021, his monograph *Earth Jurisprudence: Private Property and the Environment* was translated into Mandarin and distributed by Routledge and the Commercial

Press. In 2017, Peter published a book on Hannah Arendt and the trial of Nazi war criminal Adolf Eichmann.

Vincent Chapaux is the Research Manager of the *Maison des Sciences Humaines* of the Université libre de Bruxelles (ULB), Belgium. Before holding this position, he was a lecturer in International Law at that same university and a guest lecturer in International Animal Law at the University of Ottawa, Canada. He is currently a guest lecturer in Animal Law at the Université de Liège. He holds a PhD in political science (ULB), a *master spécialisé* in international law (ULB), and conducted his postdoctoral research at both New York University, USA (Fulbright Grantee) and Université du Québec à Montréal, Canada.

Valérie Chevrier-Marineau completed a bachelor degree in International Relations and International Law at the University of Quebec in Montreal, Canada, and at The Hague University of Applied Sciences in the Netherlands, in 2018 (BA). With bidisciplinary training and a marked interest in critical approaches to law, she completed her Master of Laws (LLM) under the supervision of Professor Hélène Mayrand at the University of Sherbrooke, Canada (2020). Her research interests focus more broadly on the contribution of critical positioning in environmental transnational law. Valérie also holds a Bachelor of Laws (LLB) from the University of Sherbrooke (2022) and is now acting as a Law Clerk at the Federal Court of Appeal in Ottawa.

Elena Cirkovic is a legal scholar currently working on her project at the University of Helsinki, Finland, and the Max Planck Institute Luxembourg for International, European, and Regulatory Procedural Law, entitled "Anthropocentrism and Sustainability of the Earth System and Outer Space (ANTARES)". The project aims to connect Earth System and Outer Space with the application of complex systems approaches, critical theory, and posthuman approaches in contemporary thought. Her work has been funded by Kone Foundation of Finland, Arctic Avenue administered by the University of Helsinki and Stockholm University, Sweden, and the Minerva Center for the Rule of Law under Extreme Conditions at the University of Haifa, Israel. Dr. Cirkovic has a transdisciplinary background, and her work focuses on international public and private law, outer space law, climate law, human rights, critical theory, and philosophy, as well as sustainable design, architecture, and the arts.

Godwin E.K. Dzah is an Assistant Professor at the Faculty of Law, University of Alberta, Canada where he teaches international law and corporate law as well as energy law and international environmental law. His research interests are in public international law, international environmental law, sustainable development and climate change, law of the sea, international law and global corporations, and critical approaches. He is interested in how legal concepts, ideas, and knowledge derived from non-dominant systems of law can help in reimagining international law. He holds political science and law degrees from the University of Ghana; Harvard Law School, USA; and the University of British Columbia, Canada, and received his qualifying certificate in law from the Ghana School of Law. He previously held a Law of the Sea fellowship at the Schulich School of Law at Dalhousie University, Canada, and the United Nations Secretariat in New York, and was recently a postdoctoral fellow at the Osgoode Hall Law School, York University, Canada.

Anna Grear is Professor of Law and Theory at Cardiff University, Wales, UK. Anna is founder and Editor-in-chief of the *Journal of Human Rights and the Environment* and was the founding Director of the Global Network for the Study of Human Rights and the Environment.

Contributors

Tim Lindgren is a doctoral candidate at Melbourne Law School, University of Melbourne, Australia. He is a member of the Institute for International Law and the Humanities (IILAH) and a Visiting Scholar at the Lauterpacht Centre for International Law, University of Cambridge, UK. He researches and teaches in international law. His research concerns international law and the environment, peoples' tribunals, colonialism, and the performance of international law in informal spaces, including histories of development and international law. He has been a returning lecturer and faculty member for the Oxford Consortium for Human Rights (2018 and 2019) and a Teaching Fellow at Melbourne Law School (2020 and 2021). He is also an editor for the *Journal of Human Rights and the Environment*. He holds a BA in Global Studies (*summa cum laude*) from Westminster University, UK, including a Master in International Law (distinction) from the School of Oriental Studies (SOAS), where he was awarded Best Overall Course Performance.

Alejandro Lorite Escorihuela is a Professor of International Law at the University of Québec in Montréal, Canada.

Vito De Lucia is a Professor of International Law at the Faculty of Law, UiT The Arctic University of Norway and a full member of the Norwegian Centre for the Law of the Sea (NCLOS). His most immediate research interests are located at the intersection of critical theory, international law, and ecology. His current research agenda focuses on ocean commons and the ongoing negotiations towards a global treaty on marine biodiversity in areas beyond national jurisdiction, where he has been an observer since 2016, and on Arctic governance. He is author of *The 'Ecosystem Approach' in International Environmental Law. Genealogy and Biopolitics* (2019).

Ugo Mattei is full Professor of Civil Law at the University of Turin, Italy and Distinguished Professor of Law and Alfred and Hanna Fromm Chair in International and Comparative Law (emeritus), at the UC Hastings Law School, San Francisco, California. He is also a full member of the International Academy of Comparative Law, a general editor for the Common Core of European Private Law since its beginning in 1993 and he serves as the Academic Coordinator of the International University College of Turin, Italy. He has published more than thirty books and one hundred other publications in numerous languages, including *The Commons. A Manifesto* (2011), *The Ecology of Law: Toward a Legal System in Tune with Nature and Community* written with Prof. Fritjof Capra (2015), and *The Turning Point in Private Law. Ecology, Technology and the Commons*, written with Professor Alessandra Quarta (2019).

Hélène Mayrand has been Professor at the Faculty of Law of the Université de Sherbrooke in Canada since 2012. She has been Vice-Dean of Research and Graduate Studies since 2022. She holds a Bachelor of Laws (LLB) from the University of Sherbrooke, a Master of Laws (LLM) from Cambridge University, UK, and a Doctor of Juridical Science (SJD) from the University of Toronto, Canada. She teaches immigration and refugee law, legal interpretation, and theories of international law and international relations. She is a co-founder of the Critical Legal Research Laboratory. Professor Mayrand's research focuses on environmental law, both international and domestic, as well as immigration and refugee law. Her research interests include the development of critical approaches to law, the relationship between law and power, and theory and practice.

Frédéric Mégret is a full Professor and William Dawson Scholar at the Faculty of Law, and the co-Director of the Centre for Human Rights and Legal Pluralism, Faculty of Law, McGill University, Canada. From 2006 to 2016 he held the Canada Research Chair on the Law of Human Rights and Legal Pluralism. Before coming to McGill, he was an assistant Professor at the University of Toronto, Canada, a research associate at the European University Institute, and an attaché at the International Committee of the Red Cross. He is the editor, with Philip Alston, of the second edition of *The United Nations and Human Rights: A Critical Appraisal* and the co-editor of the *Oxford Handbook of International Criminal Law*. His research interests are in general international law, the laws of war, human rights, and international criminal justice.

Roger Merino is an Associate Professor at the Public Management School of *Universidad del Pacífico* in Lima, Perú. He holds a PhD and a MSc in Globalization and International Policy from the University of Bath, UK. He has been a Visiting Scholar at Harvard's Institute for Global Law and Policy, USA. His research areas include political ecology, international environmental governance, and Indigenous rights. He has published research articles in leading journals, such as *World Development, Geopolitics, Third World Quarterly, Global Environmental Politics, among others*. His monograph *Socio-legal Struggles for Indigenous Self-Determination in Latin America: Reimagining the Nation, Reinventing the State* was published by Routledge in 2021.

Michael W. Monterossi is a post-doc researcher in Private Law at the University of Turin, Italy. After receiving his PhD degree in Private Law from Ca' Foscari, University of Venice, Italy, he worked as a researcher at the University of Lucerne, Switzerland, on a project on the protection of the interests of future generations and other non-human entities in Private Law. He is the author of the book *L'orizzonte intergenerazionale del diritto civile. Tutela, soggettività, azione* (The Intergenerational Horizon of Private Law. Protection, Subjectivity, Action, 2020) and of numerous publications both in national and international law reviews. His research focuses on the evolution of Private Law in light of ecological and climate mutations as well as on the Private and Commercial Law aspects of the regulation of data and new technologies.

Karen Morrow LLB, LLM was educated at the Queen's University of Belfast and King's College London, UK. She has lectured at Buckingham, Durham, and Leeds Universities, UK and at Queen's University of Belfast. She has been Professor of Environmental Law at Swansea University, UK since 2007. Her research interests focus on theoretical and practical aspects of public participation in environmental and sustainability law and policy and on ecofeminism, gender, and the environment. Her current work focuses on ecofeminist approaches to the challenges of the Anthropocene, with particular emphasis on climate change. She is a series editor for *Critical Reflections on Human Rights and the Environment* and a member of the international advisory board for the cross-disciplinary *Gender and Environment* series (Routledge). She serves on the editorial boards of the *Journal of Human Rights and the Environment*, the *Environmental Law Review*, and the *University of Western Australia Law Review*.

Usha Natarajan is Edward W Said Fellow at Columbia University, USA, and International Schulich Law Visiting Scholar at Dalhousie University, Canada. Her research is interdisciplinary, utilising Third World Approaches to International Law (TWAIL) to provide an interrelated understanding of development, environment, migration, and conflict. She is a leader in the TWAIL movement and a founding editor of the *TWAIL Review*. With over 40 publications, her research is recognised by leading disciplinary awards, fellowships, and global research grants. Prior to academia, she worked with the United Nations.

Mario Prost is Senior Lecturer and Programme Director of the LLM in International Law at the Keele Law School, UK. He obtained his first degree and LLM from the Sorbonne Law School (Université Paris 1), France, and his PhD from McGill University, Canada, where he was a Liberatore Major Fellow. His dissertation on the fragmentation of international law was awarded the Association of Quebec Law Professors annual prize for best doctoral thesis (2009).

Mario is a member of the executive board of the European Society of International Law and a founding member of its interest group on international environmental law.

Britta Sjöstedt is a Senior Lecturer in Environmental Law at the Faculty of Law at Lund University, Sweden. Her research interests relate to legal issues in regard to environmental protection, armed conflict, environmental peace building, and sustainable investments.

Valarie G. Waboose is an Anishinabe Kwe from the Bkejwanong Territory (also known as Walpole Island First Nation). She is a great-grandmother of two, grandmother of seven, and mother of two daughters. She is a second-generation survivor of two parents who attended the Shingwauk Indian Residential School. She completed her PhD dissertation (Trent University, Canada) on the impacts of the compensation processes upon residential school survivors. She is also a second-degree member of the Midewiwin Lodge on Manitoulin Island. Before completing her doctoral degree, she obtained a BA at Trent University, an LLB at Windsor Law, and LLM at Osgoode Hall Law, Canada. She is currently Assistant Professor at the University of Windsor, Faculty of Law. Her research interests include Indigenous legal traditions, orders and methodologies, child welfare, reconciliation, and the residential school legacy as well as Indigenous justice systems.

INTRODUCTION

The Complexities of Anthropocentrism in International Law

Vincent Chapaux, Frédéric Mégret and Usha Natarajan[1]

This book examines the relationship between anthropocentrism – the idea that human beings are socially and politically at the centre of the cosmos – and international law. International law has not been understood by international lawyers historically as anthropocentric, nor has the study of anthropocentrism particularly focused on the role of international law in diffusing hegemonic ideas about anthropocentrism. We believe that taking seriously the intersection between international law and anthropocentrism will aid better understanding of both. Central questions include: Is international law anthropocentric? If so, in what way? How central is anthropocentrism to international law's self-constitution? Can and should international law seek to transcend this bias? How is it helpful to understand international law by foregrounding its anthropocentrism, when compared with existing modes of critique that examine how international law privileges some humans over others, or focus on humans' relationship to the state? What are some of the risks and blind spots of a focus on anthropocentrism? How is anthropocentrism structured through international legal structures?

The study of anthropocentrism has of late gained prominence in the context of the mass extinction of species, climate change, and environmental degradation more generally, from the perspective of social theory, philosophy, or ethics (Thompson and Barton, 1994; Coward, 2006; Callicott, 1984; Norton, 1984; Plumwood, 2005; Nagel, 1974). It is also implicated in the study of the relationship between human and non-human animals.[2] Within the legal field, there have been considerable efforts to better conceptualise the general relationship of anthropocentrism to law (Gillespie, 2014; Garner, 2005; Regan, 2001; Otomo and Mussawir, 2013; Maloney and Burdon, 2014; Davies, 2017, chap. 10; Burdon, 2015; Philippopoulos-Mihalopoulos, 2011; Nedelsky, 1990; Naffine, 2009). Indeed, it has been argued that 'the separation and hierarchical ordering of the human and non-human worlds constitutes the primary assumption from which most Western legal theory begins' (Burdon, 2010).

At the same time, anthropocentrism needs to be problematised as, in part, dependent on certain legal forms, including the international legal form. Anthropocentrism is a historical construct with cultural and historical provenance in Western Europe. Neglecting that provenance and failing to connect it to the history of colonialism and imperialism may lead to vacuous critique of anthropocentrism. Indeed, such critique risks reproducing bias if used to obfuscate or distract from global inequalities of wealth and power. After all, global challenges, such as climate change, mass species extinction, and increases in economic inequality, are caused by a

DOI: 10.4324/9781003201120-1

small percentage of humanity consisting of privileged cultures, races, genders, classes, castes, and other groups.

A thorough examination of anthropocentrism in international law has been lacking. Anthropocentrism is at times addressed by those branches of international law that perceive themselves to be directly concerned with animals and the environment. For the most part, these disciplines consider themselves to be less anthropocentric than the rest of international law. Nonetheless, some scholars in these fields are attentive to the critique that international norms striving to protect the non-human may themselves do so in a human-centric way (De Lucia, 2015). International environmental laws may protect the non-human environment in ways that subtly or explicitly reinscribe, and even extend, international law's anthropocentricism. Similarly, animal rights promotion may involve a subtle double move of critiquing the restriction of rights solely to humans but not interrogating the anthropocentrism of the notion of rights.

Where the anthropocentric bias of international law has been problematised, different conversations have not been systematically connected to each other. It is hoped that this collection can be an opportunity to enable fruitful cross-conversations. For instance, the emerging field of international animal law and its anthropocentrism critique (Escorihuela, 2011; Kelch, 2017; Peters, 2016; Sykes, 2016; Peters, 2021) has been partially disconnected from efforts to transcend international environmental law's anthropocentrism (White, 2014; Benton, 1993; Varner, Light, and Rolston, 2003; Kotzé and French, 2018). This separation reproduces a similar division in domestic law in which, for example, animal rights scholars are not particularly in dialogue with deep ecology scholars.[3] This disconnection neglects the deeper potential of anthropocentrism as a common critique and foregoes the opportunity to deepen and refine each particular approach to the issue through mutual collaboration and inquiry. What is left out, moreover, is a critique of the broader and foundational anthropocentrism of international law within which some of its branches are embedded.

A critique that includes not only those branches of international law that perceive themselves to be directly concerned with animals and the environment but encompasses also the anthropocentrism of *general* international law, is one that, by contrast, can make more headway. It should include an examination of the anthropocentrism of international economic trade and investment law, international human rights law, the laws of war, the law of the sea, and other branches of international law not traditionally viewed as having a particularly strong connection to non-human animals or the environment. This collection also considers the potential pitfalls of a critique of anthropocentrism – its flattening of human diversity and inequality; its implications for hierarchies of gender, race, and class; the neo-fascist tendencies of certain ecocentric approaches; and whether and how to negotiate these shortcomings. The editors have sought to bring together authors working in different fields of international law to consider the implications of anthropocentrism in their area and the broader discipline. The book also connects the two existing strands of the critique of anthropocentrism – that inspired by the animal rights movement and that emanating from deep ecology and its broader focus on both animal and non-animal beings.

We argue that the move towards a shallower form of anthropocentrism in international law – one, for example, that typically grants a mere instrumental value to the human environment and non-human animals – is no substitute for a deeper engagement with international law's anthropocentric blind spots. Instead, we direct attention to the way in which international law obscures its own implication in the disasters it seeks to remedy (Natarajan and Dehm, 2022; Natarajan and Khoday, 2014; Alam *et al.*, 2015; Gonzalez, 2015). Such deeper engagement is especially urgent in a context where consensus is building that a simplistic approach to environmental and animal protection that does not challenge international law's broader and systemic anthropocentric biases will not solve current global environmental and animal welfare crises.

Introduction

Several reasons may explain the relative lack of scholarly attention to the issue of anthropocentrism within general international law. To begin with, one of the central disciplinary debates in international law is structured by opposition between international law as state-centred versus international law as human-centred, with the human-centeredness of international law often presented as the losing and more progressist prong of that equation (Trindade, 2010). In this context, the supposed human-centric character of international law — or at least the constitution of an international law that fundamentally serves the needs and aspirations of human beings — is always portrayed as a work-in-progress, perhaps even a project of distant realisation. Such a project, inspired as it is by humanist ideals, has long had a strong claim on the discipline of international law and is treated as the yardstick by which disciplinary progress is to be evaluated. This is evident in the concept of human rights as well as in notions such as 'crimes against humanity' or the 'common heritage of humankind'. Humanism is generally contrasted with inhumanity, with the humanist idea treated as the highest ambition of international law. That project unfolds not only in relation to and opposition with the state, but according to its own dialectics of ever-greater inclusion into the human species. Thus, defining what counts as 'human' has been described as the single most important force in the unfolding of the human rights project (Rorty, Shute, and Hurley,1993).

What is interesting in this dichotomy is what it excludes: Everything beyond human beings and their institutions. What of the possibility that international law might manifest, through and beyond its constantly reproduced human/sovereign dichotomies, a broader project of human domination? The humanist project, precisely through what it excludes, may be blind to its own inhumanity to other species and to nature. Such inhumanity ultimately may be related to the way in which humans treat each other and thus is a subject worthy of interrogation by adherents to the humanist project. The project of extending humanity's boundaries, indispensable as it may be from within an evolving humanist canon, has its own evident limits, as it remains squarely within the continuous actualisation of humanistic philosophy (Argyrou, 2005). Crucially, examining the anthropocentrism of international law provides insights into the very intraspecies distributive debates that have long preoccupied at least some international lawyers.

International law has sought over past decades to transcend many of its biases but, in some ways, anthropocentrism is the ultimate disciplinary blind spot: The one question almost never asked and that creates discomfort when it is. Where, for example, feminist critique has underlined the androcentrism of international law, and where postcolonial critique has stressed its Eurocentrism, then what might be termed an anthropocentric critique of international law seeks to question the unproblematically central position given to the human being and its institutions in international law.

This book examines thoroughly the relationship between anthropocentrism and international law. It builds on an emerging foundation: First, conversations in international law that have sought to problematise the ill treatment of animals or the degradation of the environment within processes triggered by international law itself and, particularly, a larger history of international law's anthropocentric construction of nature (Natarajan and Dehm, 2022; Natarajan and Khoday, 2014; Porras, 2014; Watson, 2011; Humphreys and Otomo, 2014). Second, efforts to conceptualise what it means to speak of a non-anthropocentric law, based in particular on attempts to theorise a non-anthropocentric philosophy and ethics, have been going on for decades but have so far left international law outside their critical focus (Maloney and Siemen, 2015; De Lucia, 2013; Burdon, 2011; Bosselmann, 2011; Cullinan, 2011; Cullinan and Falstrom, 2008; Warren, 2006). Third, actual experiences in trying to transcend the anthropocentric bias of international law (Sydes, 2014; Rogers and Maloney, 2017; Margil, 2014; Maloney, 2016; Wright, 2013; Maloney, 2015; Warren, Filgueira, and Mason, 2009; Barlow, 2010), notably those driven by Indigenous legal traditions that have been developed domestically and transnationally and that could contribute to reshaping the conversation about international law (Charpleix, 2018; Albro, 2018; Watson, 2017; Humphreys, 2017; Watson, 2014; Graham, 1999; Tomas, 2011; Black, 2010).

The goal of the book is not to duplicate these conversations but to galvanise and refine them by uniting them under the common analytical framework of anthropocentrism. As its title suggests, the collection does not simply assume that international law is anthropocentric or that such a diagnosis, in its generality, is necessarily or always useful. Indeed, it proposes that we should problematise and historicise the anthropocentric critique itself to see it fully for what it is. What matters is not just assessing whether anthropocentrism characterises international law in the abstract, but understanding where the anthropocentric critique comes from, how it operates, and what its consequences are for our understanding of international law. What does 'anthropocentrism' as a concept shed light on and what does it hide?

This edited collection proposes an exercise in analysis, critique, and reimagination. It understands anthropocentrism as the doctrine that institutions, notably legal institutions, should be directed above all to benefit human beings, and that these beings constitute the ultimate yardstick of all value. In that respect, although it may not be fully articulated, anthropocentrism is more often than not founded on a combination of notions including that: (i) Human beings are a special species based on certain unique characteristics; (ii) human beings hold a superior place in a hierarchy of the natural order; (iii) human beings are conceptually separable from nature; and (iv) human beings have special claims to its direction and governance. Specifically, as we shall see, anthropocentrism manifests itself in the law as an insistence that only human beings (or at least human institutions) can be *subjects* of the law and have a broader responsibility in the law's direction and enforcement.

There are at least three broad approaches one could take to international law and anthropocentrism. The first is that anthropocentrism is a project largely unrealised: We are simply not there in terms of meeting the needs of all human beings in such a way as to make humanity a meaningful concept. International law is only imperfectly a humanist project. The second is that, notwithstanding its limitations and the imperfectly realised nature of a humanist international legal order, anthropocentrism is still the fundamental lens through which all international law is analysed. Even as less-than-fully realised, therefore, anthropocentrism is still a problem that needs to be addressed because it has a descriptively narrow and inaccurate understanding of nature and non-human animals, and because it is normatively problematic, endlessly reproducing modernity's appropriating and instrumentalising move. The third approach, which is the one we take in this collection, is that both these approaches are interconnected: The attention to international law's anthropocentrism, even as it captures something, also serves to distract from the degree to which interspecies and environmental issues are profoundly connected to intra-human ones.

The collection understands anthropocentrism not as a universal or atemporal notion. Rather, especially as it manifests itself within the international law discipline, its core assumptions derive from Western modernity and the European Enlightenment (Last, 2018; Plumwood, 2005; Rist, 2003; Argyrou, 2005). This is not to suggest that other cultures have not produced anthropocentric narratives, but rather that Western modern culture has a unique and unrivalled imprint on the world of international law (Gathii, 2021; Chimni, 2017; Anghie, 2007; Chimni, 2003; Gathii, 1998). To discuss the history of international law, then, is to discuss the history of anthropocentrism and vice versa. If anthropocentrism sheds light on the making of international law, international law has been one of the means through which *a particular kind of* anthropocentrism has unfolded on a global scale through colonialism, imperialism, and other forms of norm-diffusion. Cognisant of this inextricable link, the collection adopts a decolonial stance.

Introduction

The book is divided into three parts. The first part explores the extent to which international law is anthropocentric. Mario Prost begins with sovereignty as an anthropocentric concept. International law is fundamentally premised on an anthropocentric worldview, in that sovereignty from the outset is premised on the transformation and subjection of nature. The exploitation of nature is, in fact, mandated by sovereignty; and those societies that failed to abide by that injunction, could be considered not to possess sovereignty.

For Frédéric Mégret, the anthropocentrism of sovereignty is a function of an even deeper anthropocentrism: That of human rights. For all their humanist mystique understood as an effort to improve the condition of humans in society, human rights are based on a deeper Western claim about the primacy of the human over all other realms. This cautions against thinking that one can extirpate international law from the global environmental crisis because the two are in a sense two facets of the same coin. Mégret hints at the need to further problematise humanity and its exclusions as a way of confronting international human rights law's inherent anthropocentrism.

Perhaps less surprisingly, international trade law is based on a commodification of the living. But Charlotte Blattner argues that efforts to protect non-human animals remain limited by their legalism and the limitations of trade law. This is because, notwithstanding such efforts, commodification of non-human animals remains the key to understanding international trade law. Non-human animals are primarily conceived as goods and non-trade concerns carefully insulated from the core trade regime. Ultimately, underlying beliefs, such as endless economic growth and the emphasis on trade law as a law made by humans and for humans, caution against breaking free from the anthropocentric mindset. The protection of non-human animals remains at best an afterthought in the broader legal economy of trade.

One might think that international environmental law would better respond to the challenge of moving beyond anthropocentrism but, as Vito de Lucia argues, its inherent anthropocentrism may explain its inability to fight global environmental degradation. This is because of a deep contradiction, or at least tension, at the heart of an international environmental law designed to limit harm to nature but simultaneously enables it. Ominously, de Lucia suggests that not even the turn to ecocentrism can fundamentally rescue international environmental law from its modernist anthropocentrism, given its indebtedness to a biopolitics of exploitation/conservation of nature.

For Godwin Dzah, the law of the sea promulgates its own form of anthropocentrism. As he puts it, 'the law of the sea has hardly ever been "of" or "for" the sea' but thoroughly embeds anthropocentric values and interests, notably the extension of human control under the guise of ocean governance. The oceans, in fact, provide merely a sort of background to a thoroughly human narrative about the sea's finality, one powerfully framed by Eurocentrism, imperialism, and colonialism. This is evident in the 1982 UN Convention on the Law of the Sea but also in more recent developments, such as bioprospecting and the exploitation of marine diversity. Dzah suggests ways in which one might move international law beyond its anthropocentrism, by acknowledging its responsibility in ecological deterioration and unrestrained resource extraction.

The last chapter of this first part sees Matilda Arvidsson and Britta Sjöstedt explore the 'human-other' ordering potential of international law in times of armed conflict and disaster. Taking as its case studies the 'natural' environment and AI-powered swarms of drones, this chapter suggests that international humanitarian law could develop in a more post-anthropocentric direction. In adopting a post-humanist feminist and post-anthropocentric lens, Arvidsson and Sjöstedt study how international humanitarian law orders the relationship of the human and its other during armed conflict. The notion of the 'war machine', borrowed from Deleuze and

Guattari, serves to tease out ways in which less anthropocentric understandings of international humanitarian law might surface under the banner of post-anthropocentrism to protect a range of entities from violent forces.

The second part of the book considers varied ways of conceptualising anthropocentrism in international law towards a fuller understanding of this phenomenon. Hélène Mayrand and Valérie Chevrier-Marineau provide an overview of diverse critical approaches to international law and anthropocentrism. From structuralism to Marxism, from Third World Approaches to International Law (TWAIL) to feminism, from environmental justice to posthumanism, they find almost all critical approaches to international law are both descriptively and normatively anthropocentric. The authors point to potential in posthuman critical approaches to law for moving beyond deconstruction and founding law on an acentric conception of human, non-human, and inhuman bodies.

Anna Grear asks crucially what the stakes are of legal anthropocentrism. Through reflecting on the colonial foundations of the international legal order, Grear describes the 'subject-at-the-centre' of legal anthropocentrism: The privileged individual, the dominant state, and the powerful transnational corporation. By situating patterns of oppression enacted by legal anthropocentrism alongside patterns of coloniality, this chapter renders visible the role of the international legal order in the emergence of the Anthropocene. Finally, Grear reflects on proposals for a new international legal imaginary in the form of Earth system law and governance, and what insights emerge when such proposals are brought into theoretical encounter with the agentic forces of the planetary.

Karen Morrow deploys ecofeminism as one lens to help us to grasp more fully the structural roots and nature of the Anthropocene predicament and to develop effective responses to it. Ecofeminism focuses on shared problematic impacts of dualistic and hierarchical thinking, and the 'othering' of both inter-human and human–environment relationships. Morrow points out that, while such attitudes and assumptions have always been problematic, now in the vanguard of the Anthropocene they are untenable because understanding and acting on the entwined inter-human and human–environment relationships will determine the future of our species. Through ecofeminism, Morrow posits that a sustainable basis for inter-human and human–environment interactions and the social ordering that embodies them is fundamentally relational in nature.

Disciplinary critique oftentimes urges prioritising knowledges outside the Eurocentric and patriarchal mainstream, but the international law response is usually an appropriative commodification of subaltern knowledges in a manner that reinscribes existing hierarchies. Roger Merino traces how the gradual recognition of Indigenous knowledges within international law has reaffirmed hegemonic ideas, institutions, and regulations. While there is some recognition of Indigenous knowledge as a human right that must be respected, as an asset to be protected under property rights rules, and as a condition to effectively implement environmental and development initiatives, there is less engagement with how Indigenous knowledges undermine and fundamentally unmake concepts such as human rights, property, environment, and development. Merino shows how international law has essentialised, commodified, and instrumentalised Indigenous knowledges, treating them as a second-level system of knowledge unable to support broader international rules to address contemporary ecological crises.

Peter Burdon and Samuel Alexander explore the meaning and uses of anthropocentrism in the literature of Earth Jurisprudence and ecological law. They break down anthropocentrism into its component parts – perceptual, descriptive, and normative – and argue that the Anthropocene compels humans to recognise descriptive anthropocentrism and their place as subjects capable of rupturing the Earth system. From this insight they draw not a hubristic normativity but rather the need to engage with the Anthropocene with humility. They caution that the dominant logic of neoliberalism has important implications for how anthropocentrism is engaged with and understood today. In this context, Burdon and Alexander point out how rights arguments such as the rights of nature can be co-opted by neoliberalism to further capital accumulation. They conclude that turning to alternative knowledges is essential for attaining a deeper disciplinary understanding of environmental justice.

Alejandro Lorite Escorihuela explores the ambivalent understanding of anthropocentrism that animates global animal law as a scholarly and policy-making project: To free animals from anthropocentric cruelties or indifference, we make them the object of further regulatory proliferation. Anthropocentrism in such a project is understood as a form of injustice characterised as the legal system's irrational preference for human interests over those of other animals, where additional rules will correct the ethical imbalance and increase welfare. However, Lorite points out that the real source of the harm done to animals is not discrimination but objectification. Animals are not the victims of speciesism, understood as irrational treatment under liberal rule. Rather, they are the victims of dominionism, understood as rational treatment under totalitarian rule. Thus, this chapter asks who we are really trying to liberate through an international law of animal liberation and whether, in the name of animal liberation, we are rather expanding the reach of anthropocentric law and rights as instruments of totalitarian terror.

The third part of this collection is prospective. It explores how international law could be conceptualised differently. Ugo Mattei and Michael Monterossi open this final part by asking what a post-anthropocentric legal system would look like. The authors show that international law is not an entirely anthropocentric system. Instead, they insist that it can be described both as an embodiment of short-term economic logics *and* a laboratory for post-anthropocentric worldviews. Relying on the increasing number of legal suits regarding climate change, they sustain that law can become a crucial tool in remodeling current international systems, particularly when inspired by the work of domestic courts.

In the subsequent chapter, Valarie Waboose relies on the epistemological views of the Anishinabe people of the Great Lakes region of the North American continent to offer two important elements that could help international lawyers and scholars go beyond the anthropocentric structure of international law. First, she lays out in detail the Anishinabe worldview, a cosmology profoundly non-anthropocentric in its foundations. Second, she explains how she drew on her non-anthropocentric Anishinabe worldview to create a distinctive methodology to conduct research and create legal knowledge in new and helpful ways. Profoundly original and prospective, her work is a gateway to new forms of legal imaginaries.

Post-anthropocentric legal propositions are often drafted to better protect non-human entities, but the latter are rarely included in the process. Vincent Chapaux explores what it would mean to rethink international law not only *for* non-human animals but to a certain extent *with* them. Relying on the work of Belgian philosopher Vinciane Despret, Chapaux tries to build a way to integrate non-human worldviews (particularly those of cats) into the construction of international legal norms, without stumbling into the pitfalls of natural law theorists and social Darwinists.

Among the current propositions towards a post-anthropocentric international law lies the idea of adding an international crime of ecocide to the 1998 Rome Statute. In the penultimate chapter of this book, Tim Lindgren offers a nuanced assessment of this call. After attending the hearings of the Australian Peoples' Tribunal for Community and Nature's Rights, he realised that reaching towards the international legal order in search for a protection against ecocidal violence is a very complicated and loaded gesture that could both import the international environmental predicaments into a domestic situation and cut local actors from their direct relationship with the land. Without rejecting the idea of an international crime of ecocide altogether, he invites us to slow down and to carefully measure what resorting to international law would mean in cases of ecocidal violence.

Anthropocentrism is the idea that human beings are socially and politically at the centre of the cosmos, but this collection is mainly focused on one tiny part of it: The Earth. In the last chapter, Elena Cirkovic tries to broaden the perspective and invites us to think about the current expanding presence of humans in outer space and its consequences in terms of environmental degradation. She invites us to rethink international law from a cosmic point of view: A perspective in which anything that has the potential to affect anything is considered agentic and should be taken into account in the construction of legal systems.

Overall, this collection demonstrates that anthropocentrism is profoundly embedded in the logic of international law. While it is less surprising that certain subfields of international law, such as international trade law, were designed for human profit at the expense of other living and non-living entities, this collection uncovers layers of anthropocentrism that run much deeper and wider. Anthropocentrism contributes to structuring legal fields usually depicted as post- or non-anthropocentric, such as international environmental law; it underpins many of the ways that scholars organise their critical approaches to international law; it provides the foundation to many of the main building blocks of international law, such as the notion of sovereignty; and, in myriad ways, constrains the ability of law to organise the world fairly and justly.

Faced with such deep disciplinary structures of power and injustice, the collection turns to alternatives from outside mainstream international law. The authorial propositions are diverse, but many share a sense of humility about the challenges of trying to see the world from the perspective of another: An 'other'. All the chapters in varied ways bring us face to face with a disciplinary hubris that refuses to allow such perspectives to emerge on their own terms and in their own languages: The perspectives of other humans whose voices are systemically silenced, living entities whose worldviews are overlooked, and non-living entities whose agency is not considered at all.

Notes

1 The authors would like to thank Isabella Spano for her invaluable help in the editing of this collection.
2 While the questions surrounding the human–animal divide are as ancient as the history of human thought (De Fontenay, 2014), current debates are usually structured around Peter Singer and Tom Regan's propositions (Singer, 1977; Regan, 1983) and their opposition to environmental ethics (Callicott, 1980). For a broader treatment, see Donaldson and Kymlicka (2011).
3 In contrast, some scholarship connects climate change, animal suffering, and ecological degradation (see Grear, 2013).

References

Alam, S., Atapattu, S., Gonzalez, C.G. and Razzaque, J. (2015). *International environmental law and the global south*. Cambridge: Cambridge University Press.
Albro, R. (2018). 'Bolivia's indigenous foreign policy: vivír bien and global climate change ethics', in Berry, E. and Albro, R. (eds.) *Church, cosmovision and the environment*. London: Routledge, pp. 99–122.
Anghie, A. (2007) *Imperialism, sovereignty and the making of international law*. Cambridge: Cambridge University Press.
Argyrou, V. (2005) *The logic of environmentalism: Anthropology, ecology and postcoloniality*. New York: Berghahn Books.
Barlow, M. (2010) 'The growing movement to protect the global water commons', *The Brown Journal of World Affairs*, 17(1), pp. 181–195.
Benton, T. (1993) *Natural relations: Ecology, animal rights and social justice*. New York: Verso.
Black, C.F. (2010) *The land is the source of the law: A dialogic encounter with indigenous jurisprudence*. London: Routledge.

Bosselmann, K. (2011) 'From reductionist environmental law to sustainability law', in Burdon, P. (ed.) *Exploring wild law: the philosophy of earth jurisprudence*. Kent Town: Wakefield Press, pp. 204–213.

Burdon, P. (2010) 'Wild law: The philosophy of earth jurisprudence', *Alternative Law Journal*, 35 (2), pp. 62–65.

Burdon, P. (2011) *Exploring wild law: The philosophy of earth jurisprudence*. Kent Town: Wakefield Press.

Burdon, P. (2015) *Earth jurisprudence: Private property and the environment*. London: Routledge.

Callicott, J.B. (1980) 'Animal liberation: A triangular affair', *Environmental Ethics*, 2 (4), pp. 311–338.

Callicott, J.B. (1984) 'Non-anthropocentric value theory and environmental ethics', *American Philosophical Quarterly*, 21 (4), pp. 299–309.

Charpleix, L. (2018) 'The Whanganui River as Te Awa Tupua: Place-based law in a legally pluralistic society', *The Geographical Journal*, 184 (1), pp. 19–30.

Chimni, B.S. (2003) 'Third world approaches to international law: A manifesto', in Anghie, A. (ed.) *The third world and international order*. Leiden: Brill Nijhoff, pp. 47–73.

Chimni, B.S. (2017) *International law and world order*. 2nd edn. Cambridge: Cambridge University Press.

Coward, M. (2006) 'Against anthropocentrism: The destruction of the built environment as a distinct form of political violence', *Review of International Studies*, 32 (3), pp. 419–437.

Cullinan, C. (2011) 'A history of wild law', in Burdon, P. (ed.) *Exploring wild law: The philosophy of earth jurisprudence*, Kent Town: Wakefield Press, pp. 12–23.

Cullinan, C. and Falstrom, A. (2008) 'If nature had rights', *Orion Magazine*, 27, pp. 26–31.

Davies, M. (2017) *Asking the law question*. 4th edn. Pyrmont: Thomson Reuters.

De Fontenay, E. (2014) *Le silence des bêtes: La philosophie à l'épreuve de l'animalité*. Paris: Fayard.

De Lucia, V. (2013) 'Towards an ecological philosophy of law: A Comparative Discussion', *Journal of Human Rights and the Environment*, 4, pp. 167–190.

De Lucia, V. (2015) 'Competing narratives and complex genealogies: The ecosystem approach in international environmental law', *Journal of Environmental Law*, 27 (1), pp. 91–117.

Donaldson, S. and Kymlicka, W. (2011) *Zoopolis: A political theory of animal rights*. Oxford: Oxford University Press.

Escorihuela, A.L. (2011) 'A global slaughterhouse', *Helsinki Review of Global Governance*, 2, pp. 25–29.

Garner, R. (2005) *The political theory of animal rights*. Manchester: Manchester University Press.

Gathii, J.T. (2021) 'The promise of international law: A third world view', *American University International Law Review*, 36(3), pp. 377–477.

Gathii, J.T. (1998) 'International law and eurocentricity', *European Journal of International Law*, 9, pp. 184–211.

Gillespie, A. (2014) *International environmental law, policy, and ethics*. 2nd edn. Oxford: Oxford University Press.

Gonzalez, C.G. (2015) 'Bridging the north-south divide: international environmental law in the anthropocene', *Pace Environmental Law Review*, 32, pp. 407–434.

Graham, M. (1999) 'Some thoughts about the philosophical underpinnings of Aboriginal worldviews', *Worldviews: Global Religions, Culture, and Ecology*, 3 (2), pp. 105–118.

Grear, A. (2013) 'Law's entities: Complexity, plasticity and justice', *Jurisprudence*, 4 (1), pp. 76–101.

Humphreys, D. (2017) 'Rights of pachamama: The emergence of an earth jurisprudence in the Americas', *Journal of International Relations and Development*, 20 (3), pp. 459–484.

Humphreys, S. and Otomo, Y. (2014) 'Theorising international environmental law', in Orford, A., Hoffmann, F. and Clark, M. (eds) *The Oxford handbook of the theory of international law*. Oxford: Oxford University Press, pp. 797–820.

Kelch, T.G. (2017) *Globalization and animal law: Comparative law, international law and international trade*. Alphen aan den Rijn: Kluwer Law International BV.

Kotzé, L.J. and French, D. (2018) 'The anthropocentric ontology of international environmental law and the Sustainable Development Goals: Towards an ecocentric rule of law in the Anthropocene', *Global Journal of Comparative Law*, 7 (1), pp. 5–36.

Last, A. (2018) 'To risk the Earth: The nonhuman and nonhistory', *Feminist Review*, 118 (1), pp. 87–92.

Maloney, M. (2015) 'Finally being heard: The great barrier reef and the international rights of nature tribunal', *Griffith Journal of Law & Human Dignity*, 3 (1), pp. 40–58.

Maloney, M. (2016) 'Building an alternative jurisprudence for the earth: The international rights of nature tribunal', *Vermont Law Review*, 41, pp. 129–142.

Maloney, M. and Siemen, P. (2015) 'Responding to the great work: The role of earth jurisprudence and wild law in the 21st century', *Environmental and Earth Law Journal*, 5, pp. 6–22.

Maloney, M.M. and Burdon, P. (eds.) (2014) *Wild law: In practice*. London: Routledge.

Margil, M. (2014) 'Building an international movement for rights of nature', in Maloney, M.M. and Burdon, P. (eds.) *Wild law in practice*. London: Routledge, pp. 149–160.

Naffine, N. (2009) *Law's meaning of life: Philosophy, religion, Darwin and the legal person*. Oxford: Hart Publishing.

Nagel, T. (1974) 'What is it like to be a bat?', *The Philosophical Review*, 83 (4), pp. 435–450.

Natarajan, U. and Dehm, J. (2022) *Locating nature: Making and unmaking international law*. Cambridge: Cambridge University Press.

Natarajan, U. and Khoday, K. (2014) 'Locating nature: Making and unmaking international law', *Leiden Journal of International Law*, 27 (3), pp. 573–593.

Nedelsky, J. (1990) 'Law, boundaries, and the bounded self', *Representations*, 30, pp. 162–189.

Norton, B.G. (1984) 'Environmental ethics and weak anthropocentrism', *Environmental Ethics*, 6 (2), pp. 131–148.

Otomo, Y. and Mussawir, E. (2013). *Law and the question of the animal. A critical jurisprudence*. London: Routledge.

Peters, A. (2016) 'Global animal law: What it is and why we need it', *Transnational Environmental Law*, 5 (1), pp. 9–23.

Peters, A. (2021) *Animals in international law*. Leiden: Brill Nijhoff.

Philippopoulos-Mihalopoulos, A. (2011) *Law and ecology: New environmental foundations*. Abingdon: Routledge.

Plumwood, V. (2005) *Environmental culture: The ecological crisis of reason*. London: Routledge.

Porras, I. (2014) 'Appropriating nature: Commerce, property, and the commodification of nature in the law of nations', *Leiden Journal of International Law*, 27 (3), pp. 641–660.

Regan, T. (2001) *Defending animal rights*. Urbana: University of Illinois Press.

Regan, T. (1983) *The case for animal rights*. Berkeley: University of California Press.

Rist, G. (2003) *The history of development: from western origins to global faith*. London: Bloomsbury Publishing.

Rogers, N. and Maloney, M. (2017) *Law as if earth really mattered: The wild law judgment project*. New York: Routledge.

Rorty, R., Shute, S. and Hurley, S. (1993) *On human rights: The Oxford amnesty lectures 1993*. New York: Basic Books.

Singer, P. (1977) *Animal liberation: A new ethics for our treatment of animals*. New York: Avon.

Sydes, B. (2014) 'The challenges of putting Wild Law into practice: Reflections on the Australian Environmental Defender's Office movement', in Maloney, M.M. and Burdon, P. (eds.) *Wild law: In practice*. London: Routledge, pp. 58–72.

Sykes, K. (2016) 'The appeal to science and the formation of global animal law', *European Journal of International Law*, 27 (2), pp. 497–518.

Thompson, S.C.G. and Barton, M.A. (1994) 'Ecocentric and anthropocentric attitudes toward the environment', *Journal of Environmental Psychology*, 14 (2), pp. 149–157.

Tomas, N. (2011) 'Maori concepts of Rangatiratanga, Kaitiakitanga, the environment, and property rights', in Grinlinton, D. and Taylor, P. (eds.) *Property rights and sustainability*. Leiden: Brill Nijhoff, pp. 219–248.

Trindade, A.A.C. (2010) *International Law for Humankind*. Leiden: Brill Nijhoff.

Varner, G., Light, A. and Rolston, H. (2003) 'Can animal rights activists be environmentalists?', in Armstrong, S.J. and Botzler, R.G. (eds.) *The animal ethics reader*. Abingdon: Routledge, pp. 410–421.

Warren, L., Filgueira, B. and Mason, I. (2009) 'Wild law: Is there any evidence of Earth jurisprudence in existing law and practice', *An international research project UK Environmental Law Association*. The Gaia Foundation.

Warren, L.M. (2006) 'Wild law – the theory', *Environmental Law and Management*, 18 (1), pp. 11–17.

Watson, I. (2011) 'Aboriginal (ising) international law and other centres of power', *Griffith Law Review*, 20 (3), pp. 619–640.

Watson, I. (2014) *Aboriginal peoples, colonialism and international law: Raw law*. London: Routledge.

Watson, I. (2017) 'Aboriginal laws of the land: Surviving fracking, golf courses and drains among other extractive industries', in Rogers, N. and Maloney, M.M. (eds.) *Law as if earth really mattered*. London: Routledge, pp. 209–218.

White, S. (2014) 'Wild law and animal law: Some commonalities and differences', in Maloney M.M. and Burdon, P. (eds.) *Wild law: In practice*. London: Routledge, pp. 247–262.

Wright, G. (2013) 'Climate regulation as if the planet matters: The earth jurisprudence approach to climate change', *Environmental and Earth Law Journal*, 3 (1), pp. 33–57.

SECTION 1

Unveiling the Anthropocentrism of International Law

1

'ONE VAST GASOLINE STATION FOR HUMAN EXPLOITATION'

Sovereignty as Anthropocentric Extraction

Mario Prost

Introduction: Fracking Lancashire

In March 2014, the British Conservative Government led by David Cameron called for the UK to go all out for shale gas. Despite the known risks of hydraulic fracturing to human health and the environment – including seismic activity, contamination of water sources and atmospheric pollution (Peduzzi and Harding Rohr Reis, 2013; Kovats *et al.*, 2014) – Cameron announced that it was time to 'roll up the sleeves' and 'get shale gas wells up and running' in the UK (Dominiczak, 2014). The announcement was faced with considerable opposition from local and environmental activists, following incidents at various test drilling sites in Lancashire. The government was not deterred, however, and went on the offensive to win over the public. Speaking to reporters after a nuclear security summit in Brussels, Cameron laid out his case for fracking: shale exploration would be good for the country and good for Europe, ushering in a new era of cheap energy and reducing dependence on foreign (and in particular Russian) gas. Closing his statement, the prime minister insisted that Britain had not just a clear economic interest but, in actual fact, a 'duty to frack' (Witte and Faiola, 2014).

Four thousand miles away, in the oasis town of Aïn Salah, some 30,000 protesters took to 'Resistance Square' early on Tuesday 24 February 2015 to oppose shale-gas exploitation planned in their community and in other parts of Algeria. They were soon joined by others in Tamenrrasset, Timimoune, Metlili, Adrar, Touggourt, Ghardaia, Ouargla, Tizi Ouzou, Bejaia and Algiers (Lamri, 2015). Despite widespread and growing discontent, the Algerian government decided to press ahead with plans to expand hydraulic fracturing for shale gas in the Sahara. Responding to the 24 February protests, Mohamed Ali Boughazi, advisor to Algerian President Bouteflika, took to national television to read a twenty-minute declaration defending the government's position. His statement included the following admonition: 'Shale gas is a gift from God, and it is our duty to exploit it' (Reid Ross, 2015).

To most international lawyers, the notion that a country has a 'duty' to exploit its natural resources will appear at once banal and peculiar. That a country 'owns' the resources lying within its borders and may do with them what it pleases has long been recognised as an essential aspect of state sovereignty. So deep-rooted is this principle that territorial sovereignty and resource ownership 'usually are linked inseparably in peoples' mind' (Bilder, 1980, p. 454). Enshrined in

DOI: 10.4324/9781003201120-3

13

the principle of permanent sovereignty over natural resources (PSNR), the inalienable right of states to freely dispose of their natural wealth for their own ends has been restated in countless international instruments and is universally recognised as one of the most fundamental organising principles of international relations (Schrijver, 1997; de Waart, 2009). Yet what the statements by Cameron and Boughazi appear to entail is something qualitatively different, not a mere right but a sovereign obligation to use nature's bounty as fully as man is able.

This chapter seeks to explain and situate Cameron and Boughazi's statements in international law's *longue durée*. Specifically, it aims to demonstrate that their views are not at odds but, in fact, entirely consistent with the doctrine of sovereignty as historically articulated in international law. Sovereignty is the 'basic constitutional doctrine' of international law, from which everything else within the system is derived (Crawford, 2012, p. 447). International law, in the formal sense at least, can be interpreted as a set of arguments about what subjects can legitimately claim to be sovereign, what legal privileges are attached to sovereignty, and how to resolve clashes of sovereignty. International law's core normative foundations are all, in one way or another, wedded to the concept of sovereignty, and most international law questions can be reconfigured as aspects of the sovereignty questions: what it is, who has it, and what can be done with it (Simpson, 2008; Jodoin, 2008; Mégret, 2012).

As will be shown, one of the most enduring features of the doctrine of sovereignty is its deep anthropocentric structure. From the sixteenth century onwards, sovereignty has been intimately tied to peoples' willingness and ability to make productive use of nature and utilise its resources to satisfy human needs. The transformation and subjection of nature's forces by man has long been posited as a primordial and constitutive act of sovereignty and, as a result, sovereignty has historically been denied to (primarily non-European) societies deemed to be failing in their duty to tame and shape the natural world to enable human development. Put differently, the exploitation of nature is not simply, as is commonly assumed, a legally protected attribute of sovereignty. It is, in more ways than one, a core requisite of sovereignty. International law is thus anthropocentric in a more profound and radical way than is perhaps usually acknowledged. It does not merely enable or facilitate the exploitation of nature, but in fact demands it.

The chapter explores these ideas by looking, first, at the origins of the sovereignty doctrine as articulated in the context of the colonial encounter in the Americas and legal debates about the justice of empire. In doing so, it focuses specifically on Locke's arguments about land property through labour and how these arguments were developed into full-fledged international law doctrines positing the cultivation of the earth as a core sovereign obligation of states under the law of nations. The last section explores the legacy of these ideas in contemporary international law doctrines and practises, including the role of resource commodification in the decolonisation process and post-war development policies. It will be seen that, despite profound transformations to the law of statehood and the emergence of environmental norms calling for the conservation of nature, the international legal ontology of sovereignty – what it means to be a sovereign state in the world – remains intimately connected to, and to an extent predicated upon, a willingness to subdue nature and bend it to human designs.

Ploughing Wilderness to Sovereignise It: Property Formation in Locke's American Wasteland

As Europeans first laid claim to the New World, they quickly became confronted with the necessity to provide a legitimate basis for the taking of lands that were patently inhabited. The justification of land appropriation in the Americas emerged as a vital question that came to preoccupy some of the most prominent political thinkers of sixteenth and seventeenth-century

Europe (Ince, 2018, pp. 42–44). It also provided the vital context out of which emerged critical ideas about sovereignty and the conceptual elaboration of the law of nations more generally (Anghie, 2005; Carty, 1996; Miéville, 2006).

The English were latecomers to the Atlantic. Lacking the papal bulls Spain and Portugal enjoyed as the basis for their colonial ambitions, they developed a distinct set of theories to justify their claims, premised on the classification of native land as legally vacant (Pagden, 1998). The English did not share the ambivalence of some of their Spanish counterparts. Salamanca theologians and jurists such as Vitoria and Soto had seen Native Americans as bearers of rights, legitimate owners, and lords of their land before the arrival of the Spaniards. As such, they were forced to develop long-winded legal arguments to defend the dispossession of the Indians, 'tak[ing] back with the left hand what they gave away with the right' (Koskenniemi, 2011, p. 10). Invoking the right to travel, dwell and be granted hospitality in the New World, the Salamancans affirmed the Christian Europeans' right to wage war against those opposing free commerce and evangelisation (Williams, 1990; Anghie, 1999; Porras, 2014).

English (and French) imperial claims were, in a sense, of a much more straightforward nature. The English took the view that it wasn't necessary to justify the dispossession of the Indians since the Indians were not, in actual fact, in lawful possession of the land upon which they lived, having failed to bring it under proper dominion as per the requirements of natural law. Indian land was thus free of any pre-existing title. Though not empty, the New World was legally vacant and 'open for occupation' (Schmitt, 2006, p. 130). It was, in other words, a juridical blank slate, waiting to be subdued by the providential hand of civilisation.

But what, then, was the mediating device by which people could bring land under proper dominion, which the Indians had failed to display? According to English colonial ideology, territorial rights could only be established through certain kinds of value-creating activities. Specifically, the land had to be settled, enclosed, and productively developed to be lawfully appropriated. Ownership of the earth was not, in other words, established through discovery, conquest or occupation, but by human toil. Land was to be treated as capital – a giant storehouse of dormant reserves – and only labour could release its value. Since the natives had barely scratched the surface of the Earth, living off nature's fruits in a state of primaeval simplicity, they could not rightfully claim ownership of the land they occupied. Their land was therefore to be treated as *vacuum domicilium*, i.e., unoccupied ground free for the taking by anyone willing and able to bestow culture and husbandry upon it (Corcoran, 2017).

In March 1630, seven hundred Puritans assembled in Southampton to take ship to Massachusetts Bay. Many of them needed reassuring, as they were about to leave behind comfortable homes and security for the harsh conditions of the New World. Among other things, they needed to know that braving the hardship ahead would earn them not just the freedom to worship but rightful individual ownership of the land (Linklater, 2013, pp. 26–27). The promise of free land was tempting. But the challenges of emigration were immense, and people needed to be convinced, among other things, that indigenous land was theirs to take and that, once settled, they would enjoy security of tenancy. To appease these anxieties, John Winthrop – an English lawyer, founding member and governor of the Massachusetts Bay colony – published a set of papers for circulation to prospective colonists in which he sought to address common causes of concern. In response to the objection that the English have 'no warrant to enter upon land which hath been so long possessed by others', Winthrop put forward what would quickly become the central defence of England's right to American soil:

> That which lies common and hath never been replenished or subdued is free to any that will possess and improve it […] by enclosing and peculiar manurance […] As for

the natives in New England, they enclose no land, neither have any settled habitation, nor any tame cattle to improve the land by, and so have no other but a natural right to those countries. So as if we leave them sufficient for their use, we may lawfully take the rest, there being more than enough for them and us.

(Vaughan, 1997, pp. 28–29)

These notions of unredeemed American land laying in a state of unproductive waste, and thus bereft of ownership, were commonplace in early colonial literature. With Locke, they received their fullest and most enduring theoretical elaboration. Locke's interest in the Americas was not merely philosophical. He was deeply involved in the administrative webs of nascent colonial capitalism and in the intellectual effort to articulate the colonial agenda of expropriation and disenfranchisement in a 'protoliberal framework' of natural law (Ince, 2018, p. 44; Ivison, 2003). Through the patronage of Anthony Ashley Cooper – 1st Earl of Shaftesbury, founder of the Whig party and Lord Proprietor of the Caroline colony – Locke secured key positions in the private and public administration of England's colonies, including secretary to the Proprietors of the Carolina colony (1669–1675) and secretary to the Board of Trade and Foreign Plantations, the main body which oversaw commerce and settlements in the Atlantic world (1695–1700). Locke also had economic investments in the English colonies: he was a merchant adventurer to the Bahamas, held shares in the Royal African Company (the English monopoly for trading in slaves), and briefly invested in the British East India Company (Armitage, 2012).

Locke's *Second Treatise of Government* – and specifically its Chapter 5 'Of Property' – is simultaneously a philosophical elaboration about the natural right to property as the basis of civil government and a robust defence of England's rights to American soil against sceptics, counterclaims by aboriginal nations, and competing claims by rival European powers. In the beginning, argues Locke, God gave all of nature 'to mankind in common' (Locke, 2003, p. 111). In this original state of nature, all men had an equal right to gather natural resources and make use of them to the best advantage of their life and convenience. Once gathered, an item became the property of the person who, 'through the labour of his body and the work of his hands', made the effort to remove it from the common state nature had placed it in (Locke, 2003, p. 112). But nature itself remained common property of mankind. One owned the acorns picked up under an oak, but not the oak tree itself. Applied to land, the labour theory of property holds that land originally held in common is rightfully appropriated by the person who, through agrarian cultivation, makes it more productive: 'as much land as a man tills, plants, improves, cultivates and can use the product of, so much is his property' (Locke, 2003, p. 113). He who clears the forest, ploughs the soil and cultivates crops can lay claim to the land and enclose it from the common.

Locke's theory up to this point is mostly concerned with explaining the original appropriation of property, i.e., how something goes from being held in common by everyone to being the private and exclusive property of a particular individual. Underlying these explanations, however, lies a theory of natural law that allowed Locke to define native systems of property away as illegitimate. For Locke, the cultivation of the earth is not just expedient. It is morally required, a natural duty imposed upon man by divine providence and the necessities of material reproduction. God, Locke argues, cannot have intended for the earth to remain forever in common and uncultivated. In a world of abundance, with 'plenty of natural provisions' and 'few spenders', living easily on the fruits which the 'spontaneous hand of nature' has brought forth of its own accord, without toil, may have been warranted for a time. But in a world of scarcity, it must be God's command that man shall subdue the earth and act industriously: the 'condition of human life' and the 'penury of his condition' require it of him (Locke, 2003, p. 113).

Sovereignty as Anthropocentric Extraction

Enclosure, cultivation, and extraction thus represent for Locke material conditions of executing God's purpose in accordance with the natural law of preservation. Nature is designed by providence to invite property and increase the common stock of mankind. It is against this background that Locke interprets America and defends England's right of settlement in the New World. For Locke, America presented itself as a vast, unredeemed wilderness in which the Indians lived in a perfect state of nature. The defining feature of this primitive state was the sheer waste caused by nature being left 'unimproved' (Neocleous, 2011, pp. 510–511). Land in America was 'left wholly to nature', with 'no improvement of pasturage, tillage, or planting'. The Indians enclosed no ground and bestowed no labour upon the land. As such, they merely possessed a right to the fruits and the beasts of the earth but 'exercise[d] very little Dominion, and ha[d] but a very moderate Sovereignty' (Locke, 2003, p. 147).

The 'very moderate sovereignty' the Indians possessed over their land was, according to Locke, limited in two fundamental ways. Spatially, it was strictly limited to the land the Indians effectively occupied and made use of. Whatever land they tilled and reaped was their peculiar right. But everything beyond this 'was still to be looked on as waste, and might be the possession of any other' (Locke, 2003, p. 116). But more significantly perhaps, Indian dominion over land was, for Locke, normatively weak and transient. The Indians only possessed a primitive or embryonic right to the land, characteristic of their pre-historic and pre-political state. The English enjoyed a superior claim to proprietorship, one that superseded any rights claimed by virtue of prior occupation. This was so not because of England's military might or some religious justification but because of its industry and its superior ability to extract value out of America's paradise of fertility. Wherever the Indians had failed to enclose nature and dutifully exploit its resources, the Englishman could claim their land for cultivation by virtue of a higher yield of goods to humanity (Arneil, 1996, p. 63).

Locke's chapter on property is replete with comparisons between English industriousness and Indian idleness. These are often expressed in terms of ratios, which Locke takes as evidence of England's superior title to the land. The Indians, he writes, are 'rich in land' yet 'poor in all the comforts of life'. Despite having been provided as liberally as other peoples with the 'materials of plenty', the Indians, for want of improving their land through labour, 'have not one hundredth part of the conveniencies' which the industrious and ingenious Englishman is able to extract out of the well-tended land of Devonshire (Locke, 2003, p. 116).

For Locke, the Indians are mere dwellers on their land, content with what 'unassisted nature offers to their necessities'. They are thus to be regarded as 'tenants', with no legal interest in a land that has yet to be turned into property. This, to Locke, is an offence against the common law of nature and against God himself, who commanded man to improve the earth for the benefit of life. From there, Locke concludes that 'different degrees of industry' give men 'possessions in different proportions' (Locke, 2003, p. 120). In other words, the degree of intensity with which land is used and cultivated 'establishes a correspondingly strong title to the land as well as more sophisticated societies' (Fitzmaurice, 2014, p. 121).

With Locke, then, subduing the earth and having dominion over it are joined together: 'the one g[ives] title to the other' (Locke, 2003, p. 114). People clinging to the lawless existence of pastoralism and foraging remain unformed, without politics or society. Having yet to undergo the process of primitive appropriation, the Indians knew few controversies over property and therefore had no need for civil laws to decide on them. Whilst carrying in them the germs of civilisation and capable in principle of entering civil society, they were, at the time of colonisation, without an overarching sovereign, living in a stateless form (Beard, 2007, p. 102).

The triumph of civilisation in the New World demanded the transformation of nature into property so that it could be converted into things useful to man. The propertisation of nature became the hallmark of progress and modernity. Europeans had it; Native Americans did not. Colonisation, in this view, 'meant installing this mechanism of progress on New World soil, where it had previously been unknown' (Greer, 2018, p. 2). The Lockean defence of empire is one that therefore expresses a distinctly anthropocentric and capitalistic worldview, demanding that inept nature be turned into an ever-expanding domain of value as a *sine qua non* for state formation and human progress, and appealing to a productive and accumulative mode of appropriation as the superior and morally exalted basis of land tenure (Ince, 2018, p. 39).

Globalising Locke: The Law of Nations and the Sovereign Obligation to Cultivate the Earth

For Locke, the purpose of government is to establish and protect property so that men can release the full potential of nature, ensuring their self-preservation and thriving as free individuals (Moore, 2015, pp. 17–18). This pre-supposes, as we have seen, waging war on land tenure systems and traditional use-rights deemed wasteful and inefficient and consolidating land's status as a fully fungible commodity: propertised, tradable, and exploitable at will. In Europe, this entailed the dispossession of self-supporting rural populations and the enclosure of the commons. In the rest of the world, this justified the taking and exploitation of regions of nature which non-European peoples were neglecting and were thus waiting to be settled and productively developed (Özsu, 2019).

Locke, to be sure, did not invent this set of ideas about waste, improvement, and legitimate territorial authority. The general thrust of the argument – that title to the land is acquired through effective occupation, and occupation means efficient use of resources – was already nascent, before him, in the writings of Grotius, Hobbes, Puffendorf, and among the colonists themselves (Fitzmaurice, 2015). But Locke was instrumental in giving these ideas their intelligible unity and credence and was 'the most explicit in linking land appropriation, enclosure, ownership and the founding of sovereign power and right' (Brown, 2010, p. 44). His views became highly influential among politicians, lawyers, and public intellectuals debating property relations, land rights, and good government in England and her colonies.

In England, they were most notoriously embraced by Blackstone in his *Commentaries*, 'the most influential law book in Anglo-American history' (Alschuler, 1996, p. 2). In the volume dealing with the 'right of things', Blackstone adopts an explicitly Lockean view of property. Addressing the matter of original appropriation – how a thing that belongs to everyone comes to be owned by someone – Blackstone begins by affirming the rule of first possession, which grants an ownership claim to the party that gains control or makes use of a thing before others (the 'first taker'). By the law of nature and reason, he posits, the right of possession goes to 'he who first began to use it' (Blackstone, 1771, p. 3). First possession, however, grants but a type of 'transient property', provisional and limited to the use and enjoyment of a thing, not the thing itself. For Blackstone, 'permanent dominion' cannot be established by mere occupation. It requires that the thing be 'improved and ameliorated by the bodily labour of the occupant'. This, he concludes, 'give[s] the fairest and most reasonable title to an exclusive property therein' (Blackstone, 1771, p. 5).

Turning to the matter of land ownership, Blackstone takes the view that occupancy gives the right to the temporary use of the soil. As the world grows more populous, however, it becomes necessary to purse regular methods of providing a constant subsistence to mankind, as unassisted nature 'w[ill] not provide her fruits in sufficient quantities without the assistance of tillage'. Permanent property in the soil must therefore go to those who make it fructify through their industry, art, and labour. Without this incentive, Blackstone asks, 'who would be at the pains of

Sovereignty as Anthropocentric Extraction

tilling it?'. By necessity, it follows that it is the 'art of agriculture' that gives a 'more permanent property in the soil' (Blackstone, 1771, p. 7).

From these general principles of property, Blackstone extrapolates a number of observations about the right of European states to occupy land in 'wild and uncultivated nations that have never been formed into civil states'. Chief among those rights is what Blackstone terms the 'right of migration'. Countries that are 'overcharged with inhabitants', he writes, must be deemed to have the right to 'send colonies to find out new habitations'. In the process, 'overcharged' nations must pay due respect to the humanity and dignity of local inhabitants. Driving them out and massacring the innocent merely because they differ from their invaders in culture, government or colour would not be 'consonant to nature, to reason or to Christianity'. Yet for Blackstone, 'the stocking and cultivation of desert uninhabited countries [is] strictly within the limits of the law of nature' (Blackstone, 1771, p. 7).

The *terra nullius* idea – i.e., the view that land and its resources ought to be used intensively and that entitlement follows such usage – became the central defence of England's sovereignty in her colonies. *Terra nullius* was of course never an idea expressing the absence of people from overseas territories. It was always used as a 'code for the absence of agricultural use of those lands' (Graham, 2011, p. 95). At the core of the *terra nullius* idea is thus a distinct view of nature as something to be appropriated and transcended by culture. That vision formed an integral part of the 'mental furniture' of the founders and defenders of England's colonies (Gascoigne, 2002, p. 8). An editorial published in 1838 in the *Sydney Herald*, for instance, typifies the intellectual coupling of cultural progress, agricultural improvement, and British sovereignty in the Australian settler ideology:

> This vast land was to them a common – they bestowed no labour upon the land – their ownership, their right, was nothing more than that of the Emu or Kangaroo. They bestowed no labour upon the land and that – and that only – it is which gives a right to property to it. Where, we ask, is the man endowed with even a modicum of reasoning powers, who will assert that this great continent was ever intended by the Creator to remain an unproductive wilderness? […] The British people […] took possession […] and they had a perfect right to do so […] Herein we find the right to the dominion, which the British Crown, or, more properly speaking the British people, exercise over the continent of New Holland.
>
> *(cited in Graham, 2011, p. 96)*

Nowhere was this idea of progress as intensively productive cultivation more prevalent perhaps than in revolutionary America, where it became the keystone of the country's imagined national mission: to redeem the continent and bring it into flourishing order. The incorporation of Locke's labour theory of property in Blackstone's *Commentaries* was adopted by judges, lawyers, and politicians seeking to justify their own belief in the agrarian basis of their title to American land (Arneil, 1996). American colonists were faced with a unique set of challenges in attempting to secure the propriety of their settlements. They had to find a way to simultaneously defend the dispossession of the natives and resist the claims of the Crown. This could not be achieved by relying on the right of conquest. If achieved through conquest, title to the land would lie with the British Crown. While the Crown was adamant that all rights in the New World came from the state, the American revolutionaries had to find a way to argue that they had established property independently of the Crown, from their own occupation of the land, including land upon which Native Americans lived (Fitzmaurice, 2014, pp. 171–173).

Locke and Blackstone furnished a uniquely well-suited set of justifications which became 'terrifically prominent in early US political and legal culture' (Purdy, 2015, p. 76). In pre-revolutionary colonies, they were used as a means of questioning the rights of both the imperial state and colonised peoples. When ordered by the Privy Council to respect the natives' right to their ancestral land, settlers in Connecticut refused to execute the Crown's orders, claiming that they had 'built upon, planted and greatly improved' the colony and, as such, should be regarded as 'absolute owners' of the land (Yirush, 2011, p. 123). The Indians, on the other hand, made 'very little use … of the Earth further than to walk upon it'. They had no utensils, no ploughs, no hoes, no axes and therefore could not till the soil. Whatever land they occupied was therefore legally vacant and could be appropriated without their consent, save for the 'few spots of it' in which they had effectively mixed their labour (Tully, 1993, pp. 166–167).

Locke and Blackstone's ideas were not popular only among the puritan pastors and landed elites of the original colonies. They were also staples of the political imagination of the American Revolution, providing essential justifications for America's right to declare independence and its policies towards the Amerindians. Founders such as Thomas Jefferson and John Adams defended American sovereignty in distinctly Lockean terms, claiming it was rooted in the sweat and tears of the plantation workers, who fertilised the great American wasteland with their labour and put it to productive use: 'their own blood was spilt in acquiring land for their settlement [and] for themselves alone they have right to hold' (Jefferson, 1774, p. 8). America, in other words, was not 'granted' by the Crown. It was conquered by the plough, redeemed by the planters 'and the treasure they expended to subdue to by cultivation' (Adams, 1819, p. 133). If it didn't belong to the Crown, America didn't belong to the Indians either, whose static, unproductive nature stood against providential design. 'Shall the lordly savage', asked Adams rhetorically,

> forbid the wilderness to blossom like a rose? Shall he forbid the oaks of the forests to fall before the axe of industry, and to rise again, transformed into the habitation of ease and elegance? Shall he doom an immense region of the world to perpetual desolation … and the fields and the valleys which a beneficent God has formed to teem with the life of innumerable multitudes, be condemned to everlasting barrenness?
>
> *(cited in Purdy, 2015, p. 81)*

Sovereignty could not belong to the Indians, who had failed to turn trees into chairs. It could only belong to those who, through their art, industry, and the galvanising power of labour, could turn the earth's 'innumerable multitudes' into conveniences of life and tradable commodities. Sovereignty was established through the dutiful exploitation of the earth and, once established, could only be maintained through perpetual control over nature. Leaving nature to its own device meant letting wilderness take over and eroding one's hard-earned sovereignty. The sovereign-deflating potentialities of stasis and unexploited nature was a recurring theme in the mythology of the American frontier. As the young American republic expanded westward, settlers were kept under a 'constant admonition' to impose human order over the world and improve nature through unfaltering activity (Chaplin, 1993, p. 108). 'Withhold the hand of cultivation' wrote John Drayton, a South Carolina Governor and US District Judge, 'and nature immediately causes weeds and plants to spring up again; and, in course of time, covers them with her dark retreats' (Drayton, 1802, p. 149).

Cultivate the earth to keep nature's dark retreats at bay, so that the welfare of nations may be enhanced and mankind be multiplied: what started as a set of defences improvised out of the colonial encounter and concerned mainly with property formation in the Americas became, by the mid-eighteenth century, a formal doctrine of the law of nations itself, a sort of sacred mandate defining what it means, under international law, to be a state, how territorial sovereignty is established, and how it ought to be exercised. Nowhere was this move more visible, perhaps, than in the work of Emer de Vattel. Considered by many the father of modern international law (Chetail and Haggenmacher, 2011; Stapelbroek and Trampus, 2019), Vattel published *The Law of Nations or the Principles of Natural Law Applied to the Conduct and to the Affairs of Nations and of Sovereigns* in 1758. The work enjoyed immense and enduring acclaim in Europe and the United States, where for more than a century it was regarded as the standard reference work on international law in universities, chancelleries, and diplomatic circles. Much of Vattel's success can be explained by the methodological innovations that underpin his work. The legal constructs and categories (sovereign equality; law of treaties; law of neutrality; *ius ad bellum*; right of responsibility, etc.) developed in Vattel's *Law of Nations* were relied on by jurists and politicians for centuries (Jouannet, 2012a).

The law of nations theorised by Vattel was liberal and pluralist. This was a law of nations based on the principles of sovereign equality and political liberty, including the freedom of states to choose their system of government and religion. Its basic tenet was simple: states are free to do whatever they wish on their own territory, provided this does not diminish the sovereign rights and independence of other nations. It isn't surprising, therefore, that his work became enormously popular with small and medium-sized states seeking to resist the hegemonic impulses of great powers. But Vattel's *Law of Nations* was also, on another level, far more complex. Though premised on a principle of formal equality and self-determination, it also promoted a singular vision of good politics and human happiness, i.e., substantive principles he believed would contribute to the welfare and perfectibility of nations. Vattel's is not so much, therefore, a liberal law as a 'liberal-welfarist' law, which simultaneously protects the freedom of the state and advances principles of good government deemed necessary to promote the happiness and well-being of its people (Jouannet, 2012b).

At the centre of Vattel's providentialist project was the notion that the end and object of civil society is 'to produce for the citizens whatever they stand in need of for the necessities, the conveniences, the accommodation of life, and, in general, whatever constitutes happiness – with the peaceful possession of property, a method of obtaining justice with security, and, finally, a mutual defence against all external violence' (Vattel, 1844, p. 4). From these aims, Vattel elaborated a number of principles of good government, including principles concerning the legitimate use by states of their territory. Since the first and principal object of a good government is to provide for the necessities of the nation and maximise public welfare, Vattel considered the matter of the proper use of a country's natural wealth to be of prime importance and dedicated an entire chapter (Ch. VII) to the 'cultivation of the soil'.

In it, Vattel claimed the intensive exploitation of a country's natural capital to be the 'surest' and 'most solid' source of utility for a nation (the other being commerce). States therefore have a duty to take all available measures to ensure that the land under their jurisdiction is tilled and cultivated in the most efficient manner possible. Every nation is, according to Vattel, not only allowed but 'obliged by the law of nature to cultivate the land that has fallen to its share' in the most advantageous way through proper incentives (including tax incentives) and by suppressing 'inimical' land tenure regimes (including communal land ownership) that do not allow the land to be enclosed, engrossed, and dutifully improved (Vattel, 1844, pp. 34–35).

Critically, for Vattel, states owe it not just to themselves but to the international community to intensively cultivate the earth, for 'the whole earth is destined to feed its inhabitants; but this it would be incapable of doing if it were uncultivated'. Nations that inhabit fertile lands but disdain to cultivate them intensively, choosing instead to live by plunder, 'are wanting to themselves [but also] injurious to all their neighbours, and deserve to be extirpated as savage and pernicious beasts'. Idleness might have been tolerable 'in the first ages of the world' when nature was capable of furnishing spontaneously and without culture sufficient support for its few inhabitants. But with humanity so greatly multiplied, concludes Vattel, those who still pursue this idle mode of life usurp more extensive territories than, with a reasonable share of labour, they would have occasion for and have, therefore, no reason to complain if other nations, more industrious and too closely confined, come to take possession of a part of those lands.

(Vattel, 1844, pp. 35–36)

The prime responsibility of the sovereign, under international law, is for Vattel to promote the intensive use of land to procure things of utility and satisfy life's necessities. Under-utilisation of the world's resources, on this view, is seen as antisocial and offensive. A country which fails to productively utilise its resources is, in a sense, committing a dereliction of duty and, as such, must be seen to forfeit part of its territorial sovereignty to other, more industrious countries that are willing and able to take over neglected land to make it fructify.

This argument is made in the most explicit manner by Vattel when he evokes the colonisation of North America. Whilst critical of the conquest of the civilised empires of Peru and Mexico and more broadly of the violence perpetrated against Native Americans, he saw the dispossession of 'erratic nations' that 'ranged' extensive tracts of land without settling or cultivating them as 'extremely lawful' (Vattel, 1844, p. 36). Those nations could only claim legal possession of those regions of the New World where their 'ecological footprint' was manifest (Mickelson, 2014). Critically, for Vattel, the use of the land had to be constant to satisfy the ecological footprint requirement:

if a North American nation relocated, providing nature time to regenerate, they automatically lost the rights over the lands from which they had moved. Only the European's more intensive approach of permanent utilisation … assured full legal rights of possession.

(Jiménez-Fonseca, 2017, p. 201)

Vattel's intervention was critical in universalising, secularising, and radicalising Lockean arguments about the appropriation of land through surplus extraction. His doctrine of sovereignty possessed unique qualities that made it enormously popular among lawyers seeking to justify occupation and conquest, from the eighteenth to the early twentieth century. With Vattel, it became possible, under the law of nations, to affirm the right to sovereign equality whilst simultaneously denying or limiting the sovereignty of non-European nations deemed to be failing in their duty to intensively utilise the resources which providence placed at their disposal. Critically, this could now be done on seemingly value-neutral grounds. The cultivation argument was free of the religious or cultural connotations that pervaded legal arguments for centuries. Occupation was no longer about conquering and subjugating the infidels, but solely about improving nature and, with it, humanity. Suddenly, it became unnecessary to argue that non-Europeans were outside of humanity or lived in a pre-political state of nature, as had long been the case. Instead, the enjoyment of sovereignty – including the right to exclude others from one's territory – became conditioned or predicated upon the observance of international standards governing, among other things, the proper use of land and nature.

In the nineteenth century, the focus of empire and of the law of nations shifted away from the formal appropriation and settlement of land to the control of non-European sovereignty, and it is to this task that nineteenth century legal positivists notoriously turned their attention. The main legal technique developed to justify imperial expansion was the postulation of a gap between civilised states – those that had reached evolutionary maturity and whose sovereignty was beyond question – and uncivilised states, standing at lower stages of human development and enjoying unequal rights and privileges under international law (Anghie, 2005). Sovereignty, in other words, became identified with specific standards of civilisation and, whilst the subject of intense speculation, the standard of civilisation was articulated first and foremost in terms of conformity with the basic tenets of capitalist modernity (Tzouvala, 2020; Parfitt, 2019).

To be admitted as members of the family of nations, with full legal personality and equal rights, non-Western countries were expected to undergo legal, institutional, and political reforms required for capitalist relations of production and exchange to thrive and expand. This included the establishment of an organised political bureaucracy and a system of law capable of guaranteeing the security and property of foreign nationals (Gong, 1984). Where those standards were imperfectly met (semi-civilised states) or lacking altogether (uncivilised states), capitulations, protectorates, and other modes of imperial governance were deemed legitimate mechanisms to bridge the civilisational gap, improve non-European polities, and enable their progress towards higher stages of human development.

Whilst the distinction between civilised and uncivilised states appears at first to represent a radical departure from Vattel's pluralist law of nations, things are in reality more complex, and Vattelian arguments were largely absorbed and re-deployed by nineteenth century jurists justifying the imperial occupation of non-European sovereignties. For colonial trade to flourish and capitalist expansion to take place, legal reforms for the protection of private property and the security of transactions were necessary but not sufficient. In vast swathes of the non-European world, the permanent transition to a capitalist mode of production required the transformation of 'natural economies' of subsistence into commodity economies. This process of 'primitive accumulation' involved, among other things, abolishing pre-capitalist forms of land appropriation and exploitation, suppressing itinerant modes of agriculture, and commodifying natural resources (Neocleous, 2012).

In justifying the occupation and control of 'waste' countries so that they could be 'improved', Vattel's standards of environmental exploitation proved most useful. Whilst prepared to concede that non-European peoples had entered civil society and lived under some form of political formation, most nineteenth century jurists took the view that, to the extent that sovereignty existed in Africa and Asia, it was a 'personal' sovereignty akin to that which prevailed in medieval Europe. Territorial sovereignty, on the other hand, was deemed to be largely absent (Fitzmaurice, 2014, p. 239). Native inhabitants living in pastoral or primitive agricultural states had yet to subdue nature through intensive extraction and consumption, as mandated by the law of nations. As such, they lacked an essential requisite of civilisation and their territorial sovereignty remained unrealised.

Johann Kaspar Bluntschli – one of the most influential jurists of the nineteenth century and a founder of the *Institut de Droit International* – was explicit in linking territorial sovereignty to civilisation and civilisation to industrial development. A nation, he claimed, 'is civilised or uncivilised, depending on whether it exhibits labour, development, progress, or only latent forces' (Bluntschli, 1883, p. 38). The end of the state must be to 'face nature boldly and independently, making use of her when she is kind and combating her when she is cruel' (Bluntschli, 1885, p. 217). A nation that demonstrates an unwavering commitment to the control, harnessing, and intensive exploitation of nature creates civilisation out of wilderness. A nation that leaves nature minimally altered and does not develop it through its arts, science, and industry does not advance and remains uncivilised.

This mode of argumentation was especially apparent when Bluntschli discussed the status of the 'Negro monarchies' of Africa. Unlike industrial and mercantile states, he claimed, those tribal nations did not display the 'expenditure of serious and persistent labour' characteristic of more advanced nations and thus 'fail[ed] to develop the hidden resources of their nature'. With 'no industry to ennoble labour' and despite a great fertility of soil, 'famine or invasion reduce[d] the careless population to misery'. Whilst land in those countries was not, strictly speaking, uncultivated, and people occasionally lived a 'happy sensual life', they were not 'proper states' but 'childish' nations that could 'offer no opposition' to occupation by more advanced nations prepared to stimulate civilisation and cultivation (Bluntschli, 1885, pp. 78–81).

Other apologists for empire legitimised occupation by reference to Vattelian ideas. Phillimore, citing Vattel, affirmed that 'the cultivation of the soil is an obligation imposed upon man' and that habitants of 'overflowing' countries had the right to make settlements 'in countries capable of supporting large numbers by cultivation' and occupied by unproductive tribes 'refus[ing] to cultivate the soil' (Phillimore, 1879, p. 347). Discussing the status of treaties concluded with tribal leaders in Africa, Westlake noted that those tribes possessed some of the attributes of civilisation (settled agricultural populations, some notions of property, trading activities) yet could not be said to possess sovereign title to their land in the full international sense of the term. Their occupation was too rudimentary and 'took no rights under international law'. There could be no cession of sovereignty, as tribal leaders could not cede what they did not possess (or understand). Likewise, 'so-called protectorates' in uncivilised regions of Africa were protectorates in name only. As a matter of law, there could be no protectorate as 'there [was] no state to be protected'. With no state and no sovereignty worthy of the name, nothing could stop the 'inflow of civilisation' wherever there was 'land to cultivate, ore to be mined, commerce to be developed' (Westlake, 1894, pp. 142–146).

The standard of civilisation, as a mode of argumentation about the unequal distribution of international rights and duties, therefore, defined the inhabitants of non-European territories as too low on the scale of human progress to be recognised as possessing title in their land. Unwilling or unable to derive any profit from their biophysical environment, leaving sources of wealth unproductive, the right of 'undeveloped' nations was not to statehood but to guardianship (Koskenniemi, 2016). To be regarded as sovereign and progress up the ladder of civilisation, these communities would have had to manifest, first, 'the propensity and capacity to exploit nature in the creation of physical and political structures' (Fitzmaurice, 2014, p. 44). Inhabiting the earth was deemed insufficient. What human progress and the welfare of nations demanded was economic improvement, through intensive cultivation, industry and 'martial spirit' (Lorimer, 1883, p. 186). The divide between civilised and uncivilised states was therefore as much an ecological as a racial or cultural one:

> those communities who lived as rational, productive economic actors, evidenced by particular forms of cultivation, were deemed to be proper subjects of law and history; those who did not were deemed to be in need of improvement as much as their waste lands were.
>
> *(Bhandar, 2018, p. 8)*

The Postcolonial Lives of Sovereignty: Independence, Development, Sustainability

Imperialism has been defined and studied in a variety of ways, as a political, economic or cultural formation. Yet, as historian Richard Drayton has observed, imperialism was (is) also, at its core, a 'campaign to extend an ecological regime: a way of living in Nature' (Drayton, 2000, p. 229). As the previous section has shown, international law played a momentous role in that campaign. By advancing specific understandings of the relationship between the human and the non-human world, it made the possession and enjoyment of sovereignty conditional upon the embrace of a certain idea of nature. Classical jurists transfigured nature into a 'storehouse of dormant reserves' destined to be harnessed and put into motion by global imperial enterprise (Bonneuil, 2019, p. 3). Nature was reduced to a pure domain of utility or, as Argyrou puts it, a 'standing reserve of resources' (Argyrou, 2005), the control and intensive exploitation of which was progressively elevated as the primordial end of the state. Polities that conformed to this end were recognised as legitimate subjects of international law and their sovereignty regarded as beyond question. For the others, subjecthood would have to be earned by civilising wilderness off the face of the earth and creating a 'smooth space' for capitalist accumulation and exchange (Hardt and Negri, 2000). Until then, sovereignty would remain 'permeable, negotiable, penetrable, vulnerable' (Charlesworth and Chinkin, 2000, p. 130).

How much remains of this vision of sovereignty in contemporary international law has been the subject of intense debate in past decades, most notably in response to the ecological crisis (Litfin, 1998; Eckersley, 2004; Folch, 2016). On one level, the concept of sovereignty appears to have been largely secularised. The modern definition of the state, as reflected for instance in the 1933 Montevideo Convention, is based on seemingly abstract and neutral categories (territory, population, government) and centres first and foremost on the requirement of effectiveness, i.e., the presence of a stable political organisation strong enough to assert itself throughout the territory of the state, to the exclusion of other entities. But contemporary international law lays no specific requirements as to the nature and extent of this control, 'except that it include some degree of maintenance of law and order and the establishment of basic institutions' (Crawford, 2006, p. 59). The productive use of resources through labour and industry no longer constitutes a formal requirement of sovereignty. Modern statehood, in other words, is rooted less in the government of nature than in the government of people.

In this respect, contemporary international law can be said to be more Kantian than Vattelian. Whilst Kant, like most of his contemporaries, saw occupation as one of the accepted legal methods of acquiring sovereignty over territory, he did not see occupation as requiring productive development of the land. For Kant, asserting sovereignty required control over land but the form of possession was for each agent to determine: 'as long as they keep within their boundaries the way they want to live on their land is up to their own discretion' (Kant, 1996, p. 107). This, to an extent, is the position under contemporary international law: sovereignty is no longer formally premised on a given ecological regime, and land is no longer classified as *terra nullius* simply because it is sparsely populated or lightly utilised.

Whilst the above is certainly true at a formal level, it is also true that contemporary international law remains deeply committed, at a normative level, to the intensive exploitation of nature. Many of the assumptions about what it means to be a modern state, forged between the seventeenth and the nineteenth centuries, were not overturned but in many ways consolidated with decolonisation. The decolonisation process itself, to begin with, was in a sense designed to ensure the continuation of colonial patterns of ecological transformation. The League of Nations' Mandate System is a case in point. Established at the end of the First World War, the mandate system was a regime of international administration of non-European territories 'not yet able to stand by themselves', purportedly aimed at guiding their transition towards sovereignty. Mandate territories were classified according to their degree of advancement as class A, B, or C mandates, with class C mandates regarded as having to undertake the most significant societal development to reach self-government.

As noted by Natarajan and Khoday, this classification was largely reflective of the mandated societies' degree of control and use of their natural environment and their preparedness for integration in world markets (Natarajan and Khoday, 2014). Advancement from one stage of progress to the next required, among other things, opening up mandated territories to transnational capital so as to extract the raw materials which 'lay wasted and ungarnered' because the natives 'did not know their use or value' (Lugard, 1929, p. 613). To achieve sovereignty, colonial peoples – deemed incapable of supplying their own and others' needs – were expected to undertake a range of legal and administrative reforms to enable the optimal use of their surroundings, transitioning from subsistence economy to mining, manufacturing and intensive agriculture. This included a shift from communal landownership to private property, the alienation of land to concessionary companies, the commodification of natural resources, and exploitation at the industrial scale (Ross, 2017).

When colonial peoples in Latin America, Africa and Asia achieved their independence, these ideological prejudices against unproductive uses of nature were not cast off but in fact largely internalised by postcolonial states 'as they yearned to prove their mettle as states deserving of membership in the international community' (Hendlin, 2014, p. 148). Newly independent states engaged in performative demonstrations of possession to concretise their title to the land according to European signifiers of sovereignty. This was especially visible in campaigns aimed at reclaiming indigenous territories as a way of cementing their authority over spaces seen as weakly occupied and thus vulnerable to foreign encroachment. In what some have called a process of 'internal colonialism', many postcolonial states engaged in aggressive campaigns of settlement and intensive development of 'unoccupied' spaces to demonstrate their sovereign possession in some visible way (Maybury-Lewis, Macdonald and Maybury-Lewis, 2009).

This process often involved the violent exclusion of indigenous peoples that did not play the part and did not in some way make an economic use of the land. From 1879–1883, the Argentinean government launched the *conquista del desierto* (conquest of the desert) project, a military campaign aimed at flexing its sovereignty in the Pampa and Patagonian Desert, a region inhabited primarily by indigenous peoples. Tens of thousands of *indios* (Indians) were killed or displaced and replaced by white settlers sent to develop the land through irrigation and agriculture. The main justification for this campaign was that the desert was empty, and that emptiness made it dangerous, a threat to civilisation and the newly united Argentinian republic that nurtured it (Hopkins, 2020). Leaving the Pampa and Patagonia in a state of unproductive waste stood in the way of national wealth and displayed the weakness of Argentinean sovereignty. The making of the modern Argentinean state thus demanded the conquest and intensive development of the under-utilised interior hinterlands.

In Brazil, a similar campaign was launched in 1964–1985 to occupy the Amazon and intensively develop its resources through extraction, ranching, logging, and roads. Indigenous Amazonians were perceived as unable to fulfil the role of *occupatio* (occupation). Their non-exploitative way of living in the world was not performing land ownership, thus weakening Brazil's claim to the land, and leaving its sovereignty vulnerable to potential control by foreign nations or environmentalists attempting to frame the Amazon as the common heritage of mankind. These anxieties about the 'sovereignty-deflating possibilities of wild lands' (Hendlin, 2014) led the Brazilian government to strengthen its sovereign hold of the region by encouraging large-scale settlement by landless workers and the intensive extraction of its natural resources.

These examples illustrate the enduring nature of the *terra nullius* logic, which recognises sovereignty as performed and demonstrated through surplus extraction, i.e., the production of greater surplus than required for subsistence, and devalues other forms of land use. This logic, as Mickelson has remarked, 'is alive and well just beneath the surface' (Mickelson, 2014, p. 639). It can be seen at work, for instance, in the recent wave of global land grabs, i.e., the large-scale land acquisitions by private and public investors of farmland for the production of agricultural commodities. The global land rush is made possible by a range of different factors, including unequal economic relations and the relatively weak tenure security of local land users in many postcolonial states where, under the *uti possidetis* rule, ownership of the land was formally transferred from former colonial powers to newly independent states as public property (von Bernstorff, 2013).

But as Cotula has observed about African land-grabs, another factor facilitating the land rush is the widespread and ingrained perception that Africa 'has plenty of land' (The Economist, 2018) that is for all intents and purposes 'empty' and available for productive development (Cotula, 2013). Speaking at an international summit in May 2011, the Ethiopian prime minister called on international investors to develop commercial farms in Africa stressing that 'we have three million hectares of unutilized land. This land is not used by anybody. This land should be developed' (Cotula, 2013, p. 80). Despite empirical evidence showing that valuable agricultural land in Africa is often already used or claimed by local groups (Chouquer, 2012), the 'empty land' discourse provides a powerful narrative to legitimise large land deals with foreign investors. This narrative is made possible by a normative framework that delegitimises and devalues non-intensive land uses for grazing, foraging or hunting, or land set aside by communities for future generations, as unproductive, without owners and thus ready for allocation. International institutions like the World Bank have been key in sustaining this vision, frequently alluding to the presence of vast 'reserves' of prime agricultural land in poor countries lying idle and waiting to be grabbed by transnational capital so that they can be productively utilised (World Bank, 1975; 1978).

The normative coupling of sovereignty and extraction was further amplified by the international community's embrace of development as the prime objective of all properly governed states after the Second World War. As is well known, the architects of the post-war order envisioned economic progress as the key to lasting peace and prosperity. This required reconstructing Europe and uplifting the Third World out of poverty by replicating the world over the characteristic features of 'advanced' societies: high levels of industrialisation, technicalisation of agriculture, and the rapid growth of material production and living standards. International law played a crucial role in writing development deep into the fabric of international life in the post-war era (Rajagopal, 2003). The UN Charter lists the promotion of 'higher standards of living, full employment, and conditions of economic and social progress and development' as the primary means of creating 'conditions of stability and well-being which are necessary for peaceful and friendly relations among nations' (Charter of the United Nations 1945, art. 55). In Europe, the Rome Treaty provided that ensuring the 'development of economic activities' and 'mitigating the backwardness of the less favoured' represented the highest aims of the Community and the surest means of promoting international peace and security (Treaty Establishing the European Community 1957, Preamble).

The concept of development embraced as part of the post-war multilateral system was a notoriously narrow one (Esteva, 1992). While there might have been differences of opinion regarding matters such as tariffs, protectionism or state controls of the economy, there was little dissension about the main objectives of the development project: the expansion of mass consumption through extractive growth. The Final Act of the UN Conference on Trade and Employment referred to 'the expansion of the production, exchange and consumption of goods' as the primary purpose of the post-war settlement (Final Act of the UN Conference on Trade and Employment 1948). The General Agreement on Tariffs and Trade drew an even more explicit link between mass consumerism and resource extraction, noting that raising standards of living required 'developing the full use of the resources of the world' with a view to 'expanding the production and exchange of goods' (General Agreement on Tariffs and Trade 1947, Preamble).

On the road to development, 'backward' states were once again admonished to increase the productivity of nature to free their societies from want and need, working through different 'stages' of economic growth to reach the age of mass consumption (Rostow, 1960). In this vision, non-intensive uses of nature were not valued as alternative models of development but interpreted as symbols of economic stagnation and obstacles to human progress. As Sachs has observed, 'in this view, Tuaregs, Zapotecos or Rajasthanis [were] not seen as living diverse and non-comparable ways of human existence, but as somehow lacking in terms of what has been achieved by the advanced countries' (Sachs, 1992, p. 3). Catching up was thus declared to be the historical task of all non-industrial societies, and this was to be achieved by emulating the Western model of extractive growth.

This narrow understanding of economic growth as the historical task of properly governed states received its 'purest and most definitive form … in the halls of international financial institutions' (Mickelson, 2015, p. 97). There, postcolonial states were 'swept up in a vision of development as the pinnacle of human progress', and that pinnacle was itself defined as requiring to leave behind traditional existences and embracing Western industrial modernity (Mickelson, 2015, p. 97). The so-called 'Green Revolution' provides a prime example of this type of intervention, aimed at reforming the primitive and stagnant nature of underdeveloped European countries through the deployment of Western extractive technology. The Green Revolution was designed as a techno-political strategy for creating abundance in poor countries 'by breaking out of nature's limits and variabilities' (Shiva, 1991, p. 11). This was pursued through the introduction of high-yielding grain varieties and the adoption of modern methods of farming, including the use of chemical fertilisers, the concentration of land holdings, hydraulic infrastructure, and mechanisation. The World Bank and agencies like the Food and Agriculture Organisation (FAO) played a leading role in funding the Green Revolution, training agricultural experts, and promoting the legal and institutional reforms needed to implement it.

The ecological, social, and political instability generated by the Green Revolution is now well-documented (Pearse, 1980; Shiva, 1991; Tilman, 1998). The main point here, however, is that the Green Revolution was the explicit articulation of a philosophy which envisions the state first and foremost as environmental exploiter and which measures human progress in distinctly extractive terms. As Bryant and Bailey (1997, p. 85) have observed about the Green Revolution and other large development projects, 'in a manner reminiscent of colonial times, IFI-sponsored "progress" was measured in terms of trees felled, valleys flooded, minerals extracted and acreage dedicated to cash crops or cattle ranging'. A certain vision of what it means to be a well-functioning state was thus re-encoded at the heart of international law through the mantra of development. The 'rational' management of natural resources and the capacity to extract abundance out of scarcity became indicators of good government. The mismanagement of natural wealth, on the other hand, and the inability to squeeze development out of abundance have become closely associated with poor governance and state fragility (Nordstrom, 2008; Collier and Venables, 2010).

The development discourse has of course gone through a series of structural changes since the post-war era. The claim that rapid expansion of gross domestic product (GDP) would automatically translate to prosperity for all has been largely discredited. Likewise, the environmental wreckage caused by industrial development has become hard to ignore. The concept of sustainable development, coined in the 1980s and since endorsed by the UN as its main political blueprint for saving the world, constitutes the latest attempt to resolve these contradictions and integrate the economic, social, and environmental dimensions of development. As has been widely observed, however, sustainable development, as articulated in the Brundtland Report or, more recently, in the UN Sustainable Development Goals, proclaims the imperative of achieving development within planetary boundaries whilst simultaneously remaining committed to the frenzied exploitation and universal commodification of nature to reach the final goal of general abundance (Rist, 2008). Despite mounting awareness that the pursuit of endless industrial growth is eating through our living planet and threatening our existence, sustainable development remains, at its core, committed to logic of 'catching up' through forceful economic growth (no less than 7% GDP growth in the least developed countries) and ever-increasing levels of extraction, production, and consumption (Hickel, 2019). At the end of the day, the thing that is meant to be sustained is 'development', and environmental considerations are only allowed to stand in the way of growth when they start affecting productivity. Sustainable development 'reconfigured and reshuffled' (Escobar, 1995, p. 195) many of the concerns of classical development by adding in progressive-sounding qualifications ('inclusive' growth, 'responsible' consumption, 'efficient' resource use, etc.). Yet the metabolic patterns of the global economy are left broadly unaffected, and other scenarios – redistribution, degrowth, frugality – are all carefully cast aside, enabling the seamless pursuit of 'binge development' (Porras, 2015).

International law, in the end, is caught up in a deep contradiction that perpetuates and extends, rather than resolves, the intensity of the environmental crisis. It is self-reflexively aware of the need to protect the environment and manage nature rationally so that it continues to provide the benefits we rely on, now and in the future, whilst remaining structurally committed to its exhaustive exploitation. The field of international environmental law may bring that contradiction to the fore of international attention and polarise the terms slightly differently, addressing some of the worst excesses of an unregulated global marketplace along the way. Yet the conservation requirement does not subvert the extractive logic which sits at the core of the international legal order. International law remains as concerned with nature's exhaustion as it is about surplus extraction and the technocratic management of nature to maintain its productivity as a supplier of goods to humanity.

Conclusion: International Law and the Extractive Imaginary

According to a recent study, just 3% of the world's ecosystems remain ecologically intact, with healthy populations of animals and undisturbed habitats (Carrington, 2021). Save for a few fragments of untouched wilderness, there simply is 'no more nature that stands apart from human beings' (Purdy, 2015, p. 3). If by nature we understand that which is independent, autonomous, and free from human influence and impact, then nature is dead; it has been 'triumphantly blotted out' (Jameson, 2019, p. 309). And it isn't neutrally 'dead'. It is being destroyed and devastated by identifiable social processes (capitalist expansion, class relations, technologies, etc.) at a rate unseen in the past 10,000 years. We have entered the Anthropocene, and this isn't the 'good Anthropocene' that some have envisioned (Meyer, 2016).

To date, international law has failed spectacularly to respond to the challenges of the Anthropocene 'in any way that really counts' (Grear, 2013). A range of reasons have been offered to explain this state of affairs, from flawed policy frameworks to corporate capture, weak institutions or the de-prioritisation of the environment against short-term economic gain (United Nations Environment Programme, 2019). Whilst these are all valid considerations, they tend to focus on somewhat exogenous factors: the market, politics, social relations and so on. What this chapter has attempted to show is that, as well as these external considerations, international law has been a central enabling element of the Anthropocene because of a set of ideas and normative expectations which are deep-rooted in the system itself.

Chief among them is a remarkably stable pattern of thought about the essence of sovereignty as the taming of spontaneous, wild, recalcitrant and unpredictable nature. To an extent at least, the current environmental crisis is the consequence of a distinctly anthropocentric understanding of sovereignty as established and maintained through the methodical exploitation of nature, and the globalisation of that understanding through international law. As this chapter has attempted to show, this globalisation was not a random or historically contingent occurrence. It was in large part determined by the legal requirements of empire and the spread of global capitalism. As others have observed (Klein, 2014; Moore, 2016), to speak of the Anthropocene – or anthropocentrism for that matter – is for that reason problematic (and inimical to action). Blaming all of humanity for ecological breakdown or attributing it to some universal properties of the human species conceals the real drivers of environmental change: 'ours is the geological epoch not of humanity, but of capital' (Malm, 2015).

Global capitalist expansion requires the fungibility of nature, its conversion into something which can be propertised, transformed and traded. The process of incorporating nature into capital has historically, and to a significant degree, been a legal one (Özsu, 2019). International law, through its central juridical concept of sovereignty, has helped normalise a worldview fully compatible with that goal. The late James Crawford once observed that 'the term sovereignty has a long and troubled history, and a variety of meanings' (Crawford, 2006, p. 32). This chapter has shown that, as far as interpretations of nature are concerned, it also possesses a reasonably well-defined core, one that deprives nature of any autonomous life force and understands it, above all else, as an economic resource.

International law is caught up in what – borrowing from Ranganathan (2019) – we might call an 'extractive imaginary'. It envisions and evaluates the natural world first and foremost as a warehouse of commodities or, as Heidegger once lamented, 'one vast gasoline station for human exploitation' (cited in Harvey, 2014, p. 250). In this extractive imaginary, progress is defined in extractive terms. The primary end of the sovereign state – one might say its principal vocation under international law – must be to take control of nature to unearth the wealth buried in it. Put simply, 'states assert sovereignty over their environment and that is what makes them sovereign states' (Kuehls, 1998, p. 31). From this point of view, unmodified nature is seen as a sovereign aberration. It is against the nature of the state to not stand against nature. That is why Britain *must* frack Lancashire and Algeria *must* frack the Sahara.

The 'epistemology of mastery' that sits at the centre of international law's extractive imaginary has, to an extent, been upset and rendered unstable by the ecological crisis. Calls are multiplying for the development of new socio-juridical imaginaries to think our way out of the wreckage, including radical shifts in the way we think about the nature, the role, and the ends of the sovereign state (Grear, 2013; Eckersley, 2004; Philippopoulos-Mihalopoulos, 2017; De Lucia, 2020). Alternative understandings of the relationship between nature and law are 'stretching legal systems everywhere' (Natarajan and Dehm, 2019, p. 7), from law reforms granting rights to nature and non-human entities, to class actions on behalf of future generations or people's climate justice summits and tribunals. Amidst the blasted landscapes of the Capitalocene, 'fresh and green sprouts of change are emerging' (Grear and Bollier, 2020, p. 13).

But the ecological re-calibration of our key legal concepts and categories, necessary and long overdue though it may be, remains a marginal one to date. In practise, international institutions continue to refer to animal populations as 'stocks'. Healthy stocks are described as 'under-exploited'. And unspoilt ecosystems are seen as wasted opportunities for exploitation. On 13 February 2020, the World Bank tweeted an image of Bhutan's untouched old-growth forests (World Bank South Asia, 2020). The only carbon-negative country in the world, Bhutan's forest cover has remained remarkably stable over the years, with a national constitution mandating that at least 60% forest cover be maintained in perpetuity. The World Bank looked at healthy forests and saw a problem requiring its intervention: '71% of Bhutan's territory is covered in forest, but with a contribution of only about 2% to GDP per year, the forest sector remains underutilized'. Bhutan is letting its forest stand, doing nothing (except of course providing habitat for millions of animals, producing oxygen, and capturing CO_2) 'when they could be turned into Ikea furniture' (Hickel, 2020).

The extractive imaginary is alive and kicking.

References

Adams, J. (1819) *Novanglus, and Massachusettensis; Political Essays, Published in the Years 1774 and 1775, on the Principal Points of Controversy, Between Great Britain and Her Colonies*. Boston: Hews & Goss.

Alschuler, A.W. (1996) 'Rediscovering blackstone', *University of Pennsylvania Law Review*, 145(1), pp. 1–55.

Anghie, A. (1999) 'Francisco de Vitoria and the colonial origins of international law', in Darian-Smith, E. and Fitzpatrick, P. (eds.) *Laws of the Postcolonial*. Ann Arbor: University of Michigan Press, pp. 89–107.

Anghie, A. (2005) *Imperialism, Sovereignty and the Making of International Law*. Cambridge: Cambridge University Press.

Armitage, D. (2012) 'John Locke: Theorist of empire?', in Muthu, S. (ed.) *Empire and Modern Political Thought*. Cambridge: Cambridge University Press, pp. 84–111.

Arneil, B. (1996) *John Locke and America – The Defence of English Colonialism*. Oxford: Clarendon Press.

Beard, J.L. (2007) *The Political Economy of Desire – International Law, Development and the Nation State*. New York: Routledge-Cavendish.

Bhandar, B. (2018) *Colonial Lives of Property – Law, Land, and Racial Regimes of Ownership*. Durham: Duke University Press.

Bilder, R.B. (1980) 'International law and natural resources policies', *Natural Resources Journal*, 20(3), pp. 451–486.

Blackstone, W. (1771) *Commentaries on the Laws of England – Book II – Of the Rights of Things*. Philadelphia: Robert Bell.

Bluntschli, J.K. (1883) *La Politique*. Paris: Librairie Guillaumin.

Bluntschli, J.K. (1885) *The Theory of the State*. Oxford: Clarendon.

Bonneuil, Ch. (2019) 'Seeing nature as a "universal store of genes": How biological diversity became "genetic resources", 1890–1940', *Studies in History and Philosophy of Science*, 75, pp. 1–14. DOI: 10.1016/j.shpsc.2018.12.002

Brown, W. (2010) *Walled States, Waning Sovereignty*. New York: Zone Books.

Bryan, R.L. and Bailey, S. (1997) *Third World Political Ecology*. London: Routledge.

Carrington, D. (2021) 'Just 3% of world's ecosystems remain intact, study suggests', *The Guardian*, 15 April [Online]. Available at: https://www.theguardian.com/environment/2021/apr/15/just-3-of-worlds-ecosystems-remain-intact-study-suggests (Accessed: 31 May 2021).

Carty, A. (1996) *Was Ireland Conquered? International Law and the Irish Question*. London: Pluto Press.

Chaplin, J.E. (1993) *An Anxious Pursuit – Agricultural Innovation and Modernity in the Lower South, 1730–1815*. Chapel Hill: University of North Carolina Press.

Charlesworth, H. and Chinkin, C. (2000). *The Boundaries of International Law: A Feminist Analysis*. Manchester: Manchester University Press

Chetail, V. and Haggenmacher, P. (eds.) *Vattel's International Law in a XXIst Century Perspective*. Leiden: Brill.

Chouquer, G. (2012) 'L'Afrique est-elle disponible? Ce que l'on voit quand on regarde', *Grain de Sel*, 57, pp. 7–8.

Collier, P. and Venables, A.J. (2010) 'Natural resources and state fragility', EUI Working Papers RSCAS, No. 2010/36.

Corcoran, P. (2017) 'John Locke on native right, colonial possession, and the concept of *vacuum domicilium*', *The European Legacy*, 26(2), pp. 225–250. DOI: 10.1080/10848770.2017.1416766

Cotula, L. (2013) *The Great African Land Grab? Agricultural Investments and the Global Food System*. London: Zed Books.

Crawford, J. (2006) *The Creation of States in International Law*, 2nd edn. Oxford: Clarendon Press.

Crawford, J. (2012) *Brownlie's Principles of Public International Law*, 8th edn. Oxford: Oxford University Press.

De Lucia, V. (2020) 'Rethinking the encounter between law and nature in the anthropocene: From biopolitical sovereignty to wonder', *Law and Critique*, 31, pp. 329–349. DOI: 10.1007/s10978-020-09281-9

de Waart, P.J.I.M. (2009) 'Permanent sovereignty over natural resources as a corner-stone for international economic rights and duties', *Netherlands International Law Review*, 24(1), pp. 304–322. DOI: 10.1017/S0165070X00016302

Dominiczak, P. (2014) 'David Cameron: Britain has a "duty" to frack', *The Telegraph*, 26 March [Online]. Available at: https://www.telegraph.co.uk/news/earth/energy/fracking/10723294/David-Cameron-Britain-has-a-duty-to-frack.html (Accessed 12 December 2020).

Drayton, J. (1802) *A View of South-Carolina as Respects her Natural and Civil Concerns*. Charleston: W.P. Young.

Drayton, R. (2000) *Nature's Government – Science, Imperial Britain, and the 'Improvement' of the World*. New Haven: Yale University Press.

Eckersley, R. (2004) *The Green State – Rethinking Democracy and Sovereignty*. Cambridge: MIT Press.

Escobar, A. (1995) *Encountering Development – The Making and Unmaking of the Third World*. Princeton: Princeton University Press.

Esteva, G. (1992) 'Development', in Sachs, W. (ed) *The Development Dictionary – A Guide to Knowledge as Power*. London: Zed Books.

Fitzmaurice, A. (2014) *Sovereignty, Property and Empire, 1500–2000*. Cambridge: Cambridge University Press.

Fitzmaurice, A. (2015) 'Sovereign trusteeship and empire', *Theoretical Inquiries in Law*, 16(2), pp. 447–471. DOI: 10.1515/til-2015-108

Folch, C. (2016) 'The nature of sovereignty in the anthropocene: Hydroelectric lessons of struggle, otherness, and economics from paraguay', *Current Anthropology*, 57(5), pp. 565–85. DOI: 10.1086/688580

Gascoigne, J. (2002) *The Enlightenment and the Origins of European Australia*. Cambridge: Cambridge University Press.

Gong, G.W. (1984) *The Standard of 'Civilization' and International Society*. Oxford: Clarendon.

Graham, N. (2011) *Lawscape – Property, Environment, Law*. Abingdon: Routledge.

Grear, A. (2013) 'Towards a New Horizon: In search of a renewing socio-juridical imaginary', *Oñati Socio-Legal Series*, 3(5), pp. 966–990.

Grear, A. and Bollier, D. (2020) *The Great Awakening – New Modes of Life amidst Capitalist Ruins*. Goleta: Punctum Books.

Greer, A. (2018) *Property and Dispossession – Natives, Empires and Land in Early Modern North America*. Cambridge: Cambridge University Press.

Hardt, M. and Negri, A. (2000) *Empire*. Cambridge: Harvard University Press.

Harvey, D. (2014) *Seventeen Contradictions and the End of Capitalism*. Oxford: Oxford University Press.

Hendlin, Y.H. (2014) 'From *Terra Nullius* to *Terra Communis*: Reconsidering wild land in an era of conservation and indigenous rights', *Environmental Philosophy*, 11(2), pp. 141–174. DOI: 10.5840/envirophil20143205

Hickel, J. (2019) 'The contradiction of the sustainable development goals: Growth versus ecology on a finite planet', *Sustainable Development*, 27(5), pp. 873–884. DOI: 10.1002/sd.1947

Hickel, J. (2020) [Twitter], 16 October. Available at: https://twitter.com/jasonhickel/status/1317028075947892736?lang=en (Accessed 1 June 2021).

Hopkins, B. D. (2020) *Ruling the Savage Periphery – Frontier Governance and the Making of the Modern State*. Cambridge: Harvard University Press.

Hutzler, A. (2021) 'Rick Santorum slammed for saying America was birthed from nothing, there "Isn't Much Native American Culture"', *Newsweek*, 26 April [Online]. Available at: https://www.newsweek.com/rick-santorum-slammed-saying-america-was-birthed-nothing-there-isnt-much-native-american-1586529 (Accessed 29 April 2021).

Ince, O.U. (2018) *Colonial Capitalism and the Dilemmas of Liberalism*. Oxford: Oxford University Press.

Ivison, D. (2003) 'Locke, liberalism, and empire', in Anstey, P. (ed.) *The Philosophy of John Locke – New Perspectives*. New York: Routledge.

Jameson, F. (2019) *Postmodernism – Or, the Cultural Logic of Late Capitalism*. London: Verso.

Jefferson, T. (1774) *A Summary View of the Rights of British America*. Williamsburg: Clementina Rind.

Jiménez-Fonseca, M. (2017) *Civilizing Nature – Revisiting the Imperialist History of International Law, 1511–1972*. PhD thesis. University of Helsinki.

Jodoin, S. (2008) 'International law and alterity: The state and the other', *Leiden Journal of International Law*, 21, pp. 1–28. DOI: 10.1017/S0922156507004700

Jouannet, E. (2012a) 'Emmer De Vattel (1714–1767)', in Fassbender, B. and Peters, A. (eds) *The Oxford Handbook of the History of International Law*, Oxford: Oxford University Press.

Jouannet, E. (2012b) *The Liberal-Welfarist Law of Nations – A History of International Law*. Cambridge: Cambridge University Press.

Kant, I. (1996) *The Metaphysics of Morals*. Cambridge: Cambridge University Press.

Klein, N. (2014) *This Changes Everything – Capitalism vs. The Climate*. London: Penguin Books.

Koskenniemi, M. (2011) 'Empire and international law: The real spanish contribution', *University of Toronto Law Journal*, 61(1), pp. 1–36. DOI: 10.3138/utlj.61.1.001

Koskenniemi, M. (2016) 'Race, hierarchy and international law: Lorimer's legal science', *European Journal of International Law*, 27(2), pp. 415–429. DOI: 10.1093/ejil/chw017

Kovats, S., Depledge, M., Haines, A., Fleming, L.E., Wilkinson, P., Shonkoff, S.B. and Scovronick, N (2014) 'The health implications of fracking', *The Lancet*, 383, pp. 757–758. DOI: 10.1016/S0140-6736(13)62700-2

Kuehls, T. (1999) 'Between sovereignty and environment: An exploration of the discourse of government', in Litfin, K.T. (ed.) *The Greening of Sovereignty in World Politics*. Cambridge: MIT Press.

Lamri, R. (2015) 'A question of sovereignty, justice and dignity: The people vs. the government on fracking in Algeria', *Open Democracy*, 4 March [Online]. Available at: https://www.opendemocracy.net/en/north-africa-west-asia/question-of-sovereignty-justice-and-dignity-people-vs-government-on-fra/ (Accessed 12 December 2020).

Linklater, A. (2013) *Owning the Earth – The Transforming History of Land Ownership*. London: Bloomsbury.

Litfin, K.T. (ed.) (1998) *The Greening of Sovereignty in World Politics*. Cambridge: MIT Press.

Locke, J. (2003) *Two Treatises of Government and A Letter Concerning Toleration*. New Haven: Yale University Press.

Lorimer, J. (1883) *The Institutes of the Law of Nations – A Treatise of the Jural Relations of Separate Political Communities – Vol. I*. Edinburgh: William Blackwood and Sons.

Lugard, F. (1929) *The Dual Mandate in British Tropical Africa*. London: William Blackwood.

Malm, A. (2015) 'The anthropocene myth', *The Jacobin*, 30 March [Online]. Available at: https://www.jacobinmag.com/2015/03/anthropocene-capitalism-climate-change/ (Accessed 1 June 2021).

Maybury-Lewis, D., Macdonald, T. and Maybury-Lewis, B. (eds.) (2009) *Manifest Destinies and Indigenous Peoples*. Cambridge: Harvard University Press.

Mégret, F. (2012) 'L'étatisme spécifique du droit international', *Revue Québécoise de Droit International*, 24(1), pp. 105–129.

Meyer, W.B. (2016) *The Progressive Environmental Prometheans – Left-Wing Heralds of a 'Good Anthropocene'*. London: Palgrave Macmillan.

Mickelson, K. (2014) 'The maps of international law: Perceptions of nature in the classification of territory', *Leiden Journal of International Law*, 27(3), pp. 621–639. DOI: 10.1017/S0922156514000235

Mickelson, K. (2015) 'International law as a war against nature? Reflections on the ambivalence of international environmental law', in Stark, B. (ed.) *International Law and Its Discontents – Confronting Crises*. Cambridge: Cambridge University Press.

Miéville, C. (2006) *Between Equal Rights – A Marxist Theory of International Law*. London: Pluto Press.

Moore, J.W. (ed.) (2016) *Anthropocene or Capitalocene? Nature, History, and the Crisis of Capitalism*. Oakland: PM Press.

Moore, M. (2015) *A Political Theory of Territory*. Oxford: Oxford University Press.

Natarajan, U. and Dehm, J. (2019) 'Where is the environment? Locating nature in international law', *TWAILR – Reflections*, 30 August [Online]. Available at: https://twailr.com/where-is-the-environment-locating-nature-in-international-law/ (Accessed: 6 October 2022).

Natarajan, U. and Khoday, K. (2014) 'Locating nature: Making and unmaking international law', *Leiden Journal of International Law*, 27(3), pp. 573–593. DOI: 10.1017/S0922156514000211

Neocleous, M. (2011) 'War on waste: Law, original accumulation and the violence of capital', *Science & Society*, 75(4), pp. 506–528.

Neocleous, M. (2012) 'International law as primitive accumulation; or, the secret of systematic colonization', *European Journal of International Law*, 23(4), pp. 941–962. DOI: 10.1093/ejil/chs068

Nordstrom, S. (2008) 'Fragility and natural resources', Danish Institute for International Studies Policy Brief, pp. 1–4.

Özsu, U. (2019) 'Grabbing land legally: A Marxist analysis', *Leiden Journal of International Law*, 32, pp. 215–233. DOI: 10.1017/S0922156519000025

Pagden, A. (1998) *Lords of All the Word – Ideologies of Empire in Spain, Britain and France c.1500-c.1800*. New Haven: Yale University Press.

Parfitt, R. (2019) *The Process of International Legal Reproduction – Inequality, Historiography, Resistance*. Cambridge: Cambridge University Press.

Pearse, A. (1980) *Seeds of Plenty, Seeds of Want – Social and Economic Implications of the Green Revolution*. Oxford: Clarendon Press.

Peduzzi, P. and Harding Rohr Reis, R. (2013) 'Gas fracking: Can we safely squeeze the rocks?', *Environmental Development*, 6, pp. 86–99. DOI: 10.1016/j.envdev.2012.12.001

Philippopoulos-Mihalopoulos, A. (2017) 'Critical environmental law as method in the anthropocene', in Philippopoulos-Mihalopoulos, A and Brooks, V. (eds.) *Research Methods in Environmental Law – A Handbook*. Cheltenham: Edward Elgar.

Phillimore, R. (1879) *Commentaries Upon International Law*. London: Butterworths.

Porras, I. (2014) 'Appropriating nature: Commerce, property, and the commodification of nature in the law of nature', *Leiden Journal of International Law*, 27(3), pp. 641–660. DOI: 10.1017/S0922156514000247

Porras, I. (2015) 'Binge development in the age of fear: Scarcity, consumption, inequality, and the environmental crisis', in Stark, B. (ed), *International Law and Its Discontents – Confronting Crises*. Cambridge: Cambridge University Press.

Purdy, J. (2015) *After Nature – A Politics for the Anthropocene*. Cambridge: Harvard University Press.

Rajagopal, B. (2003) *International Law from Below – Development, Social Movements, and Third World Resistance*. Cambridge: Cambridge University Press.

Ranganathan, S. (2019) 'Ocean floor grab: International law and the making of an extractive imaginary', *European Journal of International Law*, 30(2), pp. 573–600. DOI: 10.1093/ejil/chz027

Reid Ross, A. (2015) 'The Ain Salah uprising', *Counterpunch*, 13 March [Online]. Available at: https://www.counterpunch.org/2015/03/13/the-ain-salah-uprising/ (Accessed 28 April 2021).

Rist, G. (2008) *The History of Development – From Western Origins to Global Faith*, 3rd edn. London: Zed Books.

Ross, C. (2017) *Ecology and Power in the Age of Empire – Europe and the Transformation of the Tropical World*. Oxford: Oxford University Press.

Rostow, W.W. (1960) *The Stages of Economic Growth – A Non-Communist Manifesto*. Cambridge: Cambridge University Press.

Sachs, W. (ed.) (1992) *The Development Dictionary – A Guide to Knowledge as Power*. London: Zed Books.

Schmitt, C. (2006) *The Nomos of the Earth in the International Law of the Jus Publicum Europaeum*. New York: Telos Press.

Schrijver, N. (1997) *Sovereignty over Natural Resources – Balancing Rights and Duties*. Cambridge: Cambridge University Press.

Shiva, V. (1991) *The Violence of the Green Revolution – Third World Agriculture, Ecology and Politics*. London: Zed Books.

Simpson, G. (2008) 'The guises of sovereignty', in Jacobsen et al (eds) *Re-envisioning Sovereignty – The End of Westphalia?*. Aldershot: Ashgate, pp. 51–69.

Stapelbroek, K. and Trampus, A. (eds) *The Legacy of Vattel's Droit des gens*. London: Palgrave Macmillan.

The Economist (2018) 'Africa has plenty of land. Why is it so hard to make a living from it?', 28th April [Online]. Available at: https://www.economist.com/middle-east-and-africa/2018/04/28/africa-has-plenty-of-land-why-is-it-so-hard-to-make-a-living-from-it (Accessed 27 May 2021).

Tilman, D. (1998) 'The greening of the green revolution', *Nature*, 396, pp. 211–212. DOI: 10.1038/24254

Tully, J. (1993) *An Approach to Political Philosophy – Locke in Contexts*. Cambridge: Cambridge University Press.

Tzouvala, N. (2020) *Capitalism as Civilisation – A History of International Law*. Cambridge: Cambridge University Press.

United Nations Environment Programme (2019) *Environmental Rule of Law – First Global Report*. Nairobi: UNEP.

Vattel, E. (1844) *The Law of Nations or Principles of the Law of Nature, Applied to the Conduct and Affairs of Nations and Sovereigns*, 6th edn. Philadelphia: T. & J.W. Johnson.

Vaughan, A.T. (1997) *The Puritan Tradition in America, 1620–1730*. Hanover: University Press of New England.

von Bernstorff, J. (2013) 'The global "Land-Grab", sovereignty and human rights', *ESIL Reflections*, 2(9), pp. 1–6.

Westlake, J. (1894) *Chapters on the Principles of International Law*. Cambridge: Cambridge University Press.

Williams, R. (1990) *The American Indian in Western Legal Thought – The Discourses of Conquest*. Oxford: Oxford University Press.

Witte, G. and Faiola, A. (2014) 'Amid showdown with energy-rich Russia, calls rise in Europe to start fracking', *Washington Post*, 7 April [Online]. Available at: https://www.washingtonpost.com/world/amid-showdown-with-energy-rich-russia-calls-rise-in-europe-to-start-fracking/2014/04/07/f3616058-2c24-4683-abe3-728a5572debf_story.html (Accessed 18 December 2020).

World Bank (1975) *Land Reform*. Washington, DC: World Bank.

World Bank (1978) *World Development Report 1978*. Washington, DC: World Bank.

World Bank South Asia (2020) [Twitter]. 13 February. Available at: https://twitter.com/WorldBankSAsia/status/1227865071164022784 (Accessed 1 June 2021).

Yirush, C. (2011) *Settlers, Liberty, and Empire – The Roots of Early American Political Theory, 1676–1775*. Cambridge: Cambridge University Press.

2

THE ANTHROPOCENTRISM OF HUMAN RIGHTS

Frédéric Mégret

Introduction

This chapter considers a claim that is on one level powerfully intuitive and compelling, yet on another level perhaps so obvious as to be unintelligible from within the categories of human rights; namely, that human rights are anthropocentric. By anthropocentric I mean a philosophy of centring the human, based on the idea that humans have primary moral standing and are the only measure of justice. That human rights are anthropocentric may strike one as precisely the point. After all, human rights are the rights of humans and, therefore, how could they be anything other than anthropocentric?

The paradox, however, is that historically and philosophically, human rights are not always or specifically articulated as an anthropocentric claim. Rather, human rights are first and foremost claims made about what I will describe as intra-human justice, namely, the justice of particular social and political human formations. The foundational Western human rights instruments, for example, were adopted as an attempt to resist tyranny and the oppression of some humans by other humans (humans' 'inhumanity' to each other or themselves); even as human rights have moved from that minimal liberal script, for example through attention to economic and social or collective rights, they have highlighted the importance of the state for guaranteeing such rights for humans. The overall impression is certainly one of a project of justice for humans and by humans, and one that does not have much to say about the non-human realm. But what if human rights were understood not only as the-rights-of-humans but, in a fundamentally anthropocentric vein, as the-rights-that-only-humans-can-have? What if the right-to-have-rights (and confer them) was thereby affirmed as the defining prerogative of humanity? What is left out in a movement that entrenches the centrality of the rights of humans amidst the social orders that they create (for themselves)?

In this chapter, I ask the question of the excluded middle in this fundamental setup. I seek to highlight the possibility that, niched amidst otherwise highly legitimate concerns about the harms and goods that humans may do to each other, lies an even more fundamental implicit claim about the superiority and sacrality of all things human. I make the case that the human rights project is, in fact, more deeply, centrally, and problematically anthropocentric than the focus on intra-human justice suggests; that it is, in fact, only exclusively focused on intra-human justice because it has fundamentally bracketed and excluded questions of relations to other

DOI: 10.4324/9781003201120-4

35

animal species and, more broadly, to the environment. I suggest that, by contrast, much can be gained from a reading of human rights as not so much a political arrangement inter se (as suggested, for example, by the anthropocentric focus of the language of social contract) but at least in part as much as a political arrangement from which nature, including non-human animals, is specifically and deliberately – if implicitly and covertly – excluded.

Intellectually, the chapter is indebted to a tradition of thought that has sought to address deep questions of identity not in the cogito but in the relationship to the other. Just as for Anthony Anghie, for example, sovereignty was 'forged out of the attempt to create a legal system that could account for relations between the European and non-European worlds in the colonial confrontation' rather than the intra-European dealings of states (Anghie, 2007, p. 3), I will argue that there could not be a 'human' in international human rights law without the human encounter with its radical (animal, natural) other.

That other was for a long time and arguably still to this day a human 'other', characteristically and metaphorically framed as close to nature (primitive, savage, bestial, etc.) and therefore not partaking in those quintessential qualities that make the human *human* (women, blacks, Jews) (Rorty, Shute, and Hurley, 1993); but it is, if nothing else, the quite literally non-human, namely 'brute animals' and, beyond them, the vast realm of living and inanimate nature, as the ultimate background prop in the human morality tale. The attribution of pejorative 'natural' characteristics to excluded human minorities – however much it tells us about the treatment of those minorities – also tells us much about the underlying contempt in which those characteristics are held (Srinivasan, 2015; Deutscher, 2008; Anderson, 2017). It is in the crucible of that encounter with humanity's radical 'other' – at least as much as in the self-referential encounter of humans with other humans – that I argue human rights were fundamentally conceived.

The key question for our purposes, then, is not so much whether non-human animals or nature should have rights, but what it might mean for the human rights project itself if they were to have such rights. In turn, that debate hinges on the degree to which protections of non-human animals or nature should be anthropocentric or ecocentric. Closely related is the notion of whether the foundation for rights protection is deontological or utilitarian and, ultimately, what the basis for humans having rights is in the first place (Benton, 1998). As can be gathered, to inquire about the anthropocentrism of human rights, both as a practise and as a theory, is necessarily to re-explore the history of the human rights movement. Debates in international and human rights law on these contested issues have lagged behind debates in political theory and ethics, even as they have been largely derivative of those debates.

The chapter thus aims to contextualise the question of the anthropocentrism of human rights in some of the more salient debates on ecology and environmental ethics of the last decades on the one hand, but also to highlight, when necessary, how the specific constraints of legal form inform those debates. It begins by tracing the origin of the exclusion of the non-human from the historical rights project in fundamental assumptions about rights born with modernity. It then traces the continuation of this obfuscation all the way to contemporary human rights instruments mediated by international law and discusses how the dominance of human rights makes certain aspects of the devastation brought about by the Anthropocene fundamentally unintelligible. Finally, the chapter highlights some of the complex ways in which reinventions of human rights along less anthropocentric lines are already ongoing and need to be problematised.

Throughout the chapter, I grapple with what might be described as the anthropocentric curse: The fact that any attempts to transcend rights must contend with the fact that they are ultimately proclaimed through human imagination, language, and institutions, and at best involve the reimagining of more inclusive *human* cosmogonies. I suggest that this curse may be a blessing, and that the true incidence of moving beyond an anthropocentric conception of human rights, rather than an abandonment of what makes humans *human*, may be a renewed ability to imagine humanity. In short, decentring the human as a creature of its own making may also help envisage opportunities for a more inclusive and less alienated sense of human-ness.

The Emergence of Human Rights as an Anthropocentric Moment

Human Rights as Humanist Pursuit

Pinning down the vast worlds of human rights to one central intuition is notoriously difficult. However, perhaps all theories and practises of human rights share at least an implicit understanding that they are about the centrality of humans' becoming (to, at least, humans). Whether the foundational conceit is that of some fundamental human nature or the political notion of the 'social contract', human rights emerge first and foremost as part of an attempt to give meaning to human, particularly social and political, existence. Notably, the key questions that the human rights project asks, historically, are questions about the proper limits of a characteristically human institution, the State.

The specific details of these intra-human controversies need not detain us at this stage. It is important to note how such claims, within the larger arc of human civilisation, have been simultaneously emancipatory and oppressive. They are implicated in the human liberation from tyranny, tradition, and hierarchy; they are also deeply implicated in exclusionary, imperial, and destructive patterns. It is the same 'human rights' that bring down the tyrant and colonise the other. It is not, of course, as if the particulars of the history of human rights for humans do not matter, far from it; in fact, as I will suggest, the continued decolonisation of the human rights project may well be an opportunity to simultaneously forge new conceptions of rights and their relationship to nature (Maldonado-Torres, 2017).

The point is merely that that story, compelling as it may be on its own terms, cannot fully be understood if one ignores what it simultaneously excludes. The more interesting story, in fact, may be one hiding in plain sight amidst intra-human discussions about the justice, effectiveness and success of rights: Whatever else it may be, the human rights project is, at heart, the heir to an older *humanist* project that puts humans at the centre of the world. This is true in the narrow historical sense that Renaissance humanism prepared the ground for the human rights revolution in proclaiming 'the dignity and excellence of man' (Steenbakkers, 2014). But it is also true in the sense that the human rights movement is the culmination of a movement that has invented 'man' as its own centre of the world, away from earlier cosmologies that subordinated the human to the God-like. It is fundamentally a *philanthropic* project, a story of mankind being enamoured with mankind (Weitzenfeld and Joy, 2014).

The origins of this historically incongruous mindset can be found in antiquity, notably the Stoics, and in ideas of justice being fundamentally a cosmopolitan issue from which nature and animals are excluded except as the implicit background to 'fall from grace' narratives (where the fall from grace is nonetheless entirely understood in anthropocentric terms and not as entailing continuing duties to the animals with whom humans previously lived in peace) (Steiner, 2010). The tradition of human rights is a distant offshoot of such attempts to grapple with notions of the just society understood as the ordering of purely human affairs, albeit one that may well add an extra layer of anthropocentrism because of the way in which it further idealises humans as the primary subject of rights. Human rights as rights inter se: Granted by humans, for humans.

Theories of justice in the Western canon, from Aristotle to Rawls, then, may differ considerably in their identification of salient human issues and their inspiration (Theistic or Godless, idealist or materialistic, etc.), but otherwise tend to implicitly agree on the notion that the goal of theories of justice is to address the continuing difficulties of humans living alongside other humans. As Gary Steiner has pointed out, this means that theories of justice are for the most part not whole and that they can only be understood 'against the background of a more fundamental theory of the natural order and our place in it' (Steiner, 2013, p. 169) that is taken as so evident as to never fully need to be extrapolated.

Defining the Human Rights Project in Opposition to Nature

How, then, might one rewrite the history of the human rights project not (or at least not only) as the working through of intra-human intuitions about justice but as, more fundamentally, part of the ongoing constitution of the relationship between humans and nature? In the traditional human rights narrative, of course, it is hardly as if nature is absent. It is famously offered by liberal thinkers as a glimpse into an imagined 'state of nature' understood as more or less hellish for humans, with significant consequences for the tenor of the rights project. For those of a more Hobbesian disposition, the state of nature is ghastly and the Leviathan the only hope; for those of a more Lockean inclination, the state of nature is gentler and a reminder that humans only enter the social contract on condition of their natural rights being respected.

However, the 'state of nature' and 'nature' are not the same thing. The former is a highly stylised and contrived prequel to the social contract that disappears as soon as it has served its purpose and which, ultimately, says more about imagined relations between humans than anything about, as it were, the nature of nature. Something resembling actual nature reappears in the human rights project but in at least three, more complex ways.

First, the human rights project is premised on a fundamental intuition about the distinction of the human subject from nature and not simply, as is conventionally understood, about the equality of humans between themselves. As Lynn Hunt has shown, the modern discourse of human rights was in fact preceded by eighteenth-century efforts to distinguish humanity from both God and animals, far more than preoccupied with the by now familiar rights claims against the state (Hunt, 2008, p. 23). The claim that only humans have rights is, in turn, tied to ideas about the distinctiveness and, indeed, the superiority of human animals. The origins of such a claim in the Western tradition are clearly tied to Biblical injunction (Genesis (1:26): 'Then God said, "Let us make mankind in our image, in our likeness"'). There is an abundance of characteristics that are held to set apart humans from the non-human world (reason, morality, emotions, etc.), some of which are tied to humans' very relationship to nature, such as their ability to deliberately master and engineer it. Humans do not just live in their environment – they fundamentally transform it in their image and to suit their needs. Each of these foundations is susceptible to lead to different rights projects; but all share a sense of compelling human uniqueness.

Second, the human rights project, having proclaimed the fundamental distinguishability of humans from nature doubles down on that intuition by defining freedom as the fundamental right in a liberal order. Freedom involves the right to pursue one's uniquely human goals, to self-determine unhindered. That freedom, in the conventional modern rights narrative, is conceived principally in political terms as freedom from the State, where the only limitation is framed by the demands of collective and public order as well as the rights of other humans. In other words, human rights contain no embedded limitation to human freedoms that would not themselves be justified by the need to maximise human freedoms.

Lurking within that familiar political definition of freedom, however, is an underlying claim that does not even need to be mentioned, namely that there are no worthwhile non-human limitations to freedom – that humans, in short, should have complete freedom from nature itself. Human freedom is the freedom to continuously emancipate oneself from nature and, in the process, further cement one's humanity as fundamentally distinct from nature (Schroeder,

2002, pp. 296–297). The Prometheanism implicit in such an attitude is what gave modernity its unbounded optimism about never-ending growth and ever-higher levels of both security and material comfort.

Third and logically, claims about human rights quickly lead to claims of dominion over nature as the natural consequence of freedom from nature. Nature does not have dominion over itself. Genesis (1:26) makes it clear that the world was created 'so that they may rule over the fish in the sea and the birds in the sky, over the livestock and all the wild animals, and over all the creatures that move along the ground'. Aristotle famously considered that: 'if nature makes nothing in vain, the inference must be that she has made all animals for the sake of man'. The purely scriptural and ancient basis for man's dominion is eventually recycled through nascent liberalism, except that it is restated as a humanist philosophy: 'God, who hath given the world to men in common, hath also given them reason to make use of it to the best advantage of life and convenience. The earth and all that is therein is given to men for the support and comfort of their being' (Locke, 1961, p. 134). In fact, nature left to waste is of no value, and whatever value it acquires it acquires mostly through human labour (Locke, 1961, p. 141).

A fortiori, this means that humans can engage in killing non-human animals. Just as the end of society is self-preservation, killing 'wild savage beasts' such as lions and tigers amidst which 'men can have no society nor security' (Locke, 1961, p. 126) is fundamental to human freedom. More fundamentally, this means that humanity can 'remake the world' in its image (Genesis (28): 'fill the earth and subdue it. Rule over the fish in the sea and the birds in the sky and over every living creature that moves on the ground'). Of course, quite radically for the times, Locke makes this point about the availability of nature for *all* men (at least, of course, as he imagines them to be) and not just the Monarch. Property rights protect humans from unjust confiscation. Yet behind that equalising human ambition lies the irrepressible reality of nature's dispossession from itself: Nature is *equally* to be instrumentalised by all humans.

Natural Rights vs. Nature's Rights?

The fate of rights, moreover, continues to be deeply constituted by the relationship to nature on a more ontological level. In a cruel twist, the primacy of the human is often justified by reference to nature itself. It is as if it were nature that inspired the secret of human domination to humans. A vision of human rights as grounded in nature naturally inclines one to imagine the default of humanity to be untrammelled freedom in a pre-societal state. The goal of human rights then becomes to perpetuate, as much as possible and compatible with the exigencies of social life (but only such exigencies), those freedoms, but without making any concessions to the rights of other beings or realms. 'Natural rights', it turns out, have very little to say about the 'rights of nature' and, in fact, merely usher in an era of domination, even if tempered by the occasional hints of conservationism (Liebell, 2011).

At the same time, there is no doubt that the project of human rights is a project of distinction from the merely natural. Rights may get their cue from nature, but the ability to have them is precisely what distinguishes humans as those creatures capable of leaving behind the 'slavery' of 'mere appetite' to master their destiny (Rousseau, 1997, p. 54). At any rate, how would one know what nature 'says' in a world riddled by value conflicts? That nature has a certain order does not translate well into moral prescription. Neither animals nor natural forces seem to treat humans as special. If anything, one may be wary of what invocations of this natural order portend, given associations of nature with ideas about biological and racial superiority, and indeed 'natural' ideas have justified a range of practises such as slavery that one has every reason to condemn from a righteously anthropocentric framework. Natural law theory often only held together as a result of earlier theistic commitments that saw nature as having been divinely ordained. In a thoroughly disenchanted world, that foundation crumbles and with it the idea that nature ought to elicit any particular commitment to rights.

Over time, the foundation of rights has thus emancipated itself even from *that* tenuous connection to nature, to re-found human rights in a much more explicitly political, contractarian and self-derived fashion. If anything, this has further severed any indebtedness to nature by making human will the foundation of all rights. The idea of human rights, then, appears as the endlessly repeated move by which humans grant themselves rights, a sort of perpetual performative 'coup'. Nature is not represented in the social contract and so whatever consideration it may secure is derivative of human society rather than constitutive of it; nature stands outside the 'public sphere' of politics, its object rather than a subject. The relationship of humans to nature is then deeply over-determined by that fundamental compact. The Kantian emphasis on inherent worth, often equated with dignity, suggests an entirely distinct and sui generis basis upon which humans should treat other humans; by contrast, the relationship to non-human animals and nature is almost fundamentally conducted on instrumental terms. In fact, humans have dignity not because that dignity is inherent but by contrast with animals who have no value in themselves. The value of the environment and animals lies in their ability to sustain human life and therefore human rights.

It is worth noting that, initially at least, it is hard to wager whether nature or non-human animals are ultimately made better or worse by humans seeking justice for themselves along liberal rights lines. Admittedly, nature may be marginally more harmed by major armed conflicts, genocide, or tyranny. But nature might also occasionally benefit from major breakdowns of human society. Nature around Auschwitz was largely oblivious to the intra-human atrocity that was being perpetrated in its midst, and it has been said that the COVID pandemic gave breathing space to a range of species and the climate (Rume and Islam, 2020). The point is that the human rights project, even if it were more perfectly realised on its own terms, is certainly not thought of as being for the benefit of nature, directly or incidentally.

International Human Rights Law and the Globalisation of Modernist Approaches to the Non-Human

In this section, I suggest that international human rights law is fundamentally indebted to and not particularly critical of these earlier theoretical moves which it prolongs in an international context, making them into the backbone of the international system. In that respect, international human rights law can be seen as merely a part of a broader international legal project which has itself made its commodification of nature and resulting environmental harm invisible, even as it is constantly presented as the solution to the problem (Natarajan and Khoday, 2014). Whilst international environmental law itself can be faulted for its anthropocentrism, it can at least lay claim to the notion that it is occasionally involved in protecting ecosystems and species as such (Redgwell, 1996, p. 73). The intersection of human rights and environmentalism, such as it is, however, tends to tilt the balance further in an anthropocentric direction.

In short, the entire pedigree of international human rights law speaks to the continuation of the fundamental project of emancipating humans from and justifying domination over nature and its intensification across borders. The juridification of rights both domestically and internationally calls attention to the fact that the subject is no longer simply a notional or philosophical subject but one that needs to be able to participate in judicial processes as well as answer for its legal wrongs, something which seems to exclude non-humans. Furthermore, from the point of view of the general narrative of international law, 'humanisation' and, specifically, the emergence of humans as subjects of international law are typically associated with the general pacification and progress of international relations.

International law's implicit ideology of anthropocentrism becomes even more unassailable. Again, and with remarkable constancy, human rights are imagined, this time internationally, as resolving a particular intra-human problem (for example, the abuse of sovereignty in a Westphalian international system or the exclusion of certain humans from the promise of human rights) but in ways that are simultaneously oblivious to what this move enables in terms of the invisibility of nature. International human rights law, predictably, did not traditionally devote much if any attention to the environment. The 1948 Universal Declaration of Human Rights (UDHR) and the main human rights treaties are oblivious to its existence. They are, in fact, quite adamant about their anthropocentrism, proposing the foundation of human rights as lying in the 'inherent dignity [...] of all members of the human family' (UDHR, 1948, Preamble).

Certainly, human rights instruments have occasionally been deployed to protect the environment. If nothing else, human rights bodies such as the European Court of Human Rights (ECtHR) have recognised that the protection of the environment is a legitimate aim, susceptible to lead to limitations of rights (*Pine Valley Devs Ltd and Others v. Ireland*, paras. 54, 57). Various national policies of nature conservation have been sanctioned by the Court as a legitimate aim, in certain circumstances, even to the point of trumping the right to property of businesses or house owners (Desmet, 2010). In a sense, then, an already anthropocentric view of the environment is what makes it possible to further deploy environmental policies under a human rights framework: Anthropocentric views of rights are in a sense only so problematic as the general anthropocentric construction of the world. In a world where the environment is itself understood in anthropocentric terms, they at least ensure that the environment and human rights are a 'match' since they are conjoined by their anthropocentrism. But even on anthropocentric protection of the environment grounds, that approach already falls prey to the criticism that it is excessively individualistic (Francioni, 2010).

International Human Rights' Constitution of Nature

Developments in the rights tolerance of environmental protection should not hide that the price to pay for such relative protection has also been the continued human appropriation of lifeworlds and a certain construction of the environment as a sort of inert background to mankind's destiny. This is apparent in a series of moves embedded in the radical modernity of rights which have been, if anything, accelerated by rights' internationalisation. The non-human world, in this context, is constructed, I contend, either as a threat, a commodity, or a resource.

First, then, the environment may be considered as a threat, in ways that powerfully rearticulate age-old tropes about its menacing character and the consequent need to be protected from it. The entire arc of environmental protections in the context of human rights reproduces a powerful sense that the State in particular should, if need be, move to forcibly protect humans from the harms that can be inflicted on them by nature. A series of ECtHR cases have long confirmed this. For example, in *Lopez Ostra v. Spain* [1994], the ECtHR noted '[S]evere environmental pollution may affect individuals' well-being and prevent them from enjoying their homes in such a way as to affect their private and family life adversely, without, however, seriously endangering their health' (para. 51). The crucial finding is that harm to the environment can create harms to humans. This explains the somewhat contrived and roundabout invocation of, inter alia, the right to family life to comprehend harms that in other contexts might be deemed to be primarily harms against the environment per se. If anything, it is the interests of the community or society that must be weighed against those of private individuals (*Giacomelli v. Italy* [2006]), leaving no space for any 'third constituency'.

Many of the ECtHR cases in this vein are concerned with man-made action on the environment which, in turn, creates rights violations. However, it is also the case that states can be liable for harms produced by environmental forces themselves from which they have failed to protect persons within their jurisdiction. The ECtHR has highlighted that the obligation of the State arises not just in relation to man-made dangerous activities but also, specifically, natural disasters. Although the latter are 'beyond human control' and 'do not call for the same extent of State involvement', states 'were still required to do everything in their power to protect lives' (and, to a lesser degree, human property) (*Budayeva and Others v. Russia* [2008], para. 174).

The Human Rights Committee (2018, para. 26) has similarly arrived at the general conclusion that the duty to protect life includes 'appropriate measures to address the general conditions in society that may give rise to direct threats to life' including 'degradation of the environment'. In its first contentious case (*Portillo Cáceres v. Paraguay* [2019], paras. 7.4–7.5), it has confirmed an 'undeniable link' between environmental protection and human rights and a duty on the State to take all appropriate measures to protect persons within its jurisdiction from any 'reasonably foreseeable threat'. On an even broader level, the effects of climate change on the enjoyment of human rights have been explicitly recognised by the High Commissioner for Human Rights (HRC, 2009).[1]

Although this author knows of no cases where violations of rights have been alleged as a result of failure by the State to protect from non-human animals, it stands to reason that such cases could also implicate state responsibility. The point is that the perception of the environment as a potential threat powerfully and further reinstates the State as the guardian of human rights against forces of nature. This is in the name of human rights and in a context where, as the Strasbourg Court has been keen to note, 'no right to nature preservation is as such included among the rights and freedoms guaranteed by the Convention' (*Fadeyeva v. Russia* [2005], para. 68).

Second, the environment is conceived through a series of appropriative moves henceforth ratified on the international plane. The most obvious way in which this is so is that the relationship to nature or animals is often understood through the prism of property, whose centrality to the genesis of international law has been the object of much rediscovery of late (Koskenniemi, 2021). The human right to property of nature's products and animals is not one that stands to be limited by nature or animals' inherent rights but merely by societal objectives (which may of course, include the protection of nature or animals but not qua rights). To the extent that animals or nature appear at all in rights discourse, then, it is as having been impinged on as forms of property. For example, in the case of the *Massacres of El Mozote and nearby places v. El Salvador* [2012], the Inter-American Court of Human Rights (IACtHR), denouncing scorched earth tactics, noted that 'The right to property is a human right and, in this case, its violation is especially serious and significant' because, according to an expert witness, 'They killed or took the animals; they [...] took the cows, the hens; they took my cows, they killed two bulls: a loss of both material and affective significance in the peasant universe' (*The Massacres of El Mozote and nearby places v. El Salvador*, 2012, para. 180). Understanding the non-human world at best through a property lens makes most actual environmental or animal harm invisible.

But even beyond private property the consecration of the right to self-determination – outlining as it does what is certainly, in the human realm, a much-needed legal corrective to imperialism – is also, simultaneously, a claim to dominion over nature by particular peoples. The right to self-determination solidifies a sense that there is a human right to collectively create the structures wherein conditions of political freedom – including resource exploitation – can be nurtured. It should be seen in light of affirmations of sovereignty over natural resources or the right to development, which are themselves known for their very anthropocentric understandings.[2] Ecocentric limitations to collective freedoms, in this context, are likely to be seen as standing in the way of peoples' development, possibly in the sole interest of amorphous global priorities and benefiting Western carbon rentiers.

Note, crucially, that both the human right to property and the right to self-determination straddle otherwise persistent human ideological divides between neo-liberal and central ownership models. For example, whatever (genuine) doubts the international human rights movement has had about the place of property are almost entirely traceable to intra-human controversies between proponents of private and collectivised property (focusing on the *subjects* of property) – not about whether one or the other could be claimed over nature (the *object* of property). Similarly, the right to self-determination straddles the distinction between regimes based on private property and those of a more collectivist bent. The avowed neutrality of international human rights law in terms of the economic and political regime, such as it is, is certainly not neutral towards what is shared by these regimes, namely a commitment to limitless resource exploitation.

Third, departing from a conception of property understood as exclusive *usus* and *abusus*, the environment can be understood as a broader resource for the satisfaction of human rights. The discourse of economic and social rights adds a further dimension to the relationship to nature: Not only should human beings be protected from nature, but there is a positive obligation to exploit natural resources to meet basic needs (Recommendation on the Exploitation of Natural Resources and Human Rights Violations, 2006). Of course, this does not necessarily lead to overexploitation or abuse, but it does frame the relationship to nature as one of fundamental subservience to human needs, requiring the state at the minimum to not interfere with humans' attempts at subsistence through nature's exploitation.

It is true that guaranteeing economic rights also requires that a certain care be taken of the environment as the externality upon which the fulfilment of rights relies. The Committee on Economic, Social, and Cultural Rights (CESCR), for example, considers that environmental policies should be adopted in order to secure the right to food. Indeed, 'the right to adequate food is […] inseparable from […] environmental and social policies, at both the national and international levels' (CESCR, 1999, para. 4). That conception of the environment is clearly more in line with conceptions of sustainable development and a milder, less carefree approach to the non-human world – but it is still largely anthropocentric and foreshadows a broader move to conceptualise the environment from a human rights perspective.

Human Rights Approaches to the Environment

There is much ambiguity about whether human rights approaches to the environment are compatible with or exclusive of more ecocentric approaches. Authors such as Sumudu Atapattu have strenuously denied that to adopt a human rights approach is to give up on non-anthropocentric approaches to the environment (Atapattu, 2002). Rights such as property rights are not necessarily averse to the protection of nature; it depends what the property is used for. In a series of cases before the Inter-American system, the issue was the failure by states to recognise the property rights of Indigenous groups to their land and, instead, the concession of exploitation and mining rights to corporations. Said Indigenous groups sought to uphold their title to the land precisely as part of an attempt to

> reconfigure the right to property as a vehicle for reclaiming control of strategic resources and articulating a more complex relation between people and territory,

43

> whereby land and resources are interrelated with history, culture, way of life and sense of belonging, and with political and economic agency via free, prior and informed consent.
>
> *(Cotula, 2020)*

Thus, whilst property can be a sword it can also paradoxically be a shield, and one should be wary of conceptions of property rights as necessarily incompatible with respect for the environment. There are certainly sound bases for considering that human rights and the environment are related. Their 'close relationship' and 'interdependence', in particular, have been repeatedly emphasised by UN bodies. On the one hand, human rights risk being negatively affected by environmental degradation; on the other hand, respect for human rights is presented as a key plank of environmental protection. Much attention has been paid, for example, to the extent to which human rights might help the environmental movement. Under that perspective, human rights obligations to consult or inform populations, the rule of law, accountability, and transparency, all concur towards greater environmental protection.

These are all valid points, but they should not obfuscate the limits of even human rights approaches to the environment in a context where such approaches are deeply embedded in a certain anthropocentric matrix. It is almost as if the failure to address the deep anthropocentric infrastructure of human rights ensured that whatever international human rights law could do for the environment with one hand, it has had a tendency to undo with the other. The attention to environmental issues is less endogenous to human rights than it is the product of a belated process of catching up with international environmental legal developments, most notably in relation to climate change.

The fundamental blueprint of rights approaches to the environment is what is often referred to as the 'human right to a healthy environment'. For example, Principle 1 of the 1972 Stockholm Declaration states that: 'man has the fundamental right to freedom, equality and adequate conditions of life, in an environment of a quality that permits a life of dignity and wellbeing'. More specific proclamations of said right have followed in regional contexts.[3] The notion of a right to a healthy environment, which is of course, as far as it goes, unmistakably pro-environment and has been used to that effect, is also simultaneously, unashamedly anthropocentric in that it portrays a healthy environment as a human entitlement.

The environment is no longer presented as merely matter to be appropriated; but it is still presented, fundamentally, as an 'offering' to humanity. It only fundamentally enters the human rights conscience to the extent that it threatens conditions of equitable human existence. This is in a sense no mean feat, but it also, in turn, powerfully reinforces an agenda of state intervention to protect the environment for human ends and, in turn, shape it in humans' image. Rather than merely protecting against the environment or guaranteeing its free appropriation, the State is legitimised as the appropriate ultimate guardian of a certain continuity of a beneficial environment. The right to a healthy environment may be a critique of the worst manifestation of the Anthropocene, but it is also, ironically a powerful implicit reconduction of that model.

To the extent that a different conception of the environment enters the realm of international human rights law, it is through the very specific lens of Indigenous peoples. In that context, the state is to respect and protect 'Indigenous peoples' cultural values and rights associated with their ancestral lands and their relationship with nature […], in order to prevent the degradation of their particular way of life' (CESCR, 2019, para. 36). Note however that this different conception of nature does not per se become a foundational stone of human rights: It is merely incorporated as an element of the protection of particular peoples and their cultural specificity, whose own rapport to the environment is not understood to entail any particular consequence for the human rights edifice. The environment is, in fact, understood as a cultural artefact of Indigenous life. In provincialising Indigeneity, even in the process of protecting it, its implicit critique of the human rights mystique is marginalised.

Indeed, the debate on whether nature should be protected for its uses to humans or for its own sake reproduces arguments that have been fundamental to the emergence of human rights themselves. There, the debate was on whether there was anything inherent to humans that would deserve unqualified protection regardless of social utility. A long tradition running from Kant to Dworkin has argued that the unique grace of human rights is their reliance on a notion of human dignity, i.e., that there is something irretrievably unique and even sacred about humanity. Even on human rights grounds, such claims have long had to be nuanced through more politically inclined conceptions of human rights that see them as deeply immersed in the management of social and political reality and, in particular, the communal character of much of human life. Nonetheless, whether as trumps or whether sublimated through human utility, the exact measure of rights – whatever rights may be – has consistently been human.

A Crisis of Humanism?

Notwithstanding the prevalence and hegemony of these solidified anthropocentric constructs, it may be argued that the model on which ideas of human rights and the legal institutional machinery of international human rights law are premised has been subjected to a triple crisis of Western modernity. Although long in the making, that crisis may reach conditions of ripening in the Anthropocene that deeply problematise the anthropocentrism of human rights.

First, a relative normalisation of 'humanity'. If the shiny specificity of human rights was based on ideas about the incommensurability of the human to other living and inanimate categories and a sense of 'human exceptionalism', that myth has been deeply challenged. The Darwinian revolution had already shaken the sentiment of fundamental distinctiveness that foundational religious narratives had long instilled (after the Copernican revolution had established that humans were not at the centre of the universe). Humans may be very different, but they are, ultimately, only a late ramification of the tree of life, sharing ancestors with all living things. Palaeontology has uncovered the existence of a number of archaic human species, including Neanderthals, that suggest the early possibility of several quite distinct human families a few million years ago. Aetiology has relativised the sentiment that the use of tools, bipedalism, large brains, language, moral judgment, or mere sentience are that distinctive, even though their combination in humans is arguably quite unique. In late modernity, the possibility of hybridisation with animals and ideas about 'post humanity' suggests that life could be made of human animal and non-human animal assemblage (Jones, 2021). Finally, the Anthropocene itself calls into question the distinction between humans and nature as nature becomes fundamentally defined by the doings of humans.

The loss of a sense of specificity of the human is intimately linked to an abandonment of traditional Western ontological and epistemological dualism and a sense, by contrast, that the human and nature are co-constituted. Paradoxically, human destructiveness, whilst confirming a kind of technical human supremacy over nature, also fundamentally erodes the distinction between humanity and nature in that every breath that humanity takes stands to affect its environment. The porosity between the human and the non-human, the human and the human-sustaining, and the human and the harmed-by-humans creates a radical challenge for the human rights project: The undermining of dualism has further highlighted the circular character of rights proclamations and the impossibility of making only humans subjects of rights.

The second crisis involves a major rediscovery in the West of the vulnerability of humans to nature, following centuries of an inflated notion of human dominion over it (Barry, 2012). Amidst increasing panic about global ecological catastrophe, resource depletion, and even the possibility of human extinction, an overarching sense of doom is projected. Humanity cannot exist independently of the life systems that sustain it; in a sense, it is not only embedded into such life systems but part of them, defined by them. But because those life systems have themselves become entirely enmeshed in the dense fabric of human society and have considerable inertia, it becomes very difficult to imagine their undoing without unravelling the very conditions of modernity and perhaps even human life.

This leads to a further relativisation of the human as continuous with its life systems, and is itself merely a cog in larger planetary eco-constellations. It brings to mind ways forward that prioritise dramatic action in order to safeguard what can still be saved. It suggests, perhaps, surrendering one's collective fate to the clemency of forces that transcend humanity, and certainly individuals. But it also highlights ways in which the intertwined projects of humanism, modernity, and the market need to be rethought from the bottom up, including in their human rights components. This is all the more so that vulnerability, aside from being a universal form of human experience, is also an important moral category. It potentially stands to re-orient the human rights project in strikingly different directions than the traditional focus on 'autonomy', as humans' status as 'semi-dependent creatures' comes to the fore (Dobson, 2009).

Third, a late modern disenchantment with humanity. The Holocaust proved beyond doubt humans' cruelty to humans, but also that cruelty's specific enabling at the heart of modernity (Bauman, 2000). In a sense, moreover, that failed promise of modernity had already long been the ordinary experience of victims of slavery or colonisation. Despite this slow reckoning, the underlying promise of the modernist ideology as a project of endless control and appropriation of nature has survived, safeguarded by the modernist international legal edifice. Nonetheless, climate change and the emergence of the Anthropocene, by suggesting both the finitude of planetary resources and the ultimately self-destructive nature of human freedom alongside capitalism's deep complicity in racism, may precipitate an even deeper crisis of Western modernity. Where the Holocaust could be dismissed as a perversion of modernity, the colossal harm wreaked by the pursuit of material accumulation arises in a context where modernity is working exactly as it was supposed to (Treanor, 2019).

Attention to human self-destructiveness and the long, murderous feedback loop of modernity thus helps sow not a Weberian disenchantment with God but a disenchantment with 'man'. It should come as no surprise in that context that animal rights activists reprise the language of slavery or genocide which, whatever the outlandish character of such claims, hits modernity where it hurts most (Spiegel, 1996) by bringing attention to the connection between its 'otherization' of other humans and the non-human (Saeed, 2019). The exploitation and persecution of (notably) non-human animals and humans are one (de-humanisation as the ongoing intra-human manifestation of extra-human non-humanisation) (Nibert, 2002).

Post-Anthropocentric and Non-Anthropocentric Reimaginings

These conjoined crises, then, have created not only the desire but also the need to expand the register of rights to non-humans. As it turns out, however, and although this remained obscured for a long time, it is very difficult to extend rights to non-human realms without deeply problematising the ascription of rights to humans in the first place and, more generally, the finality of the human rights project.

Expanding the Register of Rights Beyond Humans

Much thinking has gone into renewing the category of rights by looking at how a variety of non-human entities might benefit from them. These developments, occurring relatively late in the development of rights, borrow from the register of human rights but for what seems the purpose of subverting their anthropocentrism. The discourse, then, is both critical and derivative of the anthropocentrism of rights. It is critical evidently because it denies humans the monopoly of rights (although for the most part not of *human* rights); it is derivative in that it borrows from the idea of human rights, the idea of rights, even as it dispenses with the human qualifier.

For example, the rise of animal rights has been heavily influenced by the critique of speciesism, itself evocative of classic human rights struggles against previous exclusions (sexism, racism, ableism, etc.). It borrows heavily from the traditional human rights playbook, including in terms of highlighting the fundamentally deontological and non-utilitarian nature of claims made on behalf of animals (Regan, 2004) as well as the superiority of rights over merely welfare discourse (Francione, 1996). Similarly, more recent but increasingly influential approaches to the rights of nature ('wild law', 'earth jurisprudence') seek to deploy rights discourses to endow certain natural formations or indeed earth as such with intrinsic value.

Such approaches have been criticised even from non-anthropocentric perspectives, for example, as borrowing too heavily from a register of rights that is itself human and potentially paternalistic, ascribing certain desires to nature on the basis of human second-guessing. The idea that the best and only way to protect non-human animals and nature is through the conferral of rights has been challenged on the basis that simply because they do not have 'rights' does not mean that they should not benefit from 'right' treatment by humans. A discourse of human *duties* might yield broadly similar results, whilst being more directly enforceable and politically mobilising. Such arguments are reinforced by a sense that non-human rights constantly need to be exercised through human agency anyhow, such that non-human rights risk being transformed into little more than another vernacular of human agency. The fear is that granting animals rights, as Derrida put it, 'would reproduce the philosophical and juridical machine thanks to which the exploitation of animal material for food, work, experimentation, etc., has been practiced', for example by drawing lines between different categories of animals (Derrida and Roudinesco, 2004, p. 71). This is a fear, incidentally, occasionally shared by lawyers themselves, wary that 'environmental rights' fail to tackle the deeply incomprehensible (Giagnocavo and Goldstein, 1989).

From a more anthropocentric tradition, the very idea of rights of non-human animals or nature might be found to be demeaning of historical human struggles (surely speciesism is not on a par with racism?) (Rose, 1991). It can lead to attitudes ranging from indifference to a blatant reaffirmation of the primacy of the human, animated by fears of the biologism presaged in earlier ecocentric philosophies, a renewed extolling of human liberty against 'liberticide' agendas of conservation (Machan, 2004), or left-leaning developmentalist agendas with little patience for rights that might derail broader social emancipatory goals.

Nonetheless, such approaches have begun to trickle into international human rights law with as yet unproven effects. Even though the environment of human rights bodies is already unmistakably committed to an anthropocentric lens, hints of a non-human space with the rights edifice have been dropped here and there. In a separate opinion, Judge Pinto de Albuquerque of the ECtHR has noted that 'Wild, abandoned or stray animals are also protected by the Convention as a part of a healthy, balanced and sustainable environment' (*Herrmann v. Germany* [2012]). Such intimations, although they noticeably depart from a property/resource/threat framework, still tend to see even wildlife as somehow part of the background to human prosperity (Sparks, 2020).

More spectacularly, the IACtHR has argued that:

> the right to a healthy environment [...] protects the components of the environment, such as forests, rivers and seas, as legal interests in themselves, even in the absence of the certainty or evidence of a risk to individuals [...] it protects nature and the environment, not only because of the benefits they provide to humanity or the effects that their degradation may have on other human rights, such as health, life or personal integrity, but because of their importance to the other living organisms with which we share the planet that also merit protection in their own right [...]. Thus, the right to a healthy environment as an autonomous right differs from the environmental content that arises from the protection of other rights, such as the right to life or the right to personal integrity.
>
> *(IACtHR, 2017, paras. 62–63)*

These pronouncements, sporadic and of uncertain provenance and impact, remain deeply constrained by the conditions of their emergence within the great temples of human rights. In short, they seem well-meaning but also somewhat gratuitous, relying on neither extensive interpretative work (arguably, how could they?) nor significant intellectual scaffolding or much attention to the potential radical implications of such recognition. If nature is to have an inherent dignity, moreover, one would hope that it would be rediscovered not merely incidentally through a finding on Indigenous territorial holdings.

Indeed, if broader reimaginings are to arise, they are likely to arise on the margins or even at a safe distance from the high sites of Western intellectual modernity. In that respect, it is striking that, sociologically speaking, sectors active in the rights of nature or non-human animal sectors are typically not an emanation of even the broader human rights movement. Their own reference to and absorption of the language of rights occurs through a kind of distanced and even ironic mimetism, at best challenging the human monopoly rather than proceeding from a conception of how human rights specifically might be better rethought. For example, rights of nature are described emphatically as 'environmentalism's next frontier' – not human rights (Mogensen, 2019). This is clear within the Western intellectual tradition, where defenders of the rights of nature and non-human animals have emerged at a significant distance from liberal orthodoxies (in deep ecology, post-humanity, etc.) and have involved entirely new, unorthodox sites of contestation such as the International Rights of Nature Tribunal (Maloney, 2015; 2016).

Nonetheless, ideas about the rights of nature have begun entering if not the mainstream at least the periphery of international law. The Convention on Biological Diversity's first principle was to the effect that nature has 'inherent value'. The Earth Charter (2000) mentions the need to respect and protect 'ecological integrity'. The United Nations General Assembly set up in 2016 a virtual dialogue on Harmony with Nature which has drawn on Earth jurisprudence 'to inspire citizens and societies to reconsider how they interact with the natural world in order to implement the Sustainable Development Goals in harmony with nature'. As the Secretary General put it, 'In the Earth-centred worldview, the planet is not considered to be an inanimate object to be exploited, but as our common home, alive and subject to a plethora of dangers to its health' (UNGA, 2016, para. 6).

States and international organisations, however, may be too compromised with the productivist and extractivist mindset of the international system to be more than echoes of radical conceptions of nature uttered elsewhere. The critique of anthropocentrism has, in fact, emerged most forcefully against the background of Indigenous and non-Western cosmologies (O'Donnell *et al.*, 2020) that have simultaneously stood for a civilisational critique of modernity (Deckha, 2020) and anchored non-human rights in a deeper critique of anthropocentrism (Borràs, 2016). In that context, the fact that some countries recognise the rights of nature has been heavily scrutinised for its potential in international law (Tignino and Turley, 2018) and indeed such soft international law instruments as the Universal Declaration of Rights of Mother Earth adopted in 2010 in Cochabamba.

The Problem of Integration with the Human Rights Project

Be that as it is, the movement to expand the beneficiaries of rights, particularly in its Western variant, also runs the risk of creating a series of further rights siloes without much of an integrative effort. Advocates of non-human rights (particularly of non-human animal rights) at first seemed less intent on critiquing the ontology of rights than claiming separate realms of rights. Moreover, whilst animal or nature rights activists have done much to extend the scope of rights to non-humans, human rights activists have not reflected much on the impact that this broadening could have, if any, on the human rights project. Animal or nature rights projects are rarely if ever referenced in institutional human rights environments.

The dominant thinking within the human rights community seems to be one of benevolent indifference or agnosticism about the rights of non-humans, without a sense of the intellectual crisis that such extension should evoke or thinking about what an overall posthumanist rights project might mean for the classical human rights project. At worst, the siloed conception to rights may lead to a kind of separatism wherein the internal integrity of the human rights project is not challenged but merely leads to a deferral of much-needed critical synthesis. What is currently lacking are theories that would integrate both human rights, animal rights, and nature rights. It seems as if the latter two have been deployed as counterfires to the first but with little recognition of how they might have to cede to or override each other, leaving the impression of a disjointed patchwork that skirts some of the difficult questions.

One can wonder, in particular, whether proclamations of rights of non-humans alone can offset the hegemonic propensities of an anthropocentric conception of human rights. Although the image of a triangle between human animal, non-human animal, and nature might give the impression that the three are equal poles, in truth by far the more ambitious pole if one is willing to let go of the tyranny of the anthropocentric mindset is the biocentric one (focusing on nature), in that it stands to encompass others into a theory of all beings. Non-human animal rights, particularly those traditions focusing on mammals and sentient animals, are extremely wedded to a certain view of their proximity to human-ness. Only rights of nature potentially offer a radical and holistic decentring from not just the human subject, but any sense that the particular qualities of humans (distributed and found in other near-species as they may be) ought to be the sole measure of rights (Pelizzon and Gagliano, 2015).

Indeed, the more ambitious proclamations of the rights of nature do hint not just at different subjects of rights but also at a different *ontology* of rights, one that transcends the distinction between human animals and non-human animals but also nature at large. The Universal Declaration on the Rights of Mother Earth, for example, uses the word 'being' as its core term, which it describes as comprising 'ecosystems, natural communities, species, and all other natural entities which exist as part of Mother Earth' (Article 4.1). Mother Earth, in particular, is to have 'the right to life and to exist', and in 'the right to maintain its identity and integrity as a distinct, self-regulating and interrelated being' (Article 2.1). In a way, the notion that rights flow from nature's very existence (Sheehan, 2013) is no more extravagant than the notion that rights flow from humans' existence.

Beyond this foundational discourse and its apparently wholesome embrace of the diversity of modes of being, there remain lingering questions about the ever-present – indeed, constitutive – possibility of conflicts between different orders of rights. A certain discourse of liberal commensurability of ends suggests that what is good for the environment or non-human animals is good for humans and that ultimately what is good for humans, properly understood, is good for the environment and non-human animals. There are no fundamental trade-offs involved and therefore the projects can coexist peacefully, perhaps as adjuncts to human rights or at least not as frontally interfering with their reign (UNGA, 2016, para. 36). But this discourse of reconciliation is too general for its own good and unhelpful in settling 'tactical' tensions between the projects, even assuming (a best-case scenario) a kind of long-term convergence.

Once more than one realm has been acknowledged as having intrinsic worth and our legal cosmologies no longer revolve around a single (anthropocentric) axis, we risk having merely set up a battle of absolutes which must by necessity eventually belie the notion that any realm does, in fact, have such intrinsic worth. This sets up a situation where some sort of equality across realms is affirmed, leading to conflicts that will be hard to disentangle. The human rights project itself has long battled with that broad realisation and has greatly watered down its ambitions from a project of affirming absolute human sanctity to a project of managing the human in light of evolving collective imperatives. But there is no doubt that the multiplication of rights orders stands to exponentially complicate that always fragile resolution. This is not to mention the considerable porosity of notions such as 'nature' or 'species' if and once one moves beyond the rights of individual entities therein (Eckersley, 1995, p. 190).

At best, the reconciliation of the competing pulls of human and non-human rights is yet to be settled on an ad hoc basis, unknown authority, and undecided principles. If the practise of the human rights movement's weighing of rights (between the collective and the individual; between individuals) teaches us anything in that respect, it is the highly fluid and indeterminate nature of such a resolution, as well as the tendency over time for it to merely reinforce the status quo. But we already have plenty of reason to think that there are unresolved tensions between nature and animal rights (baobab-destroying elephants, wild horses as 'feral pests', cormorant threats on wild fish). Indeed, the very justification of animal rights (that they are sentient, psychologically aware 'subjects of a life' to use Regan's term) by virtue of their proximity to humans, for example, seems (unwittingly?) to deny the quality of rights subject to the inanimate world and even many animal species. This is not to mention that 'rights of nature', 'animal rights' and 'human rights' may all presuppose, in their generic drive, that these categories are largely homogeneous when we have reason to think that all involve quite polarised interests, to the extent those can be discerned at all (Stanford, 2013).

There is no reason to think that those tensions will not be replicated on a broader scale between humans and non-human realms. Consider the case of global warming. On some level, it is unclear that global warming is necessarily an unmitigated catastrophe for the planet in the long run. Some species might be expected to prosper whilst others will suffer. It is not even clear, from a deep ecological perspective, that what might turn out to be a colossal catastrophe for the human species, would not be an opportunity for nature, in ways that underline some of the limits of an inclusive environmental ethics (Hassoun, 2011). This might then militate for weighing the interests of certain animals or plants against those of human animals. Environmental change clearly has a distributive dimension between humans of course, but also between species and realms.

And what of the possibility of nature or non-human animal challenges to certain human rights such as the right to property, especially in a context where it is precisely being subjected to ownership that is understood as denying one's (e.g., non-human animal) personhood (Francione, 2012)? What if even averting certain threats did not entirely obviate the right to be of non-human creatures or things? Should animal experimentation be banned, even when it can save human lives (Lansdell, 1988)? What of the compatibility of severe population control with the right to family life? Who will stand against that seemingly most human (albeit theologically sustained) of human impulses, namely the natalist urge to 'populate the earth'?

In short, what is missing is a sense of how different rights registers relate to each other. The overall impression is one of overlayed efforts delaying some more radical re-evaluation of the human rights project itself, one that would begin to tease out the profound contradictions and synergies involved in the project. Intellectually that project has begun as the project of reformulating human rights in light of the need to protect non-human animals and the environment (Taylor, 1997) regardless, in fact, of whether either are themselves protected by rights of their own. Nonetheless, even the international soft law approach (operating without any of the constraints that would apply if a treaty were involved) remains very scattershot. The 2010 Declaration of the Rights of Mother Earth hints at steps in that direction, for example, by noting that 'The rights of each being are limited by the rights of other beings and any conflict between their rights must be resolved in a way that maintains the integrity, balance and health of Mother Earth' (Article 1.7). It also points out that 'Every human being is responsible for respecting and living in harmony with Mother Earth'. In the same spirit, the Dialogues on Harmony with Nature had emphasised that 'humanity, which is inextricably part of the community of life on Earth, cannot continue to override the laws that maintain the homeostatic balances of the Earth system' (UNGA, 2016, para. 12).

These institutional pronouncements have much in common with Deep ecology and notably George Sessions and Arne Næss' third basic principle that 'Humans have no right to reduce (the) richness and diversity (of life forms) except to satisfy vital needs' (Naess and Sessions, 1986). Similar rules to arbitrate between the human and the non-human can be found in, for example, Article 2(2) of the 1978 Universal Declaration of Animal Rights.[4] Whilst some rules do not particularly challenge human dominance (the prohibition of wanton killing), others seem more explicitly at odds with the utility of animals to humans. These include prohibitions on animal experimentation (Universal Declaration of Animal Rights, 1978, Article 8) or exploiting animals 'for the amusement of man' (Article 10).

However helpful, these scattered attempts to litigate inter-rights questions remain a poor approximation of what might be an integrated theory of across-realm rights, one that would be a true product of dialogue between these different vernaculars. Clearly, such a theory would need to take into account some of the glaring asymmetries involved, including the fact that duties would presumably weigh mostly on human animals (an earthquake does not commit genocide) in ways that might re-tilt rights towards a kind of humanist *noblesse oblige*. Ecocentric limitations, however, could be built more clearly into general human rights language, either as indirectly protecting the rights of others, a necessary and legitimate social pursuit, or yet another, as yet inexistent, form of sui generis limitation to the rule of rights (Taylor, 1997).

Indeed, the common point of all these challenges to the human centring of rights is perhaps, first and foremost, the (further) problematisation of human freedom at least understood as untrammelled profligacy, including as it has consistently been framed internationally. But the question of 'who decides?' is likely to remain key, in a context where the question often begs the answer (in the sense that, as we have seen, human rights mechanisms almost inevitably produce human-slanted responses even with the best of environmental intentions), unless one is willing to thoroughly rethink human rights (Eckersley, 1995).

Bringing the Human Back In?

The ontological decentralisation of rights concepts might give rise to a number of claims and scenarios. On one level, it may precipitate defensive moves by human rights advocates, once the truly radical potential of a non-anthropocentric conception of human rights becomes apparent. It has already been hinted that 'Recognizing new, non-human, rights-holders […] may diminish the respect given to all rights-based claims, even as it diverts attention from the many unfinished struggles for human freedom' (Petrasek, 2018). As this chapter has argued, that is, of course, precisely the point of a non-anthropocentric perspective on rights, as one that clearly takes issue with the world-signifying character of even 'unfinished struggles for human freedom'.

But is the fear that non-anthropocentric conceptions of rights might be anti-humanist a valid one? This may depend, as it turns out, on which non-anthropocentric conception one has in mind. The spectre of eco-authoritarianism is sometimes agitated as suggesting something deeply illiberal in the environmental movement. Deep ecologists have, it is true, sometimes hinted at a preference for dramatic top-down solutions. There is more than a whiff of human 'self-hating' involved in such initiatives as the anti-natalist Voluntary Human Extinction Movement, for example, or some of the Anthropos-demonising dimensions of deep ecology (Biehl and Bookchin, 1995). These at times seem to engage in a kind if idolatry that replicates, in inverted fashion, the very Cartesian dualism of human rights; as if, having successfully argued that nature can be a subject too, they then reduced humans to mere objects – even as ecological premises suggest that humans are part of nature, and cannot easily be disowned.

It may be, then, that, in certain circumstances, human rights, reconciled with their anthropocentrism, would be called upon to stand as a bulwark against excessively post-humanist or even misanthropic readings of humanity's fate. This might seem ironic and as an inversion of who is truly the victim in that plotline – but if it was 'humans' who were threatened by naturalist fundamentalism rather than nature by fundamentalist humanism, then it is conceivable that human rights could still serve that role, for which history has prepared it well. Just as certain human rights have been presented as procedurally conducive to environmental protection, it may be that they would occasionally have to be deployed as shields to temper ecological authoritarianism or fanaticism.

If that seems a bit of red herring in our current predicament, at least a kind of reckoning with the disruptive potential of non-anthropocentric categories might have the salutary effect of drawing the intellectual battle lines and prompting an awakening that human rights, animal rights and nature rights are all political projects with definite implications for each other and not merely for themselves. It may be that in the struggle to even obtain any recognition of nature or animal rights, some of these questions had to be silenced; but it will be increasingly hard to ignore them going forward, as nature and non-human animal rights move from securing a precarious foothold in our imaginaries to spelling out the full implications of their potentially civilisation-reconfiguring ambitions.

Rather than misanthropic per se, however, ecocentric conceptions of rights may more justly incur the accusation that they are too undifferentiated in their broad condemnation of 'humanity'. The deep-ecological worldview may be part and parcel of a sort of amorphous depoliticisation of environmental degradation, one that pits nature against a broad, undifferentiated 'humanity' (Latour, 2009). It engages in a kind of middle-class, post-materialist contemplation of the savage destruction of nature whilst being oblivious to its determinative role in bringing it about. In that respect, ideas about the rights of nature are vulnerable to the environmental justice challenge that the problem is less with humanity at large than with small parts of humanity or at least certain systems of human governance and, in particular, environmentalism's own implication in patterns of ongoing colonialism (Schlosberg, 2009).

In other words, the problem is both much more specific and, in fact, intra-human than a kind of contemplative and nature-deifying route would suggest. The wrong of deep ecology may instead be that, in its excessively broad sweep, it fails to disaggregate humanity and produces a project that is as thoroughly depoliticised as it is, in the end, toothless. It is, at best, as social ecologists have long argued, a distraction from the urgent issue of addressing humanity's most dire contradictions. In fact, that may be a generous reading. Obsessed with population growth in the South rather than over-consumption in the North (Linkola, 2009), some 'eco-brutalists' or 'eco fascists' are arguably only too aware that their earth-centeredness is a pretext to call for the destruction of racialised or third-world masses (Bookchin, 1987) – 'environmentalism through genocide', as Naomi Klein put it (Klein, 2015, p. 276). A certain kind of eco-centrism, then, is not so much generally and blindly misanthropic as it is misanthropic about *certain humans*, in ways that are surely familiar from the darker recesses of the human rights project itself (the 'pseudo-humanism' denounced by Aimé Césaire).

The fear of environmental rights being co-opted by such tendencies may be magnified by some transhumanists' willingness to take advantage of the decentring of the human subject to further carve distinctions between humans. Be that as it may, the stigmatisation of 'humanity' as the single driving force behind the Anthropocene is perpetually at risk of neglecting highly differentiated responsibilities in bringing about ecological catastrophe. Its late-day, hand-wringing universalism – coming as it does, of course, on the heels of centuries of finessing the unity of humanity – may appear as little more than an opportunistic attempt to dilute the blame for our ecological predicament by awkwardly extending it to all.

The Anthropocene is not the product of wayward humanity's 'fall from grace' as much as it is embedded in modernity and the industrial revolution, colonialism, sexism, and racism (Malm and Hornborg, 2014). This then calls for much more astute problematisation of the 'anthropos' in anthropocentrism, beyond generalities about 'humans' (Grear, 2015). If anything, a chastised human rights project shares a much deeper affinity with the project of social ecology than with deep ecology, the former allowing it and even encouraging it to continue to come to terms with deep fault lines within humanity that have arguably always haunted it and being more consonant with a vision of liberal human rights as a persistently fraught project.

In extremis, the anthropocentric critique of rights might encounter the reality that human rights were never that anthropocentric in the first place, as much as fixated on a particular category of the Anthropos, endlessly projecting its particular brand of humanity onto the world. Apocalyptic discourses about human extinction, in foregrounding both a collectively helpless humanity and, potentially, an *avant-garde* showing the way to harmony with nature, may be more than a little self-serving, parochial, and privileged. Indeed, a life of environmental degradation and miserable ecological prospects has long been the consistent experience of a large part of humanity. The novelty of the current moment – which biocentric concepts of rights grapple with but can never fully account for – is that the West may well have finally brought itself down with the rest of humanity (Colebrook, 2015), creating a cascading crisis of its core modernist creed.

Indeed, non-anthropocentric views of rights have a paradoxical relationship to the imperative to diversify and provincialise international law and human rights. On the one hand, the rediscovery of more biocentric conceptions of the world is only made necessary by highly peculiar, non-inevitable civilisational developments that have made the expansion of a liberal Western mode-of-knowing nearly all pervasive. Less anthropocentric views of rights, in that context, may certainly capitalise on long-vibrant debates within the human rights tradition itself about the need for pluralism and cultural variation. In drawing on and invoking non-Western, particularly Indigenous, imaginaries of nature, they are part of a long-standing critique of international law's civilisational blind spots. In many ways, the broad, seemingly impossibly difficult excursion into what might be less anthropocentric modalities of rights described in this chapter is only a reckoning with what has been the common sense of a great variety of cultures.

On the other hand, there may be something bittersweet about the fact that Indigenous groups, for example, only manage to get traction when they frame claims in the relatively vapid terms of nature rights rather than their own rights. The vocation of ecocentric conceptions of rights is not to be exoticised into cultural or national specificities ('Indigenous peoples', the 'constitution of Ecuador') but to be understood for the decolonial critiques of the Western mindset that they are (Temper, 2019). If non-anthropocentric views of rights are either dismissed as anti-human rights, cultural relativism, or ghettoised as part of human rights legal pluralism, then 'the West' might even claim a kind of civilisational exception to biocentrism, confident that its own world-commodifying urge is neither better nor worse than any other system of human values. The point is that being dependent on nature ought not to be seen as an idiosyncratic Indigenous specificity (as a strategy focused on 'cultural rights' might suggest) as much as a much-neglected but pervasive human condition.

Note, conversely, that beyond New-Age instrumentalisation and appropriation of their beliefs, Indigenous peoples evidently have their own anthropocentric inclinations (e.g., the Ecuadorian constitution's emphasis not just on Pachamama but also on *buen vivir*, itself a sometimes decidedly modern amalgam (Hidalgo-Capitán and Cubillo-Guevara, 2017)) and should not be held hostage to a romanticised 'good savage' view of their role as stalwarts of the environment (Anaya, 1999). Holding them to standards to which the West does not hold itself, for the opportunistic goal of sanctuarising certain territories in the name of the higher interests of nature, could create a cultural 'straitjacket' at the expense of Indigenous agency and self-determination (Lalander, 2014). Indeed, even a perception of Indigenous peoples as guardians of nature could play into visions of 'primitive' people as benign but non-historical agents living on the margins of the world. It should be stressed that Bolivia and Ecuador, for all the proclamations of rights of nature, are not paragons of ecocentrism and remain societies deeply torn by tensions between different models of development (Humphreys, 2017). Moreover, the idea that rights of nature could be extended globally may smack of legal imperialism (Chaturvedi, 2019).

Indigenous peoples, in short, do and have the right to pollute too, just as maintaining the Third World in a perpetual pre-industrial low-carbon state to offset the West's own over-consumption as part of lofty appeals to 'humanity's responsibility', and the idealisation of subsistence farming is an utterly unappealing option. Focusing on rights of non-human animals and nature cannot obfuscate legitimate international demands for development, especially when those demands are incomparably more modest than those that have grown out of modernity and when they account properly for global resource distribution and stewardship. The neo-Malthusian echoes of a programme of minding ecocentric balances should not obviate the fact that the burden of degrowth should be allocated very unevenly to reflect historical responsibilities in the global environmental crisis: Malthusianism where Malthusianism is due.

Conclusion

The field of international human rights law is one where foundational assumptions about rights are tested through the categories of the international system. In that respect, it is at risk of becoming a largely conservative system that tends to replicate existing structures (Mégret, 2013), including what has been described in this chapter as human rights anthropocentrism. The field is also marked by the pragmatics of rights promotion and litigation: The best short-term outcome for one's 'client' may not necessarily be the best long-term one; victories for human applicants may or may not translate into victories for non-human animals or the environment. It is also worth bearing in mind that activists extolling non-anthropocentric conceptions of rights may at times have the enthusiasm of the neophyte and neglect well-known critiques of human rights even on anthropocentric grounds: In short, anthropocentric rights have not even actually served 'humanity' (or significant sectors thereof) that well. Much remains to be clarified in terms of how non-anthropocentric conceptions of rights would work legally and not just in the realm of environmental ethics and green politics, cautioning against a one-size-fits-all approach to decentring rights (Angstadt and Hourdequin, 2021).

In practise, some kind of reconciliation of bio-centric and anthropo-centric theories of rights is unavoidable, perhaps one that emphasises their eco-centric commonalities from some kind of meta-normative standpoint. Whatever non-anthropocentric line of argument may be available to challenge environmental disruptions, there will often be an anthropocentric one as well. For example, a major mining extractive project may well affect animals and the local environment, and there is no reason why we should give up on arguments that this is problematic to the extent that they capture something about the intrinsic value of both; but at the same time, it will most often be the case that such disruptions also harm local inhabitants even as they contribute to enrich wealthy corporations and individuals thousands of miles away who never have to live with the consequences of the degradation. Indeed, embracing ecocentric rights certainly does not commit one to deep ecology or a radical environmental ethics, and such moves have long been defended on pragmatic legal grounds to remedy the 'absence' of nature in proceedings where it is deeply implicated (Stone, 1972).

The possibility remains, it is true, that some harms are either not captured by human rights violations or at such a level of generality (e.g., global warming) as to create insolvable problems of juridical standing and of collective action (Kopnina *et al.*, 2018). Deep ecological philosophies of rights at least caution against the temptation of technological quick fixes (e.g., geoengineering) for the exclusive benefit of humans and more generally about the continuing value of that which is beyond humans. And there is the argument that, as far as non-human animals and nature are concerned, the fact that their dispossession and disruption are anchored in intra-human foibles is not particularly *their* problem: This is, unequal and barbaric inter se as it may be, the humanity with which the planetary environment has to contend (Hawkins, 2014).

Still, that environmental degradation and animal cruelty are simply another facet of 'man's inhumanity to man' is hard to resist as a conclusion. It calls into question the very 'colonial matrix' of modernity as it has expressed both in relation to nature and the dispossessed. It suggests that the search for environmental harmony may paradoxically be a much more agonistic process than bland UN appeals to the compatibility of human rights and the environment suggest ('we are all in this together'), one deeply implicated in class, gender, race, and colonial struggles. And it suggests a re-energising of the human rights project in conditions of global environmental degradation, rather than its dilution through excessive diffusion.

Finally, although much of the emphasis on nature and non-human animals is on how the human is called upon to continuously reframe the rights of both, it is important to note the extent to which both then retroact on the very intellectual project of humanity, understood as the quintessence of political modernity. In that respect, the nature rights and animal rights projects might be understood not so much as normative than as epistemological projects, challenging dominant, human-centred modes-of-knowing (Avery, 2004). This is visible in the way non-anthropocentric versions of rights take the project away from its liberal modern foundations in purely human concerns about liberty back (or forward) towards more ontological foundations involving the very definition of *being* and, in some cases, a quasi-mystical absorption of life forces; or the way in which they could yet renew the language of the social contract (Rowlands, 1997). But it is also visible in the transformation of conceptions of humanity themselves, understood as both human-ness (the quality that makes one human) and the collectivity of human beings.

As to the latter, theories of what constitutes humanity, in addition to severing humans' connection to nature, have had a perverse tendency to quickly draw lines between categories of humans. The move by humans to give humans rights by distinguishing humanity from nature, thus turned out to be rife with possibilities for exclusion, including through de-humanisation of 'others'. This has long been visible in the tendency of human rights to historically draw hierarchies between humans who most distinctly emancipate themselves from nature (e.g., men, through the exercise of Reason) and those who most fundamentally struggle to dissociate themselves from its pull (e.g., women, 'savages', the disabled, and children, dominated by primitive needs and passions). The effort to define humanity, of course, is not without its own totalitarian proclivities, potentially conditioning rights on being upstanding humans (no human rights for terrorists!) and reaching out for an essence that then serves to define some humans out of humanity. The question, then, is how a more inclusive concept of rights altogether precipitated by the recognition of the rights of animals and nature might affect the definition of humans, perhaps around a new, encompassing ontology of being.

As to the latter, non-anthropocentric conceptions of human rights might reinstate a sense that Humanity rather than the traditional individual human subjects of rights is the appropriate framework for conceptualising inter-realm relations – indeed, that the focus on individuals is what has led human rights into some of the dead ends from which global environmental degradation is born. Instead of a self-knowing, self-referring, and self-empowering Humanity, the move to nature and to a lesser extent non-human animal rights prompts a renewed framing of the human species' destiny (Flipo, 2012) and, in fact, an embrace of human diversity as it expresses itself in a pluralist patchwork of conceptions of nature and legacies of differentiated responsibilities towards it.

Whereas modernity, imperfectly universalised among others through the international law of human rights, emphasised an untrammelled right to frame nature in one's image, the Anthropocene may yet highlight conditions in which a wounded humanity's role is framed not so much by itself than by the obligations imposed upon it by its immersion into forms of existence beyond it and how those stand to be affected by continuing inequities between humans. Indeed, even the least anthropocentric visions of non-human rights may reinscribe a degree of human dependence (for example, the 1978 Declaration on Animal Rights stipulates that 'All animals have the right to the attention, care and protection of man'). The extent to which that new vocation of humanity should be informed by concepts of guardianship, stewardship, or fiduciary duties (as opposed to violence, freedom, and dominion) (Fox-Decent, 2012) is clearly a question that will test not just the basic anthropocentric worldview but also the actual meaning of anthropocentrism.

The proclamation of non-human rights might thus help spell out anew the specificity of the human rights project. Indeed, projects of non-human rights do not give up on a sense of human specificity; they arguably merely reframe it, except as a specificity that does not cut humans from the world through a relationship of dominion as much as one that reinscribes nature within the human and the human within nature. In that respect, not so much ecocentrism itself, but the ecocentric critique of rights and, as the case may be, the rights' reception of that critique

(Etchart, 2022) could be in the best tradition of critiques of rights that have, since at least Marx and extending to contemporary feminist and communitarian thinkers, underscored the poverty of liberal conceptions of rights framed around the atomistic and desocialised nature of the human (Benton, 1993).

Notes

1 See also Joint statement of the special procedure mandate holders of the Human Rights Council on the United Nations Climate Change Conference (Copenhagen, December 7–18, 2009).
2 See, e.g., Rio Declaration on Environment and Development, 1992, Principle 1: 'Human beings are at the center of concerns for sustainable development. They are entitled to a healthy and productive life in harmony with nature'.
3 See, e.g., 1981 African Charter on Human and Peoples' Rights, art. 24: 'all peoples shall have the right to a general satisfactory environment favorable to their development'; 1988 Additional Protocol to the American Convention on Human Rights, art. 11, para. 1, states: 'everyone shall have the right to live in a healthy environment'.
4 'Man, as an animal species, shall not arrogate to himself the right to exterminate or inhumanely exploit other animals'.

References

Anaya, S.J. (1999) 'Environmentalism, Human Rights and Indigenous Peoples: A Tale of Converging and Diverging Interests Human Rights, Environment & Community: A Workshop', *Buffalo Environmental Law Journal*, 7, pp. 1–14.
Anderson, B. (2017) 'The Politics of Pests: Immigration and the Invasive Other', *Social Research: An International Quarterly*, 84(1), pp. 7–28. DOI: 10.1353/sor.2017.0003
Anghie, A. (2007) *Imperialism, Sovereignty and the Making of International Law*. Cambridge: Cambridge University Press.
Angstadt, J.M. and Hourdequin M. (2021) 'Taking Stock of the Rights of Nature', in Corrigan, D.P. and Oksanen, M. (eds.) *Rights of Nature: A Re-examination*. London: Routledge, pp. 14–35.
Atapattu, S. (2002) 'The Right to a Healthy Life or the Right to Die Polluted?: The Emergence of a Human Right to a Healthy Environment Under International Law', *Tulane Environmental Law Journal*, 16 (1), pp. 65–126.
Avery, S. (2004) 'The Misbegotten Child of Deep Ecology', *Environmental Values*, 13(1), pp. 31–50. DOI:10.3197/096327104772444811
Barry, J. (2012) *The Politics of Actually Existing Unsustainability: Human Flourishing in a Climate-Changed, Carbon Constrained World*. Oxford: Oxford University Press.
Bauman, Z. (2000) *Modernity and the Holocaust*. Ithaca: Cornell University Press.
Benton, T. (1993) *Natural Relations: Ecology, Animal Rights, and Social Justice*. London: Verso.
Benton, T. (1998) 'Rights and Justice on a Shared Planet: More Rights or New Relations?', *Theoretical Criminology*, 2(2), pp. 149–175. DOI: 10.1177/1362480698002002002
Biehl, J. and Bookchin, M. (1995) 'Theses on Social Ecology and Deep Ecology', *Left Green Perspectives*, 33.
Bookchin, M. (1987) 'Social Ecology versus Deep Ecology: A Challenge for the Ecology Movement', *Left Green Perspectives*, 4–5.
Borràs, S. (2016) 'New Transitions from Human Rights to the Environment to the Rights of Nature', *Transnational Environmental Law*, 5(1), pp. 113–143. DOI: 10.1017/S204710251500028X
Budayeva and Others v. Russia, Apps nos 15339/02, 21166/02, 20058/02, 11673/02 and 15343/02 (ECHR, 20 March 2008).
CESCR (1999) 'General Comment No. 12: The Right to Adequate Food (Art. 11)' UN Doc E/C.12/1999/5.
CESCR (2009) 'General Comment No. 21. Right of Everyone to Take Part in Cultural Life (art. 15, para. 1(a), of the Covenant)' UN Doc E/C.12/GC/21.
Chaturvedi, I. (2019) 'Why the Ganga Should Not Claim a Right of the River', *Water International*, 44(6–7), pp. 719–735. DOI: 10.1080/02508060.2019.1679947

Colebrook, C. (2015) 'The Human World Is Not More Fragile Now: It Always Has Been', Aeon [Online]. Available at: https://aeon.co/essays/the-human-world-is-not-more-fragile-now-it-always-has-been (Accessed: 21 September 2022).

Cotula, L. (2020) 'Between Hope and Critique: Human Rights, Social Justice and Re-Imagining International Law from the Bottom Up', *Georgia Journal of International and Comparative Law*, 48, pp. 473–521.

Deckha, M. (2020) 'Unsettling Anthropocentric Legal Systems: Reconciliation, Indigenous Laws, and Animal Personhood', *Journal of Intercultural Studies*, 41(1), pp. 77–97. DOI: 10.1080/07256868.2019.1704229

Derrida, J. and Roudinesco, E. (2004) *For What Tomorrow: A Dialogue*. Stanford: Stanford University Press.

Desmet, E. (2010) 'Balancing Conflicting Goods The European Human Rights Jurisprudence on Environmental Protection', *Journal for European Environmental & Planning Law*, 7(3), pp. 303–23. DOI: 10.1163/161372710X540319

Deutscher, P. (2008) 'Women, Animality, Immunity – And the Slave of the Slave', *Insights*, 1(4), pp. 2–19.

Dobson, A. (2009) 'Freedom and Dependency in an Environmental Age', *Social Philosophy and Policy*, 26(2), pp. 151–172. DOI: 10.1017/S0265052509090207

Eckersley, R. (1995) 'Liberal Democracy and the Rights of Nature: The Struggle for Inclusion', *Environmental Politics*, 4 (4), pp. 169–198. DOI: 10.1080/09644019508414232

Etchart, L. (2022) 'Buen Vivir and the Rights of Nature in National and International Law', in *Global Governance of the Environment, Indigenous Peoples and the Rights of Nature: Extractive Industries in the Ecuadorian Amazon*. Cham: Palgrave Macmillian, pp. 57–86.

Flipo, F. (2012) 'For the Rights of Nature', *Mouvements*, 70(2), pp. 122–137.

Fox-Decent, E. (2012) 'From Fiduciary States to Joint Trusteeship of the Atmosphere: The Right to a Healthy Environment through a Fiduciary Prism', in Coghill, K., Sampford, C.J.G. and Smith, T. (eds.) *Fiduciary Duty and the Atmospheric Trust*. Farnham: Ashgate, pp. 253–268.

Francione, G. (1996) *Rain Without Thunder: The Ideology of the Animal Rights Movement*. Philadelphia: Temple University Press.

Francione, G. (2012) *Animals Property & The Law*. Philadelphia: Temple University Press.

Francioni, F. (2010) 'International Human Rights in an Environmental Horizon', *European Journal of International Law*, 21(1), pp. 41–55. DOI: 10.1093/ejil/chq019

Giacomelli v. Italy, App no 59909/00 (ECHR, 2 November 2006).

Giagnocavo, C. and Goldstein, H. (1989) 'Law Reform or World Re-Form: The Problem of Environmental Rights', *McGill Law Journal*, 35(2), pp. 345–386.

Grear, A. (2015) 'Deconstructing Anthropos: A Critical Legal Reflection on "Anthropocentric" Law and Anthropocene "Humanity"', *Law and Critique*, 26(3), pp. 225–249. DOI: 10.1007/s10978-015-9161-0

Hassoun, N. (2011) 'The Anthropocentric Advantage? Environmental Ethics and Climate Change Policy', *Critical Review of International Social and Political Philosophy*, 14(2), pp. 235–57. DOI: 10.1080/13698230.2011.529710

Hawkins, R. (2014) 'Why Deep Ecology Had to Die', *The Trumpeter*, 30(2), pp. 206–230.

Herrmann v. Germany, App no 9300/07 (ECHR, 26 June 2012).

Hidalgo-Capitán, A.L and Cubillo-Guevara, A.P. (2017) 'Deconstruction and Genealogy of Latin American Good Living (Buen Vivir). The (Triune) Good Living and Its Diverse Intellectual Wellsprings', *International Development Policy | Revue Internationale de Politique de Développement*, 9, pp. 23–50. DOI: 10.4000/poldev.2351

HRC (2009) 'Report of the Office of the United Nations High Commissioner for Human Rights on the Relationship Between Climate Change and Human Rights' UN Doc A/HRC/10/61.

HRC (2018) 'General Comment No. 36 (2018) on Article 6 of the International Covenant on Civil and Political Rights, on the Right to Life', UN Doc CCPR/C/GC/36.

HRC (2019) 'Views Adopted by the Committee under Article 5 (4) of the Optional Protocol, Concerning Communication No 2751/2016', (Portillo Cáceres v. Paraguay) UN Doc CCPR/C/126/D/2751/2016.

Humphreys, D. (2017) 'Rights of Pachamama: The Emergence of an Earth Jurisprudence in the Americas', *Journal of International Relations and Development*, 20(3), pp. 459–484.

Hunt, L. (2008) *Inventing Human Rights: A History*. New York: W.W. Norton & Company.

Jones, E. (2021) 'Posthuman International Law and the Rights of Nature', in Grear, A., Boulot, E., Vargas-Roncancio, I.D. and Sterlin, J. (eds.) *Posthuman Legalities: New Materialism and Law Beyond the Human*. Cheltenham: Edward Elgar Publishing Limited, pp. 82–101.

Klein, N. (2015) *This Changes Everything: Capitalism vs. the Climate*. Toronto: Alfred A. Knopf Canada.

Kopnina, H., Washington, H., Taylor, B. and Piccolo, J.J. (2018) 'Anthropocentrism: More than Just a Misunderstood Problem', *Journal of Agricultural and Environmental Ethics*, 31(1), pp. 109–127. DOI: 10.1007/s10806-018-9711-1

Koskenniemi, M. (2021) *To the Uttermost Parts of the Earth: Legal Imagination and International Power 1300–1870*. Cambridge: Cambridge University Press.

Lalander, R. (2014) 'Rights of Nature and the Indigenous Peoples in Bolivia and Ecuador: A Straitjacket for Progressive Development Politics?', *Iberoamerican Journal of Development Studies*, 3(2), pp. 148–172.

Lansdell, H. (1988) 'Laboratory Animals Need Only Humane Treatment: Animal "Rights" May Debase Human Rights', *International Journal of Neuroscience*, 42(3–4), pp. 169–178. DOI: 10.3109/00207458808991594

Latour, B. (2009) *Politics of Nature: How to Bring the Sciences into Democracy*. Cambridge: Harvard University Press.

Liebell, S.P. (2011) 'The Text and Context of "Enough and as Good": John Locke as the Foundation of an Environmental Liberalism', *Polity*, 43(2): 210–241. DOI: 10.1057/pol.2010.28

Linkola, P. (2009) *Can Life Prevail?: A Radical Approach to the Environmental Crisis*. New York: Routledge.

Locke, J. (1961) *Two Treatises of Government*. Reprint, New York: Hafner Publishing Company.

Lopez Ostra v. Spain, App no16798/90 (ECHR, 9 December 1994).

Machan, T.R. (2004) *Putting Humans First: Why We Are Nature's Favorite*. Lanham: Rowman & Littlefield Publishers.

Maldonado-Torres, N. (2017) 'On the Coloniality of Human Rights', *Revista Crítica de Ciências Sociais*, 114, pp. 117–136. DOI: 10.4000/rccs.6793

Malm, A. and Hornborg, A. (2014) 'The Geology of Mankind? A Critique of the Anthropocene Narrative', *The Anthropocene Review*, 1(1), pp. 62–69. DOI: 10.1177/2053019613516291

Maloney, M. (2015) 'Finally Being Heard: The Great Barrier Reef and the International Rights of Nature Tribunal', *Griffith Journal of Law & Human Dignity*, 3(1), pp. 40–58.

Maloney, M. (2016) 'Building an Alternative Jurisprudence for the Earth: The International Rights of Nature Tribunal', *Vermont Law Review*, 41, pp. 129–142.

Mégret, F. (2013) 'The Apology of Utopia; Some Thoughts on Koskenniemian Themes, with Particular Emphasis on Massively Institutionalized International Human Rights Law', *Temple International and Comparative Law Journal*, 27, pp. 455–497.

Mogensen, J.F. (2019) 'Environmentalism's Next Frontier: Giving Nature Legal Rights', *Mother Jones*, July/August [Online]. Available at: https://www.motherjones.com/environment/2019/07/a-new-wave-of-environmentalists-want-to-give-nature-legal-rights/ (Accessed: 21 September 2022).

Næss, A. and Sessions, G. (1986) 'The Basic Principles of Deep Ecology', *The Trumpeter* 3(4), p. 14.

Natarajan, U. and Khoday, K. (2014) 'Locating Nature: Making and Unmaking International Law', *Leiden Journal of International Law*, 27(3), pp. 573–593. DOI: 10.1017/S0922156514000211

Nibert, D.A. (2002) *Animal Rights/Human Rights: Entanglements of Oppression and Liberation*. Lanham: Rowman & Littlefield.

O'Donnell, E., Poelina, A., Pelizzon, A. and Clark, C. (2020) 'Stop Burying the Lede: The Essential Role of Indigenous Law(s) in Creating Rights of Nature', *Transnational Environmental Law*, 9(3), pp. 403–427. DOI: 10.1017/S2047102520000242

Pelizzon, A. and Gagliano, M. (2015) 'The Sentience of Plants: Animal Rights and Rights of Nature Intersecting', *Australian Animal Protection Law Journal*, 11, pp. 5–14.

Petrasek, D. (2018) 'Human and Non-Human Rights – Convergence or Conflict?', *Centre for International Policy Studies*, 10 December 2018 [Blog]. Available at: https://www.cips-cepi.ca/2018/12/10/human-and-non-human-rights-convergence-or-conflict/ (Accessed: 21 September 2022).

Pine Valley Developments Ltd and Others v. Ireland, App no 12742/87 (ECHR, 29 November 1991).

'Recommendation on the Exploitation of Natural Resources and Human Rights Violations' (2006) *The African Centre for Democracy and Human Rights Studies*, 8 May 2006 [Blog]. Available at: https://www.acdhrs.org/2006/05/recommendation-on-the-exploitation-of-natural-resources-and-human-rights-violations/ (Accessed: 21 September 2022).

Redgwell, C. (1996) 'Life, the Universe and Everything: A Critique of Anthropocentric Rights', in Anderson, M.R. and Boyle, A.E. (eds) *Human Rights Approaches to Environmental Protection*. Oxford: Clarendon Press, pp. 71–87.

Regan, T. (2004) *The Case for Animal Rights*. Berkley: University of California Press.

Rorty, R., Shute, S. and Hurley, S. (1993) *On Human Rights: The Oxford Amnesty Lectures 1993*. New York: BasicBooks.

Rose, S. (1991) 'Proud to Be Speciesist', *New Statesman and Society*, 4, pp. 21–22.

Rousseau, J.-J. (1997) *The Social Contract and Other Later Political Writings*. Cambridge: Cambridge University Press.

Rowlands, M. (1997) 'Contractarianism and Animal Rights', *Journal of Applied Philosophy*, 14(3), pp. 235–247. DOI: 10.1111/1468-5930.00060

Rume, T. and Didar-Ul Islam, S.M. (2020) 'Environmental Effects of COVID-19 Pandemic and Potential Strategies of Sustainability', *Heliyon*, 6(9), e04965. DOI: 10.1016/j.heliyon.2020.e04965

Saeed, A. (2019) 'The "Other'" in the Context of Human Slavery and Genocide and Current Nonhuman Animal Industrial Systems', *Sloth*, 5(1) [Online]. Available at: https://www.animalsandsociety.org/research/asi-publications/sloth/sloth-volume-5-no-1-winter-2019/the-other-in-the-context-of-human-slavery-genocide-and-current-nonhuman-animal-industrial-systems/ (Accessed: 21 September 2022).

Schlosberg, D. (2009) *Defining Environmental Justice: Theories, Movements, and Nature*. Oxford: Oxford University Press.

Schroeder, J.L. (2002) 'The Stumbling Block: Freedom, Rationality, and Legal Scholarship', *William & Mary Law Review*, 44(1), pp. 263–374.

Sheehan, L. (2013) 'Realizing Nature's Rule of Law through Rights of Waterways', in Voigt, C. (ed.) *Rule of Law for Nature: New Dimensions and Ideas in Environmental Law*. Cambridge: Cambridge University Press, pp. 222–240. DOI: 10.1017/CBO9781107337961.018.

Sparks, T. (2020) 'Protection of Animals Through Human Rights: The Case-Law of the European Court of Human Rights', in Peters, A. (ed.) *Studies in Global Animal Law*. Berlin: Springer Open, 153–171.

Spiegel, M. (1996) *The Dreaded Comparison: Human and Animal Slavery*. London: Mirror Books.

Srinivasan, K. (2015) 'The Human Rights Imagination and Nonhuman Life in the Age of Developmentality', *Journal of the National Human Rights Commission*, 14, pp. 289–309.

Steenbakkers, P. (2014) 'Human Dignity in Renaissance Humanism', in Mieth, D., Braavig, J., Düwell, M. and Brownsword, R. (eds.) *The Cambridge Handbook of Human Dignity: Interdisciplinary Perspectives*. Cambridge: Cambridge University Press, pp. 85–94.

Steiner, G. (2010) *Anthropocentrism and Its Discontents: The Moral Status of Animals in the History of Western Philosophy*. Pittsburgh: University of Pittsburgh Press.

Steiner, G. (2013) *Animals and the Limits of Postmodernism*. New York: Columbia University Press.

Stone, C.D. (1972) 'Should Trees Have Standing–toward Legal Rights for Natural Objects', *Southern California Law Review*, 45(2), pp. 450–501.

Taylor, P.E. (1997) 'From Environmental to Ecological Human Rights: A New Dynamic in International Law', *Georgetown International Environmental Law Review*, 10(2), pp. 309–397.

Temper, L. (2019) 'Blocking Pipelines, Unsettling Environmental Justice: From Rights of Nature to Responsibility to Territory', *Local Environment*, 24(2), pp. 94–112. DOI: 10.1080/13549839.2018.1536698

Tignino, M. and Turley, L. (2018) 'Granting Rivers Legal Rights: Is International Law Ready for Rights-Centered Environmental Protection?', *New Security Beat*, 19 June [Online]. Available at: https://www.newsecuritybeat.org/2018/06/granting-rivers-legal-rights-international-law-ready-rights-centered-environmental-protection/ (Accessed: 22 September 2022).

Treanor, B. (2019) 'Hope in the Age of the Anthropocene', *Analecta Hermeneutica*, 10, pp. 1–22.

UNGA (2016) 'Harmony with Nature, Note by the Secretary-General', UN Doc A/71/266.

Universal Declaration of Rights of Mother Earth (adopted 22 April 2010). Available at: https://www.garn.org/universal-declaration/ (Accessed: 22 September 2022).

Weitzenfeld, A. and Joy, M. (2014) 'An Overview of Anthropocentrism, Humanism, and Speciesism in Critical Animal Theory', *Counterpoints*, 448, pp. 3–27.

3

INTERNATIONAL TRADE LAW AND THE COMMODIFICATION OF THE LIVING

Charlotte E. Blattner

Introduction[1]

Human history shows that we have typically used nature as a resource to satisfy our needs – sometimes excessively, at times to a lesser degree; sometimes in a respectful manner or in a more reckless form. With the rise of market-based approaches to regulation – including the expansion of private property rights (i.e., privatisation), liberalisation, and the emphasis of voluntary compliance – nature has become the object of a steady process of commodification. Consider our use and regulation of fish, but also of genes, or even water: What we historically used merely for consumption is now appropriated, extracted, given a price tag, and sold on the market (Prudham, 2016, p. 123). The commodification of nature is a rapidly growing area of research concerned with the ways in which natural entities and processes are made exchangeable through the market, and with the study of the implications thereof. The approaches to this topic and the disciplines by which it is examined differ widely, encompassing, among others, political ecology, critical environmental studies, critical legal studies, Marxist geography, and history (Castree, 2016; Ertman and Williams, 2005; Harcourt, 2012; Harvey, 2009; Laitos, 2021; O'Neill, 2001; Richards, 2014; Smith, 2008; Tsing, 2017). There are different ways in which both 'commodification' and 'nature' are understood. Commodification, as used in the following, describes an appropriation of someone or something not born, designed, or produced for sale to be converted into an exchangeable form (Kosoy and Corbera, 2010, p. 125). Nature, too, can be understood in a variety of ways. It can be seen as the essence of things such as when we say it is the 'nature' of someone or something to be or act in a certain way. Nature sometimes is used to denote everything separate from society ('the natural world'). And yet for others, humans are an integral part of nature itself. In the following, I leave this question open as I focus on the law's treatment of nonhuman animals (or in short 'animals') as a particular instance of the commodification of the living. This specific 'case study', if you will, is symptomatic of the broader commodification of nature and allows us to uncover dimensions of the commodification process that are instructive for this broader discourse.

The chapter begins by showing why trade law is increasingly hailed as a beacon of hope for animals as no general international treaty is in sight that would as such aim to protect animals across and beyond state borders. Efforts to analyse the potential of international trade law to better protect animals are limited to the specifics of trade law (e.g., national treatment clause or

DOI: 10.4324/9781003201120-5

61

the general exceptions) and often do so in legalistic fashion. What we risk losing sight of in this process are the inherent limitations of trade law and the full spectrum of our possibilities to truly consider animals' interests.

This chapter evaluates the existence and degree of commodification of animals under trade law through four lenses. To start, the status of animals under trade law, as posited in treaty law and mediated by parties to such treaties, gives us a good picture of the extent to which animals are commodified. As I will show, animals are self-evidently accepted as goods, with little to no attention being paid to how animals are commodified and without critically evaluating such processes. I use feminist theory's well-developed notion of objectification to disentangle the multiple legal, political, social, and economic dynamics at play as animals are appropriated by trade law. This status of animals as commodified goods remains largely unchallenged due to the narrow focus on trade as a transfer of goods or services between human, corporate, or state actors, blocking out non-trade concerns that are necessarily (and often coercively) involved in these relations, including concerns about welfare, well-being, claims to life and integrity, environmental degradation, pollution, and natural resource use (Nadakavukaren, 2010). This clinical separation of trade from non-trade concerns, as I argue, leads to a siloisation that helps to create a veneer, an illusion of legitimacy. Further, there are fundamental beliefs woven into the fabric of trade law that remain unchallenged, such as endless economic growth and trade liberalisation as the gate towards ending suffering and eradicating poverty, despite contrary evidence.

Finally, international trade law is not questioned as a system of binding norms made by and for human actors alone, in which nonhuman animals figure indirectly and instrumentally only and upon which they have no bearing. It is precisely these concerns that prompt the question of whether international trade law is not so thoroughly steeped in anthropocentrism – the belief that the world rotates around the human/man, for whom all other lives and life systems exist as objects (Grear, 2015, p. 225; Probyn-Rapsey, 2018, p. 47) – that the total commodification of the living nonhuman is the natural and necessary consequence, regardless of well-intentioned but essentially superficial attempts to render trade law 'moral', 'animal-friendly', and the like. In this context, it is also important to note that 'human' is not a neutral term encompassing all members of humanity. 'Human' is instead a shorthand to denote the white, male, self-sufficient, fully abled, and rational human who needs no positive assistance from the state, or, indeed, anyone else (Deckha, 2021, p. 12).

Trade Law as a Beacon of Hope for Animals: A Premature Judgment?

Over the past decades, the body of international law has increased at a fast rate, expanding into diverse areas such as international human rights law, international humanitarian law (e.g., the 1949 Geneva Conventions), criminal law (e.g., Genocide Convention, 1948), labour relations (e.g., Fundamental Rights and Principles of Work), finances, environmental law (e.g., Kyoto Protocol 1997), and trade (e.g., General Agreement on Tariffs and Trade (GATT) 1994) (Crawford, 2019; Evans, 2018; Shaw, 2017). The considerable growth and incremental expansion of international law notwithstanding, today, there is no international treaty in force that deals with core issues of *animal law*, such as the recognition of animal sentience, prohibiting cruelty towards animals, preventing different forms of suffering, demanding humane treatment, or, let alone, establishing rights for animals (Blattner, 2019, p. 2; Bowman, Davies, and Redgwell, 2011, p. 698; Cunniff and Kramer, 2013, p. 230; Favre, 2016, p. 92; Wagmann and Liebmann, 2011, p. 279; Peters, 2021; White, 2013, pp. 391–392).

The primary concern of animal law[2] is about animals as beings with intrinsic value and as having their own lifeworlds which, in turn, imposes on us human animals certain legal obligations. To be clear, there are international treaties dealing with a minority of animals affected as a species (such as the 1948 International Convention for the Regulation of Whaling (ICRW), the 1973 Convention on International Trade in Endangered Species of Wild Fauna and Flora (CITES), the 1979 Convention on the Conservation of Migratory Species of Wild Animals (CMS) or the 1992 Convention on Biological Diversity (CBD)).

However, the concern of these treaties for animals is limited in several respects, two of which are most noteworthy. To begin with, these treaties understand and value animals as groups, in the scheme of which individual animals are not worthy of protection *qua* individuals, but are instead seen as fungible, exchangeable parts of a larger entity ('species') that is worthy of protection (Pearce, 2015, p. 157).[3] It does not matter, in this scheme, that one animal dies, as long as they are replaced by another of their species to maintain the overall number necessary to conserve their species.[4] Another limit of such treaties is that they are blind to the majority of animals, which are neither endangered nor threatened with extinction but who are, instead, bred on an industrial scale to secure human needs, including, for example, domesticated animals like cows, pigs, chickens, horses, and other animals used for farming, or macaques, beagles, frogs, and fishes used for research. In essence, the aforementioned treaties, though they touch on aspects of our relations with certain animals, are not an instrument of animal law proper, but treaties primarily environmental or trade-related in nature.

Scholars and interest groups from a range of disciplines have attempted to inscribe normative dimensions about our relations with animals in the international sphere, using different legal avenues and tools, offering varying moral and political motivations, and meeting varying degrees of public awareness and recognition. By and large, however, these proposals have not been successful to date. They are either too focused on codifying the status quo, as in the case of the 1988 International Convention on the Protection of Animals (ICPA), the 2011 Universal Declaration on Animal Welfare (UDAW), and the 2018 UN Convention on Animal Health and Protection (UNCAHP), which prompt governments to sign up for them (Blattner and Tselepy, forthcoming); or they appear, for now, too aspirational vis-à-vis the rather conservative concerns for animals currently present on the international plane, such as the 2011 Declaration of Animal Rights.

Alongside general international law, international trade law has come into focus as a potentially useful avenue to further the interests of animals. Indeed, in the past years, concern for animals has incrementally pervaded international trade matters, in terms of the number of disputes raised, the number of successfully litigated disputes, and the member states' attitudes towards animal law evolving within and outside its dispute settlement mechanism (Blattner, 2019, pp. 156 ff.). A notable case that prompted scholars to hail international trade law as a beacon of hope for animals was the *Seals* case. This was a case brought by Norway against the European Communities (EC) for its laws prohibiting the importation and sale of processed and unprocessed seal products. Norway claimed that the EC's 'seal regime' contains certain exceptions that give privileged access to the EU market to seal products originating in the EC and certain third countries but did not afford those to Norway. In doing so, as Norway argued, the EC violated key obligations of non-discrimination under trade law.[5]

On May 22, 2014, in a much-awaited decision, the Appellate Body (AB) – which is a standing body of seven persons that hears appeals from reports issued by panels in disputes brought by members of the World Trade Organisation (WTO) – issued its report holding for the first time that protecting animals is a legitimate concern of public morality that members can rely on if they violate trade obligations (WTO Seals AB Report, 2014). The earlier panel report even noted that protecting animal welfare is a 'globally recognized issue' (WTO Seals Panel Report, 2013, para. 7.420) that is 'a matter of ethical responsibility for human beings in general' (WTO Seals Panel Report, 2013, para. 7.409). Across the board, animal studies scholars welcomed the reports and considered the case 'a significant victory' (Howse, Langille, and Sykes, 2014), a 'landmark decision' (Sykes, 2021, p. 1), and even 'a watershed case in the global animal protection movement' (Lurié and Kalinina, 2015, p. 444; cf. Hennig, 2015). These positive judgments mirror the more general evaluations of trade law as possessing 'enormous potential as a tool to tackle [...] global problems successfully' (Conrad, 2011, p. 1).

The *Seals* reports were guaranteed to cause ripples throughout international trade law and international law, more generally. Nearly a decade after the reports' adoption by WTO members, however, one must ask why few to none of the WTO's members have taken seriously this new line of adjudication. One might expect that this 'softening' of trade law would, all other things being equal, encourage states to seize the opportunity to put in place domestic laws that protect animals from the most sweeping onsets of economic liberalisation (see special issue GJAL, 2022). Different hypotheses could help explain this development (or the lack thereof): One could say states might not be convinced that the *Seals* adjudication is here to stay (Mavroidis, 2015). It might also be possible that despite the legality of harsher trade restrictions put in place on the grounds of moral concern for animals, states fear economic repercussions from their trading partners. Yet another, more cynical interpretation would be that even the (seemingly) most adamant defenders of animal welfare at the WTO were, in fact, never truly concerned about animals but much more about ways to indirectly secure domestic production while blocking and limiting access for foreign products (hence, engaging in protectionism). Relatedly, one could hypothesise that members resorted to trade law to respond to demands by their people or consumers (for example, ostentatiously invoking seal protection in one isolated case) while masking that they were not truly interested in protecting animals so they could continue to engage in the economic exploitation of animals.

At the point of this writing, no empirical analysis has yet shed light on these developments, but they are part of ongoing scholarly discussions (e.g., Offor, 2020; special issue GJAL, 2022). Whatever the motivations and rationales behind states' cautious stance on protecting animals through trade law, the question remains if, and to what extent, trade law can, in principle, be used to protect, liberate, or emancipate animals on the international plane. This question has been widely addressed, often in legalistic fashion, paying attention to the specifics of trade law (e.g., national treatment clause and general exceptions under the GATT), and attending to the narrow language of trade. While such analyses are indispensable to answer the above question, they do not consider the full spectrum of possibilities and limitations inherent in trade law. This includes, most notably, the fact that animals – as sentient beings with their own lifeworlds – are self-evidently seen as 'goods' over whom members have factual and legal authority and whom they consequently can treat and trade as they please.

These and other limitations, as I argue in the following, can be directly traced to the broader project of trade law to commodify animals, which makes it intrinsically implausible that trade law would ever, whatever its marginal reform, fundamentally challenge or overcome that basic structure. Commodification is 'the action or process of treating a person or thing as property which can be traded or whose value is purely monetary' (Oxford English Dictionary, 2022a). In scholarship, 'commodification' is mostly used without definition and applied to living beings and their actions (rather than to things like sneakers or a cup of coffee) to narratively bring to the fore the oxymoronic nature of treating as an object what/who might not be an object (Collard, 2014; Gunderson, 2011; McDonald, Harrington, Moorhouse, and D'Cruze, 2021).

Some scholars have done this in more obvious ways, by terming animals 'lively commodities' (Collard and Dempsey, 2013), 'animal capital' (Shukin, 2009), 'lively capital' (Collard, 2020), 'sentient commodities' (Wilkie, 2017), even 'undead things' (Haraway, 2008) and the 'once-living commodities', 'soon-to-be-dead', 'living dead', and 'future corpses' (Gillespie, 2021). This tension of 'a sentient, dynamic, emotional being [being] *made* thing*like* when it is made a commodity, made property, through markets, the law, the state, and other institutions and mechanisms' (Collard, 2020, p. 5) is not only present in debates about our treatment of animals, but indeed, of any agents we consider ought not be commodified. In a sense, these charged terms are a discursive, normative tool against the commodification of animals.

A close and critical look at the commodification of animals under trade law is all the more necessary as trade law has grown in content and application in the past years, expanding the range of trade disciplines, creating new institutions and establishing a binding dispute settlement system (Bethlehem *et al.*, 2009, p. 1). Further, trade law's purview is exclusive, precluding states from having recourse to any other forum where animals and trade are affected (WTO, Dispute Settlement Understanding (DSU), 1994, art. 23(1)). Since a majority of animals (especially domesticated animals) is subject to trade, the issue of animal law is not left much room in general international law.

Before I begin my analysis, a note on the functioning and dynamics of trade law is due. International trade law is an aggregate of multilateral and bilateral legal rules, but the subject matter of the present analysis is limited to trade law under the tenets of the WTO, an intergovernmental organisation that regulates and facilitates international trade between nations, because WTO membership includes more than 164 member states that account for 98% of global trade flows (WTO, 2023). WTO law hence covers the majority of all legal disputes that arise in trade matters. The WTO facilitates trade in goods, services, and intellectual property among participating countries. It does so by providing a framework for negotiating trade agreements, which usually aim to further liberalise trade by reducing or eliminating tariffs, quotas, and other restrictions, and by prohibiting discrimination between trading partners.

Violations of these core commitments can be justified for environmental protection, national security, and other important goals. The WTO also has its own independent dispute resolution for enforcing participants' adherence to trade agreements and resolving disputes. The WTO oversees roughly 60 agreements, the most notably of which, in the context of animal protection, are the GATT, the 1994 Agreement on Technical Barriers to Trade (TBT) and the 1994 Agreement on the Application of Sanitary and Phytosanitary Measures (SPS). Reviewing each of these with a view to excavating the existence and degree of commodification of animals would go beyond the reach of this chapter. The following analyses will instead make functional reference to the treaties to exemplify the argumentative point made.

Animals as Goods: Necessary and Natural Objectification?

Objectification is a well-developed notion central to feminist theory, which can help elucidate the state of commodification experienced by nonhuman animals under trade law. In her early work, Martha Nussbaum (1995, p. 257) identified seven features, each of which represents one incremental step towards seeing or treating as an object what is not an object.[6] The first step towards objectifying others is to instrumentalise them (the objectifier treats the 'object' as a tool of his or her purposes). The objectified subject is then denied autonomy (the objectifier treats the 'object' as lacking in autonomy and self-determination) and treated as inert (the objectifier treats the 'object' as lacking agency and activity). The objectifier acts and interacts with objectified subject as if they were a fungible asset (the objectifier treats the 'object' as interchangeable with other 'objects'), as if they were not violable (the objectifier treats the 'object' as lacking in boundary-integrity, as something permissible to break). Next, the oppressor subjects the other to ownership (the objectifier treats the 'object' as something that is owned by another and can be bought and sold). This shows that, in the feminist theory of objectification, commodification features only as one aspect of a larger set of actions that constitute another's objectification. Finally, in Nussbaum's theory, the objectifier denies the objectified's subjectivity (the objectifier treats the 'object' as something without experience and feelings, or with experiences and feelings which need not be taken into account).[7]

Rae Langton has further developed Nussbaum's theory of objectification and added three more features: The objectified is reduced to their body, meaning that they are identified in the treatment by others as their body or body parts; relatedly, they are reduced to their looks and appearance or impression on others (reduction to appearance); and they are treated *as if* they are silent or lack the capacity to speak (Langton, 2009, pp. 228–229). Although this combined theory of objectification emerged from debates about the treatment of women in the context of prostitution and other contexts of (assumed) exploitation, Nussbaum and Langton's work seems instructive to better understand how we have come to (and continue to) objectify animals who are, in fact, sentient, conscious beings with tremendous agential capacities.

Biologically speaking, animals are all living beings excluding human animals, plants, and other biological kingdoms (Merriam-Webster, 2023a). Whilst nonhuman animals are thereby negatively defined (by reference to what they are not), they are, by this definition, not characterised by their use or utility for human purposes. For example, animals are not seen as 'livestock', 'providers of human food', 'research tools' or the like (Eisen, 2010). The above definition is quite broad, including all animals from small insects to large mammals, from invertebrates to hominoids, from the 1.95-gram bee hummingbird to the 90-foot blue whale, and from the broiler chicken to the Yorkshire pig. That trade law understands animals in such a broad fashion is exemplified in the only definition of animals in its treaty texts, namely Annex A of SPS, which provides that the term 'animals' includes fish and wild fauna. We can infer from this that wild animals and animals commonly understood as 'domesticated' self-evidently qualify as 'animals' for the purposes of trade law.

Reduction to Body and Appearance: Fungibility

In trade law, the term 'animals' denotes living or dead entire animals, and 'animal products' equates to unprocessed, fully processed, or partially processed animal parts or liquids (cf. FAO, 2013, pp. 140–144). These definitions demonstrate that animals are, in the scheme of trade law, in a fundamental sense reduced to their bodies and body mass. The most trade law cares about is whether an animal is alive or dead, and this only for classification purposes. By and large, animals in live or dead state are seen as a medium to deliver flesh, blood, milk, reproductive organs, eggs, and other objectified parts of their bodies.

Animals and their processed body parts or liquids are subject to multiple sections in the members' Schedules of Concession (which include bound duties for goods from other WTO members and form part of the GATT). Section III covers animal fats and oils and Section VIII deals with leather goods. Section I deals with 'live animals and animal products', under which most animals fall. Chapter 1 of Section I includes 'live animals', Chapter 2 'meat and edible meat offal', Chapter 3 'fish and crustaceans, mollusks and other aquatic invertebrates', Chapter 4 'dairy produce, birds' eggs, natural honey, edible products of animal origin', and Chapter 5 deals with 'products of animal origin, not elsewhere specified or included' (World Customs Organization, 2022). These classifications bring to the surface the forceful imposition of equivalence on animals and the lack of identification of them as individual beings, which in turn shows that they are treated as fungible, interchangeable with other (objectified) animals. A cow is a cow is a cow. This process is 'inherently violent in so far as its valuation requires erasure of the inherent qualities' of animals (Wadiwel, 2015, p. 165), except insofar as such qualities (e.g., consumability, edibility, reproducibility) are of use to humans.

Instrumentality

Instrumentality denotes the quality or state of being instrumental, i.e., serving as a means or tool (Merriam-Webster, 2023b). Instrumentality exists on a spectrum and comes in different forms and degrees: One can be fully instrumental to others, which often coincides with one's own needs being disregarded; or one can be partially instrumental to others, whilst one's own needs are still met (Kaufmann, 2010). To some extent, a certain level of instrumentality is, in a globalised, interconnected world where social beings communicate and interact on a daily basis, unavoidable. However, many cultures believe that animals are made for human purposes alone, i.e., that they, by their very nature, serve human needs. Cows produce milk for humans and chickens make our breakfast eggs. As such, animals are fully instrumental – indeed, without their own doing and often against their will (Hribal, 2010; Wadiwel, 2018), instrumentalised, i.e., rendered instrumental to accomplish a purpose or result (Oxford English Dictionary, 2022b). Embedded in these views is a stark rejection of the reality, notably of the influence of domestication on animals and us, animals' biological evolution, and the everyday reality they experience. Isabella La Rocca (2015) brings out the flaws of the view that 'animals are natural and necessary producers for humans' as follows:

> The belief that hens naturally produce eggs for human consumption is absurd. A wild chicken produces only 10–15 eggs per year. A domesticated hen, however, has been bred to produce hundreds of eggs during her most fertile years, which causes osteoporosis and accompanying broken bones, and reproductive disorders (tumors, egg binding, peritonitis, etc.). To stimulate maximum egg laying, artificial light is alternated with periods of darkness. She will likely be starved for extended periods until she loses 25% to 35% of her body weight, also to stimulate increased egg production. Standard industry practices include debeaking, intensive confinement, and gassing or slaughter when the hen no longer produces enough eggs to be profitable, at one to two years of age, much younger than her natural life span.
>
> …
>
> A calf in the dairy industry typically spends the first 2–3 months of life isolated in a small stall. As soon as she is old enough to become pregnant she is artificially inseminated – among dairy industry workers, the colloquial term used for the enclosure used to restrain her during this process is 'rape rack'. Her calf is taken from her within hours after birth so that her milk can be sold for human consumption, causing great distress to both mother and baby. Mastitis is common in the dairy industry. When a cow no longer produces enough milk to be profitable, she is sent to slaughter – typically, when she is only 4 to 5 years old. Her natural life span is 25 years.
>
> …
>
> Artificial insemination is also standard practice in the pork industry. For most of her adult life, a sow is kept in a metal enclosure, known as a gestation crate, so small that she cannot turn around. She will chew on the metal bars, wave her head incessantly, and exhibit other signs of insanity. The metal farrowing crate in which she and her piglets

are kept while she is lactating is also not large enough for her to turn around. Then her piglets are taken from her and she is again artificially impregnated.

Animals' status as a means to satisfy human ends is thus discursively located in intrinsic properties of animals (e.g., 'milk cows', 'egg-laying hens', etc.) (Collard, 2014, p. 155) rather than being acknowledged as an external burden put on nonhuman animals by humans (i.e., impregnating female cows once a year to produce milk, which they only do upon giving birth etc.). However, the view that animals 'are made for/there to' satisfy human needs does not 'come naturally' but has been produced by a complicated set of intermingling cultural, religious, philosophical, and psychological biases of the dominant group over the oppressed group (Caviola, Everett, and Faber, 2018; Dhont et al., 2018; Joy, 2011; Kopytoff, 1986, p. 64; Paez and Faria, 2014; Weitzenfeld and Joy, 2014; Park and Valentino, 2019). In trade law, these biases are not openly acknowledged or presented, but play out implicitly in the background. The fact that animals are self-evidently seen as objects of trade, as goods, shows the privilege and casualness with which negotiating parties assumed sovereignty, power, and predatory force over other animals (Grear, 2015; Wadiwel, 2015).

To some extent, we might view the ongoing debate in trade law about whether certain animals qualify as goods or as natural resources as an opening to recognise animals not only as means to our ends. Article XX(g) GATT, which declares legal trade violations on the basis of measures that relate to 'exhaustible natural resources', is interpreted by panels and the AB to encompass animals. In *Prohibition of Imports of Tuna and Tuna Products from Canada*, the panel determined that tuna stocks constituted exhaustible natural resources (WTO Tuna I Panel Report, 1982, para. 4.9). In *Herring and Salmon*, parties concurred that salmon and herring are exhaustible natural resources (WTO Herring and Salmon Panel Report, 1988, para. 4.4). In *Shrimp/Turtle*, however, parties argued at length over whether sea turtles conserved under US law constitute exhaustible natural resources. India, Pakistan, and Thailand argued that all living natural resources, including animals, are renewable and thus inexhaustible (WTO Shrimp/Turtle I AB Report, 1998, paras. 127 ff). The AB was not convinced by the argument and pointed to the growing evidence about the threatened status of these animals that recently prompted 'concerted bilateral or multilateral action' of the international community 'to protect living natural resources' (para. 131). Because WTO members explicitly recognised the objective of sustainable development in the preamble of the WTO Agreement, the AB reasoned, the treaty must be interpreted dynamically, responding to an informed community concerned with protecting and conserving the environment (para. 131).

To end the recurring debate on animals as *goods vs. resources*, the WTO has established a working definition of 'natural resources' in its Trade Report of 2010. According to the report, natural resources are 'stocks of materials that exist in the natural environment that are both scarce and economically useful in production or consumption, either in their raw state or after a minimal amount of processing' (WTO, 2010, 46). Such natural resources are characterised by uneven distribution across countries, exhaustibility, externalities (market failures in the form of unpriced effects resulting from consumption and/or production), dominance in output and trade, and price volatility.

Especially domesticated animals seem to fulfil these criteria: They can, theoretically, go extinct, their production causes high negative externalities (causing environmental degradation, threatening human health and world food security, and compromising animal welfare (Blattner, 2014, p. 16), they provide for output and trade, and their price as traded entities varies across time and territory. The report, however, denies them the status of natural resources, not because they are abundant or found outside ecosystems, but because production of agricultural goods requires other natural resources as input (mainly land and water).

In addition, the report argues that farmed animals are cultivated rather than extracted from natural environments (WTO, 2010, p. 46). The only farmed animals the report defines as natural resources are marine animals (e.g., fishes, shrimp, and mussels) that are tradable (i.e., not subject to conservation or preservation treaties) (WTO, 2010, p. 46). These debates show how political the classification of animals in the realm of trade law is. Indeed, members interested in freely trading in animals and their body parts do their best to keep most animals in the 'goods' category, wary of the potential ramifications that come with classifying some of them as 'natural resources'.

However, it is important to note that even if a nonhuman animal ends up being defined as a natural resource, they are, in effect, traded as regular goods. Only if animals – as natural resources – belong to an endangered species are they exempted from trade law and thus protected from being traded. Aforementioned treaties like ICRW, CITES, and CBD lay down states' common desire to considerably reduce or fully prohibit trade in endangered species. These exemptions (mirrored in Art. XX(g) GATT)[8] are useful and effective for protecting some animals, but they have no bearing on the majority of animals traded under the WTO regime.

Broadly speaking, most animals are thus fully within the ambit of the Schedules of Concessions on Goods and, as such, freely tradable. Because of their absolute subjection to trade law, there has been little debate about the degree to which animals can or should legitimately be instrumentalised as commodities. Animals are, in self-evident manner, understood as being of value to us, meant to fulfil human needs alone. Under the tenets of WTO law, animals are consequently reduced to their productive roles, which reinforces instrumental views about them (Wilkie, 2017, p. 280).

Violability

Along with this implicit right to fully instrumentalise animals, comes the privilege to violate animals' interests without the need for justification. Trade law 'relies fundamentally on the ability to sever animals from their former homes, transport them globally, and deliver them [...]' (Collard, 2014, p. 156). By framing animals as 'consumables', they become readily and legitimately exploitable and violable. To be clear, all animals – including us humans – are liable to being violated, to suffer, and to being exploited in their vulnerability. Throughout our lives, we all experience vulnerability in permanent or transient, lesser, and more expansive forms. For each of us, it is inescapable that one goes through stages of dependency, for example during infanthood, at old age, during illnesses, and in other situations where we rely on others to thrive (Oliver, 2015, p. 480). Though we share vulnerability as flesh and blood across the species line (Deckha, 2015; Oliver, 2015; Satz, 2009), there is a common agreement that human animals *ought not* be violated or made to suffer, to be kept free from violation, infraction, or assault.

Nonhuman animals, by contrast, despite sharing vulnerability with us at the core, are not subject to the same normative, protective shield of inviolability. There is no agreement that animals ought not be violated. Trade law sets up no barrier to when, how, how many, or how often (most) nonhuman animals may be injured, mutilated, killed, or severed from their families. In the case of goods and services from humans, by contrast, different reasons (the nature of goods, their social meaning, or the inequality they create) serve to curb their commodification (e.g., Satz, 2010; Walsh, 2019). Unlike human animals, nonhuman animals also do not enjoy inviolable basic rights (Donaldson and Kymlicka, 2017) that would set barriers to how animals can be treated in the trade realm. Another way in which trade law openly endorses animals' violability is by facilitating trade in them and their bodies. Animals' state as 'disposables', in turn, is integral to the functioning of trade (Collard and Dempsey, 2013; Gruen, 2014).

Ownership

Ownership is perhaps the most qualifying criterion of objectification. From the perspective of the owner, one has a bundle of rights in relation to the thing that is owned: The right to title (formally identified owner of the property), the right to use property, the right to profit by the use of property, the right to exclude others from property, the right to transfer property, and the right to destroy property (a definition which goes back to Honoré (1961)). These rights are almost uniformly recognised across the globe. As most states still treat animals as legal objects, who are owned by legal persons, it seems logical that animals qualify as goods in the trade realm.

WTO agreements that deal with trade in goods, including GATT, TBT, and SPS, crucially lack a definition of *goods*. This has presented the adjudicatory bodies of the WTO (i.e., the panels and the AB) with a problem, particularly when they were asked to define the status of formerly free-living animals. They were forced to determine, sometimes quite arbitrarily, the point at which a sentient being becomes a good subject to trade. In *US – Softwood Lumber IV*, the AB asserted that goods include tangible property, which is not yet severed, but is in principle severable, from land (WTO Softwood Lumber IV AB Report, 2004, paras. 59 ff.). The panel in *Herring and Salmon* determined that the GATT applies to restrictions on exports in fishes but not to restrictions on fish catches (WTO Herring and Salmon Panel Report, 1988, para. 3.1).

Simply asserting dominion over animals by catching or hunting them does thus not render them goods. Only once animals are introduced into the stream of trade do they qualify as goods. Based on these reports, scholars now define goods as such if they are listed as goods in the member's Schedules of Concessions on Goods, if they are processed to the level laid out in the Harmonized System Nomenclature, or if they have monetary value and are tradable (Desta, 2021). By entering the stream of commerce, animals are ascribed a certain monetary value that makes them tradable and confers on others a right to trade in them. At least from the perspective of trade law, the fact that one is traded means one is ownable – which brings to the fore its circular rationale and lack of precision in determining why, precisely, someone with an own view of the world and own life plans, should be legally, albeit often forcefully, come into the stream of commerce.

Denial of Subjectivity, Autonomy, and Agency

The WTO's treaty law is silent about the possibility of animals having agency, exercising autonomy, or being subjects. Reports have acknowledged and consider important the fact that animals suffer (as done, notably, by the AB in *Seals*); one report has also dealt with how severing family relations may result in suffering (notably when dolphin mothers are separated from their calves (WTO Tuna/Dolphin III Panel Report, 2012, para. 4.73). However, there is no indication whatsoever that animals could be regarded as anything but passive sufferers. Indeed, the above elements of objectification strongly suggest that animals' own needs, preferences, interests, agential capacities, desires, and so on do not play a role in trade law. Indeed, we do not even recognise these in the first place. This denial of subjectivity, autonomy, and agency is somewhat tied to the element of fungibility: Recognising animals individually could disrupt the view that they are inanimate things made to be traded; denying individuality, in turn, clouds judgment that might recognise that animals are indeed sentient, self-willed, and conscious beings (Wilkie, 2017, p. 289).

Separating Trade from Non-Trade Concerns:
Siloisation and Distraction

In the past decades, scholars have increasingly grappled with the question of whether and to which extent trade law can be used to protect animals. This surge of interest is to some extent owed to the fact that member states have adopted trade-restrictive measures to protect animals, such as by prohibiting inhumane transport of animals, banning the importation of meat from inhumanely slaughtered or transported animals or from inhumanely trapped animals; banning cosmetics tested on animals; limiting trade in endangered animals and wild birds as well as banning trade in products from inhumanely killed seals and seal pups and shrimps caught in a fashion that endangers turtles (Blattner, 2019, p. 88). States can protect animals through labels, taxes, tariffs, and quantitative restrictions. The legality of these means, however, is debated, as WTO agreements are lengthy, complex, and cover a wide range of activities. Indeed, the WTO treaties are said to be so technical in nature that there is almost no way for member states not to violate them (Sykes, 2004).

For example, under GATT, the legitimacy of import bans on, e.g., cruelly produced eggs, meat, and dairy, is discussed the most. In this context, it is debatable whether import bans are illegal in the first place, thus falling under article XI GATT, which eliminated any and all quantitative restrictions. The opposite argument is that they are legal to begin with and subject to the national treatment obligation of article III GATT, hence, domestic and imported products should be treated alike. But is an imported egg produced in battery cages the same as an egg produced by a domestic farmer who holds a flock of 40 chickens, tends to them daily, secures them access to the outside throughout the day, and ensures that they can socialise and engage in nesting practises and the like as they please? This is part of the process and production methods (PPMs) debate (Conrad, 2011): The traditional view is that physical characteristics alone determine the likeness of domestic and imported products. Since it is virtually impossible to determine the level of animal welfare by examining the two eggs, they are the same and hence must be treated alike under trade law (e.g., Schoenbaum, 1992). This view facilitates free trade as it enables cheap and low-welfare products to flood domestic markets and effectively precludes states from protecting animals in production contexts. Newer analyses rely on relevant treaty provisions, Note Ads, panel reports, and scholarly opinions to argue that measures which distinguish between products on the basis of animal welfare should not only be justifiable in the WTO treaty framework but also legal in the first place, pointing to intrinsically different goods, in line with the changing public opinion and public policy priorities about production methods and animal treatment (Archibald, 2008; Charnovitz, 2002; Fitzgerald, 2011; Kelch, 2017; Van den Bossche and Zdouc, 2021, pp. 426–427; for an in-depth discussion see Blattner, 2019, pp. 99 ff.).

In applying such and other measures, members likely violate GATT provisions, which can be justified by the general exception of article XX GATT. Article XX(g) GATT explicitly encompasses animals as natural resources but comes with significant limitations for most animals (as argued above). For many years, article XX(b) GATT, which allows exceptions if they are 'necessary to protect human, animal or plant life or health', was thought to be the primary exception for states interested in animal welfare. However, most scholars doubt that health can be equated with (and thus cover) welfare (Kelch, 2017; cf. Sykes, 2014). After the recent reports adopted in the *Seals* dispute, the public morals exception of article XX(a) GATT has now been considered the primary exception to enable states to justify trade-restrictive measures intended to protect animals.

This brief journey into some of the specifics of WTO trade law shows how tempting it is to lose the forest for the trees. One easily gets caught up in treaty clauses, interpretations, exceptions, and reports of panels and the AB of several hundred pages and thereby risks overlooking the ingrained structures of trade law and its limits. Though the WTO shares commitment to other values in the preambles of its various treaties, its core mandate is to liberalise trade. According to the preamble to the GATT, members are committed to 'entering into reciprocal and mutually advantageous arrangements directed to the substantial reduction of tariffs and other barriers to trade and to the elimination of discriminatory treatment in international commerce'. Trade is most prominently facilitated through the principle of non-discrimination, market access rules, and provisions to eliminate unfair trade (see esp. Van den Bossche and Zdouc, 2021). Removing the most obvious barriers to the flow of goods, services, and finances among states has thrown into bolder relief the remaining regulations that impede free trade, dealing with labour rights, environmental protection, the use of natural resources, or claims to better treat animals (so-called 'non-trade concerns').

From the perspective of trade law, these non-trade concerns do not happily co-exist with commitments to liberalise trade. Indeed, from a competitive perspective, these regulatory requirements are seen as undesirable costs and barriers:

> [Regulatory requirements are] designed to do one of two things: either to limit market access of products, services, or service suppliers by or to the target country (e.g., import or export restrictions), or to permit discrimination among similar products, services, or service suppliers to the detriment of the target country (e.g. administrative/regulatory restrictions).
>
> *(Nadakavukaren, 2010, p. 101)*

These non-trade concerns, which include most social, political, moral, and other norms that in some sense affect trade, are thus at odds with axiomatic obligations of WTO law and should, in the view of trade, ideally vanish. From the regulatory perspective, by contrast, market-friendly WTO rules undermine essential political, moral, cultural, environmental, and technical achievements and aspirations (Qureshi, 1998, p. 166). With a view to animal protection, for example, states' regulatory policies – especially when it comes to commercially used animals – are plainly ineffective if they are not able to stop certain products at the border or increase their price – cheap, low-welfare products will simply flood the market (Conrad, 2011, p. 70). Unsurprisingly, animal advocates accordingly identify global free trade agreements as the principal obstacle to progress in animal law (e.g., Thomas, 2007, p. 609).

Nevertheless, the 'trade and …' problematic does not always tip in favour of trade. In certain circumstances, the rules that govern trade allow non-trade concerns to take priority over trade obligations. The preamble to the GATT enshrines the 'social' commitment of members; it values not only economic factors but considers welfare, claiming main goals that include 'raising standards of living, ensuring full employment and a large and steadily growing volume of real income'. But the GATT's preamble does not enshrine commitment towards animals or any moral aspirations we have in relation to them. The absence of such a commitment from the prominent preambular spot has an influence on how trade law is interpreted and applied.[9] Laws that aim to better protect animals are hence *prima facie* at a greater disadvantage than laws designed to further human welfare or to protect the environment.

Such broader discussions rarely feature in scholarly writings at the intersection of trade and animal law, not to mention policy discussions on the domestic level or at the WTO. Instead, the different actors involved in these processes – states, affected corporations, panellists, experts, scholars, etc. – focus on technical debates about exceptions, PPMs, and the like. As a consequence, non-trade concerns, including concern for animals, remain mired in these narrow, technical debates. This results in an excessively narrow focus on trade as a transfer of goods or services between human, corporate, or state actors, blocking out non-trade concerns that are necessarily (and often coercively) involved in these relations, including concerns about welfare, well-being, claims to life and integrity, environmental degradation, pollution, and natural resource use.

To be clear, tweaks to WTO practise, successful re-interpretations, and courageous pushes from member states could, in theory, have a positive effect on the billions of nonhuman animals currently bought, sold, and transported under trade law. These are, however, mostly cosmetic changes that leave unaffected the fact that non-trade concerns remain the exception to the rule and that trade matters first and foremost. Further, serious debate about whether PPMs should not generally influence the likeness test, clarification of the scope of Article XX(b) GATT, or general guidance on the scope, content, and depth of morality under XX(a) GATT, have never taken place. Indeed, one is tempted to think that the WTO does not seem to have an interest in clarifying or resolving these issues as the continued lack of legal security plays into the hands of free trade: If in doubt, a measure is guaranteed to violate trade law.

The continued practise of entering non-trade debates on such technical terms, and in such isolated, siloised fashion, significantly narrows and anchors the debates, drawing attention away, indeed, distracting from the broader pillars, background factors, and assumptions that fully structure our thinking around animals in trade (Offor, 2020, pp. 242, 260).[10] Our contributions to this narrow focus, and our omission to address the broader roots of these problems, might in fact be one reason why such larger debates within WTO law are not taking place and why the WTO, as an organisation, has never felt the need to address such issues openly and in a wholesome manner.

The Enduring Myth of Eternal Economic Growth

Woven into the fabric of trade law are fundamental beliefs that remain unchallenged, notably the narrative of endless economic growth.[11] Driving factors of globalisation – including enhanced means of transportation, new communication tools, the gradual removal of barriers to trade, new waves of foreign direct investment, and intensifying animal production (e.g., reducing space per animal, using growth hormones and superfoods, mutilating animals, automatic handling and slaughter, and shortening the lifespans of animals) – have unleashed unprecedented flows of goods, accelerating economic growth, and, in turn, further boosting consumption and production (Blattner, 2019, pp. 52–53).

Though positive in the eyes of trade law, the economic globalisation of animal industries has led to a host of ethical, social, and environmental problems. The steady increase in production and consumption of animal products is one of the biggest sources of animal suffering and results in the killing of 100+ billion animals per year (Mason and Singer, 1990; Schlottmann and Sebo, 2018). Animal production is also one of the biggest threats to a liveable future. It eats up vast (and disproportionate) amounts of land, water, energy, and other resources that could be used more effectively and in a more sustainable and equitable manner. Industrial animal agriculture also puts immense pressure on the environment, by polluting land, water, and air through enormous amounts of waste (Abate, 2020; Schlottmann and Sebo, 2018). Animal agriculture is increasingly recognised as 'the new coal', as it thwarts efforts to stabilise greenhouse gas emissions at a 'safe' level (Blattner, 2020; Dorning, 2021). Finally, as the COVID-19 crisis painfully showed, animal production has become a major threat to public health, by spreading diseases and pathogens, but also by leading to widespread resistance to antimicrobial agents (Wallace, 2016). In sum, 'the planet as a whole produces and consumes beyond its biological capacity, as reflected by its ecological footprint' and 'in effect consum[es] and produc[es] at the expense of future generations' (Reuveny, 2010, p. 99).

Making things even worse, a great deal of production (and hence trade) is made not for the sake of feeding people and satisfying other immediate human needs, but for the sake of increasing production and profitability. Early forms of trade, such as pre-capitalist peasant societies producing and exchanging animal products, were geared to answer the needs and consumption patterns of people (mostly of dominant classes) (Wood, 2002, p. 78) and as such, they normalised relationships of exploitation to animals (Nibert, 2013). Animal products thus had, first and foremost, 'use value'.

In today's commercial and capitalist transactions, by contrast, 'exchange value' – the price at which a commodity can be traded – is its driving force.[12] Animal products are made and valued primarily for their 'ability to produce surplus exchange value' and in this scheme, 'the production of concrete use values becomes a mere means to this end' (Zick Varul, 2011; also, Joyce, Nevins, and Schneiderman, 2017, p. 96; Wadiwel, 2015, p. 161). In simpler terms, the primary purpose of breeding, raising, rearing, and slaughtering animals today is *not to create food* but *to accumulate capital* (Gunderson, 2011, p. 261). This helps explain why, according to the Food and Agriculture Organisation's (FAO) numbers, 1.3 billion tons of food, i.e., one-third of the foodstuffs produced for human consumption, is lost or wasted in the whole supply chain (without the retail stage, about 15% of all foods are lost and wasted (FAO, 2019, p.12)). On global average, 20% of all dairy products, 35% of all fish and seafood, and 20% of meat and meat products are lost or thrown away (FAO, 2023).

While individual consumers have a role to play in this tremendous food loss and waste, too, it is mainly food suppliers, food service providers, and retailers that engage in those practises – not because the food is perished but because they would rather see their products go to dumpsters than to lower their price (FAO, 2019, p. 13; Royte, 2014). COVID-19 has aggravated these issues, as when the *New York Times* reported that farmers were dumping as many as 3.7 million gallons of milk each day; that a single chicken processor is smashing 750,000 unhatched eggs every week; and that some producers have even considered euthanising chickens to avoid selling them at unprofitable rates (Yaffe-Bellany and Corkery, 2020; see also FAO, 2020).

A major reason for the emergence and expansion of such practises is trade law's neoliberal, capitalist ideology that 'has intensified legal and regulatory controls *in the service of extending economistic logics through all social spheres*' (emphasis added) (Blanco and Grear, 2019, p. 96), giving trade concerns primacy and displacing all non-trade concerns.[13] If we juxtapose such economic motives to their (unaccounted) consequences, such as environmental destruction and degradation, threats to habitability and public health, and overreaching planetary boundaries at the expense of oppressed (human and animal) groups and future generations, it becomes clear that endless economic growth is impossible on a finite planet – whether we want to admit it or not. While it is understandable that trade has come to evolve that way in an age where globalisation was thought to make a whole planet thrive, it is less understandable why 'the fairy tale of eternal economic growth', as Greta Thunberg put it speaking at the United Nations Climate Action Summit in 2019 (Herz, 2019), still structures our normative thinking when it comes to trade relations, especially as we now know about its unfulfilled claims, its rife negative externalities, and the limits of this planet's ability to sustain us and others.

Breaking Free from 'Law as Anthropocentrism'

If we broaden our view further, it becomes clear that 'capital's blind drive for self-expansion and self-accumulation' (Gunderson, 2011, p. 260), which trade law crucially centres on and serves, does not operate in a 'legally neutral way'. Several contributions in this volume resolutely demonstrate how international law and its specialisations are fully committed to an anthropocentric logic. Anthropocentrism is the belief that 'humanity is at the centre of existence' and 'that humanity is fundamentally different to the rest of the natural world, of which we are rational, and it, isolated and valueless, has been provided as an instrument for our benefit' (Gillespie, 2014, p. 2).

The law is a quintessential site at which this anthropocentrism manifests itself. Law is seen as an invention of humans alone, framing humans as the one and only agents on this planet (Grear, 2015, p. 228). Through law, humans seemingly have separated themselves from the other animal kingdom, transcended nature, to bring to life a cultured normativity and bind nature to their service (Grear, 2015, pp. 234–235). Law is made by humans for humans and is, therefore, 'unresponsive – at a fundamental level – to the ethical implications for the vulnerable embodied biomateriality of the living order' (Grear, 2015, p. 241). Only in the past several decades has humanity considered whether other animals, too, have claims to normative protections, but most of this discussion thus far has focused on animals as passive beneficiaries of different forms of legal protections (animal protection, animal welfare, animal rights, etc.).

This anthropocentric framework is justified with reference to a range of arguments, including self-interest, economics, religion, aesthetics, culture, and future generations (Gillespie, 2014). Specifically in relation to law, the standard claims are that humans are rational 'creatures of rules' (Sellars, 1980, p. 138), 'rule-following animal[s]' (von Hayek, 1982, p. 11) that alone have 'nomic capacity' (Nozik, 2001, p. 270), i.e., the capacity to be guided in one's behaviour by norms.[14] These views have been framed as an objective fact, rather than as a bias. As Mary Midgley put it:

> [P]eople have seen themselves as placed, not just at the relative centre of a particular life, but at the absolute, objective centre of everything. The centrality of MAN (sic) has been pretty steadily conceived, both in the West and in many other traditions, not as an illusion of perspective, imposed on us by our starting-point, but as an objective fact, and indeed an essential fact, about the whole universe.
>
> *(Midgley, 1994, p. 104, emphasis maintained.)*

In practise, consideration of animals has consequently become, mainly, a matter of charity, rendering their moral, political, and legal status fragile (Donaldson and Kymlicka, 2011). When we look at trade law, we can see this fragile status of animals in the lack of a commitment in the preambles of the WTO's various treaties and the fact that protecting animals remains a rarely tolerated exception to more important obligations, i.e., liberalising trade. In mainstream discourse, animals are mostly viewed as silent sufferers, pitiful beings who cannot take care of themselves and must be put under human 'patronage' (Corman, 2016).

These implications and the theoretical conceptions giving rise to them are now criticised from a number of perspectives, including critical animal studies, animal politics, critical disability theorists, and multispecies ethnography (Corman, 2016; Blattner, Donaldson, and Wilcox, 2020; Donaldson and Kymlicka, 2011; Gillespie, 2019; Meijer, 2019a; Meijer, 2019b; Räsänen and Syrjämaa, 2017; Taylor, 2017). Drawing on nonhuman animals' languages, cultures, emotional lives, cognitive capacities, and political practises, they demonstrate that our thoroughly anthropocentric view about animals is unjustifiably biased and leads to dire and inacceptable consequences for them. One central argument of the emerging critical literature to counter such consequences is to posit animals as self-directing, self-determining agents, within their own communities and in their relations with humans (Meijer and Bovenkerk, 2021, p. 50), for which the investigation of animal normativity is central (Danón, 2019; Lorini, 2018; Palao, 2021; Vincent, Ring, and Andrews, 2018). The hypothesis of this research is that animals act in light of rules, even without recourse to human language, because they have their own sense and mode of normativity (Danón, 2019; Lorini, 2018; Palao, 2021; Vincent, Ring, and Andrews, 2018).

It might be senseless or naïve to point this out, but the possibility of consulting other animals about their views of trade law, has never crossed trade policy experts' minds. But it is worth asking anyway, for commodification is an engrossing process that colonises most of an animal's life who happens to be born in captivity or captured in the wild for trade purposes. The fact that we see animals in these contexts as lively, sentient beings, but treat them as things, objects, and commodities is a sobering reality. However, if we put emphasis on the fact that we do in fact see them as 'as lively, sentient beings', there is an opening to consider letting go of or reversing the ongoing and total commodification of animals.

Commodification, after all, is not thoroughly stable, definitive, or irreversible (Kopytoff, 1986, 64; Wilkie, 2010, pp. 3, 123). One's status as a commodity can be changed; one can, for example, be unmade as a commodity as one is freed (albeit, as Collard notes, one will never be unchanged (Collard, 2014, p. 153)). In relation to nonhumans, anthropocentric commitments, upon which trade law and its commodification crucially rest, are not inevitable or irreversible either (Youatt, 2020, p. 140). In this sense, as Kathryn Gillespie (2021, p. 293) put it:

> Might [...] we think of [...] afterlives in a different way – not only as a framework for thinking of what forms of commodification follow from the lively commodity, but one that leaves commodity relations behind entirely, making possible forms of life orienteered around the flourishing of nonhuman lives on their own terms?

Following Gillespie's prompt, one cannot help but wonder if other animals, too, would so fervently protect trade, or if they, instead, would not also want to tame the market for their own sake just as we do when it comes to human goods and labour. If we dared to listen, we would finally see that animals, through their ubiquitous forms of resistance (Hribal, 2010; Wadiwel, 2018) are 'perhaps among the most vocal and active critics of anthropocentrism' (Probyn-Rapsey, 2018, p. 51), advocating – across the board – for 'practices of decommodification' (Joyce, Nevins, and Schneiderman, 2017, p. 96). Commodification is a process that has been fundamentally guided by our normative imaginings, which have caused others unfathomable suffering and brought the planet to the brink of its sustainability, so it may be time to let nonhuman animals partake in these imaginings and propose more viable counter-imaginings.[15]

Notes

1 This chapter builds, in part, on my previous work on animals in international trade law (Blattner, 2019).
2 Animal law, as used in this chapter, is understood as a system of more or less connected legal norms governing relations between human and nonhuman animals.
3 For instance, conservationist strategies seek to ensure the survival of the African elephant (*L. Africana*), but individual elephants still suffer from and remain vulnerable to diseases, parasitism, accidents, drought, starvation, drowning, predation, and stress (CITES, 1973, Ann. I & II).
4 Indeed, such events are often seen as a positive development, rationalised as an evolutionary advantage that ensures survival of the fittest (Dawkins, 1998, p. 308).
5 These obligations, including the most favourite nation obligation and the national treatment obligation, are discussed below in the section 'Animals as Goods: Necessary and Natural Objectification?'.
6 Note that Nussbaum's theory, as the discussion about the extent and per se permissibility of pornography and objectification of women, is highly debated in philosophy. Debates include the extent to which objectification is negative or positive, the degree to which objectified persons should be positioned to choose that lot for themselves, and larger effects on society. For an overview see Papadaki, 2019.

7 For a recent critique, see Papadaki, 2012.
8 See the discussion below, in section 'Animals as Goods: Necessary and Natural Objectification?'.
9 Note that the Appellate Body and the panels frequently struggle with the principles of treaty interpretation − fluctuating between formalism and more dynamic interpretations. Art. 3 para. 2 DSU, in combination with the Vienna Convention on the Law of Treaties (VCLT) (in particular Art. 31 para. 3 lit. c), demands that other relevant rules of international law applicable between the parties must be considered when interpreting the GATT. This leaves at least some room to consider (usually more progressive) general principles of international law and, hence, for a more dynamic interpretation of its preamble (*see further* Sykes, 2014).
10 For an analysis of how such categorised debates influence our knowledge structures in law, see Chen and Hanson, 2004.
11 The following argument is tailored to animal production and trade in its goods. However, the discussion fits into the larger debate how trade can be reconciled with non-trade concerns, which currently takes place most prominently in the context of climate change. While conservative voices have argued that mere tweaks in interpretation can bring these conflicting values into alignment, newer writings propose a turn in priority-making. Bernasconi-Osterwalder and Norpoth (2019, p. 4), for example, call for 'WTO-compliance' to be morphed into 'Kyoto-compliance'.
12 As such, capitalism can be understood as 'a system of generalized commodity production' (Longo, Clausen, and Clark, 2015, p. 32).
13 See the discussion in the section 'Animals as Goods: Necessary and Natural Objectification?'.
14 For a more in-depth analysis of how animals are framed in discussions about (legal) normativity, and a critique thereof, see Lorini, 2018.
15 In recent years, political multispecies ethnography emerged as a viable methodology to put into action such a vision. See further Blattner, Donaldson, and Wilcox, 2020.

References

Abate, R. (2020) *What Can Animal Law Learn from Environmental Law?* 2nd edn. Washington, DC: Environmental Law Institute.

Agreement on Technical Barriers to Trade (adopted 15 April 1994, entered into force 1 January 1995), 1868 UNTS 120 (TBT).

Agreement on the Application of Sanitary and Phytosanitary Measures, (adopted 15 April 1994, entered into force 1 June 1995), 1867 UNTS 493 (SPS).

Appellate Body Report, EC − Measures Prohibiting the Importation and Marketing of Seal Products (adopted 18 June 2014), WTO Doc WT/DS400/AB/R, WT/DS4001/AB/R (WTO, Seals AB Report 2014).

Appellate Body Report, United States − Import Prohibition of Certain Shrimp and Shrimp Products (adopted 6 November 1998), WTO Doc WT/DS58/AB/R (WTO, Shrimp/Turtle I AB Report 1998).

Appellate Body Report, United States − Final Countervailing Duty Determination with Respect to Certain Softwood Lumber from Canada (adopted 17 February 2004), WTO Doc WT/DS257/AB/R (WTO, Softwood Lumber Panel Report 2004).

Archibald, C.J. (2008) 'Forbidden by the WTO? Discrimination Against a Product When Its Creation Causes Harm to the Environment or Animal Welfare', *Natural Resources Journal*, 48, pp. 15–51.

Bernasconi-Osterwalder, N. and Norpoth, J. (2019) 'Is World Trade Law a Barrier to Saving Our Climate? Questions and Answers', *Center for International Environmental Law (CIEL) & Friends of the Earth Europe (FoEE)* [Online]. Available at: https://www.ciel.org/Publications/ClimateTradeReport_foee-ciel _sep09.pdf (Accessed: 25 August 2022).

Bethlehem, D., McRae, D., Neufeld, R. and Van Damme, I. (eds.) (2009) *The Oxford Handbook of International Trade Law*. Oxford: Oxford University Press.

Blanco, E. and Grear, A. (2019) 'Personhood, Jurisdiction and Injustice: Law, Colonialities and the Global Order', *Journal of Human Rights and the Environment*, 10(1), pp. 86–117. DOI: doi.org/10.4337/jhre. 2019.01.05

Blattner, C.E. (2014) '3R for Farmed Animals − A Legal Argument for Consistency', *Global Journal of Animal Law*, 1, pp. 1–16.

Blattner, C.E. (2019) *Protecting Animals Within and Across Borders: Extraterritorial Jurisdiction and the Challenges of Globalization*. New York: Oxford University Press.

Blattner, C.E. (2020) 'Just Transition for Agriculture? A Critical Step in Tackling Climate Change', *Journal of Agriculture, Food Systems, and Community Development*, 9(3), pp. 53–58. DOI: 10.5304/jafscd.2020.093.006

Blattner, C.E., Donaldson, S. and Wilcox, R. (2020) 'Animal Agency in Community: A Political Multispecies Ethnography of VINE Sanctuary', *Politics & Animals*, 6, pp. 1–22.

Blattner, C.E. and Tselepy, J.C. (forthcoming) 'For Whose Sake and Benefit? A Critical Analysis of Leading International Treaty Proposals to Protect Nonhuman Animals', *American Journal of Comparative Law*.

Bowman, M., Davies, P. and Redgwell, C. (2011) *Lyster's International Wildlife Law*. 2nd edn. Cambridge: Cambridge University Press.

Castree, N., Demeritt, D., Liverman, D. and Rhoads, B. (eds.) (2016) *A Companion to Environmental Geography*. Malden; Oxford: John Wiley & Sons Ltd.

Caviola, L., Everett, J.A.C. and Faber, N.S. (2018) 'The Moral Standing of Animals: Towards a Psychology of Speciesism', *Journal of Personality and Social Psychology*, 116(6), pp. 1011–1029. DOI: 10.1037/pspp0000182

Charnovitz, S. (2002) 'The Law of Environmental "PPMs" in the WTO: Debunking the Myth of Illegality', *Yale Journal of International Law*, 27, pp. 59–110. DOI: 10.1142/9789814513258_0012

Chen, R.C. and Hanson, J.D. (2004) 'Categorically Biased: The Influence of Knowledge Structures on Law and Legal Theory', *Southern California Law Review*, 77(6), pp. 1103–1253.

Collard, R.-C. (2014) 'Putting Animals Back Together, Taking Commodities Apart', *Annals of the Association of American Geographers*, 104(1), pp. 151–165. DOI: 10.1080/00045608.2013.847750

Collard, R.-C. (2020) *Animal Traffic: Lively Capital in the Global Exotic Pet Trade*. Durham; London: Duke University Press.

Collard, R.-C. and Dempsey, J. (2013) 'Life for Sale? The Politics of Lively Commodities', *Environment and Planning A: Economy and Space*, 45(11), pp. 2682–2699. DOI: 10.1068/a45692

Committee for the Convention for the Protection of Animals (1988) International Convention for the Protection of Animals (ICAP), Proposal [Online]. Available at: https://www.animallaw.info/treaty/international-convention-protection-animals (Accessed: 25 August 2022).

Conrad, C.R. (2011) *Processes and Production Methods (PPMs) in WTO Law: Interfacing Trade and Social Goals*. Cambridge: Cambridge University Press.

Convention on Biological Diversity (adopted 5 June 1992, entered into force 29 December 1993), 1760 UNTS 79 (CBD).

Convention on International Trade in Endangered Species of Wild Fauna and Flora (adopted 3 March 1973, entered into force 1 July 1975), 993 UNTS 243 (CITES).

Convention on the Conservation of Migratory Species of Wild Animals (adopted 23 June 1979, entered into force 1 November 1983), 1651 UNTS 355 (CMS).

Corman, L. (2016) 'The Ventriloquist's Burden', in Castricano, J. and Corman, L. (eds.) *Animal Subjects 2.0*. Waterloo: Wilfried Laurier University Press, pp. 473–512.

Crawford, J. (2019) *Brownlie's Principles of Public International Law*. 9th edn. Oxford: Oxford University Press.

Cunniff, P. and Kramer, M. (2013) 'Developments in Animal Law', in Linzey, A. (ed.) *The Global Guide to Animal Protection*. Urbana: University of Illinois Press, pp. 230–231.

Danón, L. (2019) 'Animal Normativity', *Phenomenology and Mind*, 17, pp. 176–187. DOI: 10.13128/pam-8035

Dawkins, M.S. (1998) 'Evolution and Animal Welfare', The Quarterly Review of Biology, 73(3), pp. 305–328. DOI: 10.1086/420307

Deckha, M. (2015) 'Vulnerability, Equality, and Animals', *Canadian Journal of Women and the Law*, 27, pp. 47–70. DOI: 10.3138/cjwl.27.1.47

Deckha, M. (2021) *Animals as Legal Beings: Contesting Anthropocentric Legal Orders*. Toronto: University of Toronto Press.

Desta, M. (2021) 'To What Extent Are WTO Rules Relevant to Trade in Natural Resources?', Research and Analysis of the World Trade Organization [Online]. Available at: https://www.wto.org/english/res_e/publications_e/wtr10_forum_e/wtr10_desta_e.htm (Accessed: 25 August 2022).

Dhont, K., Gordon H., Kimberly C., and MacInnis, C.C. (2014) 'Social Dominance Orientation Connects Prejudicial Human–Human and Human–Animal Relations', *Personality and Individual Differences*, 61, pp. 105–108. DOI: 10.1016/j.paid.2013.12.020

Donaldson, S. and Kymlicka, W. (2011) *Zoopolis: A Political Theory of Animal Rights*. Oxford: Oxford University Press.

Donaldson, S. and Kymlicka W. (2017) 'Universal Basic Rights for Animals', in Armstrong, S.J. and Boltzer, R.G. (eds.) *The Animal Ethics Reader*. 3rd edn. New York; London: Routledge, pp. 53–64.

Dorning, M. (2014) 'Is Beef the New Coal? Climate-Friendly Eating Is on the Rise', Bloomberg, 14 May [Online]. Available at: https://www.bloomberg.com/news/articles/2021-05-14/is-beef-the-new-coal-climate-friendly-eating-is-on-the-rise (Accessed: 25 August 2022).

Eisen, J. (2010) 'Liberating Animal Law: Breaking Free from Human-Use Typologies', *Animal Law*, 17, pp. 59–76.

Ertman, M.M. and Williams, J.C. (eds.) (2005) *Rethinking Commodification: Cases and Readings in Law and Culture*. New York: New York University Press.

Evans, M.D. (2018) *International Law*. 5th edn. Oxford: Oxford University Press.

Faria, C. and Paez, E. (2014) 'Anthropocentrism and Speciesism: Conceptual and Normative Issues', *Revista de Bioética y Derecho*, 32, pp. 95–103. DOI: 10.4321/S1886-58872014000300009

Favre, D. (2016) 'An International Treaty for Animal Welfare', in Cao, D. and White, S. (eds.) *Animal Law and Welfare: International Perspectives*. Cham: Springer, pp. 87–108.

Fitzgerald, P.L. (2011) '"Morality" May Not Be Enough to Justify the EU Seals Products Ban: Animal Welfare Meets International Trade Law', *Journal International Wildlife Law and Policy*, 14, pp. 85–136. DOI: 10.1080/13880292.2011.583578

Food and Agriculture Organization (FAO) (2013) *Statistical Yearbook 2013: World Food and Agriculture*. Rome: FAO.

Food and Agriculture Organization (FAO) (2019) *The State of Food and Agriculture: Moving Forward on Food Loss and Waste Reduction* [Online]. Available at: http://www.fao.org/3/ca6030en/ca6030en.pdf (Accessed: 25 August 2022).

Food and Agriculture Organization (FAO) (2020) *Mitigating Risks to Food Systems During COVID-19: Reducing Food Loss and Waste* [Online]. Available at: http://www.fao.org/3/ca9056en/CA9056EN.pdf (Accessed: 25 August 2022).

Food and Agriculture Organization (FAO) (2023) *Food Loss and Waste Facts* [Online]. Available at: http://www.fao.org/3/i4807e/i4807e.pdf (Accessed: 5 April 2023).

General Agreement on Tariffs and Trade (adopted 15 April 1994, entered into force 1 June 1995), 1867 UNTS 187 (GATT).

Gillespie, A. (2014) *International Environmental Law, Policy, and Ethics*. 2nd edn. Oxford: Oxford University Press.

Gillespie, K.A. (2019) 'For a Politicized Multispecies Ethnography: Reflections on a Feminist Geographic Pedagogical Experiment', *Politics & Animals*, 5, pp. 17–32.

Gillespie, K.A. (2021) 'The Afterlives of the Lively Commodity: Life-Worlds, Death-Worlds, Rotting-Worlds', *EPA: Economy and Space*, 53(2), pp. 280–295. DOI: 10.1177/0308518X20944417

GJAL (2022) Special issue 'International Law and Animal Health and Protection: Persistent Themes, New Prospects for Change'.

Global Animal Law (GAL) Association (2018) UN Convention on Animal Health and Protection (UNCAHP), Proposal 2018 [Online]. Available at: https://www.uncahp.org/ (Acessed: 25 August 2022).

Grear, A. (2015) 'Deconstructing Anthropos: A Critical Legal Reflection on "Anthropocentric" Law and Anthropocene "Humanity"', *Law Critique*, 26, pp. 225–249. DOI: 10.1007/s10978-015-9161-0

Gruen, L. (2014) *The Ethics of Captivity*. New York: Oxford University Press.

Gunderson, R. (2011) 'From Cattle to Capital: Exchange Value, Animal Commodification and Barbarism', *Critical Sociology*, 39(2), pp. 259–275. DOI: 10.1177/0896920511421031

Haraway, D. (2008) *When Species Meet*. Minneapolis: University of Minnesota Press.

Harcourt, B.E. (2012) *The Illusion of Free Markets: Punishment and the Myth of Natural Order*. Cambridge, MA: Harvard University Press.

Harvey, D. (2009) *A Brief History of Neoliberalism*. Oxford: Oxford University Press.

Hennig, M. (2015) 'The EU Seal Products Ban: Why Ineffective Animal Welfare Protection Cannot Justify Trade Restrictions under European and International Trade Law', *Arctic Review on Law and Politics*, 6(1), pp. 74–85. DOI: 10.17585/arctic.v6.77

Herz, J. (2019) 'The Fairy Tale of Eternal Economic Growth: Swedish Activist Greta Thunberg Brings Attention to the Need to Steward our Planet', *Environmental and Energy Study Institute (EESI)*, 11 October [Online]. Available at: https://www.eesi.org/articles/view/the-fairy-tale-of-eternal-economic-growth (Accessed: 25 August 2022).

Honoré, A.M. (1961) 'Ownership', in Guest, A.G. (ed.) *Oxford Essays in Jurisprudence*. Oxford: Oxford University Press, pp. 112–124.

Howse, R., Langille, J. and Sykes, K. (2014) 'Sealing the Deal: The WTO's Appellate Body Report in EC – Seal Products', *ASIL Insights*, 18(12), June 6, 2014 [Online]. Available at: https://www.asil.org/

insights/volume/18/issue/12/sealing-deal-wto's-appellate-body-report-ec-–-seal-products (Accessed: 25 August 2022).

Hribal, J. (2010) *Fear of the Animal Planet: The Hidden Story of Animal Resistance*. Petrolia: CounterPunch.

International Convention for the Regulation of Whaling (adopted 2 December 1946, entered into force 10 November 1948), 161 UNTS 72 (ICRW).

Joy, M. (2011) *Why We Love Dogs, Eat Pigs, and Wear Cows: An Introduction to Carnism*. San Francisco: Conari Press.

Joyce, J., Nevins, J. and Schneiderman, J.S. (2017) 'Commodification, Violence and the Making of Workers and Ducks at Hudson Valley Foie Gras', in Gillespie, K.A. and Collard, R.-C. (eds.) *Critical Animal Geographies: Politics, Intersections and Hierarchies in a Multispecies World*. London; New York: Routledge, pp. 93–107.

Kaufmann, P. (2010) 'Instrumentalization: What Does It Mean to Use a Person?', in Kaufmann, P. Kuch, H., Neuhaeuser, C. and Webster, E. (eds.) *Humiliation, Degradation, Dehumanization*. Library of Ethics and Applied Philosophy. Dordrecht: Springer, pp. 57–65.

Kelch, T.G. (2017) *Globalization and Animal Law: Comparative Law, International Law and International Trade*. 2nd edn. Alphen aan den Rijn: Kluwer Law International.

Kopytoff, I. (1986) 'The Cultural Biography of Things: Commoditization as Process', in Appadurai, A. (ed.) *The Social Life of Things: Commodities in Cultural Perspective*. Cambridge: Cambridge University Press, pp. 64–92.

Kosoy, N. and Corbera, E. (2010) 'Payments for Ecosystem Services as Commodity Fetishism', *Ecological Economics*, 69(1), pp. 1228–1236. DOI: 10.1016/j.ecolecon.2009.11.002

Kyoto Protocol to the United Nations Framework Convention on Climate Change (adopted 11 December 1997, entered into force 16 February 2005), 2303 UNTS 162 (Kyoto Protocol).

La Rocca, I. (2015) 'Feminism and Animal Agriculture', *Women Eco Artists Dialog* [Online]. Available at: https://directory.weadartists.org/feminism-animal-agriculture (Accessed: 25 August 2022).

Laitos, J.G. (2021) *Rethinking Environmental Law: Why Environmental Laws Should Conform to the Laws of Nature*. Cheltenham: Edward Elgar Publishing.

Langton, R. (2009) *Sexual Solipsism: Philosophical Essays on Pornography and Objectification*. Oxford: Oxford University Press.

Longo, S.B., Clausen, R. and Clark, B. (2015) *The Tragedy of the Commodity: Oceans, Fisheries, and Aquaculture*. New Brunswick: Rutgers University Press.

Lorini, G. (2018) 'Animal Norms: An Investigation of Normativity in the Non-Human Social World', *Law, Culture and the Humanities*, pp. 1–22. DOI: 10.1177/1743872118800008

Lurié, A. and Kalinina, M. (2015) 'Protecting Animals in International Trade: A Study of the Recent Successes at the WTO and in Free Trade Agreements', *American University International Law Review*, 30(3), pp. 431–487.

Mason, J. and Singer, P. (1990) *Animal Factories*. New York: Harmony Books.

Mavroidis, P.C. (2015) 'Sealed with a Doubt: EU, Seals, and the WTO', *European Journal of Risk Regulation*, 6(3), pp. 388–395. DOI:10.1017/S1867299X00004839

McDonald, D.W., Harrington, L.A., Moorhouse, T.P. and D'Cruze, N. (2021) 'Trading Animal Lives: Ten Tricky Issues on the Road to Protecting Commodified Wild Animals', *BioScience*, 71(8), pp. 846–860. DOI: 10.1093/biosci/biab035

Meijer, E. (2019a) *Animal Languages: The Secret Conversations of the Living World*. London: John Murray.

Meijer, E. (2019b) *When Animals Speak: Toward an Interspecies Democracy*. New York: NYU Press.

Meijer, E. and Bovenkerk, B. (2021) 'Taking Animal Perspectives into Account in Animal Ethics', in Bovenkerk, B. and Keulartz, J. (eds.) *Animals in Our Midst: The Challenges of Co-existing with Animals in the Anthropocene*. Cham: Springer, pp. 49–64.

Merriam-Webster Dictionary (2023a) *Animal* [Online]. Available at http://www.merriam-webster.com/dictionary/animal (Accessed: 5 April 2023).

Merriam-Webster Dictionary (2023b) *Instrumentality* [Online]. Available at: https://www.merriam-webster.com/dictionary/instrumentality#:~:text=1%20%3A%20the%20quality%20or%20state,Sentences%20Learn%20More%20About%20instrumentality (Accessed: 5 April 2023).

Midgley, M. (1994) 'The End of Anthropocentrism?', *Royal Institute of Philosophy Supplements*, 36, pp. 103–112. DOI: 10.1017/S1358246100006482

Nadakavukaren Schefer, K. (2010) *Social Regulation in the WTO: Trade Policy and International Legal Development*. Cheltenham: Edward Elgar Publishing.

Nibert, D. (2013) *Animal Oppression and Human Violence: Domesecration, Capitalism, and Global Conflict*. New York: Columbia University Press.

Nozick, R. (2001) *Invariances. The Structure of the Objective World*. Cambridge, MA: Harvard University Press.

Nussbaum, M.C. (1995) 'Objectification', *Philosophy and Public Affairs*, 24, pp. 249–291. DOI: 10.1111/j.1088-4963.1995.tb00032.x

O'Neill, J. (2001) 'Markets and the Environment: The Solution is the Problem', *Economic and Political Weekly*, 36(21), pp. 1865–1873.

Offor, I. (2020) 'Animals and the Impact of Trade Law and Policy: A Global Animal Law Question', *Transnational Environmental Law*, 9(2), pp. 239–262. DOI: 10.1017/S2047102519000402

Oliver, K. (2015) 'Witnessing, Recognition, and Response Ethics', *Philosophy and Rhetoric*, 48(4), pp. 473–493. DOI: 10.5325/philrhet.48.4.0473

Our Planet, Theirs Too (2011) *The Declaration of Animal Rights* [Online]. Available at: https://declarationofar.org/textSign.php (Accessed: 25 August 2022).

Oxford English Dictionary (2022a) *Commodification* [Online]. Available at: https://www.oed.com/view/Entry/37198?redirectedFrom=Commodification#eid (Accessed: 25 August 2022).

Oxford English Dictionary (2022b) *Instrumentalize* [Online]. Available at: https://www.oed.com/about-thisentry/97158;jsessionid=35C16722B2432C4B95C13C24E615C38F (Accessed: 25 August 2022).

Palao E. (2021) 'Social Animals and the Potential for Morality: On the Cultural Exaptation of Behavioral Capacities Required for Normativity', in De Smedt J. and De Cruz H. (eds.) *Empirically Engaged Evolutionary Ethics*. Cham: Springer.

Panel Report, Canada – Measures Affecting Exports of Unprocessed Herring and Salmon (adopted 22 March 1988) WTO Doc. L/6268–35S/98 (Herring and Salmon Panel Report 1988).

Panel Report, EC – Measures Prohibiting the Importation and Marketing of Seal Products (adopted 18 June 2013), WTO Doc. WT/DS400/R, WT/DS4001/R (WTO, Seals Panel Report 2013).

Panel Report, US – Measures Concerning the Importation, Marketing and Sale of Tuna and Tuna Products (adopted 13 June 2012), WTO Doc. WT/DS381/R (WTO, Tuna/Dolphin III Panel Report 2012).

Panel Report, US – Prohibition of Imports of Tuna and Tuna Products from Canada (adopted 22 February 1982) WTO Doc. L/5198–29S/91 (Tuna I Panel Report 1982).

Papadaki, E. (2012) 'Understanding Objectification: Is There Special Wrongness Involved in Treating Human Beings Instrumentally?', *Prolegomena*, 11(1), pp. 5–24.

Papadaki, E. (2021) 'Feminist Perspectives on Objectification', in E.N. Zalta (ed.) *The Stanford Encyclopedia of Philosophy* [Online]. Available at: https://plato.stanford.edu/archives/spr2021/entries/feminism-objectification/ (Accessed: 25 August 2022).

Park, Y.S. and Benjamin V. (2019) 'Animals Are People Too: Explaining Variation in Respect for Animal Rights', *Human Rights Quarterly* 41(1), pp. 39–65. DOI: 10.1353/hrq.2019.0002

Pearce, D. (2015) 'A Welfare State for Elephants? A Case Study of Compassionate Stewardship', *Relations*, 3(2), pp.153–164.

Peters, A. (2020) *Animals in International Law*. The Hague: Hague Academy of International Law.

Probyn-Rapsey, F. (2018) 'Anthropocentrism', in Gruen, L. (ed.) *Critical Terms for Animal Studies*. Chicago: The University of Chicago Press, pp. 47–63.

Prudham, S. (2016) 'Commodification', in Castree, N., Demeritt, D., Liverman, D. and Rhoads, B. (eds.) *A Companion to Environmental Geography*. Malden; Oxford: John Wiley & Sons Ltd., pp. 123–142.

Qureshi, A.H. (1998) 'International Trade and Human Rights from the Perspective of the WTO', in Weiss, F., Denters, E. and de Waart, P. (eds.) *International Economic Law with a Human Face*. The Hague: Kluwer Law International, pp. 159–173.

Räsänen, T. and Syrijämaa, T. (2017) *Shared Lives of Humans and Animals: Animal Agency in the Global North*. London; New York: Routledge.

Reuveny, R. (2010) 'On Free Trade, Climate Change, and the WTO', *Journal of Globalization Studies*, 1(1), pp. 90–103.

Richards, J.F. (2014) *The World Hunt: An Environmental History of the Commodification of Animals*. Berkeley: University of California Press.

Royte, E. (2014) 'One-Third of Food Is Lost or Wasted: What Can Be Done', *National Geographic*, 13 October [Online]. Available at: https://www.nationalgeographic.com/environment/article/141013-food-waste-national-security-environment-science-ngfood (Accessed: 25 August 2022).

Satz, A. (2009) 'Animals as Vulnerable Subjects: Beyond Interest-Convergence, Hierarchy, and Property', *Animal Law*, 16, pp. 65–122.

Satz, D. (2010) *Why Some Things Should Not Be for Sale: The Moral Limits of Markets*. Oxford: Oxford University Press.

Schlottmann, C. and Sebo, J. (2018) *Food, Animals and the Environment: An Ethical Approach*. London: Routledge/Earthscan.

Schoenbaum, T.J. (1992) 'Free International Trade and Protection of the Environment: Irreconcilable Conflict?' *American Journal of International Law*, 86, pp. 700–727. DOI: 10.2307/2203788.

Sellars, W. (1980) 'Language, Rules and Behavior', in J. Sicha (ed.) *Pure Pragmatics and Possible Worlds: The Early Essays of Wilfrid Sellars*. Reseda: Ridgeview Publishing Company, pp. 129–155

Shaw, D. (2017) *International Law*. 8th edn. Cambridge: Cambridge University Press.

Shukin, N. (2009) *Animal Capital: Rendering Life in Biopolitical Times*. Minneapolis: University of Minnesota Press.

Smith, N. (2008) *Uneven Development: Nature, Capital, and the Production of Space*. 3rd edn. Athens, GA: University of Georgia Press.

Sykes, A. (2004) 'The Persistent Puzzles of Safeguards: Lessons from the Steel Dispute', *Journal of International Economic Law*, 7, pp. 523–564. DOI:10.1093/jiel/7.3.523

Sykes, K. (2014) 'Sealing Animal Welfare into the GATT Exceptions', *World Trade Review*, 13(3), pp. 471–498. DOI: 10.1017/S1474745613000232

Sykes, K. (2021) *Animal Welfare and International Trade Law: The Impact of the WTO Seal Case*. Cheltenham: Edward Elgar Publishing.

Taylor, S. (2017) *Beasts of Burden: Animals and Disability Liberation*. New York: New Press.

Thomas, E.M. (2007) 'Playing Chicken at the WTO: Defending an Animal Welfare-Based Trade Restriction under GATT's Moral Exception', *Boston College Environmental Affairs Law Review*, 34(3), pp. 605–637.

Tsing, A.L. (2017) *The Mushroom at the End of the World – On the Possibility of Life in Capitalist Ruins*. Princeton: Princeton University Press.

Understanding on Rules and Procedures Governing the Settlement of Disputes (adopted 15 April 1994, entered into force 1 January 1995) 1869 UNTS 401 (Dispute Settlement Understanding, DSU).

Van den Bossche, P. and Zdouc, W. (2021) *The Law and Policy of the World Trade Organization*. 5th edn. Cambridge: Cambridge University Press.

Vincent, S., Ring, R. and Andrews, K. (2018) 'Normative Practices of Other Animals', in Zimmerman, A., Jones, K. and Timmons, M. (eds.) *The Routledge Handbook of Moral Epistemology*. New York: Routledge, pp. 57–83.

von Hayek, F.A. (1982) *Law, Legislation and Liberty: A New Statement of the Liberal Principles of Justice and Political Economy*. London: Routledge & Kegan Paul.

Wadiwel, D.J. (2015) *The War Against Animals*. Leiden: Brill.

Wadiwel, D.J. (2018) 'Chicken Harvesting Machine: Animal Labour, Resistance and the Time of Production', *South Atlantic Quarterly*, 117(3), pp. 525–548. DOI: 10.1215/00382876-6942135

Wagman, B.A. and Liebman, M. (2011) *A Worldview of Animal Law*. Durham: Carolina Academic Press.

Wallace, R. (2016) *Big Farms Make Big Flu: Dispatches on Infectious Disease, Agribusiness, and the Nature of Science*. New York: Monthly Review Press.

Walsh, A. (2019) 'Commodification', in LaFollette, H. (ed.) *International Encyclopedia of Ethics*. 2nd edn. New Jersey: John Wiley & Sons Ltd. DOI: 10.1002/9781444367072.wbiee712.pub2

Weitzenfeld, A. and Joy, M. (2014) 'An Overview of Anthropocentrism, Humanism, and Speciesism in Critical Animal Theory', in Nocella, A.J., Sorenson, J., Socha, K. Matsuoka, A. (eds.) *Defining Critical Animal Studies: An Intersectional Social Justice Approach for Liberation*. Frankfurt: Peter Lang, pp. 3–27.

White, S. (2013) 'Into the Void: International Law and the Protection of Animal Welfare', *Global Policy*, 4(4), pp. 391–8. DOI: 10.1111/1758-5899.12076

Wilkie, R.M. (2010) *Livestock/Deadstock: Working with Farm Animals from Birth to Slaughter*. Philadelphia: Temple University Press.

Wilkie, R.M. (2017) 'Animals as Sentient Commodities', in Kalof, L. (ed.) *The Oxford Handbook of Animal Studies*. Oxford: Oxford University Press, pp. 279–301.

Wood, E.M. (2002) *The Origins of Capitalism: A Longer View*. New York: Verso.

World Customs Organization (WCO) (2022) *HS Nomenclature* [Online]. Available at: http://www.wcoomd.org/en/topics/nomenclature/instrument-and-tools/hs-nomenclature-2022-edition.aspx (Accessed: 25 August 2022).

World Society for the Protection of Animals (WSPA) (2011) *Universal Declaration on Animal Welfare (UDAW), Proposal* [Online]. Available at: https://www.globalanimallaw.org/database/universal.html (Accessed: 25 August 2022).

World Trade Organization (WTO) (2010), *World Trade Report, Trade in Natural Resources* [Online]. Available at: https://www.wto.org/english/res_e/booksp_e/anrep_e/world_trade_report10_e.pdf (Accessed: 25 August 2022).

World Trade Organization (WTO) (2023) *Members and Observers* [Online]. Available at: https://www.wto.org/english/thewto_e/whatis_e/tif_e/org6_e.htm (Accessed: 5 April 2023).

Yaffe-Bellany, D. and Corkery, M. (2020), 'Dumped Milk, Smashed Eggs, Plowed Vegetables: Food Waste of the Pandemic', *The New York Times*, 11 April [Online]. Available at: www.nytimes.com/2020/04/11/business/coronavirus-destroying-food.html (Accessed: 25 August 2022).

Youatt, R. (2020) *Interspecies Politics: Nature, Borders, State*. Ann Arbor: University of Michigan Press.

Zick Varul, M. (2011) 'Value: Exchange and Use Value', in Southerton, D. (ed.) *Encyclopedia of Consumer Culture*. London: Sage Publications.

4
ANTHROPOCENTRISM AND INTERNATIONAL ENVIRONMENTAL LAW

Vito De Lucia

Introduction

International Environmental Law (IEL) is that specialised body of law that regulates the conduct of states with regards to the natural environment, particularly within the overall framework of sustainable development. It can be said that, after an initial, preliminary period where the key focus was on transboundary environmental harm, its overall goal today is the protection of the natural environment from (excessive) degradation so that any human activity may remain within sustainable parameters. As such, IEL sets rules, delineates thresholds, and creates expectations. Perhaps more importantly though, it creates a narrative, a story. This story, it is often argued by scholars, is however not about the natural environment *per se*. It is, rather, about the *human* environment, as emphatically underlined in the title of the first large multilateral environmental conference held in Stockholm in 1972, the Conference on the Human Environment. The very notion of sustainable development, which from the 1990s has become arguably the overarching framework within which IEL operates, explicitly signals the centrality of human beings (UNGA, 2012).

IEL is thus imagined, negotiated, adopted, and enforced with a particular goal in mind. And, as suggested in the introduction to this volume, the effect of this orientation is that while IEL may perceive itself as 'less anthropocentric than the rest of international law' its operation routinely, whether 'subtly or explicitly[,] reinscribe[s], and even extend[s], international law's anthropocentricism'.

Indeed, faced with the chronic worsening of the state of the environment and the burgeoning amount of international environmental agreements (something which led Edith Brown-Weiss to speak of 'treaty congestion' almost three decades ago (Brown-Weiss, 1993, see also Hicks, 1998/1999)), legal scholars have increasingly focused on understanding the reasons why IEL has remained ineffective. While some literature has explored problems related to lack of implementation or enforcement of existing legislation, many IEL scholars have increasingly focused on deep level analyses trying to understand potential fundamental problems affecting IEL. Ultimately, these scholars suggest IEL suffers from deep contradictions that prevent it from effectively addressing environmental problems (M'Gonigle and Takeda, 2013). Moreover, the 'problematic' of IEL resides to a significant extent in the way that power structures and modes or forms of knowledge combine to define both the questions that can be posed and the forms the answers must take (M'Gonigle and Takeda, 2013, p. 1019)

DOI: 10.4324/9781003201120-6

Within this context, anthropocentrism has been singled out as perhaps the *key* problem affecting IEL. This critique is well-known, and this chapter will rehearse its salient elements (see section 'Understanding Anthropocentrism') and then review key documents of IEL in that light (see section 'Anthropocentrism, International Environmental Law, and Sustainable Development'). Yet when discussing anthropocentrism in relation to IEL, it is important to keep in mind that IEL is one specialised and segmented manifestation of a broader yet historically contingent legal experience. As such, it is almost inevitable that IEL reflects this broader legal experience – which we can call legal modernity – albeit within a specialised context of reference and scope of operation. IEL then inevitably reproduces the crucial underpinnings – ontological, epistemological, and axiological – and the overall thrust of legal modernity as a particular legal experience that is steeped in particular legal ideas (see e.g., Grossi, 2001; Grear, 2011; De Lucia, 2019b). In this respect, it is also important to keep in mind that IEL is in many ways *one* element in a global discourse of environmental law and governance (Percival, 2001) that is comprised of multiple 'levels of articulations' (De Lucia, 2019b, p. 20) – domestic, supranational, international, transnational – which are increasingly porous and continuously communicate and 'interbreed' (Wiener, 2001, p. 1307).

The three underpinnings or dimensions that ground IEL and absorb it within the broader legal experience of modernity are also reflected in a particular violence against nature that IEL inevitably facilitates even as it tries to delimit and contain it, in spatial, qualitative, and quantitative terms, through concepts and tools, such as protected areas (which will be discussed in the section titled 'The Violence and Exclusion of International Environmental Law'), invasive alien species, or ecological integrity. Such violence is both material – inscribed on the bodies of natural entities – and conceptual – impressed in the very idea of nature that is constructed to be outside of history, separated from culture, and devoid of agency. Unpacking anthropocentrism further reveals an additional dimension of violence that is parallel and intertwined with the violence against nature, but that is oriented against other human communities. Both these types of violence, and how they are both complementary and conjoined, will be illustrated by way of an example (see section 'The Violence and Exclusion of International Environmental Law'), while the unpacking itself will be carried out beforehand in sections 'Unpacking Anthropocentrism I' and 'Unpacking Anthropocentrism II'.

IEL, however, is also located at what I have elsewhere described as an 'epistemological crossroad' (De Lucia, 2019b, p. 12). In this sense, IEL has also been approached as an opportunity for changing in some fundamental respects the ways in which legal modernity understands, constructs, and regulates nature. Thus, the anthropocentric critique has been often accompanied by searches for a remedy, which is usually identified as ecocentrism. An ecocentric shift would, in the view of an increasing number of commentators, environmental activists, and scholars, make it possible to overcome the shortcomings of IEL – and of legal modernity more broadly – and radically reorient its normative core. The ecocentric alternative, which has a set of central ideas such as rights of nature (Stone, 1972; Cullinan, 2002; Burdon, 2014), ecological integrity (e.g., Kim and Bosselmann, 2015), or ecosystem approaches (e.g., De Lucia, 2015; 2019b), is gaining significant traction. However, while portrayed as the remedy, ecocentrism does not change the terms of the problematic in any fundamental way, as I will try to illustrate in the concluding section.

Understanding Anthropocentrism

It is useful to start by exploring in some detail the meaning of anthropocentrism. There is by now a large literature on the subject. Most of this literature characterises anthropocentrism as both the central organising principle of environmental law in all its levels of articulation (Cecchetti, 2006; Grassi, 2007; Gillespie, 2001; 2014; Grear, 2011; De Lucia, 2019b) and as the crucial problem that affects environmental law and, more broadly, the modern legal ordering of nature (e.g., Emmenegger and Tschentscher, 1994; Gillepsie, 1997; Wilkinson, 1999; Cullinan, 2002; Bosselmann, 2010; Grear, 2011; De Lucia, 2019b). Anthropocentrism is likewise a central concern for environmental ethics (Rolston, 2010; Curry, 2011), so much so that the very essence of the discipline is often mused to be 'a set of critiques [...] of the anthropocentric worldview' (Keller, 2010, p. 62). However, such critique has become 'so familiar' that it is often 'taken almost for granted without argument' (Vogel, 1996, p. 160), and this is a point on which I will return later in this chapter. Additionally, the term anthropocentrism is arguably 'loaded and ambiguous' (Vogel, 1996, p. 160) and is a 'contentious concept' (Curry, 2011, p. 54). It will be thus useful to offer a brief and careful outline of what anthropocentrism means, and what it does *not* mean, and whether all that it means or entails should be subject to a radical environmental – or eco-centric – critique.

Taking the literal meaning of the word, anthropocentric means 'human centred'. A world-view based on such an outlook implies the centrality of human beings in the world. Such centrality may refer to the ontological difference of humans (ontological anthropocentrism), of their perspective (epistemological anthropocentrism), and of their value (axiological anthropo-centrism). This centrality is premised on a number of factors, such as the capacity for agency, the ability to reason, to communicate symbolically and through speech – all elements that lend support to the human exceptionalism thesis (Gillespie, 1997, pp. 3–18) and that have been used to articulate legal agency since the early modern era (see e.g., Grossi, 1988). However, this is only a starting point, as anthropocentrism takes many forms, not all of which reproduce literally – although they do so operationally – these initial premises.

Ontological anthropocentrism is the result of the first, key gesture from which modernity, understood as both a theoretical category and as a historical period (De Lucia, 2018b), emerged in Europe; a gesture whose gestation lasted arguably several centuries and was philosophically cemented with the Cartesian separation of mind and matter.[1] This gesture reflects the new 'anthro-pology of modernity' (Grossi, 2001), situating man and his will at the very centre of the world – and of the legal world in particular, what amounted to a transformational shift if contrasted with the *reicentrism* (the centrality and normativity of the objective world) that characterised, for both cultural and technological reasons, the European medieval period (Grossi, 2001). This engenders a conceptual earthquake in Europe that both underpins and orients all three dimensions of anthropocentrism. There is no space to follow the myriad ways in which this central ontological construction has been operationalised, nor all its implications. Suffice it to say, for the purposes of this chapter, that the European human subject found himself (and the *male* pronoun is particularly apt) as the pivot of what Esposito has called a 'logic of presupposition', that is, a presupposition of antecedence of the subject with respect to both the world of phenomena, and to itself as a body (Esposito, 2010). Agency was thus located only within the European human mind, and everything else – the famous Cartesian *res extensa*, that is material reality – was transformed into a collection of passive objects. From ontological primacy follows *epistemological* anthropocentrism, which translates the ontological shift just outlined in the domain of perception and knowledge. It is thus the human subject, which is also the centre of knowledge (Philippopoulos-Mihalopoulos, 2011), that through his mind penetrates material reality and names it and knows it, with the view of exerting control over it. Epistemological anthropocentrism rests thus on a crucial threshold where there is only a subject knower at the centre and the material world of knowable objects all around him. In turn, this epistemological threshold translates into what is commonly referred to as the epistemology of mastery, as already indicated by Descartes when he considered that 'humanity should acquire absolute ownership of the world and control everything: the natural environment, the social world, even the human psyche' (Douzinas and Gearey, 2005, p. 45), as operationalised by Francis Bacon in his technological visions aimed at seizing all the secrets of nature (Merchant, 1990). A final crucial element necessary to frame the concept in a broader sense has to do with the location of value, which translates into *axiological* anthropocentrism. An anthropocentric outlook locates value *intrinsically* – that is independently of external attributions or valuations – *only* in relation to human beings. All else, in this view, has value only *instrumentally*, on account of the benefits that such entities may provide to humanity. This view was eventually captured and developed systematically by Kant, who saw humans as ends by virtue of their reason, and thus as bearer of rights; and all other entities devoid of reason as only means. Indeed, Kant considered that 'man […] is the ultimate purpose of creation here on Earth' (quoted in Gillespie, 1997, p. 3). Axiological anthropocentrism can then be described, with Bryan Norton, as

> The view that the earth and all its nonhuman contents exist or are available for man's benefit and to serve his interests and, hence, that man is entitled to manipulate the world and its systems as he wants, that is, in his interest.
>
> *(Norton, 2014, p. 136)*

This view is in turn transformed into a critique by adding an adjective, as Curry does: anthropocentrism becomes then the '*unjustified* privileging of human beings […] at the expense of other forms of life' (Curry, 2011, p. 55) (emphasis added).

All these three dimensions find legal operationalisation and are embedded deeply, albeit in different ways and to different degrees, in IEL: its conceptual apparatus, its governing principles, and in discrete regimes, treaties and rules. They offer a general theoretical template for critique – and ammunition for its operationalisation – to those IEL scholars in their effort to re-imagine IEL on other ontological, epistemological, and axiological premises. Anthropocentrism, however, is not a monolithic concept. While grounded inevitably on the three dimensions just discussed, anthropocentrism as a practise may take different shapes and is better understood as an ensemble, a plurality of articulations that, while all regulating and organising nature as a form of human environment, are correspondingly integrated and embedded in IEL in a variety of ways.

Furthermore, and at a different level, anthropocentrism is not only about how to think, structure, and operationalise the relationship between human beings and nature or how to organise the protection of the environment. It is also, and sometimes primarily, about relations between human communities (states, in international legal discourse). As such, the notion of anthropocentrism may hide more than it reveals. It is thus necessary to unpack the notion of anthropocentrism. This will be done in two steps. First, the following section will pluralise anthropocentrism, showing how there are multiple articulations that, while remaining within the same template organised around the three central dimensions discussed above, translate into significantly different legal expressions and operationalisation. Second, the underlying referent of anthropocentrism needs to be problematised, in order to understand the double effect of anthropocentrism: against nature and against (certain sections of) humanity in ways that are intimately linked. Before proceeding with the unpacking, however, the next section reviews some central IEL documents to illustrate how anthropocentrism remains, despite mounting critiques, the prevailing approach in IEL.

Anthropocentrism, International Environmental Law, and Sustainable Development

IEL, like domestic environmental law (*mutatis mutandis*), is a branch of law deputed to reacting to and addressing the widespread environmental degradation that, from the 1960s onwards, has been identified as a central problem of industrial societies. Largely spurred, particularly in North America, by Rachel Carson's formative book 'Silent Spring' (Carson, 1962), which detailed the pervasive environmental degradation caused by chemical pollutants, this environmental awakening led to the development of an environmental movement and of environmental law as a specialised field of law. Before the birth of environmental law as a distinct branch of international law, environmental concerns had been addressed in an ad hoc manner, whenever they would arise. Typically, they would address issues of an economic nature or related to property rights (whether in the form of public territorial integrity or private property rights). Early examples now firmly settled in the legal historiography of pre-Stockholm IEL are arbitrations such as the Fur Seal case, which related to resources access, conservation and exploitation, and the 'orderly development of the fishery' (Vicuña, 2000, p. 40); and the Trail Smelter Case (*Trail Smelter Arbitration (United States, Canada)* [1941]), which related to transboundary damages caused to private property by air pollution (Bratspies and Miller, 2006); or treaties, such as the Convention for the Protection of Birds Useful to Agriculture, which distinguishes between useful and noxious birds; or the Convention on the Regulation of Whaling, whose ultimate objective, to be achieved through 'the proper conservation of whale stocks', is 'the orderly development of the whaling industry' (Whaling Convention, 1946, Preamble, Last Recital).

The post-war period was characterised by a movement, still ongoing today, towards the 'publicization' of international law, which has taken the form of a shift from traditional 'inter-state normative patterns', following a private law model, to 'common-interest normative patterns' (Hey, 2003, p. 7). IEL is certainly an important expression of one such shift. However, the anthropocentric grounding of IEL remains in place. This anthropocentric grounding has been explored and exposed by Alexander Gillespie in a sustained and detailed account (Gillespie, 1997) that, while somewhat dated in its original edition, was revised in 2014 (Gillespie, 2014). The protection of the environment, he finds, remains grounded on a fundamental anthropocentrism that may take, in different legal and regulatory contexts, a variety of more targeted, subject-matter specific articulations. Gillespie goes further and carries out an analysis of the underlying justifications that motivate environmental protection in IEL, again exploring a plurality of ways anthropocentrism manifests. He isolates six such justifications, or groundings, for environmental protection: self-interest, economics, religion, aesthetics, culture and recreation, and the rights of future generations (Gillespie, 1997; 2014). All of these are important underpinnings of the foundational events and documents of IEL, albeit at different locations along the gradient that distinguishes various forms of anthropocentrism from resourcism to stewardship.

A review of some central IEL documents illustrates how the anthropocentric vision or worldview remains, despite mounting critiques, the prevailing approach in IEL.[2] These documents possess, importantly, a decisive narrative function, which exceeds their characterisation as aspirational or political declarations with a formal legal status as 'soft' law at best. Indeed, they fulfil a crucial role for the self-identify of IEL. Within this foundational context, which sets the conceptual and, perhaps, 'constitutional' premises for more specialised regimes, IEL continues to develop idiosyncratically, caught in a 'deep contradiction' (M'Gonigle and Takeda, 2013, p. 1005), which is perhaps also tragic: while IEL, as a field of study, is self-reflexively aware of how the 'environmental problematic' can only be addressed by exploring and addressing its root causes, perhaps through a change in paradigm; as a field of practise it can only operate within the existing paradigm (M'Gonigle and Takeda, 2013, p. 1005) and thus reproduce it and even expand it.

The 1972 Stockholm Conference, the first global conference focused on the environment, contains this orientation in its title – Conference on the *Human* Environment (emphasis added) – and explicitly grounds itself in an anthropocentric worldview to the extent that Man is the barycentre of the entire Declaration,[3] and all references to ecosystems and the Earth are functional to the well-being of 'present and future generations' (principles 1 and 2), of 'all mankind' (principle 5), or to the production of resources (principle 3). Principle 21, in turn, reiterates that the exploitation of resources is a sovereign right of states, its only limitations being 'the responsibility to ensure that activities within their jurisdiction or control do not cause damage to the environment of other States or of areas beyond the limits of national jurisdiction'. Some indeed consider that '[o]ne of the great achievements of the Stockholm Declaration is that it acknowledges the strong relation between environmental protection and economic development' (Jan, 2010, p. 1535), an achievement that would be completed in the next landmark UN environmental conference held in Rio in 1992. The Stockholm Declaration is also traversed through and through by what Kotzé and French (2018, p. 18) have described as a 'masculinist ontology of anthropocentrism'.

The 1992 Rio Declaration on Environment and Development (Rio Declaration), the second founding document adopted at the Conference on Environment and Development held in 1992 in Rio de Janeiro, Brazil, reiterates with even more force this orientation. In fact, Principle 1 states that 'Human beings are at the centre of concerns for sustainable development'. The anthropocentric orientation of the Declaration is readily acknowledged by most scholarship (most recently, Dupuy, 2015, 68ff.). The Rio Declaration, in particular, enshrines the concept of sustainable development as the organising principle and as the ideological narrative for the entire environmental legal field, across all levels of articulation.[4] Sustainable development, as famously defined by the Brundtland Commission, is 'development that meets the needs of the present without compromising the ability of future generations to meet their own needs' (World Commission of Environmental and Development (WCED), 1987, ch. 1, para. 1). With its emphasis on distributional concerns between present and future generations, it has a clear anthropocentric focus. Such focus is also candidly expressed elsewhere in the report, with regards to 'the exploitation of resources, the direction of investments, the orientation of technological development [and] institutional change', all elements that should 'enhance both current and future potential to meet human needs and aspirations' (WCED, 1987, ch. 2, para. 15). Finally, sustainable development is, throughout the Burndtland Report, discussed as a strategy aimed at securing a 'new era of economic growth' (WCED, 1987).

This anthropocentric orientation is confirmed, and indeed intensified, in the context of the next two UN conferences convened as follow-ups to the 1992 Rio Conference: Rio+10 and Rio+20. In 2002, a World Summit on Sustainable Development was held in Johannesburg to mark ten years since the Rio Summit. Its goal was to further commit States to take action towards the realisation of a global legal and governance framework organised around the narrative of sustainable development, a concept whose legal status remained then – and still remains – vague, but whose conceptual, policy, and normative significance is indisputable (see e.g., Lowe, 2000). The conference outcome was captured by two main documents: a political declaration and a plan of implementation (Report of the World Summit on Sustainable Development, 2002). The political declaration, also known as the Johannesburg Declaration, reaffirmed its commitment to the Rio Declaration (para. 8). The tone of the entire declaration however, reflects, even more sharply than the Rio Declaration, the increasingly central role of sustainable development and the increasingly more instrumental role of the environment as a service provider. Paragraph 13 is the only provision dedicated to a 'global environment' that 'continues to suffer' (para. 13), yet the environment continues to 'skip and turn' (Philippopoulos-Mihalopoulos, 2011, p. 22) around the central problematic of human well-being. But the political declaration was short and lacked teeth. The main outcome of the Johannesburg Summit is the Plan of Implementation (Report of the World Summit on Sustainable Development, 2002, s. I.2).

The Plan of Implementation reiterates a decisive commitment to sustainable development. Among its 11 chapters and 13 sub-chapters, there is not a single one dedicated to the environment, while almost every single one includes the expression sustainable development. The closest reference to the environment is in the title of Chapter IV: 'Protecting and managing the natural resource base of economic and social development'. It is evident, moreover, that the environment not only surrounds a human centre but is also primarily constructed as a material resource base instrumental to human economic and social development, clearly reflecting the view of nature being arrayed as a 'standing reserve' in the Heideggerian fashion (Heidegger, 1977, p. 17). Throughout the text, references to the environment are obviously present (and plentiful), but the *focus* of the entire document is *not* on the environment, and when it is, it is only instrumentally so (that is, anthropocentrically). As an illustrative example, Chapter II links health and environment to the overall goal of addressing poverty and, thus, recognises the value of environmental protection for human ends and needs. In Chapter III, on consumption and production patterns and the need to reorient them in a sustainable direction, environmental protection becomes a question of 'decoupling', efficiency, and cost internalisation, all measures of an economic character that further buttress the idea that environmental protection, within the broad narrative of sustainable development, are instrumental to human (economic) needs.

Chapter IV offers further evidence, as the opening paragraph makes clear its anthropocentric orientation: '[h]uman activities are having an increasing impact on the integrity of ecosystems that provide essential resources and services for human well-being and economic activities' (Report of the World Summit on Sustainable Development, 2002, para. 24), where the value of the integrity of ecosystems is made contingent on the ability of ecosystems to contribute to human well-being and economic activities. The paragraphs of Chapter IV address a multiplicity of issues, from water to oceans, from climate to agriculture, from disaster management to desertification. The majority of these issues are framed around the centrality of benefits to humans: water scarcity and quality are, for example, primarily framed in relation to human health (paras. 25–29). Oceans and fisheries are primarily framed in relation to their 'critical' role 'for global food security and for sustaining economic prosperity' (paras. 30–36); the protection of the marine environment from land-based activities (para. 33) and pollution (para. 34) is also arguably framed instrumentally in relation to the general goals of food security and economic prosperity. The chapter also includes sections on climate change, agriculture, desertification, biodiversity, forests, and mining. All these themes are framed as natural resources issues – and, significantly perhaps, the last section is dedicated to '[m]ining, minerals and metals', which are 'important' resources for 'the economic and social development of many countries' and are 'essential for modern living' (para. 46).

In 2012, another summit was held in Rio de Janeiro, the United Nations Conference on Sustainable Development, popularly known as Rio+20, as it marked the 20th anniversary of the 1992 landmark Rio Summit. The two official themes of the Conference were (a) how to build a green economy to achieve sustainable development and lift people out of poverty; and (b) how to improve international coordination for sustainable development. Sustainable development remains the crucial focus and overarching narrative, but there is a new concept that appears, or re-appears, as this concept was originally coined in the late 1980s, exactly in relation to the then emerging concept of sustainable development (Pearce, 1989; Pearce, Markandya, and Barbier, 1992): the green economy.[5] As in the Johannesburg Plan of Implementation, it seems that the environment and the 'natural resource base of [...] development' are used interchangeably in the outcome document of the Rio+20 Summit, entitled 'The Future We Want' (UNGA, 2012, para. 4). Moreover, paragraph 6 recognises again 'that people are at the centre of sustainable development' (UNGA, 2012, para. 6).

Finally, the recently adopted Agenda 2030, with the Sustainable Development Goals, also remains firmly anthropocentric, however slanted towards the stewardship side of the anthropocentric gradient they may be. In fact, arguably, the Sustainable Development Goals further reinforce and cement the anthropocentrism of IEL (Kotze and French, 2018).

Albeit brief and necessarily selective, this overview illustrates how the conceptual architecture of IEL – in its increasing intimacy with the narrative of sustainable development, under which, in many respects, it has been gradually subsumed[6] – is premised on an anthropocentric approach to questions of environmental significance. This anthropocentric perspective is also reflected in the case law of the International Court of Justice (ICJ), perhaps inevitably insofar as the term 'environment' always requires a centre, and that centre is always inevitably human. The Court in fact famously understands the environment as 'the living space, the quality of life and the very health of human beings, including generations unborn' (*Legality of the Threat or Use of Nuclear Weapons* [1996], para. 29).

If the anthropocentric grounding of IEL is unequivocal, the next question examines the ways in which IEL is operationalising anthropocentrism in its three dimensions. A detailed analysis is not possible given space constraints. But it is evident that the anthropocentrism of IEL in the context of sustainable development is not radical, as it recognises the need for precaution and the limitations of human knowledge and, while still grounded on rational utilitarian considerations, offers a nuanced articulation of the anthropocentric worldview. Sustainable development however has also nudged IEL towards an ambiguous and slippery slope. It is in this respect useful to note that the distinction made in the past between exploitation conventions and conservation conventions is no longer relevant, and conservation itself, imbued with the spirit of sustainable development, is located at the ambiguous crossroad between the two, as evidenced particularly in the Convention on Biological Diversity (1992) (CBD). It is now time to unpack these various ideas of anthropocentrism.

Unpacking Anthropocentrism I: From Anthropocentrism to Anthropocentrisms

First, anthropocentrism takes a variety of forms in different contexts, as already observed by Gillespie (Gillespie, 1997; 2014). These various forms or articulations are arrayed along a gradient that encompasses wildly differing perspectives, from brutal 'resourcism', that is, 'the belief held by many people in modern industrial societies that the world gains value only as nature is transformed into goods and services to meet human demands' (Grumbine, 1994, p. 34), to weak anthropocentrism (Norton, 2009), or even to visions of stewardship (Palmer, 2006) that, while anthropocentric, are in stark contrast with resourcism and are located at the other end of the spectrum. And while the core premises remain in place, it is probably more appropriate to speak of anthropocentrisms in the plural, and then zoom in on the particular dimension one seeks to address or criticise. Indeed, to this variety of anthropocentrisms correspond significantly varying conceptual cores and practical effects. This question is discussed, for example, by Curry, who concludes by using anthropocentrism to refer specifically to the 'unjustified privileging of human beings [...] at the expense of other forms of life' (Curry, 2011, p. 55). Yet a number of *justified* positions can be easily called anthropocentric, a point to which I will return later in this section. Moreover, all those stances that can be gathered under the heading of stewardship (albeit itself not a homogenous concept) arguably retain an anthropocentric outlook (Palmer, 2006), to the extent that stewardship 'originated, both as a practice and a concept, in patriarchal, elitist, and anthropocentric social systems and ideologies' (Welchman, 2012, p. 298). Some stances however also internalise a significant element of care and empathy with the non-human world, so that some authors consider stewardship as also conceivable in non-anthropocentric terms (McIntyre-Mills, 2014), which makes the analysis of the role of anthropocentrism complex and inevitably requires nuanced considerations. Indeed, whether stewardship may offer a bridge between the anthropocentric outlook and its overcoming, remains difficult to ascertain, considering that the 'managerial' orientation towards resource optimisation arguably remains embedded even in more benign articulations of stewardship (Palmer, 2006, p. 66). One definition of stewardship that is thusly oriented reads as follows:

> Stewardship is the responsible use (including conservation) of natural resources in a way that takes full and balanced account of the interests of society, future generations, and other species, as well as of private needs, and accepts significant answerability to society.
>
> *(Worrell and Appleby, 2000, p. 269)*

This definition of stewardship is evidently far removed from Curry's definition of anthropocentrism (with its emphasis on unjustified privileging), and in this sense sufficiently illustrates the significant differences that exist across the spectrum of articulations of the notion of anthropocentrism, that again, Gillespie has already pointed to in his work (Gillespie, 1997; 2014). The emphatic reference to other species included in that definition of stewardship, however, as already indicated, still arguably proceeds from the privileged position some humans are implicitly accorded at the centre of the world (and this will be further discussed in the section 'The Violence and Exclusion of International Environmental Law'). Thus, while some scholars understand stewardship as a *tertium genus* vis-à-vis both anthropocentrism and ecocentrism, it seems to the present author that stewardship remains sufficiently, if not firmly, anthropocentric, particularly in the general understanding of stewardship as 'responsible management' (Cunningham, Cunningham, and Woodworth Saigo, 2012) – it is in fact sometimes called 'responsible anthropocentrism' (Derr, cited in Sharpe, 1998, p. 32) – and is very close to the notion of weak anthropocentrism developed by Bryan Norton.

Norton's position is particularly noteworthy in that it has provided significant inspiration for at least some environmental legal theory (Tallacchini, 1996; 2000)[7] and raises the question as to whether anthropocentrism needs in fact to be overcome or whether it should simply be re-articulated. Therefore, Norton's position requires some discussion. Norton's starting point is that there is no need to locate value intrinsically in nature (responding to those criticising anthropocentrism and promoting its opposite, ecocentrism). His aim is to rescue anthropocentrism as a valuable ethical framework for environmentalism, one that can be efficacious and more easily defended than ecocentrism. To this purpose Norton distinguishes between strong and weak anthropocentrism. Strong anthropocentrism is, in Norton's account, characterised by 'felt preferences', that is 'any desire or need of a human individual that can at least temporarily be sated by some specifiable experience of that individual' (Norton, 2009, p. 161). Weak anthropocentrism, on the other hand, is characterised as being based on what he calls 'considered preferences', that is preferences that are formed through deliberative processes and are 'rationally defensible in the context of a defensible worldview' (Norton, 2009, p. 161).

Deliberation and rationality, Norton further argues, would prevent preferences from remaining impermeable to criticism and objections, insofar as they are adopted within the bounds of a 'rational worldview [...] which includes fully supported scientific theories and a metaphysical framework interpreting those theories' (Norton, 2009, p. 161). In this sense, Norton argues, weak anthropocentrism provides the basis for a 'criticism of value systems which are purely exploitative of nature', if they are 'not consonant' (Norton, 2009, p. 161) with such a rational worldview. Weak anthropocentrism, in this sense, would be *justifiable*, and hence outside of Curry's definition of anthropocentrism. Moreover, it seems that this articulation of anthropocentrism can be easily superimposed on the general framework of sustainable development, which is central for IEL, as sustainable development also rests on a 'rational worldview' and on 'fully supported scientific theories' and is not a 'purely exploitative' system of environmental governance.

What Norton underlines as the crucial distinction is not that between anthropocentrism and non-anthropocentrism but rather the distinction between individualism and non-individualism (Norton, 2009, p. 164). Demarcating value systems along an individualist/non-individualist line is certainly attractive and useful, particularly in the context of IEL, where balancing individual and common interests is an important task, as exemplified by emerging principles such as the principle of common concern, enshrined in both the UNFCCC and in the CBD, or by principles that may play an important role in relation to environmental protection of certain common domains or resources, such as the principle of the common heritage of mankind (United Nations Law of the Sea Convention, 1982 (UNCLOS)).

The aim of this section is not to discuss each individual articulation of anthropocentrism, and their merits and demerits (for example, Norton's argument seems to offer a rather unproblematised account of rationality) but rather to show how anthropocentrism, if deployed in an unqualified manner, conceals more than it reveals given the variety of articulations that exist. Additionally, a monolithic understanding of anthropocentrism may also lead to answers of the yes or no type in relation to the question of whether IEL is anthropocentric, which, like most if not all yes or no answers, tend to glide over or ignore important complexities. A more useful question is rather *how, to what extent, in what manner* is IEL anthropocentric. What is important, in other words, is to understand how IEL may operationalise the three fundamental dimensions of anthropocentrism – ontological, epistemological, and axiological – considering that not all three dimensions are always simultaneously or equally embedded and articulated in each IEL regime or treaty. It is a question of focusing not only on definitional subsumptions but on practise.

It is perhaps evident that IEL, in the current context of sustainable development, is closer to the weak anthropocentrism articulated and supported by Norton, at least in its rhetorical dimension. Sustainable development functions, in this respect, precisely as the narrative mechanism through which IEL finds a specific identity within the context of a global capitalist world where the social, economic, and environmental dimensions shall move forward in a complementary and mutually supporting manner. Yet, sustainable development, especially in the emerging and related context of the Anthropocene, also grounds and motivates action that is *violently* anthropocentric in more than the sense thus far discussed. And it is here that the next layer of anthropocentrism must be peeled off to see how this is operationalised in IEL.

Unpacking Anthropocentrism II: Anthropocentrism and the Violence of International Environmental Law

In order to penetrate this second layer, it is necessary to unpack, or deconstruct, the very semantic structure of anthropocentrism. It is then the Anthropos that becomes the focus of our attention, as the primary semantic referent in the composite abstraction anthropocentrism. *Anthropos* is the Greek word for human being, but as a semantic root operates as a symbolic placeholder for humanity in composite words such as, precisely, anthropocentrism. The latter term remains however 'far removed' from a 'rich and inclusive' understanding of human beings, as Anna Grear suggests insightfully (Grear, 2013, p. 82). So, the manoeuvre is devious, as with all synthetic language, and acquires a particularly sharp deviousness in law. As some critical legal (e.g., Grear, 2013; 2015) and non-legal (e.g., Malm and Hornborg, 2014) scholarship has been emphasising for a while now, anthropocentrism, with its undifferentiated reference to humanity, masks very differentiated realities within which different human individuals and communities live. As

Grear further explains cogently and at length for example (Grear, 2015), the *Anthropos* which is offered as the universal representation of humanity is in fact a placeholder for a very narrow instantiation of it, namely the white, male, rational agent – and this last point raises questions also in relation to the weak anthropocentrism Norton articulates, based as it is on *rational* deliberation. Anthropocentrism then, while purporting to refer to humanity at large, really refers to a particular subset thereof that historically has colonised, exploited, and plundered both other human communities and the non-human world, through what Grear calls 'vectors of oppression linking intra-and inter-species hierarchies' (Grear, 2015, p. 233).

Malm and Hornborg similarly deconstruct the concept of the Anthropocene – where Anthropos is equally the central root, fulfilling a similar role. The Anthropocene, they observe, is another synthetic concept that is however unable to account for the role of power and power relations in the pervasive transformations that determined the geological shift the Anthropocene as an idea is set to capture (Malm and Hornborg, 2014). They argue that the standard Anthropocene narrative is premised 'for reason of logical necessity' on a species narrative – i.e., humanity – and on the notion of human nature (and its species-properties), which, in a certain sense, teleologically, underlies the Anthropocene. 'Anything less', Malm and Hornborg suggest, would turn the Anthropocene from a geology of *mankind* to a geology of *only a few* (Malm and Hornborg, 2014, p. 63). Paterson makes a similar observation, suggesting that anthropocentrism, in its imbrication with capitalism, traverses humanity and selects only *certain* human beings as the beneficiaries of current regimes of ecological accumulation (Paterson, 2010). Grear further emphasizes how 'intra-species hierarchies' are central to 'histories of eco-violation', histories which 'are directly interwoven with well-practised, patterned, and predictable distributions of egregious intra-species injustice' (Grear, 2015, p. 233).

The concept of anthropocentrism, and the narrative(s) associated with it and with the emerging frame of the Anthropocene, conceals, then, very specific mechanics of exclusion and of power relations (operating, crucially in and through law) through which *both* other humans *and* ecosystems are 'systematically disadvantaged' (Grear, 2013, p. 78) by legal modernity and its operative fields of practise, such as IEL.

Anthropocentrism is thus operationalised by *certain* humans against both *other* humans and non-human worlds through the projection of mastery that is directly constructed on all three dimensions – ontological, epistemological, and axiological – underpinning anthropocentrism. In this sense, the anthropocentrism of IEL is but a reflection of the larger legacy of international law, whose colonial mechanics have been already thoroughly revealed (e.g., Anghie, 2005). IEL but reproduces these practises of colonial exclusion and violent plunder reflecting contingent ontological, epistemological, and axiological perspectives that are however rendered rational and universal.

The Violence and Exclusion of International Environmental Law

It is at this point useful to illustrate with some examples how IEL operationalises, through its normative dimension and through its operational practises, all three dimensions of anthropocentrism. An example that may help illustrate the preceding in vivid and clarifying terms is the notion of protected areas. Protected areas are a central tool of IEL, and of the conservation practise and imaginary more broadly. Stemming from the US parks movement of the late 1800s (Nanda and Pring, 2012), the idea of protecting the natural environment through enclosing it and secluding it from human activities has become a central tenet of conservation discourse and practise of IEL. Protected areas are in fact enshrined as the key obligation in the CBD under its article 8; are a key tool in many regional or sectoral regimes (from the Convention for the Protection of the Marine Environment of the North-East Atlantic (OSPAR) to the Convention on the Conservation of Antarctic Marine Living Resources (CCAMLR)); are one of the key topics in the ongoing negotiations on marine biodiversity in areas beyond national jurisdiction, which in this respect seek to adopt a comprehensive, global legal basis – which today is lacking – for designating protected areas in such marine areas; are one of the key objectives of the sustainable development goals and of the 2030 Agenda, in particular in relation to goals 14 and 15 (Gissi *et al.*, 2022; Bertzky *et al.*, 2016); are the focus of the 30 by 30 initiative (Dinerstein *et al.*, 2019), which seeks to designate as protected areas 30 percent of the Earth's land and oceans by 2030, an objective also under consideration by the CBD for its post-2020 global biodiversity framework (CBD, 2020), and also adopted under the previous biodiversity strategy, the 2020 Aichi Targets (Bertzky *et al.*, 2016). Protected areas are a crucial tool in the current conservation landscape and, besides the legal and regulatory bases, and besides also some heated controversy within the conservation science community (De Lucia, 2018a), they are considered a *sine qua non* for the achievement of current conservation objectives, especially in the context of ecosystem approaches (CBD, 2004; Arctic Council, 2015).

However, protected areas, besides the problem of efficacy and integrity linked to the problematic of the so-called 'paper parks' (e.g., Matz-Lück and Fuchs, 2014), also present problems of violence and exclusion. Indeed, protected areas have been long criticised for being used as an instrument of colonial dispossession (within a broader international legal context, see in particular Natarajan and Dehm, 2022). As Marvier, Kareiva, and Laslaz observed, 'for 30 years, the global conservation movement has been racked with controversy arising from its role in expelling indigenous people from their lands in order to create parks and reserves' (Marvier, Kareiva, and Laslaz, 2012). This is a problem that is widespread and that involves both publicly designated protected areas and the intervention of private conservation organisations. For example, environmental NGOs acquire land in developing countries to set them aside as protected areas, while in the process evicting human communities. Protected areas may thus become sites of sovereign enclosure (Marvier, Kareiva, and Laslaz, 2012) that, under the banner of conservation, effectively exclude subsistence access to resources and displace local and Indigenous communities (Brockington and Igoe, 2006). Some commentators argue in this respect, and without hesitation, that conservation programmes impose misery on people (Agrawal and Redford, 2009, p. 1).

If this example illustrates how anthropocentrism operationalises a particular *difference* with respect to who is the human centre around which the environment – and IEL – revolves, there are many dimensions to the violent exclusions that the anthropocentric environmentalism of IEL produces and enacts. While regarded as a crucial tool for the conservation of biodiversity and the protection and preservation of ecosystems, in fact protected areas may only create illusions of environmental stewardship through the production of so-called 'ecological fixes', an expression that aims to reflect how protected areas are 'spatial solution[s]' that do nothing more than remove 'environmental barriers to the accumulation of capital' (Ramesh and Rai, 2017, p. 25). Protected areas, from this perspective, ultimately only offer compensation for the widespread distribution of ecological violence that takes place in *all other places* (De Lucia, 2022) and may indeed *enable* it. IEL may thus function as a legitimating mechanism, operating as a 'force or line designed to keep chaos […] at arm's length' (Hasley, 2011, pp. 218–219), separating 'the sacred [from] the abject' (Hasley, 2011, p. 219). In this light, the protection and the preservation of 'the more "majestic" aspects of Nature' offered by IEL rely upon acts of partition and classification in which 'islands of wildness […] are conceivable only on the basis of an ongoing and generalised ecological violence' (Hasley, 2011, p. 219).

Protected areas thus enact a particular vision of the world that reflects the particular onto-logical, epistemological, and axiological coordinates of anthropocentrism and do so in relation to both *other* humans who do not fit the abstract template of the Anthropos, and who in fact are evicted and dispossessed in the name of conservation and in relation to the natural environment, which is constructed, organised, regulated, and protected on the basis of the anthropocentric apprehension of the world, and partitioned into expendable and sacred, coherently with an instrumentally anthropocentric vision of separation of humans from nature. There are other examples that further illustrate the same mechanics by emphasising in different ways the opera-tional aspects of anthropocentrism in both its aspects, for example in relation to the regulation of invasive alien species (De Lucia, 2019a), in relation to ecosystem approaches to conservation (De Lucia, 2015; 2019b), or in relation to the deployment of the framework of ecosystem services in IEL (Sullivan, 2009; De Lucia, 2018b) but for reasons of space, I direct the reader to existing literature.

Conclusions: From Anthropocentrism to Biopolitics

As we have seen, Gillespie has suggested that, while IEL remains largely anthropocentric, there are some non-anthropocentric principles that are emerging and that may help reorient the field. Additionally, and as anticipated in my introduction, despite its fundamental anthropocen-tric grounding, IEL is also located at what I have elsewhere described as an 'epistemological crossroad' (De Lucia, 2019, p. 12) to the extent that it is receptive to, and perhaps the vehicle for, the penetration of the new insights of ecology into the field of law, however problematic these may be (De Lucia, 2017b). In this sense, IEL has also been approached as an opportunity for changing in some fundamental respects the ways in which legal modernity understands, constructs, and regulates nature. Ecocentrism has been usually indicated as the remedy against anthropocentrism, as it, in a nutshell, focuses on de-centring human beings and rather centring nature, however understood (for example ecosystems, biodiversity, and so on). The goal of an ecocentric realignment of law is to integrate human beings within a nature-centric frame-work where all living and non-living entities are approached holistically. This ecocentric shift would help radically reorient the normative core of IEL in relation to all three dimensions of anthropocentrism previously discussed in section 'Understanding Anthropocentrism'. Yet the ecocentric alternative, organised around a set of central ideas such as rights of nature, ecological integrity, or ecosystem approaches, while portrayed as the remedy, does not change the terms of the problematic in any fundamental way, as I briefly illustrate in this concluding section by way of summarising other more comprehensive explorations (De Lucia, 2013; 2015; 2017a; 2018b; 2019a; 2019b; 2020)

Besides conceptual and philosophical problems already raised in relation to the obsessive 'cen-trism' that the move from anthropocentrism to ecocentrism still reproduces (Philippopoulos-Mihalopoulos, 2011; De Lucia, 2019b), there are also a set of problems that I have captured under what I called a *biopolitical* critique of environmental law (De Lucia, 2017a).

IEL, from this critical perspective, is a crucial juncture where knowledge and power inter-sect and coalesce into a regulatory framework aimed at classifying, protecting, regularising, and optimising nature, particularly through the double goal of conservation and sustainable use of biological diversity (De Lucia, 2017a; De Lucia, 2019b) and within the context of the broad narrative of sustainable development. This is the essence of biopolitics: articulating a politics of positive interventions that overcomes the distinction between exploitation and conservation and rather unifies them within a singular framework of sustainable conservation and use. IEL focuses then on this double goal and aligns with the goals of ensuring ecological integrity and deploying ecosystem approaches, or the broad framework of *Earth-care*, in ways that cut across any distinctions that may exist – in principle or in fact – between anthropocentric and ecocentric outlooks.

An example can be gleaned by reviewing paragraph six of the Oslo Manifesto for Ecological Law and Governance, which articulates the need for a shift from 'environmental law' to 'ecological law'. This 'ecological approach to law is based', as expressed in the manifesto, 'on ecocentrism, holism, and intra-/intergenerational and interspecies justice'. Moreover, law should 'no longer favour humans over nature and individual rights over collective responsibilities' (Oslo Manifesto, 2016, para. 5). The difference between the two, the manifesto further observes, 'is not merely a matter of degree, but fundamental'. With ecological law '[e]cological integrity becomes a precondition for human aspirations and a fundamental principle of law', and the central legal principle becomes 'human responsibility for nature' (Oslo Manifesto, 2016, para. 6).

However, it is precisely this principle of responsibility that becomes the focus of a biopolitical critique, which unifies anthropocentrism and ecocentrism within the same conceptual and philosophical framework as two complementary and ultimately mutually reinforcing dimensions that are operationalised through law. This I have called *biopolitical encaring*, by which I refer to the positive inflection of the exercise of power that finds its legitimation and its goal in caring for nature in ways that nudge both anthropocentrism and ecocentrism towards the notion of stewardship.

Once nature is encompassed within the sphere of knowledge and control of IEL, a further need presents itself in the context of the Anthropocene and is turned into goal. As the Earth is recognised as a vulnerable domain in need of protection, a series of discourses, underpinned by scientific assessments (such as those regularly carried out by the Intergovernmental Panel on Climate Change and by the Intergovernmental Science-Policy Platform on Biodiversity and Ecosystem Services) need to be operationalised via IEL. In this context, I have suggested elsewhere that power is no longer exercised through sovereign command but through technical norms and scientific regimes of knowledge that need to subjugate nature in order to protect it: power – and thus law – in its biopolitical inflection becomes equivalent with *Earth-care* (De Lucia, 2020, p. 336). I suggest that, in order to care for Earth, any ecocentric discourse constantly transforms into biopolitics to the extent that IEL 'must engulf nature in its entirety under a conceptual and regulatory framework where care and subjugation, vulnerability and productivity, life and death are constantly entangled in a reciprocal and inevitable relation' (De Lucia, 2020, p. 337). A biopolitical critique then makes legible the undecidable dilemma that inhabits the empty space amid the anthropocentric–ecocentric binary, where care and subjugation are inevitably entangled.

This dilemma, and continual oscillation, is precisely what I have called *encaring*, a term that, drawing inspiration from Heidegger's concept of enframing,[8] captures the biopolitical inflection and operations of IEL as an exercise of life-affirming interventions in the world (that is, power *as* care) while, at the same time, subjugating the natural environment within its framework of control and thus inflicting material violence in the same gesture.

Thus, it becomes evident that neither anthropocentrism nor its ecocentric reversal are any longer useful parameters to reflect the conceptual and operational realities of IEL. The binary framework of critique which they set up is no longer able to deal with the complexities, ambiguities, and contradictions that inhabit IEL in the context of the Anthropocene, as illustrated by the example of protected areas. Both remain entangled in comprehensive regimes of biopolitical *encaring*. Of course, if beyond anthropocentrism and ecocentrism all that is left is biopolitical *encaring*, it becomes difficult to imagine a way out. Yet this is the task left to legal scholars intent on reimagining law in the Anthropocene.

Notes

1 As many scholars have shown, the lineage of Western anthropocentrism may go back much further than to the onset of modernity and to the very roots of western civilisation: Greek philosophy as well as Judeo-Christian theology (e.g., White, 1967; Passmore, 1974). However, given space constraints, and because both these older philosophical traditions contributed crucially to the emergence of the categories of modernity, in this chapter I refer to modernity as the central grounding of anthropocentrism.

2 It is, however, perhaps useful to recall how in some other respects, as mentioned in the Introduction to this volume, IEL is *too little* anthropocentric, to the extent that human suffering and disparities of access to environmental resources or deleterious effects from environmental degradation burden disproportionally certain human communities. This will be picked up again in the section 'The Violence and Exclusion of International Environmental Law'.

3 Indeed, the Declaration proceeds from the consideration of a need to identify a 'common outlook and for common principles to inspire and guide the peoples of the world in the preservation and enhancement of the human environment' (Declaration of the United Nations Conference on the Human Environment 1972, incipit).

4 In its broadest sense. Lowe constructs the international legal space as a field open to the influence of cultural narratives and political contestations through the notion of interstitial norms, of which the concept of sustainable development is a primary example (Lowe, 2000).

5 The central characteristics, then and now, of a green economy are its reliance on environmental markets for the resolution of environmental problems.

6 Since the 1992 Rio Conference, all major intergovernmental conferences held under the aegis of the UN have framed the environment under the narrative of sustainable development.

7 See especially Cecchetti (2006, p. 56) which speaks of an eco-compatible revision of anthropocentrism.

8 Which indicates 'making things technologically accessible', which in turn leads to the arraying of nature as a standing reserve of 'resources to be exploited as means to ends' (Salisbury, 2019, p. 8).

References

Agrawal, A. and Redford, K. (2009) 'Conservation and Displacement: An Overview', *Conservation and Society*, 7(1), pp. 1–10. DOI: 10.4103/0972-4923.54790

Anghie, A. (2005) *Imperialism, Sovereignty and the Making of International Law*. Cambridge: Cambridge University Press.

Arctic Council (2015) 'Arctic Marine Strategic Plan 2015–2025. Protecting Marine and Coastal Ecosystems in a Changing Arctic'. Available at: https://www.pame.is/arctic-marine-strategic-plan-2015-2025 (Accessed: 6 March 2023).

Bertzky, B., Munroe, R., Teperman, K., Tsang, V. and Ongige, E. (2016) 'Chapter 7 – Protected Areas and the Sustainable Development Goals', UNEP World Conservation Monitoring Centre, pp. 57–71.

Bosselmann, K. (2010) 'Losing the Forest for the Trees: Environmental Reductionism in the Law', *Sustainability*, 2(8), pp. 2424–2448. DOI: 10.3390/su2082424

Bratspies, R. and Miller, A. (eds.) (2006) *Transboundary Harm in International Law: Lessons from The Trail Smelter Arbitration*. New York: Cambridge University Press.

Brockington, D. and James Igoe (2006) 'Eviction for Conservation: A Global Overview', *Conservation and Society*, 4(3), pp. 424–470.

Brown-Weiss, E. (1993) 'International Environmental Law: Contemporary Issues and the Emergence of a New World Order', *Georgetown Law Journal*, 81, pp. 675–710.

Burdon, P. (2014) *Earth Jurisprudence: Private Property and the Environment*. New York: Routledge.

Carson, R. (1962) *Silent Spring*. Boston: Houghton Mifflin.

CBD (13 April 2004) 'Decision VII/28, "Protected areas (Articles 8 (a) to (e))"' UNEP/CBD/COP/DEC/VII/28.

CBD (29 March 2022) Report of the Open-Ended Working Group on the Post-2020 Global Biodiversity Framework on Its Third Meeting (Part II) UN Doc. CBD/WG2020/3/7

Cecchetti, M. (2006) 'La Disciplina Giuridica Della Tutela Ambientale Come "Diritto Dell'Ambiente"', *Federalismi.it, Rivista di Diritto Pubblico Italiano, Comunitario e Comparato*, 25.

Convention on Biological Diversity (adopted 5 June 1992, entered into force 29 December 1993), 1760 UNTS 79 (CBD).

Cullinan, C. (2002) *Wild Law: A Manifesto for Earth Justice*. Cape Town: Siber Ink.

Cunningham, W., Cunningham, M. and Woodworth Saigo, B. (2012) *Environmental Science: A Global Concern*, 12th edn. New York: McGraw-Hill.

Curry, P. (2011) *Ecological Ethics: An Introduction*, 2nd edn. Cambridge: Polity Press.

De Lucia, V. (2013) 'Towards an Ecological Philosophy of Law. A Comparative Discussion', *Journal of Human Rights and the Environment*, 4(2), pp. 167–190. DOI: 10.4337/jhre.2013.02.03

De Lucia, V. (2015) 'Competing Narratives and Complex Genealogies. The Ecosystem Approach in International Environmental Law', *Journal of Environmental Law*, 27(1), pp. 91–117. DOI: 10.1093/jel/equ031

De Lucia, V. (2017a) 'Beyond Anthropocentrism and Ecocentrism. A Biopolitical Reading of Environmental Law', *Journal of Human Rights and the Environment*, 8(2), pp. 181–202. DOI: 10.4337/jhre.2017.02.01

De Lucia, V. (2017b) 'Critical Environmental Law and the Double Register of the Anthropocene: A Biopolitical Reading', in Kotzé, L. (ed.) *Environmental Law and Governance in the Anthropocene*. Oxford: Hart Publishing, pp. 97–116.

De Lucia, V. (2018a) 'A Critical Interrogation of the Relation between the Ecosystem Approach and Ecosystem Services', *Review of European, Comparative & International Environmental Law*, 27(2), pp. 104–114. DOI: 10.1111/reel.12227

De Lucia, V. (2018b) 'Semantics of Chaos: Law, Modernity and the Commons', *Pólemos*, 12(2), pp. 393–414. DOI: 10.1515/pol-2018-0022

De Lucia V. (2019a) 'Bare Nature. The Biopolitical Logic of the International Regulation of Invasive Alien Species', *Journal of Environmental Law*, 31(1), pp. 109–134. DOI: 10.1093/jel/eqy016

De Lucia V. (2019b) *The Ecosystem Approach in International Environmental Law: Genealogy and Biopolitics*. Abingdon: Routledge.

De Lucia, V. (2020) 'Rethinking the Encounter Between Law and Nature in the Anthropocene: From Biopolitical Sovereignty to Wonder', *Law and Critique*, 31(3), pp. 329–349. DOI: 10.1007/s10978-020-09281-9

De Lucia, V. (2022) 'Oceanic Heterolegalities? Ocean Commons and the Heterotopias of Sovereign Legality', in Braverman, I. (ed.) *Laws of the Sea: Interdisciplinary Currents*. New York: Routledge, pp. 121–143.

Declaration of the United Nations Conference on the Human Environment (1972) UN Doc A/Conf.48/14/Rev.1.

Dinerstein, E., Vynne, C., Sala, E., Joshi, A.R., Fernando, S., Lovejoy, T.E., Mayorga, J., Olson, D., Asner, G.P., Baillie, J.E.M., Burgess, N.D., Burkart, K., Noss, R.F., Zhang, Y.P., Baccini, A., Birch, T., Hahn, N., Joppa, L.N. and Wikramanayake, E. (2019) 'A Global Deal for Nature: Guiding Principles, Milestones and Targets, Science Advances', 5(4), eaaw2869. DOI: 10.1126/sciadv.aaw2869.

Douzinas. C. and Geary, A. (2005) *Critical Jurisprudence. The Political Philosophy of Justice*. Oxford: Hart Publishing.

Dupuy, P. (2015) 'The Philosophy of the Rio Declaration' in Viñuales, J. (eds.) *The Rio Declaration on Environment and Development: A Commentary*. Oxford: Oxford University Press, pp. 65–74.

Emmenegger, S. and Tschentscher, A. (1994) 'Taking Nature's Rights Seriously: The Long Way to Biocentrism in Environmental Law', *Georgetown International Environmental Law Review*, 6, pp. 545–592.

Esposito, R. (2010) *Pensiero Vivente: Origine e Attualità della Filosofia Italiana*. Torino: Einaudi.

Gillespie, A. (1997) *International Environmental Law, Policy and Ethics*. Oxford: Clarendon Press Oxford.

Gillespie, A. (2001) *The Illusion of Progress: Unsustainable Development in International Law and Policy*. London: Earthscan.

Gillespie, A. (2014) *International Environmental Law, Policy and Ethics*, 2nd edn. Oxford: Clarendon Press Oxford.

Gissi, E., Maes, F., Kyriazi, Z., Ruiz-Frau, A., Santos, C.F., Neumann, B., Quintela, A., Alves, F.L., Borg, S., Chen, W., Fernandes, M. da L., Hadjimichael, M., Manea, E., Marques, M., Platjouw, F.M., Portman, M.E., Sousa, L.P., Bolognini, L., Flannery, W., Grati, F., Pita, C., Naṭaşa Vǎidianu, Stojanov, R., Tatenhove, J. van, Micheli, F., Hornidge, A.-K. and Unger, S. (2022) 'Contributions of Marine Area-based Management Tools to the UN Sustainable Development Goals', *Journal of Cleaner Production*, 330, 129910. DOI: 10.1016/j.jclepro.2021.129910.

Grassi, S. (2007) 'Tutela dell'Ambiente (diritto amministrativo)', in Falzea, A., Grossi, P. and Cheli, E. (eds.) *Enciclopedia del Diritto: Annali I [Accertamento – Tutela]*, Milan: Giuffrè, 2007, p. 1114.

Grear, A. (2011) 'The Vulnerable Living Order: Human Rights and the Environment in a Critical and Philosophical Perspective', *Journal of Human Rights*, 2(1), pp. 23–44. DOI: 10.4337/jhre.2011.01.02

Grear, A. (2013) 'Law's Entities: Complexity, Plasticity and Justice', *Jurisprudence*, 4(1), pp. 76–101. DOI: 10.5235/20403313.4.1.76

Grear, A. (2015) 'Deconstructing Anthropos: A Critical Legal Reflection on 'Anthropocentric' Law and Anthropocene 'Humanity'', *Law and Critique*, 26, pp. 225–249. DOI: 10.1007/s10978-015-9161-0

Grossi, P. (1988) *La Proprietà e le Proprietà nell'Officina dello Storico*. Milano: Giuffrè.

Grossi, P. (2001) *L'Europa del Diritto*. Bari: Laterza.

Grumbine, E. (1994) 'What is Ecosystem Management?', *Conservation Biology*, 8(1), pp. 27–38. DOI: 10.1046/j.1523-1739.1994.08010027.x

Hasley, M., (2011) 'Majesty and Monstrosity: Delueze and the Defence of Nature', in Philippopoulos-Mihalopoulos, A. (ed.) *Law and Ecology: New Environmental Foundations*, New York: Routledge, pp. 214–236.

Heidegger, M. (1977) *The Question Concerning Technology and Other Essays*. New York: Garland.

Hey, E. (2003) *Teaching International Law*. Leiden, The Netherlands: Brill | Nijhoff.

Hicks, B. (1998–1999) 'Treaty Congestion in International Environmental Law: The Need for Greater International Coordination', *University of Richmond Law Review*, 32, pp. 1643–1674.

International Convention for the Regulation of Whaling (adopted 2 December 1946, entered into force 10 November 1948), 161 UNTS 72 (ICRW).

Jans, J.H. (2010) 'Stop the Integration Principle?', *Fordham International Law Journal*, 33(3), p. 1533–1547.

Keller, D.R. (ed.) (2010) *Environmental Ethics: The Big Questions*. Malden: Wiley-Blackwell.

Kim, R. and Bosselmann, K. (2015) 'Operationalizing Sustainable Development: Ecological Integrity as a Grundnorm of International Law', *Review of European, Comparative & International Environmental Law*, 24(2), pp. 194–208. DOI: 10.1111/reel.12109

Kotzé, L. and French, D. (2018) 'The Anthropocentric Ontology of International Environmental Law and the Sustainable Development Goals: Towards an Ecocentric Rule of Law in the Anthropocene', *Global Journal of Comparative Law*, 7, pp. 5–36. DOI: 10.1163/2211906X-00701002

Legality of the Threat or Use of Nuclear Weapons (Advisory Opinion) [1996] ICJ Rep 226.

Lowe, V. (2000) 'The Politics of Law-Making: Are the Method and Character of Norm Creation Changing?', in Byers, M. (ed.) *The Role of Law in International Politics. Essays in International Relations and International Law*. Oxford: Oxford University Press, pp. 207–226.

M'Gonigle, M., and Takeda, L. (2013) 'The Liberal Limits of Environmental Law: A Green Legal Critique', *Pace Environmental Law Review*, 30(3), pp. 1005–1015

Malm, A. and Hornborg, A. (2014) 'The Geology of Mankind? A Critique of the Anthropocene Narrative', *The Anthropocene Review*, 1(1), pp. 62–69. DOI: 10.1177/2053019613516291

Marvier, M., Kareiva, P. and Laslaz, R. (2012) 'Conservation in the Anthropocene Beyond Solitude and Fragility', *The Breakthrough Journal*, 2, pp. 29–37.

Matz-Lück, N. and Fuchs, J. (2014) 'The Impact of OSPAR on Protected Area Management Beyond National Jurisdiction: Effective Regional Cooperation or a Network of Paper Parks?', *Marine Policy*, 49, pp. 155–166. DOI: 10.1016/j.marpol.2013.12.001

McIntyre-Mills, J. (2014) *Systemic Ethics and Non-Anthropocentric Stewardship. Implications for Transdisciplinarity and Cosmopolitan Politics*. Cham: Springer.

Merchant, C. (1990) *The Death of Nature. Women, Ecology and the Scientific Revolution*. San Francisco: Harper One.

Nanda, V. and Pring, G. (2012) *International Environmental Law and Policy for the 21st Century*. Leiden: BRILL.

Natarajan, U. and Dehm, J. (eds) (2022) *Locating Nature: Making and Unmaking International Law*. Cambridge: Cambridge University Press.

Norton, B. (2009) 'Environmental Ethics and Weak Anthropocentrism', in Clowney, D. and Mosto, P. (eds) *Earthcare: An Anthology in Environmental Ethics*. Lanham: Rowman & Littlefield Publishers.

Norton, B. (2014) *Why Preserve Natural Variety?* Princeton: Princeton University Press.

'"Oslo Manifesto" for Ecological Law and Governance' (2016), *From Environmental Law to Ecological Law: A Call for Re-Framing Law and Governance* [Online]. Available at: https://elgaworld.org/oslo-manifesto (Accessed: 6 March 2023).

Palmer, C. (2006) 'Stewardship: A Case Study in Environmental Ethics', in Berry, R. (ed.) *Environmental Stewardship: Critical Perspectives, Past and Present*. London: T&T Clark, pp. 63–75.

Passmore, J. (1974) *Man's Responsibility for Nature: Ecological Problems and Western Traditions*. London: Scribner.

Paterson, M. (2010) 'Legitimation and Accumulation in Climate Change Governance', New Political Economy, 15(3), pp. 345–368. DOI: 10.1080/13563460903288247

Pearce, D. (1992) 'Green Economics', *Environmental Values*, 1(1), pp. 3–13.

Pearce, D., Markandya, A. and Barbier, E. (1989) *Blueprint for a Green Economy*. London: Routeledge.

Percival, R. (2001) 'Global Law and the Environment', *Washington Law Review*, 86, pp. 579–634.

Philippopoulos-Mihalopoulos, A., 'Towards a Critical Environmental Law', in Philippopoulos-Mihalopoulos, A. (ed.) *Law and Ecology: New Environmental Foundations*. New York: Routledge, 2011, pp. 18–38.

Ramesh, R. and Rai, N.D. (2017) 'Trading on Conservation: A Marine Protected Area as an Ecological Fix', *Marine Policy*, 82, pp. 25–31. DOI: 10.1016/j.marpol.2017.04.020

Report of the World Summit on Sustainable Development (4 September 2002) UN Doc. A/CONF.199/20.

Rio Declaration on Environment and Development (12 August 1992) A/CONF.151/26 (Vol. I) (Rio Declaration).

Rolston, H. (2012) *A New Environmental Ethics: The Next Millennium for Life on Earth*. London: Routledge.

Salisbury, P. (2019) 'Ways of Being and Ways of Knowing: Heidegger's The Question Concerning Technology and Knowledge Organization', *Proceedings from North American Symposium on Knowledge Organization*, 7, pp. 7–15. DOI: 10.7152/nasko.v7i1.15613

Sharpe, S. (1998) *Redeeming the Time: A Political Theology of the Environment*. London: Bloomsbury Publishing.

Stone, C. D. (1972) 'Should Trees have Standing? Toward Legal Rights for Natural Objects', *Southern California Law Review*, 45, pp. 450–501.

Sullivan, S. (2009) 'Green Capitalism, and the Cultural Poverty of Constructing Nature as Service-provider', *Radical Anthropology*, 3, pp. 18–27.

Tallacchini, M. (1996) *Diritto per la Natura: Ecologia e Filosofia del Diritto*. Torino: Giappichelli Editore.

Tallacchini, M. (2000) 'A Legal Framework from Ecology', *Biodiversity and Conservation*, 9(8), pp. 1085–1098. DOI: 10.1023/A:1008926819473

Trail Smelter Arbitration (United States v. Canada), Decision, [1941], 3 RIAA 1907.

UNGA, 'The Future We Want, Report of the United Nations Conference on Sustainable Development Rio de Janeiro' (20–22 June 2012) UN Doc. A/CONF.216/16.

United Nations Law of the Sea Convention (adopted 10 December 1982, entered into force 16 November 1994) 1833 UNTS 397 (UNCLOS).

Vicuña, F. (1999) 'From the 1893 Bering Sea Fur-Seals Case to the 1999 Southern Bluefin Tuna Cases: A Century of Efforts at Conservation of the Living Resources of the High Seas', *Yearbook of International Environmental Law*, 10(1), pp. 40–47. DOI: 10.1093/yiel/10.1.40

Vogel, S. (1996) *Against Nature: The Concept of Nature in Critical Theory*. New York: SUNY Press.

Welchman, J. (2012) 'A Defence of Environmental Stewardship', *Environmental Values*, 21, pp. 297–316.

Wiener, J. (2001) 'Something Borrowed for Something Blue: Legal Transplants and the Evolution of Global Environmental Law', *Ecology Law Quarterly*, 27, pp. 1295–1372.

White, L. (1967) 'The Historical Roots of our Ecological Crisis', *Science*, 155, pp. 1203–1207.

Wilkinson, D. (1999) 'Using Environmental Ethics to Create Ecological Law', in Holder, J. and McGillivray, D. (eds.) *Locality and Identity: Environmental Issues in Law and Society*. Aldershot: Ashgate, pp. 17–50.

World Commission of Environmental and Development (4 August 1987) 'Our Common Future: Report of the World Commission on Environment and Development' UN Doc A/42/427.

Worrell, R. and Appleby, M. (2000) 'Stewardship of Natural Resources: Definition, Ethical and Practical Aspects', *Journal of Agricultural and Environmental Ethics*, 12(3), pp. 263–277. DOI: 10.1023/A:1009534214698

5

THE LAW OF THE SEA'S FLUID ANTHROPOCENTRISM

Godwin E. K. Dzah[1]

Introduction

The vast, almost limitless expanse of the sea is only comparable to its law, the 1982 United Nations Convention on the Law of the Sea (UNCLOS).[2] Wide-ranging in its regulation of the uses of the ocean, the law of the sea was birthed in great controversy, and remains one of the enduring paradoxes of international law (Koh, 1987).[3] It is named not after humans nor human conduct; yet, the law of the sea is in all forms and for all practical purposes about human society. From the early days of the law of the sea as a discipline, the ocean has been invoked as the subject of law-making. However, much of the law of the sea reveals how little effort is devoted to respecting and valuing the ocean's ecological integrity.

This chapter is inspired by one of the revolutionary moments in the development of the law of the sea – the momentous speech by Ambassador Arvid Pardo, the Permanent Representative of Malta to the United Nations General Assembly in 1967. Part of Pardo's historic speech reads thus:

> The dark oceans were the womb of life: from the protecting oceans, life emerged. We still bear in our blood, in the salty bitterness of our tears – the marks of this remote past. Retracing the past, man, the present dominator of the emerged earth, is now returning to the ocean depths. His penetration of the deep could mark the beginning of the end for man, and indeed for life as we know it on earth: it could also be a unique opportunity to lay solid foundations for a peaceful and increasingly prosperous future for all peoples.
>
> *(UNGA, 1967)*

Since then, this speech has been celebrated as a significant motivation for a renewed interest in ocean governance; and along with that, Pardo has been credited as the leading architect of the modern law of the sea. Importantly, his profound words echo three concerns: the harmony between humanity and the sea; humanity's disturbance and reversal of that harmony as demonstrated through the present patterns of excessive human domination of nature; and the Janus-face of anthropocentrism as harbinger of life and death of the planetary community (Crutzen, 2006).

DOI: 10.4324/9781003201120-7

This tripartite analysis of Pardo's statement unveils the spectre of extensive human intervention in the ocean space. But it also provides glimpses into the possibility of reimagining the human–ocean connection, beginning with respect for ecological integrity. Respect for the ocean's integrity revolves around recalibrating human-centric interests and their relationship with the law of the sea. In this respect, the guiding principle behind anthropocentrism as used in this chapter is the inordinate human-centredness of international law and how it is instrumentalised to serve human interests to the detriment of non-human entities. Here, the law of the sea's anthropocentrism functions as a 'distinction between a "rational subject" at the centre of the epistemological paradigm and a distinct "other"-object, which finds validity and legitimacy through the definition of the former' (Ollino, 2019, p. 205).[4] Therefore, this chapter investigates how the law of the sea has been ingeniously adapted to 'other' the sea and elevate human interests by subordinating the ocean and its ecological integrity.

Accordingly, the final words in the excerpt warrant a closer reading. Pardo's speech hinted at human interests in harnessing the sea and its vast resources for human benefit and a seeming indifference to the ecological dimensions of human uses of the sea (Bosselmann, 2015).[5] This can be seen in the bewildering uses to which humans have put the sea – ranging from over-fishing to dumping of radioactive waste, indiscriminate offshore petroleum operations, unregulated plastic waste disposal and effluent discharge, weapons testing and nuclear exercises, deep seabed mining, and even building mega-cities on land reclaimed from the seas. The list is frighteningly endless, as the sea and its related human uses have been a rallying point for significant developments in international law.

The central argument in this chapter thus emerges from the anxiety that the law of the sea mirrors the anthropocentric dominance that characterises its disciplinary origins. The more international law scholars talk about the law of the sea, the more what is meant in fact is the extension of human control over the sea, which is nothing more than a veiled exploitation regime disguised as ocean governance (Vidas, 2011). If, along the way, there are glimpses of marine environmental awareness and subsequent ecological action, these remain incidental to the principal purpose of human exploitation. As a result, this chapter adopts the view that 'by moving beyond this anthropocentric approach and placing the ocean-space at the centre of this study, the ocean-space can therefore become known for its own characteristics' (Lalöe, 2016, p. 4).

Beyond this introduction, this chapter comprises three substantive parts and a conclusion. 'The Imperial (and Colonial) History of the Law of the Sea' canvasses a historical discussion of the law of the sea by drawing upon the primary doctrines of free seas and closed seas and their connection to Eurocentrism, imperialism, colonialism, and the crystallisation of the law of the sea's anthropocentrism. 'UNCLOS: A Three-Dimensional Analysis' utilises this history in analysing how UNCLOS continues the law of the sea's anthropocentrism in contemporary times. 'Beyond Historical Legacies: Future of the Law of the Sea' delves into new areas within the regulatory sphere of the law of the sea including bioprospecting and the exploitation of marine biodiversity beyond national jurisdiction by demonstrating how the law of the sea's present and future efforts carry on its human-centric registers, hence 'fluid' anthropocentrism. The fifth section concludes this contribution by offering some key insights into retrenching the law of the sea's anthropocentrism and ensuring ecological integrity becomes the centrepiece of the discipline.

The Imperial (and Colonial) History of the Law of the Sea

It is critically important to situate the law of the sea as integral to the anthropocentric character of European imperialism and colonialism. Instead of merely accepting international law as objective, the law of the sea's anthropocentric orientation shares in international law's imperial and colonial heritage. These concerns and anxieties invite a review of the uncritical acceptance of international law doctrines and how these doctrines eventually became central to the law of the sea. The examination of the doctrines of free seas and closed seas reveals their connection to European expansionism. In this sense, the historical origins of the discipline's anthropocentrism provoke an analysis of these two doctrines and how they 'have advanced the economic exploitation of the oceans; distributing great benefits to a few and great costs – including ecological costs – to many' (Ranganathan, 2021, pp. 167–168). These doctrines foster a particular appreciation of the seas as naturally conditioned for human exploitation. Thus, the contemporary pathological anthropocentrism of the law of the sea was long formalised in the crucible of the prevailing political and legal developments of European imperialism and colonialism, which were premised on an extractivist logic.

To begin, international law links the concept of freedom of navigation to the inauguration of the free seas doctrine. This doctrine is credited to the Dutch jurist Hugo Grotius who wrote his *Mare Liberum* in 1609. This book arose out of his legal defence of the Dutch East India Company upon its seizure of the Portuguese carrack, *Santa Catarina*, off the coast of present-day Singapore, and the carrack's subsequent sale in Amsterdam. Since then, Grotius has been described as the father of the law of the sea, although the focus of his legal argument (the concept of unimpeded ocean navigation) has always been a key feature of all seafaring societies, with Europe being no particular exception (Anand, 1983).

In his magnus opus, *Mare Liberum*, the inspiration for freedom of the seas as theorised by Grotius was hardly ever about maintaining the ocean's ecological integrity. Recent accounts even suggest the longer subtitle of *Mare Liberum*, which translates from Latin to read 'The Right Which Belongs to the Dutch to take Part in the East Indian Trade', better explains the Grotian ambition than the main title. The longer subtitle explains that the Grotian doctrine was concerned with grounding a legal basis for the Dutch contention over the Portuguese–Spanish duopoly on fishing and navigational access in primarily 'newly discovered' territories and waters. This action was justified by Grotius for breaking up the monopolisation of sea routes by Spain and Portugal.

At the time Grotius wrote this treatise, Portugal was Europe's preeminent naval power, exercising near-exclusive control over ocean navigation in Europe. It is out of this history that the Eurocentric origins of the law of the sea emerge; hence, the misattribution of its origins to Europe even though non-European societies were equally traversing the oceans at the time. As such, this so-called Europe-centred naval dominance needs to be nuanced by the fact that non-European peoples in Africa and Asia were equally skilled ocean navigators, and upon whom European seafarers heavily relied to navigate those littoral lands (Duarte, 2012, pp. 68–69). Situating this history in context, Grotius was interested in Dutch navigational access to so-called newly discovered territories and soon-to-be-established colonies. Therefore, the Grotian free seas doctrine constructed around the notion that '[t]he sea is by nature, open to all' (Grotius, 1609, p. 91) simply disguised Dutch nationalist ambitions, as ecological integrity would not have been a major consideration for commentators (like Grotius) at this time – neither would a modern conception of nature.

Unsurprisingly, with the European imperial and colonial project on the horizon, this politico-legal position advanced by Grotius was quickly embraced by other European maritime powers (Somos, 2020). In this respect, *Mare Liberum* extended European political and legal ideas beyond Europe by linking colonialism to commerce between the European centre and non-European periphery yet inspired by a Eurocentric vision of navigational freedom. It is little surprise then that the theoretical and practical constructions around the law of the sea, even now, refract these imperial and colonial imprints of the free seas theory which has been instrumentalised by non-European states too seeking to claim and assert sovereignty over the oceans, 'whether in the South China Sea or elsewhere' (Boer, 2021, p. 150).

The popularity of the Grotian *Mare Liberum* provoked a notable antithesis to the free seas doctrine. It inspired John Selden's *Mare Clausum Sive de Dominio Maris* which was written in 1635.[6] While Selden is lesser known compared to Grotius, Selden's rejoinder was not any less significant, considering he had the daunting task of justifying English claims over the sea. Selden's response arose from the English apprehension that if the Grotian free seas policy gained (European) prominence, England risked losing its control over 'the high seas to the south and east of England, as well as to undefined regions of the north and west' (Fletcher, 1933, p. 8). Selden's response was therefore written against the backdrop of England's maritime ascent, as the English contemplated the Grotian doctrine could impose limits on future English naval accomplishments and imperial interests.

Selden's *Mare Clausum* made two major arguments, both contradicting the free seas doctrine. Relying on history and state practice, he argued that first, the seas were not part of the commons, as they had historically been treated as private property (Lesaffer, 2016). Second, Selden contended, the seas abutting England were English territory. He argued that, as territorial extension, a state could exercise dominion over the sea in a way that closes off its communal use by other states. This was the opposite of the Grotian free seas position. With that, English claims over the seas could not be limited to its immediate coastal waters, but also a wider maritime dominion was possible subject to the extent of English sovereignty and control (Thornton, 2006).

However, while his work is an enduring critique of the Grotian free seas doctrine, *Mare Clausum* shares the Grotian exclusionary trait that jettisoned non-Western concepts of ocean governance. The extension of the principles in *Mare Clausum* beyond England into newer geographies through the expansion of British imperialism and colonialism, aided by English naval conquests, intensified the anthropocentrism that Selden's closed seas thesis, much like the Grotian free seas, was conditioned to produce. Thus, these doctrines conditioned the law of the sea's fluid anthropocentrism in both ideational and material terms.

Accordingly, the foundations of the law of the sea were premised on human-centredness. For its capacity to unite Europe around a common navigational philosophy that promoted an exploitative regime, the Grotian *Mare Liberum* effectively 'laid the legal groundwork for Dutch colonialism' (Mawani and Prange, 2021, p. 3). Paradoxically, while *Mare Clausum* critiqued the free seas theory, Selden's closed seas doctrine equally extended the European expansionist project as English imperial interests were supported and progressively advanced by its naval exploits beyond Europe (Somos, 2012). It is unsurprising then that neither the Grotian free seas nor Selden's closed seas paid considerable attention to the sea's inherent value, a situation that is still prevalent today. Little wonder when UNCLOS even provided for fisheries regulation through the 1995 United Nations Agreement on Straddling and Highly Migratory Fish Stocks, it lacked the necessary protection for fish ecological integrity as the agreement focused on fishery conservation for human-centred interests.

Grotius and Selden advanced their legal arguments, the Dutch and the English quickly followed them self-servingly, and other states and some legal scholars simply accepted and applied them. More than four hundred years later, these antiquated ideas still governed much of the law of the sea. It would take the Third Conference on the Law of the Sea (UNCLOS III) negotiations for there to be a purposeful attempt at introducing non-anthropocentric considerations into the discourse on the law of the sea (Steinberg, 1996). The UNCLOS III process incorporated ideas from the foundational work of the 1972 United Nations Conference on the Human Environment (Stockholm Conference) which discussed the concept of sustainable development as a significant international issue, by linking it to environmentally destructive activities in the ocean space (Hafetz, 2000). A review of the imperial and colonial origins of the discipline thus helpfully reframes the boundaries of the ensuing analysis by bridging the gap between these histories and intellectual projects dating back to the seventeenth century and how these ideas came together to presage the modern law of the sea under UNCLOS. Overall, the economic foundations of these early intrusions of international law into the ocean space overshadowed ecological concerns even to this day, as the ocean became a site of accumulating and perpetuating global unequal exchange by means of navigational prowess or through a race-to-the-bottom ocean-based extractivist approach (Porras, 2011).

UNCLOS: A Three-Dimensional Analysis

While it comes across as a grim description, UNCLOS is not post-anthropocentric in any substantive sense of the term. Far from curtailing the human-centredness of the law of the sea, it has rather entrenched this phenomenon. State parties structured UNCLOS around the idea that society can exploit the oceans at will. In fact, the central question UNCLOS seeks to answer is not if humans *can* exploit the oceans but which parts of the sea humans should have access to, the rights a state can exercise in exploiting the ocean's resources, and the duties that states might owe other states in the context of such exploitation. In this light, UNCLOS is a profoundly anthropocentric treaty in the sense that it privileges the needs of humankind and is rarely interested in the intrinsic value of the ocean's ecological integrity. Even when UNCLOS is concerned with non-human entities, whether under the common heritage of mankind or the protection and conservation of the marine environment, a closer scrutiny demonstrates that, at its core, the organising principle is still a sustained exploitation of the non-human rather than limiting ecological despoliation.

This section highlights the anthropocentricity of UNCLOS. It offers an analysis of the link between UNCLOS and anthropocentrism. For after all, UNCLOS is the constitution of the oceans and the largest and most comprehensive treaty law in modern history. Yet, it is important to point out that, fundamentally, UNCLOS guarantees oceanic resource exploitation which is now a vast and seemingly limitless enterprise. However, I limit the discussion to three related aspects of UNCLOS: common heritage of mankind; protection, preservation, and conservation of the marine environment; and the interface between UNCLOS and the language of rights in the context of the legal personhood of the sea.

Common Heritage of Mankind

The common heritage of mankind (CHM) is an ever-present reminder of the Grotian free seas doctrine. In his early works, Grotius relied on CHM's key principle, namely the oceans as 'global commons', to construct his doctrine (Schrijver, 2016). In its contemporary manifestation, CHM is a fundamental concept of international law, as it emerged as an alternative view on the exercise of control over natural resource exploitation. Viewed through the prism of the global commons, CHM was founded on the notion that 'the resources of a particular territory be held and utilised on behalf of the international community as a whole, with special attention to the needs of its most vulnerable members' (Mickelson, 2014, p. 622). Stripped of technical jargon, CHM implies all states coming together to exploit and manage deep seabed resources beyond national jurisdiction and distributing the resulting proceeds and benefits among themselves, even if some states do not (or are unable to) participate directly in the actual extraction.[7]

Pardo is once again credited with kindling global interest in CHM.[8] In his proposal for the regulation of the seabed beyond national jurisdiction, Pardo stated that:

> In ocean space however, the time has come to recognise as a basic principle of international law the overriding common interest of mankind in the preservation of the quality of the marine environment and in the rational and equitable development of resources lying beyond national jurisdiction.
>
> *(Pardo, 1975, p. 176)*

Pardo's words remained significant especially in the final rounds of the UNCLOS III negotiations.[9] Thus, the concrete manifestation of these words in Article 136 of UNCLOS provides that '[t]he Area and its resources are the common heritage of mankind'.[10]

Regrettably, Pardo's formulation did not clarify some aspects of CHM (Baxi, 2019). Conspicuously missing from his statement was CHM's scope under UNCLOS. During the UNCLOS negotiations, there was little divergence over proposals for the peaceful use of the deep seabed including for medical and research purposes. However, there was controversy over seabed mineral resource exploitation which invariably engaged the extraterritorial extension of the theory (and practice) of land grabs and a land-based resource race to the deep seabed floor and its invaluable resources (Wolfrum, 1983). Not even CHM's incorporation in the final UNCLOS text resolved the decades-long North–South contestation over CHM's relevance. As a result, its inclusion in UNCLOS only renewed ideological contestations over its legal scope (Mickelson, 2019).

Due to heightened state interests in deep seabed resources, even though UNCLOS had been adopted in 1982, it only entered into force in 1994. The initial deep seabed extraction arrangement under UNCLOS had caused international disagreement. UNCLOS thus became functional only after states from both the Global North and the Global South had adopted an implementation agreement specifically on seabed resource utilisation (Deep Seabed Implementation Agreement, 1994).[11] This agreement revised the exploitation arrangements under UNCLOS, distribution of proceeds, private participation, and technology transfer rules. The concessions underpinning the implementation agreement mainstreamed economic considerations into the law of the sea, this time designed around the International Seabed Authority (ISA). Under this regime, the Enterprise, an administrative organ under the ISA, was mandated to exploit deep seabed mineral resources on behalf of all humankind and distribute the proceeds to state parties (UNCLOS, 1982, artt. 153 and 170, Annex III and Annex IV). It was only after these compromises were made that the Global North was willing to sign on to UNCLOS, effectively ending a bitter and protracted ideological debate between the North and the South.[12] Thus, the motivation for accepting the international seabed regime under UNCLOS was directly related to the Area's enormous economic potential (Holmila, 2005).[13]

Yet, the economic prospects of the human uses of deep seabed resources have brought in tow a demand for resource exploitation accompanied by ecological risks in the Area. The ISA has a favourable disposition towards undersea mining, despite the associated ecological uncertainties (Drazen *et al.*, 2020). For instance, the Solwara 1 deep seabed project undertaken by Nautilus, a Canadian underwater mining company, in Papua New Guinea's Bismarck Sea provides a sneak preview of the ecological precarity of this novel frontier of anthropocentrism. While the Solwara 1 project has been largely unsuccessful, the cost–benefit analysis undertaken by states in respect of ocean mining suggests such decisions weigh in favour of economic interests and not ecological costs (Krutilla *et al.*, 2021). Similar undersea mining activities in Namibia and other states in the Global South confirm the uncertain character of this new form of extractivism even as marine scientists are warning us of irreversible ecological harm.[14] The ecological challenges associated with deep seabed mining include disruption in the marine environment's microbiology, contamination of marine fisheries, coral reef damage, and release of mining-induced sediment plumes (Ochoa, 2021).

Ongoing litigation also demonstrates courts are beginning to see more UNCLOS-related undersea mining cases. For example, the Supreme Court of New Zealand recently dismissed an appeal seeking to challenge the Court of Appeal's affirmation of an earlier decision of the High Court quashing the grant of mining consents issued by New Zealand's Environmental Protection Authority (EPA), which had authorised deep seabed mining in its exclusive economic zone (EEZ) (*Trans-Tasman Case* [2021]).[15] However, this reprieve may be short-lived since the New Zealand Supreme Court's decision did not fully halt deep seabed mining but only remitted the matter to the EPA for a reassessment of the environmental issues. Thus, this opens prospects for the mining corporation to reapply to New Zealand's EPA for reconsideration of its grant of mining consents (Makgill, Gardner-Hopkins, and Coates, 2020).

Trapped in this complicated politics of anthropocentrism and its tendency to overlook ecological risks, deep seabed mining continues to receive favourable reviews from states, private interests, and other interest groups (Reid, 2020). At the same time, actual implementation of environmental protection vacillates between the ISA and mining corporations. However, despite the perceived benefits of mining in the Area, these supposed profits may be easily 'offset by high degrees of uncertainty regarding its environmental impacts' (Childs, 2020, p. 197). Yet, it is unsurprising that despite this caution, human interests continue to feed the frenzied resource exploitation narrative as the inevitable circumstance of human existence.

Thus, CHM's ecological shortcomings reflect human domination with little care for ecological effects, even in the name of otherwise progressist and laudable attempts at resource exploitation. If a durable solution for protecting and safeguarding the Area's ecology is to emerge as an ethical endeavour, it must be informed by an understanding of how 'the politics of resource extraction' obscures environmental concerns (Childs, 2019, p. 4). Flowing from the above, environmental remediation can only emanate from a mindfulness to reorganise the present sociotechnical exchanges with the ocean. An extensive resource extraction agenda as an unavoidable outcome of this engagement is plainly unacceptable. Rather, protection of the ocean's integrity must assume a significant role in deep seabed mining (Kim, 2017).

To conclude this point, while some scholars look within CHM for answers, others have urged an expanded approach founded on CHM's promise (and its perils) by encouraging a broader reconsideration of the concept (Balsar, 1998). CHM's anthropocentrism must yield place to the idea of humans as stewards of the oceans. Here, the term 'common' as it occurs in CHM presents a unique opportunity for an expansive undertaking to rehabilitate UNCLOS by encouraging an ecological turn in ocean uses. Perhaps, the notion of 'commonness' might inform the integration of humans and non-human entities by igniting ethical and legal factors in assessing the ecological vulnerabilities associated with deep seabed mining.

Anthropocentrism via the Protection and Resource Conservation Regime of the Marine Environment under UNCLOS

The inclusion of the protection and the conservation of marine living resources under UNCLOS suggests the law of the sea is moving in a post-anthropocentric direction. The protection and preservation of the marine environment is under Part XII of UNCLOS, while conservation of marine living resources is under Section 2 of Part VII of UNCLOS (Scott, 2015, pp. 481–488). But as a matter of functional analysis, I discuss these two dimensions together, as they substantively speak to each other. As well, for practical purposes, the fluid flow of the oceans renders a conceptual distinction redundant, considering the present appeal for an integrated ecosystem-based approach to ocean governance (UNCLOS, 1982, artt. 192–196). Thus, the ensuing discussion of key UNCLOS provisions relative to sovereign rights and duties on the protection of the marine environment and conservation of marine life spotlights how these rights and duties apply to descriptions, arguments, and sources relevant to the anthropocentric character of the law of the sea.

Protecting the Marine Environment for Humans

Despite UNCLOS providing for the protection of the marine environment, the treaty does little to address the consequences of over-exploitation. More specifically, its general environmental obligations do not expressly itemise environmental principles (UNCLOS, 1982, Part XII). This gap has pushed law of the sea regulators, lawyers and scholars alike to look for inspiration for specific environmental guidance for ocean governance from other fields including international environmental law (Johnston and VanderZwaag, 2000). This task is complicated by the holdover of the discipline's vaunted regard for the Grotian free seas doctrine which has become a socio-technical and organisational obstruction to the healthy pursuit of the ocean's ecological integrity (Mossop, 2018).

Some scholars have revisited this topic by restating 'the law of the sea consolidated an extractive imaginary of the ocean' (Ranganathan, 2021, p. 164). Here, UNCLOS is a controlling force in the resource extraction-focused character of the ocean space. In response, it is argued that the law of the sea still provides for the protection of the marine environment (Kojima, 2017). Even so, UNCLOS is deeply anthropocentric and interested in the ocean for the benefit of human interests. In this respect, the deplorable state of the ocean's ecology is worsened by how interventions focusing on oceanic ecological matters are fiercely resisted by states seeking to subsume maritime concerns like piracy, robbery at sea, and similar threats, under the broad umbrella of marine environmental issues (Ventura, 2020). This action by some states undercuts and weakens the uniqueness of ecological damage, as it places the marine environment in keen competition with other important yet non-ecology specific matters.

Conservation of Marine Resources: In Whose Interest?

The conservation of marine living resources under UNCLOS is similarly anthropocentric since its organising ambition is still exploitation. The following discussion addresses two dimensions of this aspect of the law of the sea's anthropocentrism: maximum sustainable yield and the conservation of marine mammals. Either way, the law of the sea adopts a human-first approach by privileging resource exploitation over the maintenance of the sea's ecological integrity.

To begin, maximum sustainable yield is a deeply human-centric concept. Under this concept, a coastal state sets allowable limits of fish catch within its EEZ and the high seas (UNCLOS, 1982, artt. 61(3) and 119(1)(a)). This determination is the subject of the concept of maximum sustainable yield which is defined as the allowable fish catch at the highest level of extraction. The concept is linked to the free seas doctrine as it is premised on permissibility. This point is confirmed by Judge Higgins who observed that,

> [i]t has always been assumed that the freedom of the high seas also entailed a freedom to remove the resources that were found in those waters. Fishing on the high seas requires no one's permission because no one has title over the high seas.
>
> *(Higgins, 1995, p. 130).*

The concept of maximum sustainable yield was popularised in the 1958 Geneva Convention on Fishing and the Conservation of Living Resources on the High Seas (a predecessor treaty to UNCLOS). It set the course for the international life of this present anthropocentric method to marine fisheries conservation by embedding a human-centred focus over the marine ecosystem's integrity. Despite the responsibility UNCLOS places on a state to conserve fish stocks within that state's EEZ, it is apparent UNCLOS continues the 1958 Convention's human-centred essence, as it is concerned with the conservation of marine fisheries for a state's economic interests and not for the fish's intrinsic value nor for fish ecological integrity (UNCLOS, 1982, art. 116). But, beyond the EEZ, the major difficulty lies with conservation efforts on the high seas, since at international law, no state exercises exclusive jurisdiction on the high seas. This issue opens the concept of maximum sustainable yield up to differentiated state-to-state approaches to fishery resource utilisation. Here again, the stranglehold of anthropocentrism is palpable.

The anthropocentric character of maximum sustainable yield is further complicated by its single-fish focus. The conservation of marine fisheries under this concept siloed as the allowable catch is designed around a specific fish among several other fishes (Kumar, Pitcher, and Varkey, 2017). By reference to a particular fish, maximum sustainable yield adopts a linear conservation outlook which fails to consider the overall fishery ecosystem and the integrated character of marine life. By maximising the utilisation and extraction of a specific fish, states (and in practice, private capitalist interests) engage in unsustainable fishing practices that disrupt the ecological balance in the marine environment as these states and private economic interests contemplate specific fishery resources that are, to them, commercially viable or useful for human interests without considering the consequential oceanic disequilibrium.

The second concern lies with human–cetacean interrelations. The law of the sea enjoins states to conserve marine mammals (UNCLOS, 1982, artt. 65 and 120). Marine mammals face serious challenges including noise pollution, climate disruption, and death or injury occasioned as by-catch or through maritime accidents such as entanglement in fishing nets and undersea cables (Jefferies, 2016). Beyond human consumption of marine mammals and related uses, cetaceans are held in captivity for research and recreation. Thus, the treaty injunction in UNCLOS that seeks specially to protect marine mammals must be critically examined against a background of centuries of degeneration in a relationship that can best be described 'as grossly exploitative on the part of humans and negative in outcomes for cetaceans' (Allen, 2014, p. 31).

In tracking the genealogy of this human-cetacean interrelationship under international law, marine mammal conservation, since at least the adoption of 1946 International Convention for the Regulation of Whaling, has been described as an emergent principle of customary international law (Schiffman, 1996). Yet, this is not exactly sufficient progress. For example, attempts at restricting cetacean conservation to mere imposition of anti-whaling moratoria is a reductionist approach. Cetaceans in captivity remains a major concern as dolphins, sea lions, and whales are confined to small pens and cages for human recreation and for captive breeding. Regrettably, scientific evidence suggests marine mammals become aggressive in captivity and have, in some cases, injured or killed humans as has happened in the SeaWorld incidents in the United States (Anderson, Waayers, and Knight, 2016).

Marine mammals face other issues that are deemed not harmful to their health and socialisation. These activities include whale and dolphin watching at sea which is romanticised as eco-tourism and considered less invasive of the natural circumstances of these marine mammals. Today, countries like Canada and states like California in the United States have passed laws to address the cruel treatment of cetaceans, or directed freeing marine mammals from captivity, and are rewilding them with some degree of success. Yet still, these countries are encouraging new forms of human intervention with cetacean life under the guise of eco-tourism. If such activities, like whale watching, intensify, they may soon replace captive recreation as a new form of marine mammal disturbance with long-term consequences for marine life.

Flowing from the above, it is necessary to sustain ongoing efforts to scrutinise the connection between UNCLOS and human exploitation and how that relationship conditions the conservation of marine living resources for human interests. This reflection is critically important in the context of recent jurisprudence and subsequent action on the subject. For example, an analysis of the 2014 whaling decision of the International Court of Justice (ICJ) and later action by Japan demonstrates that, even after the ICJ's judgment, Japan justified its relentless resumption of commercial whaling under the guise of research whaling, which Japan has repeatedly described as 'sustainable exploitation of whale resources' (*Whaling in the Antarctic* [2014], para. 57; Fitzmaurice, 2015, p. 114).

Accordingly, marine conservation practices must be subjected to constant scrutiny and continually queried on why and how they routinely marginalise marine life. On this subject, some scholars argue it is not enough to focus on reforming the law of the sea – and that what is needed is emancipating the discipline from the binary moulds of 'subject/object and replace it with a broader space of plural subjectivities' (De Lucia, 2015, p. 114). This concern is seen in expressions like sustainable resource use, as in the case of maximum sustainable yield, or the human (ab)uses of cetaceans that are instilled with a 'human-centred development paradigm' (Kotzé and French, 2018, p. 28). Thus, an ethical imperative that is informed by a non-anthropocentric, multispecies justice framework which incorporates non-human entities into human relationships with the sea is plainly necessary.

Law of the Sea and Rights of Nature: The Sea as a Legal Person?

The analysis here focuses on the meaning attached to the concept of rights under UNCLOS and how that understanding fosters the law of the sea's anthropocentric outlook (UNCLOS, 1982, artt. 56, 77, 87, 90, 112, 142, 193, and 238). Like most treaties, UNCLOS is a bundle of rights and state duties at international law relative to the exploitation of the sea. While these rights are attached to states and not easily enforceable by or against an individual, the concept of rights under UNCLOS operates in practice at the intersection of 'human rights and the underlying concepts of community rights in general, or common cultural or environmental rights, common heritage of mankind under international law which gives rise to what may be properly called the right of the individual' (Cacciaguidi-Fahy, 2007, p. 87). In this sense, the notion of rights, in its diverse manifestations correspondence to UNCLOS, includes non-state-centric rights and is equally applicable to ocean governance.

Yet still, the mere mention of rights under UNCLOS evokes the inordinate power states wield over the oceans. This rights-based focus of UNCLOS is determined by human-centric dimensions that are not informed by ecological considerations. The rights-centric character of UNCLOS invites a critical review of the limits that the language of rights imposes on the ecological aspirations of ocean governance (Vidas *et al.*, 2015). Today, there is increasing support for a new vision of rights: that is, rights of nature. Rights of nature rejects the anthropocentricity of human rights and endows nature with legal personality. While there are some variations across what is described as rights of nature including 'Earth Jurisprudence', 'Wild Law', or 'Earth System Law', their diversity is all bound together by a common thread aimed at extending the concept of legal subjecthood to non-human entities and integrating ecological integrity into law's intervention in the environmental space.

In what has been termed a legal revolution, nature including trees, rivers, and mountains are now endowed as legal persons with rights. The world's oceans have also been contemplated in the context of the rights of nature (Harden-Davies *et al.*, 2020). This fundamental shift in legal thought challenges the anthropocentricity of environmental rights by substituting the objectification and commodification of nature with legal subjecthood for the sea. Rights of nature reject the perceived sufficiency of environmental rights by arguing environmental rights are rooted in human personhood. On the other hand, rights of nature invite a transformed understanding of the independent character and legal normativity of ecological integrity as a legal person, an idea that can be extended to the law of the sea (Pain and Pepper, 2021–22).

Importantly, rights of nature demand a rereading of neglected (often non-Western) ocean histories. This process involves recalling valuable yet abandoned concepts and lessons for the reconstruction of the human–sea interface in a bid to elevate ecological considerations to the disciplinary core of the law of the sea. Taking this understanding into account, a revival of these ecological concepts informs alternative modes of maintaining marine ecological integrity. These alternate pathways to securing the ocean's integrity do not reject the role of rights in this disciplinary discourse. Rather, the idea has been to adapt the language of rights in advancing ecological integrity. Accordingly, in the move to return peripheralised notions of legal normativity to law's centre, Indigenous influences on a future right of the sea are urgently needed to broaden the discourse (De Lucia, 2019; Harden-Davies *et al.*, 2020; Mortiaux, 2021).

Drawing on these ideas, some international law scholars even go further by urging a complete remake of international law. They argue for reviving sources of traditional knowledge capable of renewing environmental governance (Kotzé, 2014). This view also invites an interaction between the language of rights and Indigenous legal perspectives. However, this normative exploration is still at a very early stage, as mere incorporation of a right of the sea into rights of nature without matching political commitment to mainstreaming ecological integrity is insufficient to impel much-needed change. Yet still, there is good reason to explore neglected legal impulses in Indigenous (and traditional) knowledge and customary law and how these historically marginalised sources of law can breathe a new lease of life into the law of the sea through a re-interpretation of UNCLOS or through new legal instruments for implementing the treaty.

These developments suggest the sea is a new frontier for this novelty of attaching rights to nature. Yet it is noteworthy that this development is not altogether novel considering in some non-Western societies the sea has always enjoyed legal normativity as part of the social reality of those polities. In those societies, Indigenous legal ontologies have long cloaked non-human phenomena with legal subjecthood (Enyew, Poto, and Tsiouvalas, 2021). For example, the sea is envisioned as a living being deserving of reverence and social protection by the Indigenous peoples of Ecuador, New Zealand, and across Africa (Tănăsescu, 2020). These Indigenous legal impulses can also be found in the Global North as is observed in places like Canada, where the ethical ties between Indigenous legal systems and ocean governance are being integrated into national truth and reconciliation efforts (Paul, 2019).

Despite these normative advances, the ecological integrity of nature including the oceans need not necessarily be styled as a right before it is capable of being recognised as a legal subject. The future of the ocean's ecological integrity must transcend the bifurcated anthropocentric–ecocentric rights dichotomy and move ahead of 'ecological timescales in order to align human affairs with Earth's geological timescales' (Kotzé and Kim, 2019, p. 7). It is in this sense that a future-facing outlook on ecological integrity in the law of the sea must envisage an inherent right of the sea as *only* one of many pathways to achieving the purposes of Part XII, Section 2 of Part VII, and other UNCLOS provisions on marine environment as well as related implementation agreements adopted under the guidance of UNCLOS and allied treaties.

Beyond Historical Legacies: Future of the Law of the Sea

The law of the sea's history and its many legacies demonstrate its inherent anthropocentrism even in contemporary times. If the law of the sea must shed its historical and metaphorical deadweight, it is only appropriate to revisit its origins to derive some guidance on how to retrench its deep human-centredness. As a result, a critical review of the law and politics of ocean governance, and how that anthropocentric heritage manifests in the contemporary life of the discipline, are more urgent than ever.

A starting point in querying this history and accompanying legacies is to examine the resistance to the exclusive genealogy of Grotius as the originator of the free seas doctrine. On this subject, some Third World Approaches in International Law (TWAIL) scholars challenge the theoretical inspiration of the Grotian *Mare Liberum*, which excluded the maritime histories of non-European littoral polities (Boczek, 1984; Gathii, 2011; Enyew, 2022). In this respect, Ram Prakash Anand argued that:

> All through these centuries, there is no doubt about the freedom of navigation and commercial shipping which was experienced by various countries and peoples in the eastern seas, and which led to the development of a number of *entrepós* and trade centres. In south-east Asia, although the Sri Vijaya kings remained the strongest power in the area and more or less lords of the ocean, freedom of the seas and navigation was never interfered with, controlled or monopolised by anybody.
>
> *(Anand, 1981, p. 444)*

From Anand's viewpoint, an alternative argument is that the doctrine existed in practice prior to the Grotian construction. For Anand and other critical Third World scholars, *Mare Liberum* merely crystallised the practice within the European intellectual tradition; but, in doing so, it served as the basis for extending Eurocentrism into the realms of anthropocentrism and ultimately into the modern law of the sea.

In this reflexive consideration of the law of the sea's disciplinary history, it is apparent that ever since the formal processes for adopting the modern law of the sea begun in the early 1960s, legal scholars have kept pace with the progress and development of the discipline. During this time, critical legal scholars including those of the TWAIL tradition provided much-needed impetus for the Global South's UNCLOS III negotiations as the South joined significant aspects of the law of the sea's design to the principal plank of its primary international negotiation platform, the New International Economic Order (NIEO) (Fidler, 2003). Even long after the United Nations General Assembly adopted the historic resolution on the NIEO, TWAIL scholars continued to review the Grotian doctrine as the historical defence for the ecological despoliation of the oceans, even to this day (UNGA, 1974; Esmeir, 2017). Two strands of that critique deserve attention as they emerged as new areas of interest under UNCLOS: that is, the exploitation of marine genetic resources (bioprospecting) and biodiversity beyond national jurisdiction (BBNJ). These two concerns engage the attention of both classical and critical scholars considering their potential anthropocentric impacts.

The first concern is with bioprospecting. Defying easy definition, bioprospecting refers to commercial exploitation of biological and genetic resources in the marine environment. No legal conundrum arises with bioprospecting in territorial waters, as it is regulated under the jurisdiction of a coastal state (UNCLOS, 1982, art. 2). It is bioprospecting in areas beyond national jurisdiction, notably on the deep seabed and to an extent the superjacent water column (that is, the ocean waters above the deep seabed) that raise concerns since UNCLOS does not expressly provide for this kind of exploitation (Scovazzi, 2013).

Two different schools of thought seem to address this issue. First, bioprospecting may be regulated as CHM under Part XI of UNCLOS; or second, under Part VII of UNCLOS, subject to the freedom of the high seas (Tladi, 2019). In the first instance, bioprospecting within the remit of Article 143 on marine scientific research denotes anthropocentric purposes (Leary, 2018–19). But Part XI's reference to 'benefit of mankind as a whole' reignites concerns over anthropocentrism and revives ideological anxieties over the law of the sea; this time, the concern is that CHM discourse is revived through bioprospecting (UNCLOS, 1982, art. 136; Yu, 2020). Second, the regulation of marine areas beyond national jurisdiction under Part VII of UNCLOS once more puts the exploitation of these marine resources within the domain of freedom of the high seas. While Article 118 of UNCLOS enjoins state parties to protect, preserve, and conserve marine living resources beyond national jurisdiction, its core interest is to secure marine biodiversity for further human uses. Notably, both schools of thought seem inadequate to the task, since marine resources within CHM's purview do not expressly reference bioprospecting; and neither does UNCLOS directly define bioprospecting within the context of the freedom of the high seas (UNCLOS, 1982, artt. 87 (1) and 133(1)). Nonetheless, both approaches emphasise a common objective suggesting bioprospecting can be regulated under UNCLOS, possibly within the context of marine scientific research (UNCLOS, 1982, artt. 238–240).

On the question of BBNJ, the focus is on the conservation and sustainable use of marine biodiversity beyond areas of national jurisdiction. The resources contemplated under BBNJ are particularly valuable as they account for nearly forty percent of the total planetary surface (Burnett and Carter, 2018). Under the Grotian free seas doctrine, marine biodiversity was exploited freely in much the same way as the oceans were navigated freely. In response to increasing marine biodiversity loss, a new implementation agreement has just been negotiated, in March 2023, under the auspices of UNCLOS to regulate marine biodiversity exploitation in areas beyond national jurisdiction. While the agreement is yet to be adopted by states, it is important to track the coverage of human interests in the negotiation of the treaty text. For example, some scholars have raised concerns over the Food and Agriculture Organisation's participation in the negotiations over fears that it was there to ensure the final text of the BBNJ implementation agreement contemplates and caters for the economic interests implicated under international fisheries law (Telesetsky, 2020). So, even where the draft BBNJ implementation agreement provides for stewardship, such efforts are directed at securing prospective uses for future (human) generations. An example is seen in how the guiding principles of the draft agreement highlight CHM while simultaneously joining 'freedom of marine research' to 'other freedoms of the high seas' (Draft BBNJ Implementation Agreement, 2023, Preamble, art. 5 (b) and (b) bis). Clearly, the proposed BBNJ implementation agreement only continues the law of the sea's anthropocentric pedigree.

The present task then is how the BBNJ agreement interacts with bioprospecting, given the jurisdictional contestations between states who support bioprospecting regulation as CHM and those who favour its regulation under the freedom of the high seas (Rabone *et al.*, 2019; Vadrot, Langlet, and Tessnow-von Wysocki, 2022). To this end, a broadly construed environmental stewardship constructed on the common obligation of state parties to preserve and protect the environment under Article 192 of UNCLOS has been suggested as an alternative route for reconciling these two extremes of the debate without jeopardising the existing uses of marine biodiversity (Ridings, 2018). Yet still, before the BBNJ agreement becomes binding treaty, the United Nations General Assembly prospectively limited the scope of this agreement in a 2015 resolution, which requires the BBNJ agreement not to 'undermine existing relevant legal instruments and frameworks and relevant global, regional, and sectoral bodies' (UNGA, 2015). Evidently, these are not ecological considerations but human interests being protected in advance.

If the BBNJ agreement will eventually elevate ecological concerns, its implementation must radically change the prevailing rules of the law of the sea as relates to marine biodiversity beyond national jurisdiction. Regrettably, the language of this 2015 United Nations General Assembly resolution and the text of the BBNJ agreement suggest the agreement is not likely to be any different from its predecessor implementation agreements that focused on human uses like deep seabed mining and utilisation of straddling and migratory fish stocks. In recalibrating the contours of bioprospecting regulation, or as the world moves closer to regulating marine bioviersity under the BBNJ agreement, the sea's ecological integrity faces an existential threat from human activities in an already fragile situation impacted by human-induced climate change. Thus, the overarching concern is that the law of the sea might well have ended up with yet another implementation agreement that is nothing more than a set of self-interested human objectives styled as progress and development in international law.

Conclusion

The geographical constitution of international law is more pronounced and observable through the sea, as the oceans cover a greater part of the Earth's surface. Relating this observation to the subject matter of this chapter, international law needs to acknowledge and address its complicity in the ocean's ecological deterioration. As the constitution for the oceans, the task should be determining how the law of the sea can be afforded new interpretations or methods of implementation that help it resonate with contemporary approaches to less anthropocentric impulses and is more attuned to ecological governance. To do so effectively, the law of the sea must quickly move ahead as a 'technology in prompting, sustaining, and potentially managing' the environment (Viñuales, 2018, p. 14). Taking this concern into account, international lawyers and scholars must reconsider the purpose of the law of the sea by first querying its deep anthropocentrism. An alternative proposal for the law of the sea must outdistance human-focused interests and centre cultural and Indigenous legal imperatives in reorganising and privileging the ocean's ecological integrity (Lixinski, McAdam, and Tupou, 2022).

A better way of transforming the law of the sea involves a fundamental shift from unrestrained resource extraction while emphasising the centrality and immanent value of the ocean's ecological integrity. In this respect, international law contemplates new forms of law-making for the benefit of the ocean's integrity. However, due consideration must be accorded the complications associated with the prospects of a total remake of UNCLOS. The complex negotiation history of UNCLOS, its adoption, and coming into force are all marked by many difficult compromises. Thus, an attempt to quickly thrust aside UNCLOS entirely might be plainly naive and a failure to acknowledge and appreciate the decades-long struggle it took state parties to finalise the text of that treaty. Taking this history into consideration, it is unlikely that a new, comprehensive treaty like UNCLOS can be made or UNCLOS even amended substantially without risking potentially chaotic negotiations that may never be concluded or that may completely unravel the many compromises that would be harder to reach today, and further risk introducing new technologies of anthropocentric governance that are so often disguised as global commitment to the protection of the sea.

A more practical pathway lies with implementation agreements and creative interpretation of existing treaty provisions. But these agreements must also propose substantial changes for directing an effective and radical transformation of the law of the sea guided by a disposition to elevate oceanic ecological care as a global concern. Another route to mainstreaming ecological integrity is through international jurisprudence. For example, the International Tribunal for the Law of the Sea (ITLOS) places great emphasis on scientific evidence in its interpretive function. This could well be grounds for introducing evidence in support of the ocean's ecological integrity as an inherent value. However, international lawyers and scholars must be wary of international law's power to immobilise progress even as it pursues change. For example, while the ICJ has held that the protection and preservation of the marine environment are international obligations, and such pronouncements seem to support marine environmental objectives by declaring them as obligations *erga omnes partes*, these are state-to-state obligations and do not create direct duties relative to the ocean's integrity (*Legality of the Threat or Use of Nuclear Weapons* [1966]; *Seabed Mining Advisory Opinion* [2011]).

A failure by international lawyers, judges, and other law of the sea experts to unpack these judicial interventions through critical reviews to ensure these decisions are interpreted for the benefit of marine ecology itself will only deepen the law of the sea's anthropocentrism. Additionally, international lawyers, judges, arbitrators, Indigenous communities, and their leaders, as well as other experts must collectively expand the interpretations of relevant terms and expressions used in the law of the sea such as the 'substantive content of obligations to "conserve", "preserve", or "protect" as well as obligations to make "wise" or "sustainable" use of certain resources' to support and advance the sea's ecological sensibilities (Vordermayer, 2015). These terms and expressions should be adapted to challenge the ecological issues facing the oceans. Thus, the jurisprudence on the law of the sea must be reorganised from an ecological integrity perspective where a primary duty is first owed to the sea itself before other (including human) considerations are made (De Lucia, 2019).

More than anything else, international law scholars and lawyers, together with allied specialists in fields like marine sciences, must all break out of their siloed approaches and start becoming more of each other and working together for better and healthier oceans. For example, law of the sea scholars, blue humanities scholars, and other experts must work together and with Indigenous communities for a deeper and more wholesome appreciation of the ecological interrelations in the ocean space (Reid, 2020). It is this shared understanding that should guide human society to return to the original purpose of the modern law of the sea – a stewardship that elevates the commitment to protecting, preserving, and conserving marine life and its ecology over the perils of an unbridled human exploitation.

Human society bears an ethical responsibility to protect the ocean for its ecological value. This duty is premised on the principled foundation for a renewed (yet radical) approach to safeguarding the ocean's ecological integrity. After all, if the law of the sea is truly *about* the sea, then human action must reflect on *our relationship* with the ocean not as an exploitable resource but in terms of our connection with the ocean as 'unified ecological commons' of which humans are only a small part (Taylor, 2019, p. 166). The maintenance of the ocean's ecological integrity is a primary duty – *our* law of the sea. Human well-being is thus only optimised when humans commit first to guaranteeing ecological well-being. I surmise Arvid Pardo would equally agree efforts to address the law of the sea's anthropocentrism must be influenced by the renewed interrelations between humans and the ocean if we are to avert irreversible ecological harm to the sea, and ultimately to ourselves.

Notes

1 Assistant Professor, Faculty of Law, University of Alberta, Edmonton (Canada). BA, LLB (University of Ghana); QCL (Ghana School of Law); LLM (Harvard); PhD (University of British Columbia). Many thanks to Natasha Affolder, Asad Kiyani, Cameron Jefferies, and other reviewers for their invaluable comments on earlier drafts, and to Gideon Gabor for his excellent research assistance.
2 I use the 'sea' and 'ocean' interchangeably.

3 According to European sources, the history of the law of the sea is traced to the controversial conceptual dividing line drawn through the Atlantic Ocean by Pope Alexander VI in his infamous papal bull, *Inter Caetera*, issued in 1493. The Pope placed all seas and 'newly-discovered' lands westward under Spanish control and eastward under Portugal. Spain and Portugal formalised this papal edict in the Treaty of Tordesillas in 1494. This singular act by the Pope in the fifteenth century, which was endorsed by Portugal and Spain, underlies much of the historical foundations of the law of the sea.

4 This chapter adopts a nuanced view of anthropocentrism. Whereas it is broadly used to encompass human action, it is crucial to keep in mind that a blanket approach can be misleading, as anthropocentrism treats different humans differently based on several categories, like race, gender, and social class. As Ollino (2019, p. 205) has rightly suggested:

> Rather than putting humanity as a whole at the centre of the epistemological discourse, anthropocentrism selects only certain human beings as the beneficiaries of the quality of humans. Those who do not hold up to the standard are placed at the periphery of the system and treated as 'the others'. This is why any critical engagement with anthropocentrism and the humanist core underpinning the tenets of the international legal system must be attentive to feminist and post-colonial arguments, which help clarify the distinctiveness of the centred-human subject.

5 While not excusing Pardo for this anthropocentric disposition, Klaus Bosselmann argues that Pardo was speaking against the backdrop of a specific socio-political context characterised by decolonisation and the rise of the Global South in the international system. As was evident through decolonisation, these new newly independent states were intent on achieving formal state equality at international law through equal participation in resource utilisation. I will address much of this in later sections of this chapter.

6 Simply known as *Mare Clausum* (the long title means, 'A Closed Sea Follows from Maritime Sovereignty').

7 In relation to the Area, CHM was envisaged as the basis for equitable resource exploitation between the North and South.

8 Wolfrum challenges this widely held view that Pardo is the originator of CHM by suggesting Ambassador Cocca of Argentina deserves credit for this development.

9 Pardo's statement provided much-needed inspiration for the Global South in the context of the wave of decolonisation. For these Global South states, CHM was yet another frontier it had to overcome and decolonise in its mission to reorient the prevailing unequal global economic order.

10 Article 1(1) of UNCLOS defines the Area as 'the seabed and ocean floor and subsoil thereof, beyond the limits of national jurisdiction'.

11 Undersea mining has been a subject of interest for states and international law at least since the 1945 Truman Proclamation, which unilaterally extended American territory into the sea space with an eye for seabed mineral resources like polymetallic (manganese) nodules.

12 This subject has generated some interesting debates. Some argue the 1994 implementation agreement compromised Global North interests while others contend the implementation agreement was substantially altered to accommodate the interests of states of the Global North and allied private entities. Whichever way one looks at it, the agreement was very much about Global North interests.

13 Holmila estimated an annual turnover of USD 1 trillion.

14 Quite apart from the ecological risks, a greater part of undersea mining exploration is taking place in the Global South. This poses another level of risk to these countries that are already struggling with historical ecological problems arising from land-based pollution and climate change.

15 With regards to maritime zone delimitation, Article 55 of UNCLOS defines the EEZ as, 'an area of the beyond and adjacent to the territorial sea'. The economic and regulatory significance of the EEZ is enormous as states have greater access to fishing, exploitation of petroleum and other mineral resources, international shipping regulation, laying of undersea cables, and application of environmental laws, and the conduct of marine research.

References

Allen, S.J. (2014) 'From Exploitation to Adoration: The Historical and Contemporary Context of Human–Cetacean Interactions', in Higham, J., Bejder, L. and Williams, R. (eds.) *Whale-Watching: Sustainable Tourism and Ecological Management*. Cambridge: Cambridge University Press, pp. 31–47.

Anand, R.P. (1981) 'Maritime Practice in South-East Asia until 1600 A.D. and the Modern Law of the Sea', *International and Comparative Law Quarterly*, 30(2), pp. 440–454. DOI:10.1093/iclqaj/30.2.440.

Anand, R.P. (1983) *Origin and the Development of the Law of the Sea*. The Hague: Martinus Nijhoff.

Anderson R., Waayers, R. and Knight, A. (2016) 'Orca Behaviour and Subsequent Aggression Associated with Oceanarium Confinement', *Animals*, 6(8), pp. 49–63. DOI: 10.3390/ani6080049.

Balsar, K. (1998) *The Concept of the Common Heritage of Mankind in International Law*. The Hague: Kluwer Law.

Baxi, U. (2019) 'Intergenerational Justice, Water Rights, and Climate Change', in Cullet, P. and Koonan, S. (eds.) *Research Handbook on Law, Environment and the Global South*. Cheltenham; Northampton: Edward Elgar Publishing, pp. 2–13.

Boczek, B.A. (1984) 'Ideology and the Law of the Sea: The Challenge of the New International Economic Order' *Boston College International and Comparative Law*, 7(1), pp.1–30.

Boer, R. (2021) 'Hugo Grotius: On Freedom of the Seas and Human Nature', in Slotte, P. and Haskell, J. D. (eds.) *Christianity and International Law: An Introduction*. Cambridge: Cambridge University Press, pp. 139–152.

Bosselmann, K. (2015) *Earth Governance: Trusteeship of the Global Commons*. Cheltenham; Northampton: Edward Elgar Publishing.

Burnett, D.R. and Carter, L. (2018) *International Submarine Cables and Biodiversity of Areas Beyond National Jurisdiction*. Leiden: Brill Research Perspectives.

Cacciaguidi-Fahy, S. (2007) 'The Law of the Sea and Human Rights', *Sri Lanka Journal of International Law*, 19(1), pp. 85–108.

Childs, J. (2019) 'Greening the Blue? Corporate Strategies for Legitimising Deep Sea Mining', *Political Geography*, 71, pp. 1–12. DOI: 10.1016/j.polgeo.2019.102060.

Childs, J. (2020) 'Extracting in Four Dimensions: Time, Space, and the Emerging Geo(-)Politics Deep Seabed Mining', *Geopolitics*, 25(1), pp. 189–213. DOI: 10.1080/14650045.2018.1465041.

Crutzen, P. (2006) 'The "Anthropocene'", in Ehlers, E. and Krafft, T. (eds.) *Earth System Science in the Anthropocene: Emerging Issues and Problems*. Heidelberg: Springer, pp. 13–18.

De Lucia, V. (2015) 'Competing Narratives and Complex Genealogies: The Ecosystem Approach to International Environmental Law', *Journal of Environmental Law*, 27(1), pp. 91–117. DOI: 10.1093/jel/equ031.

De Lucia, V. (2019) 'Ocean Commons, Law of the Sea, and Rights of the Sea', *Canadian Journal of Law and Jurisprudence*, 32(1), pp. 45–57. DOI:10.1017/cjlj.2019.2.

Draft Agreement under the United Nations Convention on the Law of the Sea on the Conservation and Sustainable Use of Marine Biological Diversity of Areas Beyond National Jurisdiction (06 March 2023) [Online]. Available at: https://www.un.org/bbnj/sites/www.un.org.bbnj/files/draft_agreement_advanced_unedited_for_posting_v1.pdf. (Accessed: 07 March 2023).

Drazen J.C., Smith, C.R., Gjerde, K.M., Haddock, S.H.D., Carter, G.S., Choy, C.A., Clark, M.R., Dutrieux, P., Goetze, E., Hauton, C., Hatta, M., Koslow, J.A., Leitner, A.B., Pacini, A., Perelman, J.N., Peacock, T., Sutton, T.T., Watling, L. and Yamamoto, H. (2020) 'Midwater Ecosystems Must Be Considered When Evaluating When Evaluating Risks of Deep-Sea Mining', *Proceedings of the National Academy of Sciences*, 117(30), pp. 17455–17460. DOI: 10.1073/pnas.2011914117.

Duarte, R.T. (2012) 'Maritime History in Mozambique and East Africa: The Urgent Need for the Proper Study and Preservation of Endangered Underwater Cultural Heritage', *Journal of Maritime Archaeology*, 7, pp. 63–86. DOI: 10.1007/s11457-012-9089-6.

Enyew, E.L. (2022) 'Sailing with TWAIL: A Historical Inquiry into Third World Perspectives on the Law of the Sea', *Chinese Journal of International Law*, 21(3), pp. 439–497. DOI: 10.1093/chinesejil/jmac028.

Enyew, E.L., Poto, M.P. and Tsiouvalas, A. (2021) 'Beyond Borders and States: Modelling Connectivity According to Indigenous Cosmovisions', *Arctic Law Review on Law and Politics*, 12, pp. 207–221. DOI: 10.23865/arctic.v12.3290.

Esmeir, S. (2017) 'Bandung: Reflections on the Sea, the World, and Colonialism' in Eslava, L., Fakhri, M. and Nesiah, V. (eds.) *Bandung, Global History, and International Law: Critical Pasts and Pending Futures*. Cambridge: Cambridge University Press, pp. 81–94.

Fidler, D. (2003) 'Revolt Against or from Within the West? TWAIL, the Developing World, and the Future of International Law', *Chinese Journal of International Law*, 2(1), pp. 29–76. DOI: 10.1093/oxfordjournals.cjilaw.a000472.

Fitzmaurice, M. (2015) *Whaling and International Law*. Cambridge: Cambridge University Press.

Fletcher, E.G.M. (1933) 'John Selden (Author of Mare Clausum) and His Contribution to International Law', *Transactions of the Grotius Society*, 19, pp. 1–12.

Gathii, J. T. (2011) 'TWAIL: A Brief History of Origins, Its Decentralised Network, and a Tentative Bibliography', *Trade, Law and Development*, 3(1), pp. 26–64.

Grotius, H. (1609) *The Free Sea*. Armitage, D. (ed.). Translated by R. Hakluyt. Indianapolis: Liberty Fund, 2004.

Hafetz, Jonathan L. (2000) 'Fostering Protection of the Marine Environment and Economic Development: Article 121(3) of the Third Law of the Sea Convention', *American University International Law Review*, 15(3), pp. 586–637.

Harden-Davies, H., Humphries, F., Maloney, M., Wright, G., Gjerde, K. and Vierros, M. (2020) 'Rights of Nature: Perspectives for Global Ocean Stewardship', *Marine Policy*, 122, pp. 104059–104069. DOI: 10.1016/j.marpol.2020.104059.

Higgins, R. (1995) *Problems and Process: International Law and How to Use It*. Oxford: Oxford University Press.

Holmila, E. (2005) 'Common Heritage of Mankind in the Law of the Sea', *Acta Societatis Martensis*, 1, pp. 187–205.

Jefferies, C.S.G. (2016) *Marine Mammal Conservation and the Law of the Sea*. Oxford: Oxford University Press.

Johnston, D.M. and Vander Zwaag D. (2000) 'The Ocean and International Environmental Law: Swimming, Sinking and Treading Water at the Millennium', *Ocean and Coastal Management*, 43(2–3), pp. 141–161. DOI: 10.1016/S0964-5691(99)00070-8.

Kim, R.E. (2017) 'Should Deep Seabed Mining Be Allowed?', *Marine Policy*, 82, pp. 134–137. DOI: 10.1016/j.marpol.2017.05.010.

Koh, T.T.B. (1987) 'The Origins of the 1982 United Nations Convention on the Law of the Sea', *Malaya Law Review*, 29(1), pp. 1–17.

Kojima, C. (2017) 'South China Sea Arbitration and the Protection of the Marine Environment: Evolution of UNCLOS Part XII Through Interpretation and the Duty to Cooperate', *Asian Yearbook of International Law*, 21, pp. 166–180. DOI: 10.1163/9789004344556_010.

Kotzé, L.J. (2014) 'Rethinking Global Environmental Law and Governance in the Anthropocene', *Journal of Energy and Natural Resources*, 32(2), pp. 121–156. DOI: 10.1080/02646811.2014.11435355.

Kotzé, L.J. and French, D. (2018) 'The Anthropocentric Ontology of International Environmental Law and Sustainable Development Goals: Towards an Ecocentric Rule of Law in the Anthropocene', *Global Journal of Comparative Law*, 7, pp. 5–36. DOI: 10.1163/2211906X-00701002.

Kotzé, L.J. and Kim, R.E. (2019) 'Earth System Law: The Juridical Dimensions of Earth System', *Earth System Governance*, 1, pp. 1–12. DOI: 10.1016/j.esg.2019.100003.

Krutilla, K., Good D., Toman M. and Arin T. (2021) 'Addressing Fundamental Uncertainty in Benefit-Cost Analysis: The Case of Deep Seabed Mining', *Journal of Benefit-Cost Analysis*, 21(1), pp. 122–151. DOI: 10.1017/bca.2020.28.

Kumar, R., Pitcher, T. J. and Varkey, D. A. (2017) 'Ecosystem Approach to Fisheries: Exploring Environmental and Trophic Effects on Maximum Sustainable Yield Reference Point Estimates', *PLoS One*, 12(9), pp. 1–19. DOI: 10.1371/journal.pone.0185575.

Lalöe, A-F. (2016). *The Geography of the Ocean: Knowing the Ocean as a Space*. Abingdon: Routledge.

Leary, D. (2018–19) 'Marine Genetic Resources in Areas Beyond National Jurisdiction: Do We Need to Regulate Them in a New Agreement?', *Maritime Safety and Security Law Journal*, 5, pp. 22–47. DOI: 10.3389/fmars.2021.667274.

Legality of the Threat or Use of Nuclear Weapons (Advisory Opinion) [1996] ICJ Rep 226.

Lesaffer, R. (2016) 'Selden: The Closure of the Sea', in Dauchy, S., Martyn, G., Musson, A., Pihlajamäki, H. and Wijffels, A. (eds.) *The Formation and Transmission of Western Legal Culture: 150 Books that Made the Law in the Age of Printing*. Cham: Springer, pp. 190–194.

Lixinski, L., McAdam J. and Tupou P. (2022) 'Ocean Cultures, The Anthropocene and International Law: Cultural Heritage and Mobility Law as Imaginative Gateways', *Melbourne Journal of International Law*, 23(1), pp. 1–22.

Makgill, R.A., Gardner-Hopkins, J.D. and Coates, N.R. (2020) 'Trans-Tasman Resources Limited v. Taranaki-Whanganui Conservation Board', *International Journal of Marine and Coastal Law*, 35(4), pp. 835–845. DOI: https://doi.org/10.1163/15718085-BJA10036.

Mawani, R. and Prange S. (2021) 'Unruly Ocean: Law, Violence, and Sovereignty at Sea', *TWAILR Reflections*, 27, pp. 1–8.

Mickelson, K. (2014) 'The Maps of International Law: Perceptions of Nature in the Classification of Territory', *Leiden Journal of International Law*, 27(3), pp. 621–639. DOI: 10.1017/S0922156514000235.

Mickelson, K. (2019) 'Common Heritage of Mankind as a Limit to Exploitation of the Global Commons', *European Journal of International Law*, 30(2), pp.635–663. DOI: 10.1093/ejil/chz023.

Mortiaux, R. (2021) 'Righting Aotearoa's Coastal Marine Area: A Case for Legal Personhood to Enhance Environmental Protection', *Griffiths Law Review*, 30(3), pp.413–437. DOI: 10.1080/10383441.2021.2003743.

Mossop, J. (2018) 'Can We Make the Oceans Greener? The Successes and Failures of UNCLOS as an Environmental Treaty', *Victoria University of Wellington Law Review*, 49(4), pp. 573–593. DOI: 10.26686/vuwlr.v49i4.5341.

Ochoa, C. (2021) 'Contracts on the Seabed', *Yale Journal of International Law*, 46(1), pp. 103–153.

Ollino, A. (2019) 'Feminism, Nature and the Post-Human: Toward a Critical Analysis of the International Law of the Sea Governing Marine Living Resources Management', in Papanicolopulu, I. (ed.) *Gender and the Law of the Sea*, Leiden: Brill, pp. 204–228.

Pain, N. and Pepper, R. (2021–22) 'Can Personhood Protect the Environment? Affording Legal Rights to Nature', *Fordham International Law Journal*, 45, pp. 315–378.

Pardo, A. (1975) *The Common Heritage: Selected Papers on Oceans and World Order 1967–1974.* Valetta: Malta University Press.

Paul, K. (2019) 'First Nations, Oceans Governance and Indigenous Knowledge Systems', in International Ocean Institute – Canada (ed.) *The Future of Ocean Governance and Capacity Development: Essays in Honor of Elisabeth Mann Borgese (1918–2002).* Leiden: Brill, pp. 46–52.

Porras, I. (2011) 'Appropriating Nature: Commerce, Property, and the Commodification of Nature in the Law of Nations', *Leiden Journal of International Law*, 27(3), pp. 641–660. DOI: 10.1017/S0922156514000247.

Rabone, M., Harden-Davies, H., Collins, J.E., Zajderman, S., Appeltans, W., Droege, G., Brandt, A., Pardo-Lopez, L., Dahlgren, T.G., Glover, A.G. and Horton, T. (2019) 'Access to Marine Genetic Resources (MGR): Raising Awareness of Best-Practice Through a New Agreement for Biodiversity Beyond National Jurisdiction (BBNJ)', *Frontiers in Marine Science*, 6, pp. 1–22. DOI: 10.3389/fmars.2019.00520.

Ranganathan, S. (2021) 'Decolonisation and International Law: Putting the Ocean on the Map', *Journal of the History of International Law*, 23(1), pp. 161–183. DOI:10.1163/15718050-12340168.

Reid, S. (2020) 'Solwara 1 and the Sessile Ones', in Braverman, I. and Johnson, E. R. (eds.) *Blue Legalities: The Life and Laws of the Sea.* Durham: Duke University Press, pp. 25–44.

Responsibilities and Obligations of States Sponsoring Persons and Entities with Respect to Activities in the Area (Advisory Opinion) [2011] 50 ILM 458.

Ridings, P. (2018) 'Redefining Environmental Stewardship to Deliver Governance Frameworks for Marine Biodiversity Protection Beyond National Jurisdiction', *ICES Journal of Marine Science*, 75(1), pp. 435–443. DOI: 10.1093/icesjms/fsx122.

Schiffman, H.S. (1996) 'The Protection of Whales in International Law: A Perspective for the Next Century', *Brooklyn Journal of International Law*, 22(2), pp. 303–360.

Schrijver, N. (2016) 'Managing the Global Commons: Common Good or Common Sink?', *Third World Quarterly*, 37(7), pp. 1252–1267. DOI: 10.1080/01436597.2016.1154441.

Scott, K.N. (2015) 'Integrated Oceans Management: A New Frontier in Marine Environmental Protection', in Rothwell, D., Elferink, A.O., Scott, K. and Stephens, T. (eds.) *Oxford Handbook of the Law of the Sea.* Oxford: Oxford University Press, pp. 463–490.

Scovazzi, T. (2013) 'Open Questions on the Exploitation of Genetic Resources in the Areas Beyond National Jurisdiction', *ASIL Proceedings of the Annual Meeting*, 107, pp. 119–122. DOI: 10.5305/procannmeetasil.107.0119.

Somos, M. (2012) 'Selden's *Mare Clausum*: The Secularisation of International Law and the Rise of Soft Imperialism', *Journal of the History of International Law*, 14(2), pp. 287–330. DOI: 10.1163/138819912X 13333544461551.

Somos, M. (2020) 'Open and Closed Seas: The Grotius-Selden Dialogue at the Heart of Liberal Imperialism', in Cavanagh, E. (ed.) *Empire and Legal Thought: Ideas and Institutions from Antiquity to Modernity.* Leiden: Brill, pp. 322–361.

Steinberg, P.E. (1996) 'Three Historical Systems of Ocean Governance: A Framework for Analysing the Law of the Sea', *World Bulletin: Bulletin of the International Studies of the Philippines*, 12(5–6), pp. 1–19.

Taylor, P. (2019) 'The Common Heritage of Mankind: Expanding the Oceanic Circle', in International Ocean Institute – Canada (ed.) *The Future of Ocean Governance and Capacity Development: Essays in Honor of Elisabeth Mann Borgese (1918–2002).* Leiden: Brill Nijhoff, pp. 142–150.

Tănăsescu, M. (2020) 'Rights of Nature, Legal Personality, and Indigenous Philosophies', *Transnational Environmental Law*, 9(3), pp. 429–453. DOI: 10.1017/S2047102520000217.

Telesetsky, A. (2020) 'UN Food and Agriculture Organisation: Exercising Legal Personality to Implement the UN Convention on Law of the Sea', Ribeiro, M. C., Bastos, F.L. and Henriksen, T. (eds.) *Global Challenges and the Law of the Sea*. Cham: Springer, pp. 203–220.

Thornton, H. (2006) 'John Selden's Response to Hugo Grotius: The Argument for Closed Seas', *International Journal of Maritime History*, 18(2), pp. 105–128. DOI: 10.1177/084387140601800206.

Tladi, D. (2019) 'An Institutional Framework for Addressing Marine Genetic Resources Under the Proposed Treaty for Marine Biodiversity in Areas Beyond National Jurisdiction', *International Environmental Agreements*, 19(4), pp. 485–495. DOI: 10.1007/s10784-019-09449-4.

Trans-Tasman Resources Limited v. Taranaki-Whanganui Conservation Board [2021] NZSC 127.

UNGA (1 November 1967), 1st Committee, 22nd Session, 1515th Meeting, UN Doc A/C.1/PV.1515.

UNGA Res 3201 (S-VI) (1 May 1974) UN Doc A/RES/S-6/3201.

UNGA (1994) 'Agreement Relating to the Implementation of Part XI of the United Nations Convention on the Law of the Sea of 10 December 1982', UN Doc A/RES/48/283.

UNGA Res 69/292 (19 June 2015) UN Doc A/RES/69/292.

United Nations Law of the Sea Convention (adopted 10 December 1982, entered into force 16 November 1994) 1833 UNTS 397 (UNCLOS).

Vadrot, A.B.M., Langlet. A. and Tessnow-von Wysocki, I. (2022) 'Who Owns Marine Biodiversity? Contesting the World Order through the 'Common Heritage of Mankind' Principle', *Environmental Politics*, 31(2), pp. 226–250. DOI: 10.1080/09644016.2021.1911442.

Ventura, V.A.M.F. (2020) *Environmental Jurisdiction in the Law of the Sea: The Brazilian Blue Amazon*. Cham: Springer.

Vidas, D. (2011) 'The Anthropocene and the International Law of the Sea', *Philosophical Transactions of the Royal Society*, 369, pp. 909–925. DOI: 10.1098/rsta.2010.0326.

Vidas, D., Fauchlad, O.K., Jensen, O. and Tvedt, M.W. (2015) 'International Law for the Anthropocene: Shifting Perspectives in the Regulations of the Oceans, Environment and Genetic Resources', *Anthropocene*, 9, pp. 1–13. DOI: 10.1016/j.ancene.2015.06.003.

Viñuales, J.E. (2018) *The Organisation of the Anthropocene in Our Hands?* Leiden: Brill Research Perspectives. DOI: https://doi.org/10.1163/9789004381360.

Vordermayer, M. (2015) '"Gardening the Great Transformation": The Anthropocene Concept's Impact on International Environmental Law Doctrine', *Yearbook of International Environmental Law*, 25(1), pp. 79–112. DOI: 10.1093/yiel/yvv063.

Whaling in the Antarctic (Australia v. Japan, New Zealand intervening) (Judgment) [2014] ICJ Rep 226.

Wolfrum, R. (1983) 'The Principle of the Common Heritage of Mankind', *ZaöRV*, 43, pp. 312–337.

Yu, C. (2020) 'Implications of the UNCLOS Marine Scientific Research Regime for the Current Negotiations on Access and Benefit Sharing of Marine Genetic Resources in Areas Beyond National Jurisdiction', *Ocean Development and International Law*, 51(1), pp. 2–18. DOI: 10.1080/00908320.2019.1677018.

6

ORDERING HUMAN–OTHER RELATIONSHIPS

International Humanitarian Law and Ecologies of Armed Conflicts in the Anthropocene

Matilda Arvidsson and Britta Sjöstedt

Introduction

Drawing on recent scholarship on (post)anthropocentrism, law, and the environment (e.g., Birrell and Dehm, 2021; Davis, 2017; Gillespie, 2014; Grear, 2015; Grear, 2020; Philippopoulos-Mihalopoulos, 2017), this chapter maps international humanitarian law (IHL) and its 'legal ordering' (Lindahl, 2019) of human and other relationships during armed conflict and disaster. Our chapter focuses on two examples of human–other legal ordering during armed conflict: first, human–environment ordering, and second, human–artificially intelligent (AI) swarming drone ordering.

Our first example is covered by a well-established field of international law and protection, yet one that has rarely been prioritised in IHL: in contrast to the many man-made objects of legal ordering in IHL – such as weapon systems for example – our first example concerns the 'natural' world. Our second example is an emerging technology embedded in both military and civilian environments, only to some degree recognised – and, as with our first example, as an *object* rather than as an intelligent *subject* with legal agency of its own – and made part of IHL's legal ordering: it is an example of human-scientific 'artificial' construction.

Taking existing criticism against IHL's excluding, binary, and hierarchical modes of legal order into account, we argue that IHL has some potential in developing in a post-anthropocentric direction, specifically in reorienting its focus from armed conflicts to violent outbursts. We make use of the Deleuze-Guattarian notion of 'war-machines' (Deleuze and Guattari, 2013; Bar-On Cohen, 2011; Robison, 2010) – for example, armed groups (including state armies, although with the caveat that this involves state appropriation of war-machines), volcanoes, packs of wolves, and viruses – to capture how we believe IHL could offer protection on a less anthropocentric and more inclusive and equal basis in a shared posthuman ecology, protecting environments – inclusive or not of humans – from violent war-machines, war-machines from state violence and appropriation, as well as protecting from state violence by war-machines.

The environment (our first example) emerged only as a specific reference and separate object of protection under IHL – as the 'natural environment' – in 1977, in the wake of the Vietnam War. The impact of war on the environment is regulated by Articles 35(3) and 55 in Additional

122

DOI: 10.4324/9781003201120-8

Protocol I to the 1949 Geneva Conventions (Additional Protocol I). The need to protect the environment in times of armed conflict has since then become more prevalent, in particular after the topic 'Protection of the Environment in relation to Armed Conflict' was put on the UN International Law Commission's (ILC) work agenda in 2011 (ILC, 2011, para. 7), and the International Committee of the Red Cross (ICRC) updated Guidelines on the Protection of the Natural Environment in Armed Conflict.[1] IHL has, in recent years, been expanded to also partly apply to protect humans from the forces of the environment (and humans). In 2016, with regard to a different topic, the ILC adopted draft articles protecting persons (humans) in the event of a disaster, both natural and man-made. These draft articles largely reflect the legal ordering characteristic of core IHL principles applying to armed conflicts, with an emphasis on the protection of humans and an explicit reference to non-humans – e.g., animals, water, and machines – as 'equipments and tools' for human and humanitarian relief.[2]

The AI swarming drone (our second example) is an emerging technology used in warfare, intelligence surveillance, and disaster management, as well as in a range of civilian-commercial settings. Based on insect swarming technologies, such as that of the common mosquito, operating with distributive decision-making algorithms, and designed for human–swarm interaction, AI swarming drones have raised concerns regarding the ordering of human–technology–others relations in armed conflict. Although swarms exist as forceful phenomena in a range of planetary areas – both in and outside armed conflicts – there is no particular recognition or regulation of swarms in IHL. Almost by default, AI swarming drones are regarded by scholars as an emerging weapons system governed by Article 36 of *Additional Protocol I*, meaning that they must pass a review concerning their lawfulness as weapons before they are deployed in combat. Their use and relations as part of military operations at large, intelligence surveillance, and as part of human–swarm capacity during disaster has hitherto remained outside the scope of IHL debates.

Our analysis of IHL and its legal ordering of relations between humans and 'others' reflects, on the one hand, the state-of-the-art in contemporary IHL and environment scholarship (Hulme, 2004; Bothe *et al.*, 2010; Sjöstedt, 2020; Bruch, Payne, and Sjöstedt, 2021) as well as that of critical IHL scholarship on emerging technologies and AI (Arvidsson, 2020; Arvidsson, 2021; Kalpouzos, 2020; Johnson, 2020; Kallenborn and Bleek, 2018; Wilcox, 2017a). On the other hand, our analysis expands through posthumanist feminist and post-anthropocentric scholarship (Arvidsson and Jones, 2023; Braidotti, 2013; Braidotti, 2019; Braidotti and Bigall, 2019). The feminist question of who the human is or can be in a posthuman ecology is, in our chapter, translated into an analysis of the ecology of bodies, objects, technologies, life and death emerging through IHL's ordering of the human and its other during armed conflict.

Dichotomies unpacked by posthumanist feminist scholarship, such as human/other, man/woman, nature/culture, and more, are integral to that which we describe as IHL's anthropocentric ordering of relations in armed conflict-ecologies. Through Gilles Deleuze and Felix Guattari's notion of 'the war-machine' we consider how contemporary IHL orders human–human, human–other, as well as other–human violent relations during armed conflicts. It is our ambition to suggest a reorientation of IHL and its scholarship towards a posthumanist and post-anthropocentric ethos. Rosi Braidotti summarises this as an ethical 'rule':

> [I]t is important to be worthy of our times, the better to act upon them, in both critical and a creative manner. It follows that we should approach our historical contradictions not as some bothersome burden, but rather as building blocks of a sustainable present and an affirmative and hopeful future, even if this approach requires some drastic changes to our familiar mind-sets and established values.
>
> *(Braidotti, 2019, p. 3)*

Flowing from such an ethical position and especially its ethos of creativity, hope, and responsibility, and in contrast to a conventional state-focused, 'military necessity' and 'proportionality'-oriented legal ordering found in IHL, we suggest that the ordering of violent relations as such, between a non-exhaustive set of entities – humans, horses, gorillas, plants, viruses, cars, volcanos, AI, and more – would be a way forward for IHL and its scholarship: this would, in our analysis, be a post-anthropocentric IHL less considerate of the human, humanism, humanity, humanitarianism, or any such derivatives. Instead, it would recognise and protect the volcano's or the wolf pack's existence while seeking to protect human and other entities from its violent forces; it would recognise human existence while protecting marine life from human violent expansions and extractions from the sea; it would recognise swarms (precisely as swarms and not as 'weapons systems') of fish shoaling as potential lethal forces.

In short, while such an IHL would seem to claim an ordering of relations everywhere and at all times, we believe that a good start would be to emphasise the ordering function IHL (potentially) could have during disasters. IHL already has a bearing on and relations to disaster law and management, not the least through its connection with humanitarianism via the close relationship between the ICRC, which oversees the implementation of IHL internationally while simultaneously organising humanitarian relief during warfare, and the International Federation of Red Cross and the Red Crescent Societies (IFRC) and its national movement, which organises humanitarian relief during warfare but also in other humanitarian disasters at a national and local level. Our suggestion would be a move away from a human–centred ordering of violence during armed conflict, to instead focus on ordering violent outbursts of that which in a Deleuze-Guattarian mode would be called 'war-machines' in a posthuman ecology.

The chapter starts with a brief introduction to IHL's main principles and norms. We then go on to offer another way – our own – of reading IHL. In doing so we explain how 'the posthuman' sits within contemporary armed conflicts, IHL, and the military-industry complex. We introduce the Deleuze-Guattarian idea of the 'war-machine' as a central tenet of our rethinking of IHL. In the section that follows, we go on to look at our two examples – first 'the environment' and then the swarm and 'the AI swarming drone' – after which we conclude with our main point on how IHL could offer protection on a less anthropocentric and more inclusive and equal basis in a shared posthuman ecology.

International Humanitarian Law and Its Ordering of Armed Conflicts

Like many other fields of international law, IHL operates on the basis of a set of central legal conventions, including the four Geneva Conventions (1949) and its two Additional Protocols (1977), as well as the Hague Regulations (1907) and the many weapons conventions.[3] In addition, IHL includes a broad range of norms on targeting issues further specifying and to some degree expanding the scope of the law (Fleck, 2021). IHL also includes a broad set of customary international legal rules and principles (Henckaerts, 2005; Henckaerts and Doswald-Beck, 2005; Wilmshurst and Beau, 2011). Central principles include those of distinction, proportionality, military necessity, prohibition of superfluous (human) injury or unnecessary (human) suffering, and humanity. Decisions and evaluation of the legality of actions under IHL are to be measured against these fundamental principles as well as in relation to specific conventions guiding the field.

IHL operates with a 'humanizing mission' through which warfare is said to become more humane whilst still enabling successful human-military operations. Humanised deaths and the act of killing are integral to the principles of distinction, proportionality, military necessity, prohibition of superfluous injury or unnecessary suffering, humanity, as well as to IHL in general.

Successful military operations, while minimizing 'collateral damage' – the unproportional killing of civilians – as well as avoiding unnecessary (human) suffering and superfluous (human) damage, is often said to be the law's main objective. Who will live, who will die a 'humanised death', and who will have certain privileges in captivity during armed conflict is a matter of how IHL orders relations between its subjects (humans) and objects (non-humans) through its treaties, customs, and principles.

The anthropocentrism at work in IHL is, on the one hand, obvious: *human*itarian laws *human*ise warfare, making it more *human*e (Arvidsson, 2020). One cannot get around 'the human', taking the shape of a combatant with special privileges or as an innocent civilian, as the top predator of armed conflict as ordered by IHL. Additionally, once an individual human being becomes selected as a lawful target within a specific armed conflict, that individual human is turned into lawful prey. Military necessity, in this latter case, outweighs the value of the individual human's life. IHL's organizing rules and principles are, thus, at once celebrating humanity and infusing categories of difference between humans and other humans. IHL further orders these different humans in relations to their other non-human objects such as the environment or the AI swarming drone – the two examples we focus on in our chapter. IHL's anthropocentrism comes with a logic of separation and difference on an inter- as well as an intra-species level. Without any further qualifications, 'the human' cannot be said to be IHL's telos: there is no central logic of a 'universal human' in IHL.

While the specific anthropocentrism of IHL may seem 'natural' or outright 'good' – serving the 'right' purposes – to many IHL scholars trained in the logics of separation and hierarchisation of IHL, our contention is that IHL neither serves 'humanity', nor offers norms and guidance through which human–other relations are best arranged. IHL is, firstly, blind to armed conflicts being but one of many forms of violent eruptions. One effect of this is that armed conflicts are treated as separate events in which other fundamental legal and political principles – such as human rights and democratic governance – are largely put aside. IHL as *lex specialis* provides for derogations of several norms of human rights law (e.g., *Legal Consequences of the Construction of a Wall* [2004]; *Legality of the Threat or Use of Nuclear Weapons* [1966]; Hampson, 2008). In practice many or most laws are put aside during armed conflicts. Another effect, following on the first, is that IHL may order one part of a violent eruption – acts and actors recognised as part of an armed conflict – while other areas of international, transnational, and national law order other parts of the conflict. Examples of the latter are international environmental law protecting wildlife and national law and transnational agreements ordering disaster relief. This causes what is known as legal fragmentation, overlapping norms as well as conflict of norms: ultimately it becomes difficult for the various actors to navigate such a legal framework, as well as one in which fundamental issues and questions can be – and are – largely ignored (such as that of the environment).

Secondly, IHL focuses on humans as its subject of legal ordering, assuming that 'the human' is wholly other to, and distinctly separatable from, 'non-human' aspects and entities. While this separation can be dismissed as simply incorrect on the bases of a range of different scientific evidence – including evidence from natural sciences such as medicine, biology, and ecology – as well as through a variety of theoretical and methodological considerations – including legal dogmatic and posthuman feminist theoretical considerations – the effect of the error is what should worry legal scholars the most: legal ordering becomes ineffective once it fails to describe its subjects and objects of ordering in a convincing manner (Orford, 2012). In a way, this is the same problem as that we described as our first. While a turn to a less anthropocentric IHL certainly does not solve all issues or disperse all concerns, the epistemological and ontological change involved in such a turn is crucial.

Feminist international legal scholarship has already shown how international law in its various forms speaks of a single universal human subject (Otto, 2009) while it, as Dianne Otto has put it, normalises:

> a multitude of intra-human hierarchies – which work together to advantage the autonomous, white, able-bodied, middle class, heterosexual Man and marginalise all those who do not fit within this privileged category – rendering them not fully human – or 'exiled' within the law.
>
> *(Jones and Otto, 2020, p. 2)*

IHL is, in this regard, no exception. Yet, it differs in that its categorisations and hierarchisation are explicit parts of its legal ordering with 'humanity' acting as its overarching principle: civilians are distinguished from combatants as part of the legal ordering of killing through the principle of distinction. In the operationalisation of that distinction, civilians (unlawful targets) are mainly conflated with women, while combatants (lawful targets) are essentially conflated with military age males (MAM) (Arvidsson, 2018; Wilcox, 2017b). 'The woman' becomes a model object of protection – as long as she performs her gender along certain performative and hetero-stereotypical norms. Moreover, wounded human combatants (unlawful targets) are distinguished from non-wounded human combatants (lawful targets); pregnant women (objects of special protection) from non-pregnant women (objects of protection if recognised as civilians); and children (child soldiers as well as those who are yet to be recognised as MAM) from adults (recognised as MAM, non-civilians, and non-women) (Dinstein, 2007, pp. 145–156). IHL further recognises a range of objects of protection beyond various human forms: livestock (private property or human food security assets) is distinguished from wild animals; essential food-growing resources (human food security assets) from simply living plants; human and livestock fresh-water supplies from simply water; cultural heritage objects from their cultures; and natural environments from just any given environment.[4] In each pair, the object of greater protection gains its hierarchical status by its attachment to humankind.

It is noteworthy that, despite a strict categorisation between combatants and civilians, 'lawful' combatants and 'other fighters',[5] military targets and 'civilian objects' (all more or less clearly defined in IHL), the reality of the battlefields is different: although the '*human*itarian' and '*human*izing' suggests that humans matter first and foremost, at times non-humans, once in captivity, receive better treatment than humans. The treatment is, however, not sanctioned by IHL. An illustrative example dates from February 2014, when a military dog was shown on a video released by Taliban warfighters in Afghanistan, claiming the dog to be a US prisoner of war (POW) – POW being a central category of special protection under the third Geneva Convention applicable to human combatants only – captured by them the previous year. The dog had been carrying weapons, a GPS, a camera and other equipment for military use and purposes. The individual dog was later confirmed to be a dog-warfighter working for the British Forces as part of the International Security Assistance Force Missions (ISAF) (BBC News, 2014).

Although military dogs – dog-warfighters – due to their inability to pass as human warfighters, lack protection under IHL since they would qualify as military targets, the dog was held under favourable conditions, allegedly fed with chicken and kebab meat, and guarded by a team of Taliban warfighters (Stilwell, 2020). Similarly, in the long and violent conflict taking place in the Democratic Republic of the Congo (DRC), international humanitarian attention seems to be directed towards the protection of the mountain gorillas, at times even more than to the civilian population suffering immensely due to the armed conflict.[6] As a consequence, several armed groups known for their brutal warfare, especially directed at the civilian population, have issued statements claiming to not have harmed the endangered mountain gorillas when operating in the gorilla sector located in the World Heritage site, Virunga National Park, DRC (e.g., Mars Daily, 2007). Yet, the better protection of non-human animals over human ones in warfare is an exception rather than a rule: many livestock and pet animals who depend on humans to care for them are abandoned or let loose during hostilities, unable to feed themselves and therefore starving to death as silent and unrecognised victims of war (Peters and de Heptinne, 2022, p. 3).[7] The latter was seen, for instance, when Iraq occupied Kuwait in the 1990s, and more than 80 percent of Kuwait's cattle, sheep, and goats died (Peters and de Heptinne, 2022, p. 4).

Another Way of Thinking about International Humanitarian Law

As a matter of historical development, non-human technological entities have always been a major aspect of IHL ordering, especially when compared to other fields of international law. IHL is thus, due to how human warfare has developed over time, relatively open to emerging forms of technology and other non-human objects and aspects. The legal history of warfare is, as Joanna Bourke notes, closely intertwined with the emergence of trans- and posthumanist history and scholarship (Bourke, 2014, p. 29; Ferrando, 2018). Warfare has been the catalyst of new industrial-technological breakthroughs, and the extension of human warfighting capacity – ranging from an extension of range, precision, and force by the bow and arrow to enhancement of endurance of human warfighters through the use of exoskeletal battle uniforms (based on biological insect exoskeleton technology) (Heathcote, 2018). This has historically engendered a debate over where the human warfighter ends, where its technological extensions begin, and whether or not a human–technological distinction can ever be meaningfully made (Amoroso and Guglielmo, 2020; Arvidsson, 2018).

Moreover, non-human animals have served as an integral part of armed forces throughout history. For example, more than 10 million animals including horses, mules, donkeys, camels, dogs, and racing pigeons served during the First World War (Wishermann, 2021). Both Russia and the US military have trained beluga whales, dolphins, sea lions, and seals for military purposes. Ukraine has even established a military dolphin centre in Crimea. By combining technology with superior animal characteristics, enhanced versions of the animals can be used for combat to search for naval mines. For example, the beluga whale, thanks to strong echolocation capabilities, can dive up to 700 meters deep, which is deeper than most military submarines.[8] Sharks, dolphins, mosquitos and other animals have, in turn, inspired technological advancements in the development of new weapon systems and other military technologies – as have plants.[9] 'Human' warfare is thus best described as a multi-species and 'zoo/geo/techno-oriented' lethal affair (Braidotti and Bignall, 2019, p. 1).

Yet, even as both the natural environment (our first example) – including flora and fauna – and swarming technologies (our second example) have long since been part and parcel of warfare, IHL is far from treating non-human entities on an equal basis with humans. At best, IHL provides a framework through which to curb certain violence and 'unnecessary' human suffering in warfare, with an aim to make belligerent parties respect at least 'the principles of humanity and … the dictates of public conscience', as the Martens Clause spells it out.[10] The difference between a posthuman ecology and an armed conflict governed by IHL can best be illustrated as that between the 'war-machine', in the Deleuze-Guattarian sense, and a state appropriated war-machine: both are relationally organised by norms through which life and death become realised. The former is a mode of existence in contestation of – or pure ignorance of – the nation state and its modes of legal ordering. A state-appropriated war-machine is the military of a nation state operating within IHL's legal ordering of armed conflict ecologies.

The Deleuze-Guattarian term 'war-machine' denotes 'social (nomadic) assemblages constituting a combination of forces or elements diffusing power and, in particular, breaking down concentrated power, having war not as its goal but as one of its possible consequences' (Arvidsson, 2020, p. 123). A war-machine, in the Deleuze-Guattarian sense, can be almost anything: a pack of wolves, a cyclone, guerrilla fighters, or a fungus. Its distinct feature is that it is a grouping or assembly of many (more than one individual, thing, or element); that it functions according to an internal organisation of its collective force – a force that is usually not prone to becoming organised by a human-centred law of hierarchical orderings with a 'human' telos – that is, it is spontaneous or ephemeral; and that it moves across – rather than along – patterns and borders formed by law in its current form (including IHL).

The war-machine can be understood as a form of social assemblage 'directed against the state, and against the coalescence of sovereignty' (Robinson, 2010, p. 6). At least so in that a war-machine swarm of bees, a war-machine general strike, or a war-machine pandemic virus has the potential to unveil the weaknesses of sovereign law and power: if the unfathomable powers and violence of war-machines cannot be controlled and ordered by the state, the latter loses its monopoly on (the exercise of) violence – the foundation and continuous force of all law-making and law-preservation, in Walter Benjamin's analysis (1978). When it comes to war, to *control*, rather than to render *illegal*, is a major concern of all states. Remember, warfare was never made illegal by international law, and killing innocent civilians may be considered lawful collateral damage. Hence, the goal of the state is not to eliminate or render illegal a war-machine, but to appropriate a war-machine with the purpose of consuming its powers and curbing its potential to institute its own power as a principle of ordering (wolf-pack ordering, swarming) within a given territory. It is, as we find it in this chapter, the main concern for IHL's ordering of relations during armed conflicts and beyond.

When a state appropriates a war-machine, the state thwarts the machine's force and its 'anti-purpose' (that which Deleuze and Guattari (2013, pp. 277–278) call the 'involution', and the emancipatory 'becoming-pack', swarm, and war-machine). Yet, despite this, a 'war machine is always exterior to the State, even when the State uses it, appropriates it' (Deleuze and Guattari, 2013, p. 283). 'War-machines end up in conflict with states', as Andrew Robinson explains, because their goal is the 'deterritorialization' of the rigid fixities of state space, often to create space for difference or for particular ways of life' (Robinson, 2010, p. 6). There is thus dynamic ambivalence at work between, on the one hand, the juridical form and jurisdictional telos of states (the primary subjects of IHL) and, on the other hand, war-machines (which may be other subjects of IHL, such as resistance movements, but which might also be non-recognisable to IHL).

In order to see in greater detail how IHL orders human–other relations, as well as imagine how it can be ordered otherwise, we now look at our two examples – 'the environment' and 'the swarm and AI-powered swarm of drones'. The understanding of how IHL recognises and orders its objects and subjects will be the focus, as well as the question of how the two examples could be understood as conventional objects of IHL legal ordering as well as war-machines of a new kind of IHL legal ordering in our shared posthuman ecology.

The Environment as a Non-Human Other

What is or can 'the natural environment' be in the ecology of an armed conflict? A meadow, a swamp, an urban environment, or a bacteriological war-machine? How do certain parts of the environment become military targets, means of warfare, or objects of protection, and how are they related to humans and non-human animals? Or just receptacles of collateral harm?

First of all, it should be noted that within IHL, the environment is referred to with the qualifier 'natural', signalling that it is the untouched environment – understood as external to the human and worthy of human protection. In this sense, IHL paints a romanticised view of a pristine environment far from the vibrant and violent conditions of life and death known to any ecosystem. A more accurate view would be to define the environment as ecosystems interacting with each other in which the human is one of many entities, agents, or materials, singled out by law as culpable for violations and environmental damage and obligations to protect ecosystems as such.[11] Because of its vitality and forces, the environment can be regarded as both a planetary and a multilevel ecosystem war-machine: any given war-machine has environmental aspects, including those appropriated by states.

By way of example, belligerent state armies move in and through rural as well as urban landscapes. In doing so they interact with and change each ecological system they encounter, briefly becoming part of it, changing it, and being changed themselves: military vehicles produce dust clouds as they move in convoys through sandy deserts, making humans involuntary inhale the fine dust particles. Parts of the sandy desert ecology will, as a consequence, travel as microscopic particles stuck in the lungs of warfighters and other military personnel, cross oceans and become part of decomposing human bodies in wholly different parts of the globe. In turn, the original sandy desert ecology may receive human emissions – inclusive of hormones and medicinal drugs intended to enhance human warfighting capacity and resilience – during a pee break. IHL, in contrast, does not account for environments as constitutive for warfare, or for ecologies emerging through inter-species and multi-aspects encounters. Its scale is larger, and its legal ordering works through distancing distinctions. 'The environment' is, to be clear, conceptually set apart as an 'other' in relation to the human and humanitarian concerns.

As a consequence of the distinction between humans and the environment in IHL, the environment only has weak and inadequate legal protection in armed conflict ecologies. The result is, in most cases, that substantial environmental damage caused by warfare is regarded as lawful under IHL (Sjöstedt, 2020; Bothe *et al.*, 2010, pp. 569–592; Dinstein, 2001; Falk, 2000; Verwey, 1995; Simonds, 1992–1993; Hulme, 2010; Orellana, 2005). 'Military necessity' is, when weighed against 'environmental protection', found to have the upper hand in most cases. After all, IHL's central principles primarily aim at protecting humanity conceptually understood as distinct and set apart from 'nature'.

Since all military conduct must be carried out in a given environment, restrictions applicable on the belligerents to protect the environment could nonetheless have considerable impacts on how they can conduct warfare. There are two specific provisions protecting the environment within IHL: Articles 35(3) and 55(1) of Additional Protocol I (1977). The provisions contain an absolute protection of the natural environment from 'long-term, widespread and severe' damage even if the environment or parts of it constitutes a military target. The cumulative requirements exclude most damage inflicted towards the environment using conventional weapons in armed conflict. In fact, the two articles have never been applied to a concrete case of environmental harm during an armed conflict. Given that Additional Protocol I (1977) was adopted in the aftermath of the Vietnam War, the type of environmental damage that may be covered by the articles is most likely to be of the scale of harm inflicted during that war. As part of US military tactics during the Vietnam War, chemical agents – the infamous 'Agent Orange' being the best known – were sprayed over hectares of Vietnamese soil causing massive acute environmental damage having both immediate and long-term implications on public health and a variety of ecological systems. At the time of the war, however, this type of war on the environment was yet to be outlawed by IHL.

The two articles through which IHL orders human–environmental relations are different in scope and aims. Article 35(3) protects the environment 'irrespective of its context and relationship' with the human population (Koppe, 2014, p. 66), while Article 55 focuses on preventing adverse effects for the sake of the human population. In fact, the adoption of Article 35(3) is based on the new threats directed at humanity that environmental degradation poses. As warfare has become more technologically and chemically advanced, components of certain war remnants can have permanent harmful effects on humans, animals, vegetation, water, land, and the ecosystem as a whole (Sandoz, Swinarski, and Zimmerman, 2017, pp. 410–411).[12] In addition, Article 35(3) is not limited to the environment of the enemy but recognises the global character of the environment and the transnational adverse effects on it. Schmitt (2000, p. 128) states that, '[A]rticle 35(3) is the sole provision that operates in isolation of anthropocentric values'.

The division of the articles was mainly a result of two camps having conflicting views during the ICRC Conference (between 1974 and 1977). One camp advocated a provision protecting the environment for its importance for humans, while the other camp wanted protection for the intrinsic value of the environment unqualified by the human factors. The other camp suggested adopting articles that inclined towards treating the environment, from an instrumental perspective, based on its value for humans, which merits protection because of the harm to the population that can be caused by environmental damage. As a result, to accommodate both these efforts, two articles were adopted.

The protection is however outdated, as it does not incorporate the modern concerns about a less resilient environment due to a changing climate, increased pollution, collapsing ecosystems with mass extinction of species, and so on. As a response, the ILC has attempted to address this issue. In an early suggested draft principle that was later deleted during the drafting process, the first Special Rapporteur of the topic Protection of the Environment in Relation to Armed Conflict in the ILC proposed that 'the environment is civilian in nature' and thus should be spared from the destructive forces of armed conflicts; the latter understood as human–military (ILC, 2015). The indication here is, again, that 'the environment' is primarily seen as a romanticised passive object worthy of protection from human aggression. At the same time, the environment contains strong vital as well as deadly forces, likely for states to appropriate and try to control in warfare as part of state-appropriated war-machines. The war–machinic forces encapsulated by the IHL notion of the environment may be employed as weapons, or as part of weapons systems in warfare, although its explicit use – in terms of human manipulation of earthquakes, tsunamis, and lightning – is highly unusual. The war-machinic force of the atom, released through nuclear fission in the atomic bomb, is rarely used, yet its mere existence haunts modern warfare and IHL. In addition, the environment and the human dependence on it can also be used in warfare. Scorched earth tactics are a common feature in warfare where belligerents destroy crops, water wells, and marshlands; slaughter livestock; and so on to deny enemy forces as well as civilians food, water, and energy and thereby creating a hostile environment that challenges human existence.[13]

One of the most notorious examples of employing scorched-earth policies in modern warfare took place during the Vietnam War, when the United States directly attacked and destroyed vegetation as a response to guerrilla warfare conducted by the North Vietnamese army and the Viet Cong guerrillas using the surrounding jungle as a cover. In that context, the United States developed a strategy aiming at destroying forest and farmland. Between 1961 and 1972, the United States sprayed approximately 70 million litres of herbicides and other chemical substances, notably Agent Orange, as part of its military operations to achieve these aims. As a result, allegedly, 43 percent of the farmland and 44 percent of the total forest area of South

Vietnam were contaminated. All of these military activities have had a severe impact on the Vietnamese environment, destroying entire biotic communities. In addition, the United States army manipulated the weather patterns through cloud seeding in order to prolong the monsoon season to interfere in the North Vietnamese army and Vietcong's warfare (Westin, 1976, pp. 221–222).

The United States and New Zealand conducted secret tests during the Second World War that aimed to create a tsunami, although they never employed the technique during wartime. The so-called 'tsunami bomb' was designed to attack coastal cities by using underwater explosions to generate enormous tidal waves (Pearlman, 2013; Leech, 1950). Furthermore, several tests of using bacteria and viruses, as part of biological warfare, have been conducted by several states. Most of these activities are part of secret programs but tests made by the United States, Germany under the Nazi regime, and the Soviet Union are now known.[14] In 1942, the British tested an agent causing anthrax at Gruinard Island. Since the test, the island remains uninhabitable, primary because the microorganism that was introduced has become a permanent part of the ecosystem (Schafer, 1989).

The 1972 Biological Weapons Convention has restricted all types of acts related to all types of bacterial weapons. Also, the 1976 Convention on the prohibition of military or any other hostile use of environmental modification techniques (ENMOD Convention) has been adopted that prohibits any interference with environmental forces for military purposes. This convention prohibits manipulating forces of nature in states' warfare, such as creating tsunamis, earthquakes, lightning, and so on. Moreover, man-made structures can be used in warfare to cause destruction harming humans as well as non-humans. Dams, dykes, oil and gas wells, as well as nuclear power plants are examples of installations containing dangerous forces that have been granted special protection under IHL, given their destructive abilities (Additional Protocol I, 1977, art. 56; Additional Protocol II, 1977, art. 16).

While the environment is often invoked as an object of protection, not the least in order to secure a sustainable future, IHL has – as we have tried to show here – only little to offer towards such aims. The categorical separation of the environment from the human, as well as from central humanitarian concerns in IHL's humanizing mission has, instead, resulted in an array of dispersed IHL norms with only some degree of 'success' in offering any protection. Braidotti's suggestion for 'drastic changes to our familiar mind-sets and established values' (Braidotti, 2019, p. 3) requires, in our view, a more 'drastic' reconfiguration of IHL in relation to the environment. Concretely, this could be carried out on both normative and practical levels, including an expanded consideration for non-human aspects – such as the environment – in proportionality assessments and in assessments of military necessity under IHL.

Moreover, if violent outbursts of 'the natural environment' – including volcanic eruptions, tsunamis, and swarms of locusts – rather than (human-made) 'armed conflicts', such as IHL applies to in its contemporary form, were considered *subjects* rather than objects of protection under IHL, the environment in its various forms and figurations would be in a more promising position: it would be less of a 'romantic' and passive object 'worthy' (or, as is the case most often in contemporary IHL, unworthy) of human protection, and potentially lethal violent eruptions and relations between humans and others would instead, and more importantly, be a central concern for legal ordering. Having no specific priority to humankind – as detached from other species, elements, and aspects of our shared posthuman ecology of life – death would, in such a reconsideration of IHL, be a 'drastic' move for IHL and its scholarship, yet a necessary move in a posthuman reality. It would be a move in which environmental concerns could be more adequately fitted as both objects and subjects of legal ordering.[15]

The Swarm and the Swarm of AI-Powered Drones as the Non-Human Other

What is a swarm of AI-powered drones, or what can it be, in the ecology of an armed conflict? How does IHL seek to order emerging technologies such as the AI swarming drones? Especially, how does IHL recognise and order technologies emulating features of 'natural' war-machines – such as a swarm of birds or bees – once these technologies work autonomously within the military 'mission command', as well as once they appear as integrated human–other collaborations in a swarm-based state military war-machine?

AI-powered drones with a capacity to swarm are mathematically modelled on 'behavioural modes' of swarming insects, such as the common mosquito. In their 'natural environment' the latter 'aggregate together, hover in the same place, or migrate as a moving collective' (Dublon and Sumpter, 2014). Insects swarm either to mate or migrate: the common imagery of a swarm of angry bees chasing and attacking a cartoon character is a rare sight outside the realm of popular culture (Dublon and Sumpter, 2014). More than anything, swarms *swarm* because the movements and relations involved in swarming creates the swarm as a multiplicity, having its own making as its primary goal (mating, migration for food or shelter). Death – the death of others (it happens that swarms attack and kill others) as well as of individuals within the swarm – is only one possible outcome and not a goal in itself (Deleuze and Guattari, 2013, pp. 277–278). Deleuze and Guattari gesture towards the swarm as the war-machine par excellence, its force residing in numbers, versatility, adaptivity, sudden and unpredictable movements and eruptions, and its ability to reconfigure itself and move across – or rather regardless of – state borders. Swarms are not defined by singular-species characters or by single-materials: a swarm becomes a swarm through its swarming behaviour – its doing 'the swarm'. In a conventional IHL register, the *levée en masse* – in other words, people spontaneously taking up arms to resist an approaching (enemy) force – is a swarming war-machine in the Deleuze-Guattarian sense. The spontaneous, versatile, and unpredictable eruption of violence performed by what is known to IHL as civilians, makes it an object recognisable to the legal ordering. Yet, as we will return to shortly, the state military is the most lethal human–'zoo/geo/techno-oriented' swarm known on the planet (Braidotti and Bignall, 2019, p. 1).

Swarming is a well-known military tactic recorded as part of Mughal and Byzantine modes of warfare, considered especially suitable for guerrilla warfare or surprise attacks (Edwards, 2000; Arquilla and Ronfeldt, 2000; US Army UAS Center of Excellence, 2010). Its tactical drawback is the primary technical problem of (a lack, hitherto, of technological development suitable for) instantaneous communication during swarming – in insects and AI swarming drones such communication is known as 'sensing'. The advantage of swarming as a military tactic is its force of multitude, the strategic military advantage of sudden and irregular patterns of movement (withdrawals, outbursts), and the system of distributed decision-making throughout the entire swarm which is particular to swarming. The latter makes military decision-making instantaneous and collectively distributed for implementation: it is fast and, if correctly coordinated, extremely forceful.

AI-powered swarms – understood as tech-entities of similar size that 'aggregate together, hover in the same place, or migrate as a moving collective' (Dublon and Sumpter, 2014) – have been developed with some intensity for commercial, military, and mixed purposes during the last decade. Such swarms emulate the 'functionally versatile and powerful, and highly distributed' sensory systems of insects, meaning that they operate by 'decentralized control policies that can cope with limited local sensing and communication abilities of the agents' (Hüttenrauch, Sošic, and Neumann, 2018). Simply put, AI swarming drones are able to take on military missions as a

Ordering Human–Other Relationships

collective, communicate relevant information amongst themselves, distribute tasks, and 'reconfigure themselves, autonomously changing direction in response to sensor input to achieve the mission at hand' (Arvidsson, 2020, p. 126).

In contrast to conventional drones, AI swarming drones thus have the capacity to act within the framework of a military mission without a central command or human commander – the latter often referred to as 'the human in the loop' whose function it is to make sure that 'reasonable human control' is exercised (Skarkey, 2016, p. 23). If, or when, an individual swarm member malfunctions, is taken out by enemy fire or gets lost in terrain, the swarm automatically reconfigures itself according to a swarming protocol in order to carry out the mission. In other words, AI swarming drones act not unlike conventional (human) warfighters. The difference is that they are exceptionally good – much better than no/low-tech human–other configurations deployed in armed conflicts – at communicating, distributing tasks, making decisions, and completing the military mission successfully (at a low-cost, as compared to the potential loss of human warfighters). This is, at least, how they are being 'marketed' as part of military, state-appropriated war-machines.[16]

AI drones are already operative as part of human–machine swarming – also known as human–drone 'teaming' – for surveillance, military operative engagement, medical evacuation, and disaster response search missions (Johnson, 2020).[17] In order to fit AI swarming drones into the ecology of armed conflict, IHL practice and its scholarly debates have focused on the drones' capacities as new lethal weapons systems capable of autonomous and distributive decisions (e.g., Kalpouzos, 2020). This, as we noted in the introduction, renders them recognisable for the purpose of the Article 36 review, ordering AI swarming drones alongside (other) weapons systems: the 'obligation to determine whether its employment would, in some or all circumstances, be prohibited by this Protocol or by any other rule of international law applicable to the High Contracting Party'.[18] The review process of Article 36 can be understood, as David Roden argues, as a question of assessing the 'human-like[ness]' and the 'rational "self-mastery"' of an agent (AI entity, human, or both) (Roden, 2017, p. 99; Arvidsson, 2020, p. 128), meaning that the aim of the test is to ascertain that the weapons system is not explicitly forbidden by international law and that the central principles of IHL – distinction, proportionality, and humanity – can be upheld through the use of it (Goussac, 2019). What amounts to a 'new weapons system' (or a weapons system in the first place) is not properly defined in IHL. Scholars have addressed the problem primarily within the context of autonomous weapon systems (AWS) – pointing to the difficulties in defining what constitutes autonomy (in relation to humans), taking for granted that a weapons system is already a known definition (Kalpouzos, 2020; Jones, 2018).

The assumption in Article 36, as well as in the scholarly and otherwise recognition of AI swarming drones as weapons systems in the ecology of an armed conflict, is that such swarms are categorically distinct from and therefore other to humans: they are supposed to be, within the ecology of armed conflict as ordered by IHL, distinct from non-weapons, including humans – human military commanders (whose tools – or, indeed, extensions – they are), human warfighters, and human civilians. They are also seen as distinct from the 'natural environment', as well as the 'vast planetary network, fuelled by the extraction of non-renewable materials, labour, and data' (Crawford and Joler, 2018), which enables AI-geared tech-entities to function (regardless of their civil or military status). AI swarming drones may be 'like' humans in certain ways and capacities during armed conflicts, their 'self-mastery' even better-than-humans as fully autonomous AI swarming drones can execute 'intuitive action [that] may enable the swarm to cease an attack within nanoseconds' (Grimal and Sundaram, 2018, p. 119). Yet, it seems as if they can never become recognised as fully human – in IHL terms, neither civilians nor combatants. They are, in other words, not subjects of legal ordering.

Human warfighters – combatants – are, in contrast to the 'training-through-self-learning protocols' and AI programming through which AI swarming drones incorporate IHL's central principles as well as learn how to execute command intent within a military mission, conventionally put through military training programs. These programs teach human warfighters to embody what Roden (2017) has described above as a 'human-like', 'rational "self-mastery"'. The central pinnacles of such 'self-mastery' are to subjugate oneself to superior military command, to carry out any mission command, and to do so in accordance with the central principles of IHL – most importantly distinction, proportionality, precaution, and humanity. As noted elsewhere, Additional Protocol I (1977), Article 43, requires a responsible (human) command as well as an internal military disciplinary system that enforces individual warfighters', as well as collective, compliance with IHL. Any individual intent of a human or other warfighter must be subjugated to the intent of the (human) military commander: 'the warfighter learns how to advance the intent of its military commander and through that relationship becomes "the warfighter": no longer fully human but fully warfighter and part of/the flesh of the military swarm and [state appropriated] war machine' (Arvidsson, 2020, p. 130). In other words, once a human individual enters the military system, her humanness is transformed as she becomes a unit in a chain-of-military-command and responsibility: she becomes part of the military state-appropriated war-machine. A human warfighter is thus never really fully or only human, but always something else (too). IHL recognises her as a combatant within the ecology of the armed conflict, and as part of the military machinery as a whole.

It seems as if IHL – or at least its practitioners and scholars – recognise AI swarming drones as non-human-yet-human-like parts of state-appropriated war-machine-swarms, largely understood as human–'zoo/geo/techno-oriented' lethal forces (though their terminology may look different, this is what they mean). The ordering of human–artificially intelligent (AI) swarming drone relations undertaken by IHL is thus less a distinct separation and more a matter of recognition as different agents or individuals within a swarm – each of which are interrelated and entangled.

The latter observation invites the question: to whom, and under which circumstances, does a distinction between human and 'other' warfighters in IHL matter? What if IHL, its practitioners, and scholars were to understand and recognise AI swarming drones simply as one of many varieties of warfighters and thus subjects to legal ordering? Warfighters primarily (but not exclusively) technological and non-human in character – just as the dog-warfighter mentioned above, caught, and treated as a POW by Taliban forces – and embedded in a military system that is already operating on a swarm-logics basis? To consider a warfighter in relation to its performance, rather than in terms of speciesist hierarchies of dichotomies, would make sense not only as a gesture towards a more equal and inclusive IHL – an (admittedly small) step towards a posthuman and post-Anthropocene ethics – but also in practical terms. As species, entities, and aspects become further entangled in warfare and beyond, legal ordering would fare better – in terms of its capacity to adapt to changes, controlling as wide a set as possible of aspects of lethal violence – from dropping inter- and intra-species dichotomies, hierarchisation, and exclusions.

Or, even more straightforward, what if we were to view AI swarming drones as war-machines – as multitudes of entities, relations, minerals, forces – recognisable for IHL's ordering if and when posing a violent, lethal threat to others? Within the framework of armed conflict ecologies, they would be required to perform as any other warfighters who are subject to IHL's legal ordering; in accordance with IHL, in particular its central principles of distinction, proportionality, military necessity, and humanity. The distinction between the operative use of AI swarming drones and their deployment in combat as warfighters would, perhaps, still be one of an Article 36 review. Yet, such a review should apply to all new warfighters (and not just the ones who

Ordering Human–Other Relationships

are considered as primarily technological in design). The 'drastic changes' required, as Braidotti (2019) puts it, are thus concrete and applicable in already existing situations, making the move we envision for IHL one that better describes our contemporary existence of violent eruptions and forces.

Conclusions

In order to be worthy of our times, and to better act upon them in both a critical and creative manner, we have used this chapter as an exercise through which to critically map IHL and some of its legal ordering of human–other relationships during armed conflict and disaster. Our ambition has been to creatively move beyond mere iterations of IHL, environmental legal protection, and drone-warfare debates in contemporary practice and scholarship. In our suggestion, this means reconfiguring the conventional understanding of IHL: instead of referring only or primarily to the legal ordering of armed conflicts, a more cross-species and multi-aspect, embracive, and non-exclusionary IHL legal ordering would be one engaging with violent outbursts as such. The point would not be to save humankind from forces of violent outbursts or to conserve a pristine and romanticised 'natural' environment. Rather, the aim would be to order violent relations between humans and others during eruptions of violence originating from armed conflict or natural disasters, with no specific priority to humankind as detached from other species, elements, and aspects of our shared posthuman ecology of life as well as death.

Making use of the Deleuze-Guattarian notion of 'war-machines', we have considered how our two examples of the environment and AI swarming drones already invite a move towards fewer dichotomies in IHL's legal ordering, as well as a greater attention to a variety of violent forces – inclusive of armed groups (including state armies, although with the caveat that this involves state appropriation of war-machines), volcanos, packs of wolves, and viruses. In fact, such a move is long overdue. The move towards an IHL covering disasters as well as armed conflicts is a small step in the right direction. In the Anthropocene epoch, we are more likely to encounter further disasters due to an environment made more vulnerable. The law thus needs to be able to protect environmental aspects from further human degradation as well as to enable human protection from harmful forces – in times of armed conflict and peacetime alike.

Yet, in order to arrive at a less anthropocentric and more inclusive and equal basis in a shared posthuman ecology, in which we can exercise 'more inclusive practice[s] of becoming-human' (Braidotti and Bignall, 2019, p. 1), the real shift is ontological and ethical. Protecting environments, inclusive or not of humans, from violent war-machines, and war-machines from state violence and appropriation, requires of IHL an ability to recognise war, disaster, and other violent outbursts as multi-species and 'zoo/geo/techno-oriented', while at the same time mediating and insisting on an ethical and political distribution of human responsibility (Chakrabarty, 2009; Philippopoulos-Mihalopoulos, 2017, p. 133).

The armed conflict ecologies ordered by IHL offer relations of 'distinction', 'proportionality', 'military necessity', and 'humanity', as well as 'successful military operations' with minimal 'collateral damage'. Yet, the violence curbed and suffering avoided by such legal ordering is distributed along a set of excluding categorisations and dichotomies. The universal 'human' is, in relation to other sets of international law, less obvious, as there are several intra-human distinctions in IHL through which a hierarchy of humanitarianism unfolds: at times offering humans protection, yet at other times prioritising successful military operations or the protection of the natural environment over human lives. The anthropocentrism at work is, as it were, as messy as humankind itself.

The sustainable present and hopeful future that we want to affirm through this chapter calls on IHL scholars to challenge familiar mind-sets and established values engraved in IHL legal ordering as well as in its practice and scholarship. This, we find, is the least we can do in order to better act upon our times in critical and creative ways.

Notes

1 The ICRC Guidelines on Protection of the Natural Environment were published in 2020, available at https://shop.icrc.org/guidelines-on-the-protection-of-the-natural-environment-in-armed-conflict-rules-and-recommendations-relating-to-the-protection-of-the-natural-environment-under-international-humanitarian-law-with-commentary.

2 Article 3(a) offers a definition of disaster as: 'a calamitous event or series of events resulting in widespread loss of life, great human suffering and distress, mass displacement, or large-scale material or environmental damage, thereby seriously disrupting the functioning of society, providing a non-exhaustive list of 'equipment and goods' in Article 3(g): 'supplies, tools, machines, specially trained animals, foodstuffs, drinking water, medical supplies, means of shelter, clothing, bedding, vehicles, telecommunications equipment, and other objects for disaster relief assistance'. *Draft articles on the protection of persons in the event of disasters*, adopted by the International Law Commission at its sixty-eighth session, in 2016, and submitted to the General Assembly as a part of the Commission's report covering the work of that session (A/71/10), para. 48, available at https://legal.un.org/docs/?path=../ilc/texts/instruments/english/draft_articles/6_3_2016.pdf&lang=EF.

3 Central conventions include: *Geneva Convention (I) for the Amelioration of the Wounded and Sick in Armed Forces in the Field (12 August 1949); Geneva Convention (II) for the Amelioration of the Wounded and Sick in Armed Forces in the Field (12 August 1949); Geneva Convention (III) for the Amelioration of the Wounded and Sick in Armed Forces in the Field (12 August 1949); Geneva Convention (IV) Relative to the Protection of Civilian Persons in Time of War (12 August 1949); Additional Protocol I; Protocol Additional to the Geneva Conventions of 12 August 1949, and relating to the Protection of Victims of Non-International Armed Conflicts (Protocol II) (8 June 1977); Convention (IV) respecting the Laws and Customs of War on Land and its annex: Regulations concerning the Laws and Customs of War on Land. The Hague, 18 October 1907 (the Hague Regulations)*; and concerning weapons: *Declaration (IV,3) concerning Expanding Bullets. The Hague, 29 July 1899; Protocol for the Prohibition of the Use of Asphyxiating, Poisonous or Other Gases, and of Bacteriological Methods of Warfare. Geneva, 17 June 1925; Convention on the Prohibition of the Development, Production and Stockpiling of Bacteriological (Biological) and Toxin Weapons and on their Destruction. Opened for Signature at London, Moscow and Washington. 10 April 1972; Convention on Prohibitions or Restrictions on the Use of Certain Conventional Weapons Which May be Deemed to be Excessively Injurious or to Have Indiscriminate Effects. Geneva, 10 October 1980; Protocol (II) on Prohibitions or Restrictions on the Use of Mines, Booby-Traps and Other Devices. Geneva, 10 October 1980*; the *Convention on the Prohibition of the Use, Stockpiling, Production and Transfer of Anti-Personnel Mines and on their Destruction, 18 September 1997*; and *Convention on Cluster Munitions, 30 May 2008.*

4 Article 53 of *Additional Protocol I* prohibits attacks on cultural objects. Article 54(2) of the 1954 Hague Convention on Cultural Property prohibits attacks on, the destruction, removal, or rendering useless of objects indispensable to the survival of the civilian population. An inexhaustive list enumerated in the articles include: foodstuffs, agricultural areas for the production of foodstuffs, crops, livestock, drinking water installations and supplies, and irrigation works. For customary legal principles regarding 'Attacks against Objects Indispensable to the Survival of the Civilian Population', see Rule 54, in the ICRC Handbook on Customary IHL. Article 55 prohibits attacks on the 'natural environment'.

5 Despite the fact that this distinction does not appear in IHL, there is a distinction between privileged combatants defined in accordance with Article 4 Geneva Convention III and Article 43 Additional Protocol I and other fighters falling outside this definition.

6 The example is further discussed at https://www.orwelltoday.com/nkundagorillasafe.shtml.

7 Another example, which caught media attention at the time, of difficulties and confusion arising from IHL's difficulty in recognizing animals is drawn from the invasion of Iraq by US troops, in 2004, when US warfighters encountered a tiger allegedly belonging to Udday Hussein (one of Saddam Hussain's sons) moving about in the old palace buildings-turned-headquarters of the occupation: the soldiers, reportedly, did not know if or how to apply the Geneva Convention to the big cat (Arvidsson 2011, p. 76).

8 Submarine and animal questions are further discussed, for instance, here: https://nationalinterest.org/blog/buzz/real-submarines-check-out-russia%E2%80%99s-combat-dolphins-spy-whales-and-killer-seals-55667, https://www.theguardian.com/environment/2018/may/16/ukraine-claims-dolphin-army-captured-by-russia-went-on-hunger-strike.

9 The US Defense Advanced Research Projects Agency (DARPA) has a number of ongoing research projects in which animal behaviour is emulated as well as – although this is rarer – biologically modified with technological enhancements, in order to develop new weapons systems, surveillance, and information technologies. See, for example, the 'Insect Allies' program https://www.darpa.mil/program/insect-allies; and the Persistent Aquatic Living Sensors (PALS) program https://www.darpa.mil/program/persistent-aquatic-living-sensors. See also the 'Advanced Plant Technology' (ATP) program presented under the heading 'Nature's Silent Sentinels Could Help Detect Security Threats: New Program Envisions Plants as Discreet, Self-Sustaining Sensors Capable of Reporting Via Remotely Monitored, Programmed Responses to Environmental Stimuli' at https://www.darpa.mil/news-events/2017-11-17.

10 The Martens Clause first appeared in the preamble to the 1899 *Hague Convention (II) with respect to the laws and customs of war on land* and has formed part of IHL since then. A modern version is included in Article 1(2) Additional Protocol I. It was also reproduced in Draft Principle 12 (with a specific focus on the environment) adopted at first reading by the International Law Commission, 'Protection of the Environment. Text and Titles of the Draft Principles Provisionally Adopted by the Drafting Committee on First Reading' UN doc. A/CN.4/L.937 (2019).

11 One problem with the legal design of such 'universal human culpability' is that it does not account for the unequal distribution of interference, extraction, and irreversible damage inflicted by primarily industrial economies in the Global North. See further: Parikka, 2018, pp. 51–53, at 53. For a discussion of human responsibility from a post-anthropocentric ethics of becoming-with the environment, see Philippopoulos-Mihalopoulos, 2017.

12 This is but one of the many ways in which the human has made an increasing harmful planetary impact in the Anthropocene. It is also emblematic of how modern warfare is 'zoo/geo/techno-oriented' and posthuman, yet not in the emancipatory and ethically hopeful-constructive sense offered by posthuman theorists, such as Rosi Braidotti.

13 During the Second World War, the German occupation forces destroyed the human settlements in Northern Norway to escape Russian soldiers. All domestic animals were slaughtered; the buildings burned; the roads, bridges, and fishing boats ruined; all communications and utilities damaged; and the terrain and the harbours mined in order to hinder the advancement of the Russian forces. Another use of scorched-earth policies occurred in 1953, when the United States bombed five dams in North Korea. These attacks were undertaken for the purpose of weakening the important rice production in the state, pressuring North Korea to sign a peace agreement. Leaning, J., 'War and the Environment: Human Health Consequences of Environmental Damage of War', Critical Condition: Human Health and the Environment (1993), 127, available at http://mitpress.mit.edu/sites/default/files/titles/content/9780262531184_sch_0001.pdf.

14 See https://www.gao.gov/assets/250/242279.pdf. For instance, according to declassified documents from the Pentagon, the United States sprayed bacteria over the Hawaiian island of Oahu to simulate a biological attack on an island compound, and to develop tactics for such an attack. The test was part of Project 112, a military program in the 1960s and 1970s to test chemical and biological weapons and defences against them. The test used Bacillus globigii, a bacterium believed at the time to be harmless. Researchers later discovered the bacterium, a relative of the one that causes anthrax, could cause infections in people with weakened immune systems (see https://www.cbsnews.com/news/us-admits-bio-weapons-tests/). On 20 September 1950, a US Navy ship just off the coast of San Francisco used a giant hose to spray a cloud of microbes into the air and into the city's famous fog. The military was testing how a biological weapon attack would affect the 800,000 residents of the city. The unsuspecting residents of San Francisco certainly could not consent to the military's germ-warfare test; and there's good evidence that it may have caused the death of at least one resident of the city, Edward Nevin, and hospitalised ten others. See https://www.businessinsider.com/the-military-tested-bacterial-weapons-in-san-francisco-2015-7?r=US&IR=T.

15 The consideration of the environment as a subject may seem like an imaginary picture. However, in 2019, the Special Jurisdiction for Peace (in Spanish: Jurisdicción Especial para la Paz, JEP) formally recognised the environment as a 'silent victim' of the armed conflict in Colombia. The JEP is a parallel legal system and tribunal established after the conclusion of the Colombian peace agreement in 2016.

The petition to JEP to include the environment was made by an indigenous group and embodies a different way of how to consider the environment as a subject (*JEP, Case 02 – Territorial Situation of the Tumaco, Ricaurte, and Barbacoas Municipalities (Nariño)*).

16 See, for example, the US Defense Advanced Research Agency (DARPA) 'Gremlins' initiative: https://www.darpa.mil/program/gremlins; and *U.S. Army Roadmap for UAS 2010–2035: The Eyes of the Army* (2010).

17 For drones on search-missions without human interference, see https://www.discovermagazine.com/technology/this-swarm-of-search-and-rescue-drones-can-explore-without-human-help.

18 The Article 36 review recalls the prohibition of a range of specific weapons systems – including, for example, the *ENMOD Convention: Convention on the prohibition of military or any other hostile use of environmental modification techniques*, New York, 10 December 1976 and the *Convention on Prohibitions or Restrictions on the Use of Certain Conventional Weapons Which May Be Deemed to Be Excessively Injurious or to Have Indiscriminate Effects as amended on 21 December 2001* – with its Additional Protocols, as well as weapons prohibited by customary international law. Yet, the most important feature of the review aims at determining if a swarm of drones, operating through distributive decision-making algorithms enabling them to delegate amongst themselves various tasks within the mission on which they are sent by their (human) military commander, comply with the general requirement of distinction (Article 51(4)(c)) and proportionality (Article 51(5)(b)). See further, the ICRC 'A Guide to the Legal Review of New Weapons, Means and Methods of Warfare: Measures to Implement Article 36 of Additional Protocol I of 1977', *International Review of the Red Cross* (2006) 88:864, 931–956.

References

Amoroso, D. and Tamburrini G. (2020) 'Autonomy in Weapons Systems and its Meaningful Human Control: A Differentiated and Prudential Approach', in Giacomello, G., Moro, F.N. and Valigi M. (eds.) *Technology and International Relations: The New Frontier in Global Power*, Cheltenham: Edward Elgar Publishing, pp. 45–66.

Arvidsson, M. (2011) 'Who Happens Here? Ethical Responsibility, Subjectivity, and Corporeality: Self-accounts in the Archive of the Coalition Provisional Authority (CPA) of Iraq', *No Foundations - Journal of Extreme Legal Positivism*, (8), pp. 71–122.

Arvidsson, M. (2018) 'Targeting, Gender, and International Posthumanitarian Law and Practice: Framing the Question of the Human in International Humanitarian Law', *Australian Feminist Law Journal*, 44(1), pp. 9–28. DOI: 10.1080/13200968.2018.1465331.

Arvidsson, M. (2020) 'The Swarm that We Already Are: Artificially Intelligent (AI) Swarming 'Insect Drones', Targeting and International Humanitarian Law in a Posthuman Ecology', *Journal of Human Rights and the Environment*, 11(1), pp. 114–137. DOI: 10.4337/jhre.2020.01.05.

Arvidsson, M. (2021) 'Who, or What, Is the Human of International Humanitarian Law?', in Pahuja, S. and Chalmers, S. (eds.) *Routledge Handbook of International Law and the Humanities*. London: Routledge 2021, pp. 422 –431.

Arvidsson, M. and Jones, E. (eds.) (2023) *International Law and Posthuman Theory*. Abingdon: Routledge.

Arquilla, J. and Ronfeldt, D. (2000) *Swarming and the Future of Conflict*. Santa Monica: RAND.

Bar-On Cohen, E. (2011) 'Events of Organicity: The State Abducts the War Machine', *Anthropological Theory* 11(3), pp. 259–282. DOI: 10.1177/1463499611416719.

BBC News (2014) 'Afghan Taliban Capture British Military Dog', *BBC News*, 6 February [Online]. Available at: https://www.bbc.com/news/world-asia-26062679 (Accessed: 26 August 2022).

Benjamin, W. (1978) 'Critique of Violence', in Benjamin, W., Jephcott, E.F.N. and Demetz, P. (eds.) *Reflections: Essays, Aphorisms, Autobiographical Writings*. New York: Harcourt Brace Jovanovich, pp. 277–300.

Birrell, K. and Dehm, J. (2021) 'International Law & the Humanities in the 'Anthropocene'', in Pahuja, S. and Chalmers, S. (eds.) *Routledge Handbook of International Law and the Humanities*. London: Routledge 2021, pp. 407–421.

Bothe, M., Bruch, C., Diamond, J. and Jensen, D. (2010) 'International Law Protecting the Environment During Armed Conflict: Gaps and Opportunities', *International Review of the Red Cross*, 92 (879), pp. 569–592.

Bourke, J. (2014) 'Killing in a Posthuman World: The Philosophy and Practice of Critical Military History', in Blagaard B. and van der Tuin, I. (eds.) *The Subject of Rosi Braidotti: Politics and Concepts*. London: Bloomsbury, pp. 29–37.

Braidotti, R. (2013) *The Posthuman*. Cambridge: Polity Press.

Braidotti, R. (2019) *Posthuman Knowledge*. Cambridge: Polity Press.

Braidotti, R. and Bigall S. (eds.) (2019) *Posthuman Ecologies: Complexity and Process after Deleuze*. London: Rowman & Littlefield.

Bruch, C., Payne, C. and Sjöstedt B. (2021) 'Armed Conflict and the Environment', in Rajamani, L. and Peel, J. (eds.) *The Oxford Handbook of International Environmental Law*. 2nd edn. Oxford: Oxford University Press, pp. 865–883.

Chakrabarty, D. (2009) 'The Climate of History: Four Theses', *Critical Inquiry* 35, pp. 197–222.

Convention on the Prohibition of Military or Any other Hostile Use of Environmental Modification Techniques (adopted 10 December 1976, entered into force 5 October 1978) 1108 UNTS 151 (ENMOD Convention).

Convention on the Prohibition of the Development, Production and Stockpiling of Bacteriological (Biological) and Toxin Weapons and on their Destruction (adopted 10 April 1972, entered into force 26 March 1975) 1015 UNTS 163 (Biological Weapon Convention).

Crawford, K. and Joler, V. (2018) Anatomy of an AI System: The Amazon Echo as An Anatomical Map of Human Labor, Data and Planetary Resources [Online]. Available at: https://anatomyof.ai (Accessed: 26 August 2022).

Davis, D. (2017) *Asking the Law Question*. 4th edn. Pyrmont: Thomson Reuters.

Deleuze, G. and Guattari F. (2013) *A Thousand Plateaus: Capitalism and Schizophrenia*. Translated by B. Massumi. London: Bloomsbury.

Dinstein, Y. (2001) 'Protection of the Environment in International Armed Conflict', *Max Plank Yearbook of United Nations Law*, 5, p. 523–549. DOI: 10.1163/187574101X00141.

Dinstein, Y. (2007) 'The System of Status Groups in International Humanitarian Law', in Heintschel Heinegg W. and Epping W. (eds.) *International Humanitarian Law Facing New Challenges*. Berlin: Springer, pp. 145–156.

Dublon, I. and Sumpter D.J.T. (2014) 'Flying Insect Swarms', *Current Biology* 24, pp. 828–830. DOI: 10.1016/j.cub.2014.07.009.

Edwards, S.J.A. (2000) *Swarming on the Battlefield: Past, Present, and Future*. Santa Monica: Rand.

Falk, R. (2000) 'Evaluating the Adequacy of Existing International Law Standards', in Austin, J. and Bruch, C. (eds.) *The environmental consequences of war: Legal, economic, and scientific perspectives*. Cambridge: Cambridge University Press, pp. 137–155.

Ferrando, F. (2018) 'Transhumanism/Posthumanism',' in Braidotti, R. and Hlavajova, M. (eds.) *Posthuman Glossary*. London: Bloomsbury, pp. 438–439.

Fleck, D. (ed.) (2021) *The Handbook of International Humanitarian Law*. Oxford: Oxford University Press.

Geneva Convention for the Amelioration of the Condition of the Wounded and Sick in Armed Forces in the Field (adopted 12 August 1949, entered into force 21 October 1950) 75 UNTS 31 (First Geneva Convention).

Geneva Convention for the Amelioration of the Condition of Wounded, Sick and Shipwrecked Members of Armed Forces at Sea (adopted 12 August 1949, entered into force 21 October 1950) 75 UNTS 85 (Second Geneva Convention).

Geneva Convention Relative to the Treatment of Prisoners of War (adopted 12 August 1949, entered into force 21 October 1950) 75 UNTS 135 (Third Geneva Convention).

Geneva Convention Relative to the Protection of Civilian Persons in Time of War (adopted 12 August 1949, entered into force 21 October 1950) 75 UNTS 287 (Fourth Geneva Convention).

Goussac, N. (2019) 'Safety Net or Tangled Web: Legal Reviews of AI in Weapons and War-Fighting', *Humanitarian Law & Policy Blog*, 18 April. Available at: https://blogs.icrc.org/law-and-policy/2019/04/18/safety-net-tangled-web-legal-reviews-ai-weapons-war-fighting/ (Accessed: 26 August 2022).

Gillespie, A. (2014) *International Environmental Law, Policy, and Ethics*. 2nd edn. Oxford: Oxford University Press.

Grear, A. (2015) 'Deconstructing Anthropos: A Critical Legal Reflection on "Anthropocentric" Law and Anthropocene "Humanity"', *Law and Critique*, 26, pp. 225–249. DOI: 10.1007/s10978-015-9161-0.

Grear, A. (2020) 'Legal Imaginaries and the Anthropocene: "Of" and "For"', *Law and Critique*, 31, pp. 351–366. DOI: 10.1007/s10978-020-09275-7.

Grimal, F. and Sundaram, J. (2018) 'Combat Drones: Hives, Swarms, and Autonomous Action?', *Journal of Conflict and Security Law*, 23(1), pp. 105–135. DOI: 10.1093/jcsl/kry008.

Hamson, F.J. (2008) 'The Relationship between International Humanitarian Law and Human Rights Law from the Perspective of a Human Rights Treaty Body' (2008) *International Review of the Red Cross*, 90(871), pp. 549–572. DOI: 10.1017/S1560775508000114.

Heathcote, G. (2018) 'War's Perpetuity: Disabled Bodies of War and the Exoskeleton of Equality', *Australian Feminist Law Journal*, 44(1), pp. 71–91. DOI: 10.1080/13200968.2018.1470447.

Henckaerts, J.-M. (ed.) (2005) *Customary International Humanitarian Law, Volume I: Rules*. Cambridge: Cambridge University Press.

Henckaerts, J.-M. and Doswald-Beck L. (eds.) (2005) *Customary International Humanitarian Law, Volume II: Princples*. Cambridge: Cambridge University Press.

Hulme, K. (2010) 'Taking Care to Protect the Environment Against Damage: A Meaningless Obligation?', *International Review of the Red Cross*, 92(879), pp. 675. DOI: 10.1017/S1816383110000512.

Hulme, K. (2004) *War Torn Environment: Interpreting the Legal Threshold*. Leiden: Hotei Publishing.

Hüttenrauch, M., Šošić, A. and Neumann, G. (2018). 'Local Communication Protocols for Learning Complex Swarm Behaviors with Deep Reinforcement Learning', in Dorigo, M., Birattari, M., Blum, C., Christensen, A., Reina, A. and Trianni, V. (eds.) *Swarm Intelligence. ANTS 2018. Lecture Notes in Computer Science*. Cham: Springer, pp.71–83.

ILC (2015) 'Second Report on the Protection of the Environment in Relation to Armed Conflicts' (28 May 2015), UN Doc A/CN.4/685.

ILC (2015) 'Report of the International Law Commission to the Sixty-Third Session, Annex E. Protection of the Environment in Relation to Armed Conflicts' (26 April–3 June and 4 July–12 August 2011) UN Doc A/66/10.

Johnson, J. (2020) 'Artificial Intelligence, Drone Swarming and Escalation Risks in Future Warfare', *The RUSI Journal*, 165, pp. 26–36. DOI: 10.1080/03071847.2020.1752026.

Jones, E. (2018) 'A Posthuman-Xenofeminist Analysis of the Discourse on Autonomous Weapons Systems and Other Killing Machines', *Australian Feminist Law Journal*, 44(1), pp. 93–118. DOI: 10.1080/13200968.2018.1465333.

Jones, E. and Otto, D. (2020) 'Thinking through Anthropocentrism in International Law: Queer Theory, Posthuman Feminism and the Postcolonial: A Conversation between Emily Jones (University of Essex) and Dianne Otto (University of Melbourne)' *LSE Woman, Peace and Security*, Working Papers Series 2020, 1. Available at: https://www.lse.ac.uk/women-peace-security/assets/documents/2020/Final-Jones-and-Otto-Anthropocentrism-Posthuman-Feminism-Postcol-and-IL-LSE-WPS-Blog-2019-002.pdf (Accessed: 26 August 2022).

Kallenborn, Z. and Bleek, P.C. (2018) 'Swarming Destruction: Drone Swarms and Chemical, Biological, Radiological, and Nuclear Weapons', *The Nonproliferation Review*, 25(5–6), pp. 523–543. DOI: 10.1080/10736700.2018.1546902.

Kalpouzos, I. (2020) 'Double Elevation: Autonomous Weapons and the Search for an Irreducible Law of War', *Leiden Journal of International Law*, 33(2), pp. 289–312. DOI: 10.1017/S0922156520000114.

Koppe, E.V. (2014) 'The Principle of Ambituity and the Prohibition Against Excessive Collateral Damage to the Environment during Armed Conflict', in Rayfuse R. (ed.) *War and the Environment: New Approaches to Protecting the Environment in Relation to Armed Conflict*. Leiden: Brill Nijhoff, pp. 59–90.

Leech, T.D.J. (1950) 'The Final Report of the Project "Seal"', 18 December [Online]. Available at http://www.wanttoknow.info/documents/project_seal.pdf (Accessed: 26 August 2022)

Legal Consequences of the Construction of a Wall in the Occupied Palestinian Territory (Advisory Opinion) [2004] ICJ Rep 136.

Legality of the Threat or Use of Nuclear Weapons (Advisory Opinion) [1996] ICJ Rep 226.

Lindahl, H.K. (2019) 'Inside and Outside Global Law', *Sydney Law Review* 41(1), pp. 1–34.

Mars Daily (2007) 'Congo Rebels Agree to Stop Killing Rare Mountain Gorillas', *Mars Daily*, 24 January [Online]. Available at: https://www.bbc.com/news/world-asia-26062679 (Accessed: 26 August 2022).

Orellana, M. (2005) 'Criminal Punishment for Environmental Damage: Individual and State Responsibility at a Crossroad', *Georgetown International Environmental Law Review*, 17(4), pp.673–696.

Orford, A. (2012) 'In Praise of Description', *Leiden Journal of International Law*, 25(3), pp. 609–625. DOI: 10.1017/S0922156512000301.

Otto, D. (2009) 'The Exile of Inclusion: Reflections on Gender Issues in International Law over the Last Decade', *Melbourne Journal of International Law*, 10(1), pp. 11–26. DOI: 10.3316/informit.981522747525571.

Parikka, J. (2018) 'Anthropocene', in Braidotti R. and Hlavajova M. (eds.) *Posthuman Glossary*. London: Bloomsbury, pp. 51–53.

Peters, A. and de Heptinne, J. (2022) 'Animals in Wartime: A Legal Research Agenda', in Peters, A. de Heptinne, J. and Kolb R. (eds.) *Animals in the International Law of Armed Conflict*. Cambridge: Cambridge University Press, pp. 3–27.

Pearlman, J. (2013) "Tsunami Bomb' Tested Off New Zealand Coast', *The Telegraph*, 1 January [Online]. Available at: https://www.telegraph.co.uk/news/worldnews/australiaandthepacific/newzealand/9774217/Tsunami-bomb-tested-off-New-Zealand-coast.html (Accessed: 26 August 2022).

Philippopoulos-Mihalopoulos, A. (2017) 'Critical Environmental Law as Method in the Anthropocene', in Philippopoulos-Mihalopoulos, A. and Brooks, V. (eds.), *Research Methods in Environmental Law*. Cheltenham: Edward Elgar Publishing, pp. 131–155.

Protocol Additional to the Geneva Conventions of 12 August 1949, and Relating to the Protection of Victims of International Armed Conflicts (adopted 8 June 1977, entered into force 7 December 1979) 1125 UNTS 3 (Additional Protocol I).

Protocol Additional to the Geneva Conventions of 12 August 1949, and Relating to the Protection of Victims of Non-International Armed Conflicts (adopted 8 June 1977, entered into force 7 December 1979) 1125 UNTS 609 (Protocol II).

Robinson, A. (2010) 'Why Deleuze (Still) Matters: States, War-Machines and Radical Transformation', *Ceasefire Magazine*, 10 September [Online]. Available at: https://ceasefiremagazine.co.uk/in-theory-deleuze-war-machine/ (Accessed: 26 August 2022).

Roden, D. (2017) 'On Reason and Spectral Machines: Robert Brandom and Bounded Posthumanism', in Braidotti R. and Dolphijn R. (eds.), *Philosophy After Nature*. London: Rowman and Littlefield, pp. 99–119.

Sandoz, Y., Swinarski, C. and Zimmerman, B. (2017) *Commentary on the Additional Protocols*. Cambridge: Cambridge University Press.

Schafer, B.K. (1989) 'The Relationship Between the International Laws of Armed Conflict and Environmental Protection: The Need to Reevaluate What Types of Conduct are Permissible During Hostilities', *California Western International Law Journal*, 19(2), pp. 287–325.

Sharkey, N. (2016) 'Staying in the Loop: Human Supervisory Control of Weapons', in Bhuta, N., Beck, S., Geiβ, R., Liu, H.-Y. and Kreβ, C. (eds) *Autonomous Weapons Systems: Law, Ethics, Policy*. Cambridge: Cambridge University Press, pp. 23–38.

Simonds, S. (1992–1993) 'Conventional Warfare and Environmental Protection', *Stanford Journal of International Law*, 29(1), pp. 165–221.

Sjöstedt, B. (2020) *The Role of Multilateral Environmental Agreements: A Reconciliatory Approach to Environmental Protection in Armed Conflict*. Oxford: Hart Publishing.

Stilwell, B. (2020) 'This Canine Prisoner of War is Still Held by Taliban Captors', *We Are the Mighty* 29 April [Online]. Available at: https://www.wearethemighty.com/military-culture/working-dogs-captured-by-taliban?rebelltitem=3#rebelltitem3 (Accessed: 29 August 2022).

U.S. Army UAS Center of Excellence (2010) *U.S. Army Roadmap for UAS 2010–2035: The Eyes of the Army*.

Verwey, W. (1995) 'Protection of the Environment in Times of Armed Conflict: In Search of a New Legal Perspective', *Leiden Journal of International Law*, 8, pp. 7–40. DOI: 10.1017/S0922156500003083.

Westing, A. (1976) *Ecological Consequences of the Second Indochina War*. Taylor & Francis.

Wilcox, L. (2017a) 'Drones, Swarms and Becoming Insect: Feminist Utopias and Posthuman Politics', *Feminist Review*, 116, pp. 25–45. DOI: 10.1057/s41305-017-0071-x.

Wilcox, L. (2017b) 'Embodying Algorithmic War: Gender, Race, and the Posthuman in Drone Warfare', *Security Dialogue*, 48(1), pp. 11–28. DOI: 10.1177/0967010616657947.

Wishermann, C. (2022) 'Historical Perspectives of Animal Involvement in Wartime', in Peters, A. de Heptinne, J. and Kolb R. (eds.) *Animals in the International Law of Armed Conflict*. Cambridge: Cambridge University Press.

Wilmshurst, E. and Beau, S. (eds.) (2011) *Perspectives on the ICRC Study on Customary International Humanitarian Law*. Cambridge: Cambridge University Press.

SECTION 2

Conceptualising the Anthropocentrism of International Law

7

ANTHROPOCENTRISM AND CRITICAL APPROACHES TO INTERNATIONAL LAW

Hélène Mayrand and Valérie Chevrier-Marineau

Introduction

Critical approaches to international law have been on the rise since the 1990s. While the meaning of 'critical' is itself contested, critical approaches refers to approaches that reject law's alleged objectivity and neutrality, as well as shared political projects aimed at analyzing and denouncing power relations and exclusions enshrined in international law (Bachand, 2013; Frankenberg, 2010; Bianchi, 2016). To put it more simply, for critical approaches, international law is politics (Koskenniemi, 1990).

Critical approaches to international law refer to a variety of approaches, with diverse theoretical underpinnings that mobilize concepts and methodologies from other disciplines. For example, the roots of some critical approaches to law are often traced back to Marx and Weber, the Frankfurt School (Horkheimer, Marcuse, Adorno and later Habermas), the French philosophers Foucault, Bourdieu and Derrida, and American legal realism (Holmes) (Bachand, 2013; Frankenberg, 2010; Bianchi, 2016). Other approaches have been developed as part of social movements, including the decolonisation, feminist and environmental justice movements. Critical Legal Studies (CLS), which arose in the American domestic context in the 1970s, has been particularly influential to instill a critical turn in the international legal scholarship. Associated with the Left, CLS are not homogenous and include subcategories of approaches taking different perspectives (class, race, gender and intersectional approaches) to analyze power relations (Unger, 1986; Bianchi, 2016). CLS's challenge to law's objectivity and the indeterminacy thesis applied to rules travelled to the international sphere from the late 1980s. New Approaches to International Law (NAIL), neomarxist approaches, feminist approaches, Third World Approches to International Law (TWAIL) and various other intersectional approaches, taking into account different factors of social vulnerability, have emerged to contest the dominant representation of international law as value-neutral and progressive.

More recently, critical approaches to international law and the environment – broadly defined as approaches aimed at deconstructing human–nature power relations in international law – have also made their way into legal scholarship. Following a first wave of environmentally inclined scholars who focussed on developing a new branch of international law, greening existing regimes and finding new tools (legal or economic) to address environmental problems, some started to question whether international law, including international environmental law, was in

DOI: 10.4324/9781003201120-10

fact part of the problem. As environmental issues became more pressing even in the context of the multiplication of international environmental agreements to address them, critical scholars pointed out that international law failed to contribute significantly to better environmental protection. To formulate their critique, some were inspired by and adopted concepts and methodologies from more established critical approaches to international law, including NAIL, TWAIL and feminist approaches. Others have incorporated theories and methods borrowed from other disciplines, including from philosophy, history, sociology, anthropology, critical geography and geology. This has given rise to a rich and diverse critical literature on international law's continued implications in the constitution and degradation of the environment.

This chapter is aimed at analyzing the relationship between critical approaches to international law and anthropocentrism. While there are different understandings of anthropocentrism as a concept, we look in particular at whether scholars formulating critical approaches to international law have adopted positions of *descriptive anthropocentrism* and/or *normative anthropocentrism*, as described by Ben Mylius (2018), or have attempted to move beyond anthropocentrism altogether (both descriptive and normative).

Descriptive anthropocentrism refers to research that 'begins from, revolves around, focusses on, takes as its reference point, is centered around, or is ordered according to the species Homo sapiens or the category of "the human"' (Mylius, 2018, p. 168). Descriptive anthropocentrism can take various forms. In this chapter, we look at its more common illustrations in the critical legal literature. *Descriptive anthropocentrism by omission* limits the analysis to certain human-centred concepts and removes the human from larger geological, ecological and/or evolutionary contexts. Such an approach is implicit in social sciences that restrict the object of their studies to human behaviour and institutions (Mylius, 2018, p. 171). *Descriptive anthropocentrism by extrapolation* refers to research that universalizes concepts developed in the context of the study of human beings. Such research presents its concepts as applicable to all relations, including with nonhumans and inhuman components of the environment, either explicitly or implicitly (Mylius, 2018, p. 175). *Descriptive anthropocentrism by separation* refers to research that distinguishes human beings from nonhumans based on some capacity or feature (Mylius, 2018, p. 181).

Normative anthropocentrism can take two different forms: *active normative anthropocentrism* or *passive normative anthropocentrism*. An approach that is actively normatively anthropocentric argues that humans are superior to nature, the most valuable of beings, the only ones that should have legal rights, and so on. An approach that is passively normatively anthropocentric is the other side of the coin of descriptive anthropocentrism. Relying on a human–centred ontology or methodology constrains thinking in certain ways, limits our understandings of nature and perpetuates anthropocentrism (Mylius, 2018, p. 183).

In this chapter, we present the main critical approaches to international law that have been developed since the 1980s in the academic writings of *legal* scholars. In the first part of the chapter, we focus on critical approaches relying on structuralism to deconstruct the international legal discourse and reveal its political dimension. These approaches encompass NAIL, as well as scholars taking the structural endeavour to deconstruct the narrative surrounding the positive evolution of international law towards better environmental conditions and protection. In the second part of the chapter, we present Marxist and neomarxist approaches to international law. Third, we move to TWAIL, including more recent research underlying the relationship between colonialism and the domination of nature. Fourth, we present feminist and ecofeminist legal scholarship before turning to environmental justice. Finally, we present posthuman critical approaches to law.

Of course, we do not pretend that the chapter is exhaustive, nor that scholars are confined to specific categories. Indeed, the research of some scholars can fit in more than one approach.

Nonetheless, we argue that this broad categorization, imperfect as it may be, helps to see that some concepts are used and shared among scholars engaged in critical approaches to international law in ways that shed light on the relationship of these concepts to anthropocentrism. Critical approaches to international law presented below are almost all descriptively anthropocentric (by omission, extrapolation or separation) and passively normatively anthropocentric. The focus of their critique has been on human beings and using human-centred concepts including law as a structure, a language or a discourse, human domination, capitalism, colonialism, patriarchy, injustice, and so on. Their ontological and methodological choices (whether implicit or explicit) constrain our understandings of nature and perpetuate anthropocentrism in international law. We do not argue, however, that their projects are not worth pursuing to achieve human emancipation. We only stress that the anthropocentric bias is also dominant in critical approaches to international law. In contrast, posthuman critical approaches to law are aimed at bringing a fundamental paradigm shift, presenting non-anthropocentric concepts and methods to comprehend the human–nature relationship in a much broader context.

Structuralist Approaches to International Law

Critical approaches to international law, like many other approaches, have been developed through a sort of *bricolage*, using vocabularies and concepts 'lying around' and combining them to make new arguments.[1] One of those vocabularies has been that of structuralism, which can be broadly defined as 'thinking about the world which is predominantly concerned with the perception and description of structures' (Hawkes, 2003, p. 6). Structuralism considers that there is no true nature of things which can be objectively determined. 'Reality' is inherently influenced by biases which determine how we construct and perceive the relationship between these things (Hawkes, 2003). Structuralism has been applied in linguistics, for example, distinguishing between *langue* (linguistic structure of language) and *parole* (individual acts of speech) (Saussure, 1966). Drawing an analogy between law and language, structuralism was transposed to law notably by scholars associated with CLS (Heller, 1984). Structuralism as applied to law implies that we can understand law as a structure, often referred to as a language or discourse (Koskenniemi, 2016). While this language has its own rules, principles and procedures, the application of these rules, principles and procedures do not mandate a particular outcome. They are not chosen from randomly, nor do they result from rational choices by autonomous individuals. Outcomes are, instead, determined by structural biases (Koskenniemi, 2016).

In this section, we present scholars who have been influenced by structuralism and have understood international law as a language or discourse, deconstructing its apparent neutrality and objectivity. In the first subsection, we focus on the work of David Kennedy and Martti Koskenniemi, who have been associated with NAIL. In the second subsection, we present scholars who have also understood international law as a language or discourse to reveal that nature or the environment as subordinated due to certain biases. As explained below, the use of structuralism to deconstruct international legal discourse only focusses on one sociological aspect of human relations through the human concept of structure, making such an approach descriptively and passively normatively anthropocentric.

New Approaches to International Law

NAIL, also referred to as the New Stream (Frankenberg, 2010; Bianchi, 2016; Kennedy, 1988), arose in the late 1980s and early 1990s under the leadership of two scholars, Kennedy and Koskenniemi. NAIL was particularly influential in introducing critical theory into international legal scholarship. As Klabbers put it, 'the "newstream" has become the mainstream, if not politically then at least academically' (Klabbers, 2015, p. 471). They instilled renewal not only in international legal theory but also in international practise through an understanding of the role of international law as a 'political project' (Koskenniemi, 2007, p. 29). They directly or indirectly inspired more specific critical approaches to international law adopting a specific angle for their critique, including that of the environment.

Kennedy and Koskenniemi argued that international law can be understood through structuralism, by analogy with linguistics (Kennedy, 1987; Koskenniemi, 1989). The structure of international legal discourse, as described in Koskenniemi's formative book *From Apology to Utopia*, is characterized by a dialectic mode of argumentation which oscillates between two opposite poles: the concrete and the normative. Each argument is subject to the corresponding critique of apology or utopia, i.e., being too close to state practice and power structures or being idealistic and too removed from the actual social context. While international law has its own internal structure or grammar, Kennedy and Koskenniemi stress that it is intrinsically indeterminate. Rather than being an 'order' or a set of coherent rules and principles separate from their historical, social and political contexts, international law is based on contradictory premises, tensions and dichotomies. Kennedy and Koskenniemi have devoted their work to deconstructing mainstream assumptions of international law's objectivity and linear progress embedded in liberal ideology.

While international law is indeterminate, Kennedy and Koskenniemi reveal that its technical and sophisticated vocabulary in fact hides ideologies, attitudes and structures (Kennedy, 1980) or structural biases (Koskenniemi, 1989). It is possible to justify all kinds of practises with sound legal arguments that are valid from a legal point of view, but there is always a bias in international institutions and experts influencing these institutions (Kennedy, 2016) that favour certain outcomes. The managerial mindset in favour of compliance, effectiveness, legitimacy and technical management of international problems is a further attempt to hide the fact that international law is a site for political struggles and contestations (Koskenniemi, 1989; Kennedy, 2013 p. 7; Bianchi, 2016, pp. 155, 172).

Kennedy and Koskenniemi also deconstructed the mainstream historical narrative surrounding international law, especially the claims of a progressive and linear evolution and advancement of international law towards an improving human condition and greater justice. Rather, international law, including international lawyers, is 'part of the problem' (Kennedy, 2000 p. 460; Kennedy, 2013) and has, from the start, contributed to the production of inequality and injustice. For example, Kennedy has shed light on the 'dark sides' of global governance (Kennedy, 1999; Kennedy, 2008), international humanitarianism (Kennedy, 2004; Kennedy, 2006) and international human rights (Kennedy, 2002; Kennedy, 2012; Kennedy, 2013). Koskenniemi has described the evolution of international law resulting from the influence of 'people with projects', this history of the past shedding light on the injustices enshrined in contemporary international law (Koskenniemi, 2001; Koskenniemi, 2021; Bianchi, 2016, p. 168).

The work of Kennedy and Koskenniemi is descriptively and passively normatively anthropocentric. Nature is generally absent from their analysis, or only referred to in passages presenting particular arguments surrounding legal concepts, such as dominium or freedom of trade with an obvious anthropocentric focus (Koskenniemi, 2021). They suggest that structuralism can explain what international law really is but omit its human-specific context. In doing so, they do not come far from falling under the category of descriptively anthropocentric by extrapolation, applying structuralism as a universal concept. Moreover, they limit their analysis to law as a discipline, and to political projects and biases and are thus descriptively anthropocentric by omission. While their critique is particularly powerful in revealing the political and human constructed nature of law, rules, principles and procedures, the boundaries set by their ontological and methodological choices limit the possibilities for rethinking the human–nature relationship in international law.

Structuralism and the Environment

In this subsection, we present the research of three scholars – Ileana Porras, Hélène Mayrand and Vito de Lucia – who have relied on concepts and methodologies inspired by structuralism but have sought to problematize the relationship of law with nature. These scholars have focussed on international law as a language or discourse in which nature or the environment is subordinated as a result of biases. They have taken a historical perspective on the development of international law generally, international environmental law and specific environmental concepts.

Porras has drawn attention to the historical conception of nature that still inhabits the discipline today (Porras, 2014). She has focussed on the work of the so-called 'founding fathers of international law' Francisco de Vitoria, Alberico Gentili, Hugo Grotius and Emer de Vattel. She shows how nature was conceptualized by European founders of the discipline for the benefit and use of Europeans, especially when it comes to the fundamental right to engage in commerce and the notions of scarcity and plenty. Nature in places distant from Europe was conceived as resources available to appropriation and exploitation. She explains that

> [d]espite repeated experiences of local scarcity or exhaustion of resources elsewhere resulting from over-exploitation, Europeans remained convinced that there was plenty somewhere else for the taking. [T]he natural world that was valued was that which was productive, placed under cultivation, or otherwise exploited. Nature in its natural state, unmodified or lightly used by human beings, was viewed as a wasteland in need of an industrious owner.
>
> *(Porras, 2014, p. 660)*

While drawing attention to the political choices in favour of anthropocentric concepts that were at the heart of the development of international law, Porras' approach nonetheless remains descriptively and passively normatively anthropocentric. Indeed, she limits the object of her research to human discourses in human-defined structures (descriptive anthropocentrism by omission), which in turn does not hint at how one might change these structures through non-anthropocentric concepts.

Porras (2015) has also critically analyzed the principle of sustainable development, again from a descriptively and passively normatively anthropocentric perspective, her ontology being centred on humans and human institutions analyzed through the lens of structuralism. She looks at how the concepts of scarcity, need and consumption developed in economics are impacted by hyper and irrational human consumption patterns. She draws a distinction between humans and animals through the 'capacity to generate new material needs and [the human] ability to devise the means to fulfil those needs' (Porras, 2015, p. 34). By doing so, she also adopts an approach that is descriptively anthropocentric by separation. She argues that the discourse surrounding sustainable development has fed a fear of scarcity, which in turn has led to binge development. She put her findings in a historical context, underlining the special role scarcity has played in international law, especially in relation to the right to engage in commerce at the heart of the European international law project. She sheds light on how international law 'is structurally oriented towards enabling trade and promoting consumption-based economic growth' (Porras, 2015, p. 71). She suggests some human-centred ways to address environmental and social exploitation resulting from binge development, which include replacing the principle of sustainable development by that of environmental justice and encountering scarcity without fear (Porras, 2015, pp. 81–83).

Mayrand (2020), one author of this chapter, has also relied on structuralism as well as a methodology inspired by neomarxism to show the influence of ideologies of classical liberalism, welfarist liberalism and neoliberalism on key legal texts. She deconstructs through a neomarxist methodology of ideology critique the prevailing Western understanding of the positive evolution of international environmental law towards better environmental protection. She argues that instead of changing the problematic understandings of nature in international law, international environmental law relies on classical liberal (anthropocentric) understandings of nature. These understandings are further exacerbated under the influence of neoliberalism, with the focus on economic rationality favouring cost–benefit analysis, deformalization of law, deregulation and self-regulation by private actors, management by experts, and market mechanisms to address environmental problems. Like Porras, Mayrand's research is descriptively and passively normatively anthropocentric. She seeks to show how nature is conceived from an anthropocentric perspective embedded in international law, but she limits her research to human-centred ontology and methods and does not provide non-anthropocentric concepts to move beyond such perspective.

Vito De Lucia (2015; 2019) has analyzed the 'ecosystem approach' as part of international environmental law. Acknowledging the influence of CLS on his research, he mobilizes the methodology of 'analytics of biopolitics' inspired by Foucault (De Lucia, 2019, p. 23). Such methodology allows him to apply a genealogical method to reveal the historical contingencies and contradictions involved in the various understandings of the ecosystem approach, supported by different ideological projects. He identifies, in particular, two competing logics underlying the ecosystem approach, those of anthropocentrism and ecocentrism. He deconstructs the anthropocentric/ecocentric binary opposition as a starting point 'to articulate a biopolitical understanding of law as it relates to the environment' (De Lucia, 2019, p. 255). Relying on Foucault and his followers – Agamben, Hardt and Negri, and Esposito – De Vito expands the conceptual framework of biopolitics applied in the context of human bodies and populations to nature. By doing so, his approach is descriptively anthropocentric by extrapolation, applying a concept developed via the study of human beings to nonhumans. He argues that biopolitics create the aporetic mechanisms that '[l]ife is subjected and subjugated even as it is fostered and enhanced' (De Lucia, 2019, p. 150). He then illustrates those contradictory mechanisms at play in the context of the ecosystem approach and the *Convention on Biological Diversity*. De Lucia's conceptual tools and methodology remain, however, anthropocentric and centred on the genealogy of human ideologies surrounding the ecosystem approach. De Lucia's approach is aimed at rethinking law in the context of the Anthropocene and moving beyond legal modernity and the current biopolitical focus 'entailing *more* control, *more* management, *more* subsumption of life under the care of biopower' (De Lucia, 2019, p. 256). His research remains nonetheless passively normatively anthropocentric, as his normative endeavour is based on a framework (analytics of biopolitics) which remains centred on human concepts and institutions.

Thus, Porras, Mayrand and De Lucia have revealed, through their structuralist analysis, problematic understandings of nature in international law, international environmental law and the concepts of sustainable development and the ecosystem approach. The strength of structuralism is to illustrate how what we perceive as true or natural, including the anthropocentric understanding of nature in international law, is not objectively determined but based on political preferences and biases. Such approaches remain descriptively and passively normatively anthropocentric, in ways that limit our ultimate understanding of the human–nature relationship through human politics. Nonetheless, as De Lucia puts it, such approaches open a path for 'a conceptual displacement that *unthinks* the categories of legal modernity' (De Lucia, 2019, p. 252).

Marxist and Neomarxist Approaches to International Law

Marxism has been instrumental to critical thinking, including critical approaches to international law. Marx has developed key concepts which have been relied upon by scholars for advancing their critique of international law and institutions. This section presents these key Marxist concepts as well as new or neomarxist approaches to international law, which have been developed since the 2000s with the work of Susan Marks, Robert Cox and TWAIL scholar Bhupinder S. Chimni.

This section shows that while Marxist concepts and methods, such as historical materialism, dialectical thinking and ideology critique, are at the heart of critical approaches to international law, they remain centred on humans and significantly limit our understanding of nature, which is reduced to material conditions. Moreover, apart from Mayrand's research presented above using ideology critique as her methodology, no projects from *legal* scholars under the umbrella of neomarxism were aimed at denouncing power relations in relation to nature.

Marxist Concepts and Methods

One key method advanced by Marx and relied upon by critical approaches to international law is historical materialism, based on the idea that 'History was not produced by such abstractions but by concrete human individuals acting within material conditions that enabled such events to take place' (Koskenniemi, 2004, p. 232). Marx was particularly critical of human rights, which represents 'yet another personification of something transcendental over human species–nature' (Koskenniemi, 2004, p. 234), in which the abstract individual was in fact the bourgeois egoistic man understood in isolation from other men and the community. The ontological focus of historical materialism is on human societies, and accordingly, it is descriptively anthropocentric by omission. Moreover, it is based on a premise that nature and its components are part of 'material conditions', which are considered from an instrumental and economic value, especially through relations of production and exchange. Using historical materialism leads critical scholars to implicitly integrate and promote the bias that humans should dominate nature and that it has no intrinsic value. Such an approach is in Mylius' terms, actively normatively anthropocentric (Mylius, 2018, p. 185).

Marx also introduced dialectical thinking according to which any social reality is comprised of inherent tensions between opposite elements, which 'shows the historical contingency of the social' (Koskenniemi, 2004, p. 236; Bianchi, 2016, p. 81). Again, by focussing on social reality and excluding the nonhuman from the inquiry, the dialectic method is descriptively anthropocentric by omission, as well as passively normatively anthropocentric.

Ideology critique as method is also associated with Marx (and Engels) (Marx, 1845). Ideology refers to ideas and beliefs presented as objective and natural 'to establish and sustain relations of domination' (Knox, 2016, p. 320). According to Marks, ideology used as a critical concept can reveal how certain ideas in fact dissimulate and legitimize domination through different modes of operation (Marks, 2003, p. 19; Thompson, 1990). Ideology critique as method can be used 'to analyse international legal processes through which asymmetrical power relations are legitimated, obscured, denied, reified, naturalized, or otherwise supported' (Marks, 2003, p. 19). Ideology critique has been mobilized by neomarxist approaches (also referred to as neo-Gramscian), notably through the idea of social forces that are shaping the construction of hegemonic projects (Cox, 1981). Indeed, neomarxists have emancipated themselves from the orthodox Marxist approach. They maintain that ideology has a significant cultural component and that it allows structures to reproduce themselves internationally (Cox, 1983, p. 169). Such method remains centred on an anthropocentric ontology. Nonetheless, as illustrated through the work of Mayrand presented above, it can be used to deconstruct how ideologies have promoted an economic and instrumental understanding of nature in international law and international environmental law.

Another important Marxist insight is that the world should be grasped as a totality, as a global social theory. However, neomarxist scholars have set aside the concept of economic determinism to take into consideration 'the larger framework' of which international is part (Marks, 2011, p. 71). For example, Chimni puts forward the role that international law plays in this socially constructed world, which relied on 'the vision of globalised capitalist' and how a *transnational counter-hegemonic project* could emerge in such context (Chimni, 1999, p. 346; Chimni, 2004, p. 27). However, this 'totality' or 'larger framework' is in fact limited to its *human* social and political dimensions and is framed from an anthropocentric perspective. It extrapolates concepts developed for the study of human relations to all relations (descriptively anthropocentric by extrapolation).

Neomarxism and International Law

While dialectic reasoning and ideology critique can be used to deconstruct the anthropocentric understanding of nature in international law, nature or the environment are absent from current neomarxist legal literature. Marks and Chimni have dedicated some of their work to argue for the relevance of Marxism and Marxist concepts to critically analyze international law. Marks (2003; 2011; 2019) applied her critical tools, including some inspired by structuralism and NAIL, to the critique of constitutional and human rights. For example, she deconstructed the prevailing representations of human rights violations, which hide the systematic context of abuses and vulnerabilities from view (Marks, 2011). Also part of the TWAIL movement, Chimni has focussed on the transnational capitalist class and how international institutions, including international law, serve the interests of this class to the disadvantage of subaltern classes (Chimni, 2004).

Therefore, neomarxist projects from legal scholars are aimed at achieving human emancipation from human domination understood through social relations (class). As such, they are descriptively and passively normatively anthropocentric. Nature or the environment remains, at best, absent, or reduced to material conditions.

Third World Approches to International Law

Third World Approaches to International Law (TWAIL) is a *coalitionary movement* or a *loose network* (Mutua, 2000, p. 38; Gathii, 2011, p. 37) that emerged at the end of the 1990s. Despite divergent adherence to several critical approaches, TWAIL is inclusive of scholars in all hemispheres (including the West) having in common the analysis of the social, economic, political and legal injustices that affect the most vulnerable and marginalized people in the Third World (Mutua, 2000, p. 38; Anghie and Chimni, 2003, p. 78). Scholars associated with TWAIL include Antony Anghie, Bhupinder S. Chimni, Vasuki Nesiah, Obiora C. Okafor, Balakrishnan Rajagopal, James Thuo Gathii and Makau Wa Mutua. Usha Natarajan, Kishan Khoday and Karin Mickelson use TWAIL concepts and methods to demonstrate through examples that domination of nature is intertwined with colonialism and domination of peoples in the Third World.

As presented in this section, TWAIL projects are descriptively and passively normatively anthropocentric, their ontological focus being on the domination of peoples of the Third World, even if nature also emerges as a victim of domination in recent projects.

TWAIL Concepts and Methods

In line with structuralism, TWAILers understand international law as a language or discourse. They denounce international law's source in colonialism and its tendency to favour the most powerful at the expense of Third World peoples. They seek above all to challenge the imperial character of the established order and to assert colonialism as a central element in the history of international law. As Mutua puts it, international law 'is a predatory system that legitimizes, reproduces and sustains the plunder and subordination of the Third World by the West' (Mutua, 2000, p. 31). In so doing, they wish to foreground the will and aspirations of Third World peoples, who constitute most of the world, as an integral part of international law. Imperialism and colonialism as a mechanism of domination are therefore key TWAIL concepts to account for international law's inherent problems: '[t]he vocabulary of international law, far from being neutral, or abstract, is mired in this history of subordinating and extinguishing alien cultures' (Anghie, 1996, p. 333).

By focussing on historical and cultural dichotomies within international law and the power relations inherent in them, TWAIL scholars have mobilized methodological and epistemological strategies, such as a hermeneutics of suspicion. As is the case for many critical approaches, the historical perspective is central to the critical positioning of TWAIL, as they constantly seek to question and develop a radical critical analysis of notions considered as cornerstones of international law, including state sovereignty and the principle of universality (Sunter, 2007, p. 494). The boundaries set by TWAIL's ontological and methodological choices make this approach descriptively anthropocentric by omission, and passively normatively anthropocentric.

TWAIL and the Environment

While most of TWAIL scholarship has not been concerned with environmental issues (Natarajan, 2017), some TWAILers have been among the first to denounce how international law is constructed by and for developed countries, resulting in a greater environmental burden on Third World countries and justifying, with the help of international law, interventions and control of their resources under the pretext of better environmental protection. For example, Anghie (1993) took the *Nauru v. Australia Case* (1992) to illustrate how the enduring conception of international law as a 'civilizing mission' embedded in colonialism has impacted the doctrines of self-determination and permanent sovereignty over natural resources to allow exploitation and destruction of Nauru through mining. Such an account, however, remains anthropocentric with an instrumental understanding of nature as 'natural resources'.

Since the 2000s, Natarajan, Khoday and Mickelson developed an approach centred on nature as part of the broader TWAIL umbrella. They have dedicated much of their writing to denounce environmental injustice as the means by which the West has managed to keep Third World peoples in a state of subalternity (Mickelson, 2007; Mickelson, 2014; Khoday and Natarajan, 2012; Natarajan, 2012; Mickelson and Natarajan, 2017). They stress the problematic understandings of nature in international law, which has its roots in the European philosophy of the Enlightenment and which has been imposed on Third World states as part of the civilizing mission. They point to how international law has discredited Indigenous laws and customs and how civilizing Third World peoples included the appropriation and exploitation of nature according to a European industrial development model (Mickelson, 2014; Khoday and Natarajan, 2012, p. 426). Recognizing that these understandings of nature are deeply embedded in international law, they ask for their radical transformation.

TWAILers adopting an environmental perspective have especially focussed on the question of fairness and how 'just as the wealthy receive disproportionate benefit from natural resources and development, the poor bear a disproportionate burden of resource insecurity, toxic impacts, and other environmental crises' (Natarajan, 2017, p. 230). They stress the role that international law and international environmental law play in sustaining such environmental injustices. This focus on fairness and justice in the environmental context is also at the heart of the environmental justice movement as analyzed below. What generally distinguishes TWAILers from environmental justice scholars, apart from their explicit association with one of the movements, is TWAILers' method of relying on history to emphasize that these injustices have their roots in colonialism.

In a paper aimed at 'locating nature' in international law, Natarajan and Khoday deconstruct the narrative surrounding the positive and linear progress towards better environmental protection. Through a historical perspective, they highlight the problem of how nature is conceived from an impoverished perspective, as reflected in key principles of international law – sovereignty, development, property, economy, human rights – and the specialized field of international environmental law (IEL) (Natarajan and Khoday, 2014). While international environmental law has put at the forefront injustice and inequality among states through the principles of sustainable development and common but differentiated responsibilities, Natarajan and Khoday argue that these principles 'barely scratch the surface of the developing world's longstanding conviction that IEL is essentially unfair to them' (Natarajan and Khoday, 2014, p. 579). They underline how, in international law, nature is conceptualized as resources, that economic development should have no limit and that addressing environmental degradation can be achieved through a managerial approach. Their research remains, however, descriptively anthropocentric by omission, and passively normatively anthropocentric due to the concepts (principles of international and international environmental law, fairness) and method (deconstruction from a historical perspective) they mobilize.

Mickelson analyzes South–North relations in the context of climate change and international environmental law more broadly, critically evaluating the diverse conceptions of justice from a legal as well as a political perspective (Mickelson, 2000; Mickelson, 2009). She inquires how rethinking the South–North divide might promote a real engagement with inequality. Mickelson has also focussed on the doctrines of *terra nullius, res communis* and the principle of common heritage of mankind, that have legitimized the appropriation and exploitation of nature as well as military, political and economic interventions in Third World states (Mickelson, 2014). With respect to the principles of sustainable development and common but differentiated responsibilities, Mickelson stresses that, while these principles appear to address concerns of Third Word states, they should be questioned through broader historical, social and political contexts. As was the case for Natarajan and Khoday, Mickelson adopts an anthropocentric ontology, linking ecological harm to the domination over Third World peoples, as well as human-centred concepts and methods to conduct her research.

In terms of emancipatory avenues, the focus of TWAILers remains on human emancipation, namely Third Word peoples. They underline, for instance, that peoples, social movements, scholars, as well as states in the Third World are in a strategic position to resist the status quo and influence international environmental law in more equitable directions (Khoday and Natarajan,

2012; Natarajan, 2012; Natarajan, 2017). Natarajan, Khoday and Mickelson acknowledge that international law rests on anthropocentric principles (Natarajan and Khoday, 2014, p. 593; Mickelson, 2014, p. 637). Natarajan and Khoday also stress that the anthropocentricity of international law is another area of much needed deconstruction and reconstruction to develop new understandings of nature. While pointing to the need for this normative shift, the political project behind TWAIL constrains the inquiry in a way that implicitly privileges human beings.

Feminism

The further development of critical approaches to international law in the 1990s has also given rise to 'Feminist Approaches to International Law', as described in the groundbreaking article written by Hilary Charlesworth, Christine Chinkin and Shelley Wright (Charlesworth, Chinkin, and Wright, 1991; Otto, 2016, p. 488; Chinkin, 2010). According to these approaches, 'international law is a thoroughly gendered system, because its normative and institutional structures and practices privilege men and male power while masquerading as universal, making the law unreceptive to the voices and experiences of women' (Otto, 2016, p. 494; Charlesworth, Chinkin, and Wright, 1991, p. 615).

In this section, we discuss some of the main concepts and methods developed under these approaches through the writings of Diane Otto, Hilary Charlesworth, Christine Chinkin, Sandra Harding, among others. It is relevant to note that feminist approaches to international law are not monolithic and contain various traditions or schools. For the purpose of this chapter, we focus on the critical tradition. Such tradition is mainly wary of liberal feminism and some of its limitations (MacKinnon, 1987; Arruzza, Bhattacharya, and Fraser, 2019; Charlesworth and Chinkin, 2000, p. 32). Like NAIL, Marxism and neomarxism and TWAIL, feminist approaches have been key in developing concepts mobilized by critical approaches to law and the environment, more broadly, and also inspired ecofeminism as presented below.

As it is the case for other critical approaches taking a particular human standpoint as the centre of their research, feminist approaches to international law have been descriptively and passively normatively anthropocentric. Even new concepts developed in the context of ecofeminism remain centred on human concepts (domination and vulnerability), which are extrapolated to nonhuman beings.

Feminist Concepts and Methods

One key feminist concept is 'gender', which has to be understood through the processes of essentialism and naturalization. As a social construction, gender represents 'a systematically analytical category that refers to constructions of (privileged) masculinity and (devalorised) femininity and their ideological effects' (Petterson, 2004, p. 40). Such construction involves essentialism, a process according to which women are understood as having 'a fixed "essence" or set of characteristics' (Charlesworth and Chinkin, 2000, p. 52). As Elisabeth Grosz explains, 'essentialist theories rationalize and neutralize the prevailing sexual division of social roles by assuming that these are the only, or the best, possibilities, given the confines of the nature, essence, or biology of the two sexes' (Grosz, 1990, p. 332).

Such essentialized understanding of women is then naturalized in the system, making it seem 'natural'. Feminists have been particularly alert to such moves to explain the domination over women and other forms of domination. As Petterson explains, 'feminists theorize that not only gender hierarchy but domination more generally is naturalized (depoliticized, legitimated) by denigration of the feminine, and it is the feminisation of "others" that links multiple oppressions' (Petterson, 2004, p. 40). By positioning gender at the core of their analysis, they limit their research to human-centred concepts (descriptively anthropocentric by omission). As explained in the next section, they also rely on the assumption that all relations, including with nature, can be understood through the deconstruction of the social concept of gender, extrapolating their findings in the context of nonhuman beings (descriptively anthropocentric by extrapolation).

Feminist theories have also emphasized that traditional science has been understood through a Western and patriarchal mode of reasoning. Such reasoning has created fundamental conceptual dichotomies according to which all phenomena are understood, for example, through the dichotomies of reason vs. emotion, culture vs. nature, objectivity vs. subjectivity, mind vs. body and so on (Harding, 1993, p. 165). As feminist researchers underline, the feminine part of these dichotomies is considered inferior (Charlesworth, Chinkin, and Wright, 1991, p. 617; Harding, 1993, pp. 165, 170). Traditional research has also taken man as the model to conduct inquiries, where 'man' is understood as autonomous and rational (Western and white). Such deconstruction of traditional research and science centred on human constructed dichotomies can be mobilized to develop non-anthropocentric research, as presented in the next subsection on ecofemenism.

An important epistemological insight of feminist approaches is the standpoint epistemology developed by Sandra Harding. It implies '"starting off thought" from the lives of marginalised peoples' instead of 'thought that begins from dominant group lives'(Harding, 1993). While aimed at putting the focus of research on women's knowledge (Harding, 1993; Hartsock, 2004), the epistemological concept of 'standpoint' has been mobilized by other critical researchers to reject that research is and can be value-neutral and objective and to recognize one's own subjectivity when conducting research (Bachand, 2018, p. 33). The extrapolation of the standpoint epistemology in the context of the nonhuman is another concept that can be mobilized to develop non-anthropocentric approaches to international law. It helps in acknowledging our own biases and how our understanding of nature is limited by our own perspective. It is the first step to fully acknowledge the inherently anthropocentric limits of our research. It also points to the humble possibilities to speak for nonhuman beings.

Ecofeminism

First making its appearance as a political movement in the 1970s, ecological feminism or ecofeminism was initially led by Françoise d'Eaubonne (Warren, 2000, p. 21; d'Eaubonne, 2018). While there are different 'ecofeminisms', the approach's main postulate is 'that there are important connections between how one treats women, people of color, and the underclass on one hand and how one treats the nonhuman natural environment on the other' (Warren and Erkal, 1997, p. xi). Accordingly, ecofeminism aims to show that women and nature share a common societal status, looking at how environmental impacts particularly affect women and how they can, in turn, be powerful agents of change (Morrow, 2013; Rochette, 1998, p. 152). While ecofeminism initially emerged through the work of white Western women, it now fully engages with intersectional considerations to take into account the particularities of Indigenous women, women of colour, and women located in Third World states (Morrow, 2013).

Ecofeminists have been particularly interested in deconstructing philosophical dualisms, especially that of nature and culture (Morrow, 2013). They shed light on the constructed association of women and nature, as being 'other' to the male and irrational. Ecofeminists denounce that such association has resulted in the objectification, instrumentalization and exploitation of both. However, they also underline that such dualism has also led to a positive association of nature as a nurturing mother, as well as supported an ecocentric view according to which humanity is an integrated part of nature, as opposed to an anthropocentric perspective putting the human as superior or separate from nature. Unlike general feminist approaches to international law, which are based on an anthropocentric ontology, ecofeminists' understanding of the human–nature relationship beyond dualism moves away from anthropocentrism towards ecocentrism. At the same time, their focus remains on the problem of androcentrism as well as its impact on women as humans, not fully disengaging from anthropocentrism.

Ecofeminist scholars take the feminist critique of traditional (Western) science a step further to reveal how it contributes to the domination of nature. They rely on the feminist critique of Western science, which is not universal and neutral, but rather favours masculinized knowledge and the subordination of feminized knowledge. Karen Morrow, for example, argues for the inclusion of women and their experience in addition to scientific and technical knowledge to address global environmental problems (Morrow, 2013).

In the context of international environmental law more specifically, Annie Rochette argues that it is based on the dominant social paradigm of 'capitalist patriarchy' (Rochette, 1998, pp. 20, 35; see also Mickelson, 2007, p. 268). In her view, a paradigm shift through a 'restructuring of society in decentralised communities, where all dominations are ended and economic and social hierarchies dismantled' is necessary to achieve an ecological society (Rochette, 1998, p. 20; Merchant, 1996, pp. 13–14, 223). She stresses that women should be conceived as powerful agents to achieve such change.

Morrow has been particularly interested in underlining the inclusion and exclusion of women in building international environmental law. She raises the issue of the vulnerability of women in the context of climate change, often hidden behind intersectional discrimination in the environmental context. She acknowledges that attention has been paid to gender issues from the perspective of women's participation in the formation of the major environmental instruments, notably the United Nations Framework Convention on Climate Change or within the UN's agenda on sustainable development. However, such attention has been peripheral, and the instruments have not taken into account the systematic exclusion of women, their particular vulnerability to environmental impacts or the role they can play to induce change (Morrow, 2005, p. 191; Morrow, 2011, p. 132-133; Morrow, 2013, p. 384).

Even if they refer to ecological concepts, however, Rochette and Morrow's research is descriptively and passively normatively anthropocentric, due to their ontology centred on women and the concepts and methods they mobilize, which were developed in the context of the study of human relations.

Environmental Justice

Environmental justice is a grassroots social movement that emerged in the United States in the late 1970s and early 1980s (Mickelson, 2007; Grear, 2020). This social movement was first aimed at challenging mainstream environmentalism which did not take into account the racially unequal exposure to environmental hazards. Members of the movement revealed that it is among the most marginalized, indigenous and poor populations that there is a greater proportion of hazardous waste, environmental contamination, climate change impacts, etc.

At the international level, legal scholars associated with the environmental justice movement turned to the study of international environmental law to show that it fails to take into account not only that adverse environmental impacts are borne disproportionately by vulnerable groups and individuals but also by countries located in the Global South, revealing the environmental injustices in relation to the North–South divide. (Alam *et al.*, 2015) They point at the Western capitalist organization of the economy and the 'universalization of the Northern consumption-oriented development model' (Gonzalez and Atapattu, 2017, p. 232) as the source of inequality, failure to meet basic human needs and ecosystems collapse.

The focus of the environmental justice movement is to bring back the 'human' and justice at the centre of environmental considerations, in this case in the context of international law. As a result, their approach is explicitly anthropocentric. While their approach is descriptively anthropocentric by omission, environmental justice scholars do not argue that humans are superior to nonhumans. On the contrary, they argue that social systems of subordination are intrinsically linked to environmental harm. Their approach remains, nevertheless, passively normatively anthropocentric, but tends to move beyond anthropocentrism as it engages with posthuman critical approaches, presented below.

The issue of climate change has particularly attracted attention among environmental justice scholars, as illustrated in the work of Anna Grear, Julia Dehm and Sumudu Atapattu, among others (Grear and Dehm, 2020; Dehm, 2020; Atapattu, 2020). While the differences between developed and developing countries with respect to their contribution to greenhouse gas emissions and the impact of climate change are generally accepted, environmental justice scholars stress that climate injustices are not accidental to international law. Instead, they are a 'manifestation of a *structural pathology* in which law itself is central' (Grear, 2014, p. 106). They point to how international law is structured to displace the responsibility to adapt and address the risks associated with climate change to the most vulnerable populations (Dehm, 2020, pp. 4–5). They also underscore that the terminology to describe 'climate refugees' is not neutral, and it makes invisible the injustices made to racialized and poor communities directly impacted by climate change (Atapattu, 2020).

In terms of emancipatory proposals, different issues ask for diverse forms of environmental justice. Robert Kuehn developed a taxonomy to identify the forms of justice envisaged by the movement, in order to 'offer a method of collapsing the seemingly broad scope of environmental justice and identifying common causes of and solutions to environmental injustice' (Kuehn, 2000, p. 10682). Environmental justice can take various forms, such as distributive justice, procedural justice, corrective justice and social justice. All these forms of justice, however, only focus on justice for human beings.

Building upon Kuehn's taxonomy, Grear (2020) proposes to move away from environmental justice's anthropocentric ontology. Influenced by posthuman scholarship as presented below, she adds a fifth category to the four forms of environmental justice, that of 'ontological justice'. In her view, this new way of conceptualizing the human–nature relationship and expanding 'ethical consideration for the more-than-human victims of injustice' has started to emerge in the environmental justice scholarship and has become a necessity, considering the multiple layered injustices resulting from the global impacts of the predatory ideology of neoliberalism.

While broadening the ontological focus on 'more-than-human victims' as suggested by Grear (2020), environmental justice research using the lens of 'justice' remains framed by a concept developed in the context of human relations, which is extrapolated to nonhuman beings. Such research is, in Mylius' terms (2018), descriptively and passively normatively anthropocentric. Nonetheless, this is not necessarily problematic, as long as the limits of considering nonhuman injustices from a human standpoint are fully acknowledged. Posthuman critical approaches to law offer interesting concepts and methods to move away from the human-centred subject.

Posthuman Critical Approaches to Law

Since the 2020s, posthuman critical approaches as applied to law have emerged. These approaches have shifted from the classical anthropocentric ontology to develop a new way of understanding the human–nature relationship enshrined in law. Such approaches are aimed at rejecting the Western conception of 'the environment' and the legal responses to address environmental issues. They bring us to fundamentally rethink the human role on Earth and the modernist and humanist foundations of our legal systems. Like other critical approaches to international law, the concepts and methodology are taken from different disciplines. Such interdisciplinary theorization is not only aimed at challenging dominant conceptions in law but also at creating a 'supra-disciplinary' critical approach applicable beyond the legal framework through *radical inter-disciplinarity* (Philippopoulos-Mihalopoulos, 2017, pp. 133, 135, 148). As such, posthuman critical approaches to law attempt to reinvent the legal discipline and its ontological, epistemological and methodological foundations. In this section, we present the work of Andreas Philippopoulos-Mihalopoulos, who has identified his approach as 'critical environmental law' (Philippopoulos-Mihalopoulos, 2011, p. 19), and Anna Grear, who shares many of Philippopoulos-Mihalopoulos' assumptions but adds the analytical tool of the 'vulnerable subject' (Grear, 2017b, p. 3).

Philippopoulos-Mihalopoulos and Grear rely on posthuman approaches as developed in the social sciences to radically challenge law's ontology relying on the modern conception of human bodies, as separate from the mind and other bodies (human and nonhuman). Relying on a Deleuzian understanding of the body, posthumanism seeks 'to displace the traditional human-istic unity of the subject' (Braidotti, 2013). Bodies have to be understood as 'assemblages, namely aggregations of human and nonhuman bodies that are contingent upon the conditions of their emergence and which do not presuppose the centrality, and certainly not the exclusive presence, of the human' (Philippopoulos-Mihalopoulos, 2017, p. 138; see also Grear, 2017b, p. 15; Grear, 2015, p. 226).

Moreover, posthuman critical approaches to law have been influenced by the theoretical turn of 'new materialism'. Epistemologically, this leads to understanding knowledge not only as a language and discourse (linguistic turn), but also as matter. Bringing materiality to law means moving beyond the textual. As Philippopoulos-Mihalopoulos (2017, p. 134) puts it:

> Just as words have to be placed in a certain grammatically correct way in order for sensible sentences to emerge, in the same way bodies need to be thought of in their spatiotemporal relations in order for critical environmental law to come up with the kind of radical positions that are needed in the face of the current challenges.

Ontologically, it moves beyond descriptive anthropocentrism and recognizes that material forces can also manifest agency (Philippopoulos-Mihalopoulos, 2017, p. 141; Grear, 2017b, pp. 20–21; Grear, 2017a, p. 92; Coole and Frost, 2010, p. 7).

Such an ontological move deconstructing the human–nature divide has to be understood in the context of the geological concept of the Anthropocene as the proper framework for rethinking law. As Grear puts it, the Anthropocene is 'an increasingly popular term for a "new" geological era in which "humanity" has emerged as a geological force shaping the lifeworld of the planet itself' (Grear, 2015, p. 226). The concept of the Anthropocene helps to recognize that 'humans are everywhere, affecting the geology and future history of the Earth even after our extinction' (Philippopoulos-Mihalopoulos, 2017, p. 141). It points to how anthropocentrism enshrined in law has resulted in a '*crisis of human hierarchy*' (Grear, 2015, p. 227). At the same time, it relocates humans 'within a natural environment whose material forces themselves manifest certain agentic capacities and in which the domain of unintended or unanticipated effects is con-siderably broadened' (Philippopoulos-Mihalopoulos, 2017, p. 141; Coole & Frost, 2010, p. 10).

Understanding bodies from a posthuman perspective leads to the deconstruction of 'the environment', the etymology of the word suggesting that the subject (humans) is surrounded by an object (nature). It also challenges the binary opposition of anthropocentrism and eco-centrism. As Grear and Philippopoulos-Mihalopoulos point out, such a binary understanding of our relationship with nature has proven unhelpful to reimagining international environmental law and bringing about a radical worldview shift (Grear, 2011; Grear, 2017b, p. 17; Philippopoulos-Mihalopoulos, 2011). There is a continuity between humans and the rest of the planet and 'any distinction is often arbitrary and on the basis of partial interests' (Philippopoulos-Mihalopoulos, 2017, p. 136).

Philippopoulos-Mihalopoulos suggests understanding humans and nature in an *acentral* and *multi-agentic continuum* in which there are *ruptures*, cutting the links between bodies, and recognizing that humans are not separate nor independent from animal, vegetable or mineral bodies but are nonetheless different from material and immaterial bodies (Philippopoulos-Mihalopoulos, 2017, pp. 136–137). Even among themselves, humans cannot be understood through a monolithic understanding of humanity, some having more or less responsibility with respect to historical and geological injustices (Philippopoulos-Mihalopoulos, 2017, p. 147).

Such posthuman critical approaches to law offer the most radical solutions for reconceptualizing the human–nature relationship in international law. They require moving beyond the anthropocentric, short-term, narrow and solution-oriented mindset dominant in environmental law. It also mandates departing from the founding principles of our legal systems, including those related to property and responsibility. The approach developed by Philippopoulos-Mihalopoulos suggests a 'differential understanding of space and time in terms of depth, a need for strategic withdrawal for hyper-inclusion, and a reconfiguration of human exceptionalism' (Philippopoulos-Mihalopoulos, 2017, p. 148). Such a radical shift would imply new ontological categories: 'human (as assemblage, without contours); nonhuman (assemblages where the biological – animals and plants – but not the human prevails); inhuman (assemblages where the mineral prevails)' (Philippopoulos-Mihalopoulos, 2017, p. 150). These categories would be understood through their interactions in time (geological time) and space (planetary). 'Human exceptionalism' should be envisaged through the need to create 'the conditions for assuming responsibility, regardless of proof of causal link' (Philippopoulos-Mihalopoulos, 2017, p. 151).

While sharing Philippopoulos-Mihalopoulos' posthuman assumptions, Grear adapts the 'vulnerable subject' developed by Fineman (Fineman, 2008) as the main ontological concept for a radical environmental law (Grear, 2017b, p. 3). Inspired by feminism, this concept deconstructs the gendered ontology of environmental law, in particular the myth of the 'autonomous man', according to which disembodied rational human beings are separate and superior to nature and other human beings. Grear expands the 'vulnerable subject' concept to embrace human and nonhuman vulnerable subjects. In her view, the 'vulnerable subject' brings to the forefront substantive questions of justice. Indeed, it takes account of the actual lived experience of different bodies, being universal in scope while directed at particular vulnerabilities (Grear, 2017b).

While changing the founding principles of environmental law is indeed challenging, reconceiving the human subject through a radical acentric lens 'hold[s] out hope for environmental law as a conduit of more liveable futures' (Grear, 2017a, p. 94). It nonetheless remains to be seen how this new paradigm can concretely make its way into law.

Conclusions

Critical approaches to international law introduced concepts and methodologies aimed at revealing the politics behind apparently objective and neutral rules and principles. This chapter presents a sample of these approaches, contextualizing their relation to *descriptive anthropocentrism* and *normative anthropocentrism* as defined by Mylius (2018).

Critical approaches to international law have been developed by bricolage, influencing one another. One concept shared in many of these approaches is the structuralist notion that international law is a discourse or language that enables particular political projects by those in power but can also serve as an emancipatory tool for vulnerable individuals and groups. Such an understanding of international law and the critical ability to reveal its politics behind apparent neutral and objective rules has nonetheless tended to limit the ontology of critical approaches to international law to human beings, drawing on anthropocentric concepts and methods. Accordingly, most critical approaches to international law are descriptively and passively normatively anthropocentric.

More recent literature in line with structuralism, TWAIL, ecofeminism and environmental justice, however, has started to deconstruct the human–nature dichotomy and the anthropocentric understanding of nature enshrined in international law. They have shown that anthropocentrism is deeply embedded in international law and works in conjunction with capitalism, patriarchy, colonialism as well as Western science and Enlightenment rationality to maintain the domination of Western men over other humans and nature. These approaches have not, however, fully moved away from anthropocentrism, relying on an anthropocentric ontology or concepts and methods developed in the context of the study of human beings. Only posthuman critical approaches to law move beyond the centrality of human beings by founding law on an *acentric* conception of human, nonhuman and inhuman bodies, which cannot be fully separated and can show some form of agency.

That most critical approaches to international law are descriptively and passively normatively anthropocentric is not necessarily problematic or surprising in light of the political projects behind their critique. However, if one wants to move beyond anthropocentrism and rethink the human–nature relationship, one has to be aware that ontological and methodological choices are not neutral. They do constrain our understandings of nature, limiting or enabling a potential paradigm shift.

Note

1 See Koskenniemi (2021, p. 2), using Claude Lévi Strauss' concept of *bricolage.*

References

Alam, S., Atapattu, S., Gonzalez, C.G. and Razzaque, J. (eds.) (2015) *International Environmental Law and the Global South*. Cambridge: Cambridge University Press.

Anghie, A. (1993) '"The Heart of My Home": Colonialism, Environmental Damage, and the Nauru Case', *Harvard International Law Journal*, 34(2), pp. 445–506.

Anghie, A. (1996) 'Francisco De Vitoria and the Colonial Origins of International Law', *Social and Legal Studies*, 5(3), pp. 321–336. DOI: 10.1177/096466399600500303.

Anghie, A. and Chimni, B.S. (2003) 'Third World Approaches to International Law and Individual Responsibility in Internal Conflicts', *Chinese Journal of International Law*, 2(1), pp. 77–103. DOI: 10.1093/oxfordjournals.cjilaw.a000480.

Arruzza, C., Bhattacharya, T. and Fraser, N. (2019). *Feminism for the 99%: A Manifesto*. New York: Verso.

Atapattu, S. (2020). 'Climate Change and Displacement: Protecting "climate refugees" Within a Framework of Justice and Human Rights', *Journal of Human Rights and the Environment*, 11(1), pp. 86–113. DOI: 10.4337/jhre.2020.01.04.

Bachand, R. (2013). 'Pour une théorie critique en droit international', in *Théories critiques et droit international*. Bruxelles: Bruylant, pp. 115–132.

Bachand, R. (2018). *Les subalternes et le droit international : Une critique politique*. Paris: Pedone.

Bianchi, A. (2016). *International Law Theories: An Inquiry into Different Ways of Thinking*. Oxford: Oxford University Press.

Braidotti, R. (2013). *The Posthuman*. Cambridge: Polity.

Certain Phosphate Lands in Nauru (Nauru v. Australia) (Judgment) [1992] ICJ Rep 240.

Charlesworth, H. and Chinkin, C. (2000). *The Boundaries of International Law: A Feminist Analysis.* Manchester: Manchester University Press.

Charlesworth, H., Chinkin, C. and Wright, S. (1991). 'Feminist Approaches to International Law', *American Journal of International Law*, 85(4), pp. 613–645. DOI: 10.2307/2203269.

Chimni, B.S. (1999) 'Marxism and International Law: A Contemporary Analysis', *Economic and Political Weekly*, 34(6), pp. 337–349.

Chimni, B.S. (2004) 'International Institutions Today: An Imperial Global State in the Making', *European Journal of International Law*, 15(1), pp. 1–37. DOI: 10.1093/ejil/15.1.1.

Chinkin, C. (2010) 'Feminism, Approach to International Law', in *Max Planck Encyclopedia of Public International Law.* New York, Oxford University Press.

Coole, D. and Frost, S. eds. (2010). *New Materialisms: Ontology, Agency, and Politics.* Durham: Duke University Press.

Cox, R.W. (1981) 'Social Forces, States and World Order: Beyond International Relations Theory', *Millenium: Journal of International Studies*, 10(2), pp. 126–155. DOI: 10.1177/03058298810100020501.

Cox, R.W. (1983) 'Gramsci, Hegemony and International Relations: An Essay in Method', *Millenium: Journal of International Studies*, 12(2), pp. 162–175. DOI: 10.1177/03058298830120020701.

De Lucia, V. (2015) 'Competing Narratives and Complex Genealogies. The 'Ecosystem Approach' in International Environmental Law', *Journal of Environmental Law*, 23(1), pp. 91–117. DOI: 10.1093/jel/equ031.

De Lucia, V. (2019) *The "Ecosystem Approach" in International Environmental Law, Genealogy and Biopolitics,* New York: Routledge.

Dehm, J. (2020) 'Climate Change, "Slow Violence" and the Indefinite Deferral of Responsibility for "Loss and Damage"', *Griffith Law Review*, 29(2), pp. 1–33. DOI: 10.1080/10383441.2020.1790101.

d'Eaubonne, F. (2018) *Écologie et féminisme: Révolution ou mutation ?.* Paris: Libre & solidaire.

Fineman, M. (2008) 'The Vulnerable Subject: Anchoring Equality in the Human Condition', *Yale Journal of Law and Feminism*, 20, pp. 1–23.

Frankenberg, G. (2010) 'Critical Theory', in *Max Planck Encyclopedia of Public International Law.* New York: Oxford University Press.

Gathii, J.T. (2011) 'TWAIL: A Brief History of Its Origins, Its Decentralized Network, and a Tentative Bibliography', *Trade Law & Development*, 3(1), pp. 26–64.

Gonzalez, C.G. and Atapattu, S. (2017) 'International Environmental Law, Environmental Justice, and the Global South', *Transnational Law and Contemporary Problems*, 26(2), pp. 229–242.

Grear, A. (2011) 'The Vulnerable Living Order: Human Rights and the Environment in a Critical and Philosophical Perspective', *Journal of Human Rights and the Environment*, 2(1), pp. 23–44.

Grear, A. (2014) 'Towards 'Climate justice'? A Critical Reflection on Legal Subjectivity and Climate Injustice: Warning Signals, Patterned Hierarchies, Directions for Future Law and Policy', *Journal of Human Rights and the Environment*, 5, pp. 103–133. DOI: 10.4337/jhre.2014.02.08.

Grear, A. (2015) 'Deconstructing Anthropos: A Critical Legal Reflection on 'Anthropocentric' Law and Anthropocene 'Humanity'', *Law Critique*, 26(3), pp. 225–249. DOI: 10.1007/s10978-015-9161-0.

Grear, A. (2017a) 'Anthropocene, Capitalocene, Chthulucene: Re-encountering Environmental Law and its 'Subject' with Haraway and New Materialism', in Kotzé, L. (ed.) *Environmental Law and Governance for the Anthropocene.* Oxford: Hart Publishing, pp. 77–95.

Grear, A. (2017b) 'Foregrounding Vulnerability Materiality's Porous Affectability as a Methodological Platform', in Philippopoulos-Mihalopoulos, A. and Brooks, V. (eds.) *Research Methods in Environmental Law: A Handbook.* Cheltenham, Northampton: Edward Elgar Publishing, pp. 3–28.

Grear, A. (2020) *Environmental Justice.* Northampton: Edward Elgar Publishing.

Grear, A. and Dehm, J. (2020) 'Frames and Contestations: Environment, Climate Change and the Construction of in/justice', *Journal of Human Rights and the Environment*, 1(1), pp. 1–5. DOI: 10.4337/jhre.2020.01.00.

Grosz, E. (1990) 'A Note on Essentialism and Difference', in Gunew, S.M. (ed.) *Feminist Knowledge: Critique and Construct.* London: Routledge, pp. 332–344.

Harding, S. (1993) 'Rethinking Standpoint Epistemology: What is Strong Objectivity?', in Alcoff, L. and Potter, E. (eds.) *Feminist Epistemologies.* New York; London: Routledge, pp. 49–82.

Hartsock, N.C.M. (2004) 'The Feminist Standpoint: Developing the Ground for a Specifically Historical Materialism', in Harding, D. (ed.) *The Feminist Standpoint Theory Reader.* New York; London: Routledge, pp. 35–54.

Hawkes, T. (2003) *Structuralism and Semiotics.* London; New York: Routledge.

Heller, T. (1984) 'Structuralism and Critique', *Stanford Law Review*, 36, pp. 127–198.

Kennedy, D.W. (1980) 'Theses About International Law Discourse', *German Yearbook of International Law*, 23, pp. 353–391.

Kennedy, D.W. (1987) *International Legal Structures*. Baden-Baden: Nomos.

Kennedy, D.W. (1988) 'A New Stream of International Law Scholarship', *Wisconsin International Law Journal*, 7, pp. 1–49.

Kennedy, D.W. (1999) 'Background Noise? – The Underlying Politics of Global Governance', *Harvard International Review*, 21(3), pp. 52–57.

Kennedy, D.W. (2000) 'When Renewal Repeats: Thinking Against the Box', *New York University Journal of International Law & Politics*, 32, pp. 335–500.

Kennedy, D.W. (2002) 'The International Human Rights Movement: Part of the Problem?', *Harvard Human Rights Journal*, 15, pp. 101–125.

Kennedy, D.W. (2004) *The Dark Sides of Virtue: Reassessing International Humanitarianism*. Princeton: Princeton University Press.

Kennedy, D.W. (2006) 'Reassessing International Humanitarianism: The Dark Sides', in Orford, A. (ed.) *International Law and its Others*. Cambridge: Cambridge University Press, pp. 131–155.

Kennedy, D.W. (2008) 'The Mystery of Global Governance', *Ohio Northen University Law Review*, 34(3), pp. 827–860. DOI: 10.1017/CBO9780511627088.003.

Kennedy, D.W. (2012) 'The International Human Rights Regime: Still Part of the Problem?', in Dickinson, R., Katselli, E., Murray, C. and Pedersen, O.W. (eds.) *Examining Critical Perspectives on Human Rights*. Cambridge: Cambridge Univ. Press, pp. 19–34.

Kennedy, D.W. (2013) 'The International Human Rights Movement: Part of the Problem?', in Darian-Smith, E. (ed.) *Laws and Societies in Global Contexts: Contemporary approaches*. Cambridge: Cambridge University Press, pp. 265–281.

Kennedy, D.W. (2016) *A World of Struggle: How Power, Law and Expertise Shape Global Political Economy*. Princeton: Princeton University Press.

Khoday, K. and Natarajan, U. (2012) 'Fairness and International Environmental Law from below: Social Movements and Legal Transformation in India', *Leiden Journal of International Law*, 25(2), pp. 415–441. DOI: 10.1017/S0922156512000118.

Klabbers, J. (2015) 'Whatever Happened to Gramsci? Some Reflections on New Legal Realism', *Leiden Journal of International Law*, 28(3), pp. 469–478. DOI: 10.1017/S0922156515000230.

Knox, R. (2016) 'Marxist Approaches to International Law', in Orford, A. and Hoffmann, F. (eds.) *The Oxford Handbook of the Theory of International Law*. Oxford: Oxford University Press, pp. 306–326.

Koskenniemi, M. (1989) *From Apology to Utopia: The Structure of International Legal Argument*. Cambridge: Cambridge University Press.

Koskenniemi, M. (1990) 'The Politics of International Law', *European Journal of International Law*, 1, pp. 4–32.

Koskenniemi, M. (2001) *The Gentle Civilizer of Nations: The Rise and Fall of International Law 1870–1960*. Cambridge: Cambridge University Press.

Koskenniemi, M. (2004) 'What Should International Lawyers Learn from Karl Marx', *Leiden Journal of International Law*, 17 (2/2004), pp. 229–246. DOI: 10.1017/S0922156504001803.

Koskenniemi, M. (2007) 'The Fate of Public International Law: Between Technique and Politics', *The Modern Law Review*, 70(1), pp. 1–30. DOI: 10.1111/j.1468-2230.2006.00624.x.

Koskenniemi M. (2016) 'What Is Critical Research in International Law? Celebrating Structuralism', *Leiden Journal of International Law*, 29(3), pp. 727–735. DOI: 10.1017/S0922156516000285.

Koskenniemi M. (2021) *To the Uttermost Parts of the Earth: Legal Imagination and International Power, 1300–1870*. Cambridge: Cambridge University Press.

Kuehn, R.R. (2000) 'A Taxonomy of Environmental Justice', *Environmental Law Reporter*, 30, pp. 10681–10703.

Luke, T.W. (1995) 'On Environmentality: Geo-Power and Eco-Knowledge in the Discourses of Contemporary Environmentalism', *Cultural Critique*, 31, pp. 57–81.

Mackinnon, C. (1987) *Feminism Unmodified: Discourses on Life and Law*. Cambridge: Harvard University Press.

Marks, S. (2003) *The Riddle of All Constitutions: International Law, Democracy, and the Critique of Ideology*. Oxford: Oxford University Press.

Marks, S. (2011) 'Human Rights and Root Causes', *Modern Law Review*, 74(1), pp. 57–78. DOI: 10.1111/j.1468-2230.2010.00836.x.

Marks, S. (2019) 'Three Liberty Trees', *London Review of International Law*, 7(3), pp. 295–319. DOI: 10.1093/lril/lrz011.

Marx, K. (1998 [1845]) *The German Ideology*. New York: Prometheus Books.

Mayrand, H. (2020) 'From Classical Liberalism to Neoliberalism: Explaining the Contradictions in the International Environmental Law Project', *Revue générale de droit*, 50, pp. 57–85. DOI: 10.7202/1071277ar.

Merchant, C. (1996) *Earthcare: Women and the Environment*. New York: Routledge.

Mickelson, K. (2000) 'South, North, International Environmental Law, and International Environmental Lawyers', *Yearbook of International Environmental Law*, 11(1), pp. 52–81. DOI: 10.1093/yiel/11.1.52.

Mickelson, K. (2007) 'Critical Approaches', in Bodansky, D., Brunnée, J. and Hey, E. (eds.) *The Oxford Handbook of International Environmental Law*. Oxford: Oxford University Press, pp. 262–290.

Mickelson, K. (2009) 'Beyond a Politics of the Possible? South-North Relations and Climate Justice', *Melbourne Journal of International Law*, 10(2), pp. 411–423.

Mickelson, K. (2014) 'The Maps of International Law: Perceptions of Nature in the Classification of Territory', *Leiden Journal of International Law*, 27(3), pp. 621–639. DOI: 10.1017/S0922156514000235.

Mickelson, K. and Natarajan, U. (2017) 'Reflections on Rhetoric and Rage: Bandung and Environmental Injustice', in Eslava, L., Fakhri, M. and Nesiah, V. (eds.) *Bandung, Global History, and International Law: Critical Pasts and Pending Futures*. Cambridge: Cambridge University Press, pp. 465–480.

Morrow, K. (2005) 'Not So Much a Meeting of Minds as a Coincidence of Means: Ecofeminism, Gender Mainstreaming, and the United Nations', *Thomas Jefferson Law Review*, 28(2), pp. 185–204.

Morrow, K. (2011) 'Perspectives on Environmental Law and the Law Relating to Sustainability. A Continuing Role for Ecofeminism?', in Philippopoulos-Mihalopoulos, A. (ed.) *Law and Ecology: New Environmental Foundations*. New York: Routledge, pp. 126–152.

Morrow, K. (2013) 'Ecofeminism and the Environment: International Law and Climate Change', in Munro, V.E. and Davies, M. (eds.) *The Ashgate Research Companion to Feminist Legal Theory*. London: Taylor and Francis, pp. 377–393.

Mutua, M. (2000) 'What is TWAIL?', *American Society of International Law Proceedings*, 94, pp. 31–40.

Mylius, B. (2018) 'Three Types of Anthropocentrism', *Environmental Philosophy*, 15:2, pp. 159–194.

Natarajan, U. (2012) 'TWAIL and the Environment: The State of Nature, the Nature of the State, and the Arab Spring', *Oregon Review of International Law*, 14(1), pp. 177–202.

Natarajan, U. (2017) 'Third World Approaches to International Law (TWAIL) and the Environment', in Philippopoulos-Mihalopoulos, A. and Brooks, V. (eds.) *Research Methods in Environmental Law: A Handbook*. Cheltenham: Edward Elgar Publishing, pp. 207–236.

Natarajan, U. and Khoday, K. (2014) 'Locating Nature: Making and Unmaking International Law' *Leiden Journal of International Law*, 27(3), pp. 573–593. DOI: 10.1017/S0922156514000211

Otto, D. (2016) 'Feminist Approaches to International Law', in Orford, A. and Hoffmann, F. (eds.) *The Oxford Handbook of the Theory of International Law*. Oxford: Oxford University Press, pp. 488–504.

Petterson, V.S. (2004) 'Feminist Theories Within, Invisible To, and Beyond IR', *The Brown Journal of World Affairs*, 10(2), pp. 35–46.

Philippopoulos-Mihalopoulos, A. (2011) *Law and Ecology: New Environmental Foundations*. New York: Routledge.

Philippopoulos-Mihalopoulos, A. (2017) 'Critical Environmental Law as Method in the Anthropocene', in Philippopoulos-Mihalopoulos, A. and Brooks, V. (eds.) *Research Methods in Environmental Law: A Handbook*. Cheltenham: Edward Elgar Publishing, pp. 131–155.

Porras, I. (2014) 'Appropriating Nature: Commerce, Property, and the Commodification of Nature in the Law of Nations' *Leiden Journal of International Law*, 27(3), pp. 641–660. DOI: 10.1017/S0922156514000247.

Porras, I. (2015) 'Binge Development in the Age of Fear: Scarcity, Consumption, Inequality and the Environmental Crisis', in Stark, B. (ed.) *International Law and Its Discontents: Responding to Global Crises*. Cambridge: Cambridge University Press, pp. 25–83.

Rochette, A. (1998) *Rape of the World: An Ecofeminist Critique of International Environmental Law*. Vancouver: University of British Columbia.

Saussure, F. (1966) *Course in general linguistics*. New York; Toronto: McGraw-Hill.

Sunter, A.F. (2007) 'TWAIL as Naturalized Epistemological Inquiry', *Canadian Journal of Law & Jurisprudence*, 20(2), pp. 475–507. DOI: 10.1017/S084182090000429X.

Thompson, J.B. (1990) *Ideology and Modern Culture: Critical Social Theory in the Era of Mass Communication*. Cambridge: Polity Press.

Unger, R.M. (1986) *The Critical Legal Studies Movement*. Boston: Harvard University Press.

Warren, K. (2000) *Ecofeminist Philosophy: A Western Perspective on What it Is and Why it Matters*. New York: Rowman & Littlefield Publishers, Inc.

Warren, K. and Erkal, N. (1997) *Ecofeminism: Women, Culture, Nature*. Bloomington: Indiana University Press.

8

INTERNATIONAL LAW, LEGAL ANTHROPOCENTRISM, AND FACING THE PLANETARY

Anna Grear

Introduction

This chapter argues that the foundations of the international legal order rest on a profoundly anthropocentric set of assumptions – and that international law can rightly, in a pervasive and fundamental sense, be called 'anthropocentric'. However, this anthropocentrism is deeply uneven: legal anthropocentrism does not centralise the human being – or even the human species – *per se*. Instead, it elevates a universalised but highly particularistic Eurocentric template of 'man'/'human' as the 'universal' and paradigmatic 'anthropos' at the centre of the juridical order (Grear, 2015).

Law's 'subject-at-the-centre', this chapter argues, is deeply implicated in the persistent, predictably patterned injustices of the international legal order. International law, it turns out, in various ways assumes the enduring priority of this subject, and this remains true whether law's subject-at-the-centre takes the form of the individual or of the nation state or of the business corporation – particularly in the international legal context, the transnational corporation (TNC).[1]

Legal anthropocentrism emerges from critical analyses – including the present one – as a system of (overt and covert) power exercised against non-dominant humans, non-human animals, ecosystems and other lively[2] 'natural' systems and entities. Moreover, the Eurocentric capitalist coloniality of law's subject-at-the-centre and its eco-violence ties it to key inflection points in the emergence and intensification of the 'Anthropocene' (Malm and Hornborg, 2014; Lewis and Maslin, 2015; Kanngieser and Beuret, 2017). Given these convergences, it seems all the more important to interrogate the complex presence of the subject-at-the-centre in emergent Anthropocene-facing international law and governance aspirations (Biermann *et al.*, 2010) and to see what insights emerge when such aspirations are brought into critical theoretical encounter with 'the planetary'.[3]

The chapter unfolds in three stages: first, the dominant and (incompletely) submerged subject-at-the-centre of legal anthropocentrism will be traced out and positioned in relation to international law. Second, patterns of oppression enacted by legal anthropocentrism will be linked to patterns of coloniality visible in the international legal order and in the emergence of the Anthropocene. Finally, the chapter brings 'facing the planetary' into an encounter with the

DOI: 10.4324/9781003201120-11

165

Anthropocene-facing emergent international juridical imaginary known as Earth system law and governance.

Legal Anthropocentrism

Legal anthropocentrism rests on binary subject–object relations central to law's onto-epistemic foundations.[4] These subject–object relations – which radically objectify body, nature and world – have ongoing traction for law, and it is their very persistence (and their apparent power as plausibility structures for capitalist accumulative logics based on the objectification of nature (Nibert, 2002; Ibrahim, 2007) that, to an important extent, explains the eco-social destructiveness of legal anthropocentrism.

Putting the matter more explicitly, legal anthropocentrism ultimately mediates the juridical dominance – and appropriative interests – of the idealised white, property-owning European male template of law's central actor.[5] Law's notion of the fully agentic person[6] – and its related conception of the 'human' – are decisively shaped, not only by property-centred liberal capitalist law's ideological priorities but also by its underlying Eurocentric onto-epistemology (Davies and Naffine, 2001). Law, in other words, rotates persistently – and despite resistance and counter-movements to this very reality and its implications – around the subject-at-the-centre for which/whom *all* other life forms (and the world itself) exist as objects and/or quasi-objects. The submerged contours of law's central subject provide a quasi-clandestine template structuring contemporary juridical relations – including (and perhaps especially) in their neoliberal configurations (Blanco and Grear, 2019).

The precise contours of this subject emerge most clearly from critical legal scholarship exposing the closures, exclusions, marginalisations and long-standing patterns of slow violence (Nixon, 2011) performed by structures of its privilege under the cover of law. In 2015, in 'Deconstructing Anthropos', I argued that there are 'dense continuities between the Anthropocene and the patterned imposition of hierarchies operative within the "anthropocentrism" of law' (Grear, 2015, p. 227). I argued then, as now, that these patterns reflect the dominance of this 'systemically privileged juridical "human" subject whose persistence subtends – to a significant and continuing extent – the neoliberal global order as a whole' (Grear, 2015, p. 227), and that both the 'genesis of the Anthropocene predicament and the tilted foundations of international law can be related to [this juridical] ... construct of a paradigmatic "rational human subject"' (Grear, 2015, p. 230).

Despite the increasing *implausibility* of this subject's underlying broadly Cartesian subject–object relations, the juridical order remains structurally committed to a fundamental dichotomy between a central (and fully agentic, rational) 'actor' ('hu/man') and an 'acted upon' (world) (Halewood, 1996; Ibrahim, 2007). The subject–object split at the heart of legal onto-epistemology is far from neutral. It is not 'humanity' in any substantively inclusive sense that stands as law's subject-at the-centre, notwithstanding either the subject's alleged universalism or its putatively species-referent 'anthropocentrism' (Grear, 2015). Indeed, the subject's problematic particularity is thoroughly exposed by multiple critiques of the exclusions of the 'human' in law (including of the universal of international human rights law (Otto, 2005; Kapur, 2006; Matua, 2001)) and by related critiques of the objectified 'nature' reduced to 'environment' in international environmental law (Philippopoulos-Mihalopoulos, 2011; Code, 2006). Legal anthropocentrism is unmistakably gendered and raced and assumes an impossibly disembodied subject – a complex convergence between the Cartesian *cogito* and the Kantian moral reasoner (Halewood, 1996). The *cogito* forms the 'epistemological Panopticon' (Jung, 2007, p. 239)[7] for which *all* else, including the human body, exists as *res extensa*, while Kantian moral agency can have 'nothing whatsoever to do with the body' (Richardson, 2000, p. 128). And while many other disciplines have rejected such onto-epistemic reductivism, for law, the subject-at-the-centre remains, like the *cogito* itself, a dominant 'canonical institution', which is 'by necessity disembodied, mono-logical/narcissistic and ocularcentric' (Jung, 2007, p. 239). Law's subject-at-the-centre is thus a quintessentially (and impossibly) disembodied knower – the epistemological master viewer, the 'eye in the sky', the possessor of an 'objective' 'view from nowhere' – who through 'his' supreme rationality and by force of 'his' rational will, acts upon the world as the objectified field for the operation of 'his' unrivalled agency (Adelman, 2015). Social and political struggles for legal recognition betray the multiple lived patterns of alienation/othering – the matrices of material and semiotic violence – produced by law's privileging of this, the *only* subject thought perfectly to embody the disembodied 'rational' master actor (Naffine, 2003; Grear, 2013): the classical (social) contractor and quintessential citizen-owner of property (Naffine, 2001; Naffine, 2003). This subject primarily materialises 'his' will territorially (Nedelsky, 1990; Green, 1998) in the *extensa* of 'his' land as a mode of civic attachment expressing 'his' fealty to both polity and nation (Green, 1998, p. 252).[8] 'He' is, moreover, systemically advantaged by a juridical order in which liberal 'neutrality' itself functions as a mode of power.[9] Legal anthropocentrism – which 'centres' around *this* 'anthropos' – is thus deeply complicit in the production of multiple, networked, mutually reinforcing vectors of oppression.[10] Historically, the hierarchies fundamental to law's subject-at-the-centre, expressed in the public dominance of white, property-owning European men 'at home' (including over European women (Pateman, 1989)), were also expressed by the automatic civilisational priority assumed to belong to such men 'abroad', where gendered, rac-ist, class hierarchies legitimated the dispossession of Indigenous peoples and 'authorised' the European colonial priority that remains structurally central to the contemporary international legal order (Anghie, 2005; Quijano, 2007). Indeed, the very notion of world 'progress/history' has a 'first in Europe, then elsewhere' structure (Chakrabarty, 2007, p. 7), and the historical pri-ority of law's subject-at-the-centre and the 'civilising' mission of European international legal doctrines are twin expressions of Eurocentric juridical rationalism (Quijano, 2007).

International Law, Eurocentrism and Coloniality

Ultimately, it comes as no surprise that the objectifications enacted by legal anthropocentrism continue to subtend oppressive colonial constructions legitimating the subordination of non-dominant humans and non-human animals and the plunder of eco-systems (Nibert, 2002; Ibrahim, 2007; Koch, 2012). Given the pervasive influence of the subject-at-the-centre, it is also no surprise that traditional notions of the state should reflect some of the self-same ideological vectors and onto-epistemic assumptions.

Not only were national citizenship and the right to vote once archetypically tied to the power of European property-owning men,[11] but the construct of the nation state as a legal person replicates (albeit complexly) the subject-at-the-centre. While, as Knop points out, 'States are not like individuals in the significant respect that they are not unified beings, they are not irreducible units of analysis' (Knop, 1993, p. 320), there remains 'plenty of room for … examin-ing, in Kirsti McClure's words, "the complicity between the sovereign subject and the sovereign state in modern political theory"' (Elshtain, 1991, p. 1375).

The traditional foundations of international law are emphatically European and pervasively Eurocentric (Anghie, 2005; Quijano, 2007). As a systematic body of rules (as is well known), international law has its intellectual roots in the work of Hugo Grotius, whose *De Jure Belli ac Pacis* (1625) became 'the foundation for all later developments in the field of international law [and] … remains the intellectual font of international law as currently conceived' (Bethlehem, 2014, p. 13). At the heart of the dominant imaginary of international law stands the 'sovereign prince', the idealised projection of a bounded, territory-possessing, masculine subject. The parallel development (alongside the theories of Grotius) of the early international peace treaties, Bethlehem notes, 'resulted in the recognition of the rights of each sovereign prince to determine the internal elements of his state': 'And so was born the Westphalian system of inter-state law – a system of competing and interacting sovereign entities whose discourse and interaction was to be regulated by law' (Bethlehem, 2014, p. 13). It is not difficult to discern a continuity here between the bounded, sovereign, masterful masculine subject whose citizenship is imbricated with his stake in the land of the nation and the sovereign, territory-controlling state itself as an emanation of the will of the national sovereign prince.

Wendt underlines the pervasive anthropomorphism in the conception of the state when he points out that '[t]o say that states are "actors" or "persons" is to attribute to them properties we associate first with human beings – rationality, identities, interests, beliefs and so on' (Wendt, 2004, p. 289). These attributions, moreover, are *ubiquitous*:

> They are found in the work of realists, liberals, institutionalists, Marxists, constructivists, behaviouralists, feminists, postmodernists, international lawyers, and almost everyone in between. To be sure, scholars disagree about which properties of persons should be ascribed to states, how important state persons are relative to other corporate persons like MNCs or NGOs, whether state persons are a good thing, and whether 'failed' states can or should be persons at all. But all this discussion assumes that the idea of state personhood is meaningful and at some fundamental level makes sense. In a field in which almost everything is contested, this seems to be one thing on which almost all of us agree.
>
> *(Wendt, 2004, p. 289)*

The ubiquitous nature of these attributions and assumptions underlines the depth and centrality of the construct of the 'person' in law. Meanwhile, the sheer adhesiveness of the Eurocentric subject–object relations underwriting law's central notion of the person means that it is the European template of the rational, contract-making 'man' that provides its ultimate instantiation, howsoever imagined, even if the continuity between the individual and the sovereign state is not straightforward. The nation state – as the prime international legal person – instantiates in its own unique way these underlying commitments – not least because the 'individual and the collective (e.g. the state) Self are (philosophically) intertwined [and t]his is self-evident [in the way in which] the individualist, subjectivist perspective has marked the deep structure of international law' (Nijman, 2007, p. 26). It is entirely unsurprising, then, when all is said and done, that 'the international doctrine of State sovereignty bears an obvious resemblance to the domestic-liberal doctrine of individual liberty' (Koskenniemi, 2005, p. 224).

The state as a legal person ultimately reflects the assumptions of the Eurocentric juridical order, which is ultimately an assemblage built upon an order of overt and covert privilege granted to the rationalistic white male property-owning actor long central to European law, politics, economics and philosophy (Halewood, 1996).[12] If the entire phenomenon of 'political modernity' (namely the rule by modern institutions of the state, bureaucracy and capitalist enterprise) is, as Chakrabarty has argued, impossible to think of anywhere in the world without invoking certain categories and concepts the genealogies of which go deep into the intellectual traditions of Europe (Chakrabarty, 2007, p. 4), then it is inevitable that the subject-at-the-centre – mediated by the centrality of the 'person' for European law and political philosophy – conditions the deep structure of international law – including its notion of the state as a corporate person.

Legal anthropocentrism also underwrites European coloniality, which has further significant implications for the way in which the anthropocentrism of international law is inexorably tilted towards the structural privilege of the global North over the global South. Legal anthropocentrism, in short, is inexorably Eurocentric, and Esmeir has noted (2012, p. 1) how exclusory the notion of the Eurocentric 'universal' imported into the juridical template of the person is by pointing out that a 'central tenet of the anticolonial tradition locates the power of colonialism in the exclusion of the colonized from the realm of "universal humanity," in their "thingification"' (Esmeir, 2012, p. 1). And – inevitably – accompanying this reduction of the colonised to quasi-objects is the plunder of their equally reductively conceived 'natural resources'. Simons has noted that the entire 'underlying purpose of international law that was developed in the context of the colonial and postcolonial eras was precisely the promotion and protection of economic interests of the North' (Simons, 2012, p. 21) and that:

> as newly independent states emerged from colonial rule as sovereign entities and attempted to assert their sovereignty and establish control over their natural resources, Northern states responded using legal doctrines such as state succession, acquired rights, contracts and consent to protect the interests of their corporate nationals in these states and to resist the attempts by these new sovereign actors to establish a new international economic order which included their own sovereignty over their natural resources.[13]
>
> *(Simons, 2012, p. 21)*

Nor did these patterns disappear with the emergence of the 'post'-colonial international legal order. If anything, capitalist colonial ambition merely shape-shifted into more complex imperial dynamics and formations (Quijano, 2007). As Wood notes:

> capitalism is distinctive among all social forms precisely in its capacity to assert its domination by purely economic means ... capital's drive for relentless self expansion depends on this unique capacity, which applies not only to class relations between capital and labour but also to relations between imperial and subordinate states.
>
> *(Wood, 2005, p. 12)*

This imperialistic shape-shifting was, and remains, juridically mediated. The entire international legal order, despite geopolitical shifts and changing dynamics,[14] remains complexly legible as a Eurocentric juridical assemblage structurally dedicated to the privilege of the global North. Appropriative dynamics taking the form of blatant 'grabs' of global South 'natural resources' (McMichael, 2012; Corson and MacDonald, 2012), and more insidious practises of 'appropriation by spoilation' (Pottage, 2019), continue apace and have, if anything, intensified with the later entrance of China, India, Russia and Brazil into the capitalist geo-economy (Connolly, 2017, ch. 5). All the while, the subject-at-the-centre underwrites, in multiple ways, patterns of ongoing eco-violence and violation (both 'slow' and more episodic). European/Westphalian constructs of possession, territorial control and notions of sovereignty continue to dominate the working assumptions of international legal order – thoroughly shaped by the onto-epistemology of legal anthropocentrism. Law's underlying subject–object relations persist, as do its nested associative binaries (person/property; owner/owned; rationality/emotion; civilised/primitive; culture/nature; action/passivity and so forth), which consistently privilege the same central referent and master agent while maintaining an entire universe of 'others' constructed as marginal/complexly objectified.

A centrally important iteration of these patterns emerges in the form and power of the TNC. The early TNC was the main juridical mechanism for European mercantile colonial appropriation from the 16th century onwards (MacLean, 2004), and its centrality to capitalist colonialism is such that although

> [a] basic history of international law might treat the East India Company's rule over most of the Indian peninsula from 1757 as an aberration − [...] it was merely the most conspicuous case of the basic forms of English and early French colonial expansion.
> *(Koskenniemi, 2016, p. 109)*

Legal anthropocentrism is thus not only key to continuities between the liberal legal actor and the nation state but is also key to geographies of injustice operationalised through corporate legal personhood from the earliest phases of European colonialism. Corporate legal personhood was operatively central to the 'authorised' injustices enacted in service of the 'man'/'human' underlying the legal order, while its systematic 'otherings' included the feminisation/objectification of nature and of non-white, non-dominant peoples alike (Plumwood, 1993). As noted above, 'thingification'/objectification was central to the exclusion of the colonised from 'universal humanity' (Esmeir, 2012, p. 1), a move which is entirely predictable once legal anthropocentrism stands fully exposed.

Just as predictable is the fact that the corporation, the 'personification of capital' (Neocleous, 2003), rests on the self-same underlying idealisation of the European subject-at-the-centre and is an almost seamless expression of disembodied, rationalistic, appropriative, raced, masculine power.[15] Like that other corporation, the state, the capitalist business corporation idiosyncratically embodies the subject-at-the-centre writ large: the corporation was, at decisive points in its origins and development, emphatically Euro-masculine (Lahey and Salter, 1985, p. 555) and unmistakeably white.[16] As Lahey and Salter (1985) point out 'the business corporation is a perfection of the masculinist vision of self − existence as property, separation of accountability and enjoyment, abstract rules as justice, domination as ownership' (Lahey and Salter, 1985, p. 555). Moreover, the 'corporate legal form ... perfects and depoliticizes domination' (Lahey and Salter, 1985, p. 555), as does juridical personhood more generally. Juridical personhood, in this light, emerges as a central mechanism of juridical privilege and marginalisation,[17] a mutable ideological tool for law's mediation (and even 'justification') of uneven power relations (Blanco and Grear, 2019). It is not then, any surprise, that the template of anthropos as 'Man' (the subject-at-the-centre) conditions legal personhood as a mechanism for the privileging of capitalist corporations in ways that increasingly infiltrate orders of rights originally granted to human beings (Grear, 2010; Harding, Kohl, and Salmon, 2008).

From the very earliest periods of European colonial appropriation, the TNC was frequently deployed as the formal juridical arm of trade as appropriation, enabling European states to plunder the colonised world while simultaneously allowing capitalists to evade accountability for their predatory actions. These dynamics marked mercantile colonialism from very early on: addressing the notorious Dutch East India Company of 1602, for example, Dangerman and Schellnhuber point out that:

> Its owners neither wished nor expected to be confronted with the legal consequences of the actions and omissions of the economic entity in which they invested. In fact, they were not supposed to bear these consequences − these were colonial times. In an era when Europeans were discovering the existence of entire continents, the world appeared to have an endlessly receding horizon, and the indigenous people of the

> colonies were not necessarily considered important, let alone equal ... Then and now, companies serve as legal vehicles for generating immediate profits for shareholders and allowing investors to reap the benefits of expansion ... without being confronted by the consequences of such harvesting dynamics.
>
> *(Dangerman and Schellnhuber, 2013, E556)*

McLean (2004, p. 364) reaches the related conclusion that 'the history of colonial expansion is [also] a history of the corporate form'. The TNC is, in important ways, the *ultimate* instantiation of the disembodied European masculinist trope of the juridical subject – the ultimate colonizing force for the *imperium* of capital – the perfected appropriative mechanism of legal anthropocentrism and its orders of othering. Contemporary patterns of TNC dominance rely, inevitably, on the self-same subject–object/person–property binary, and the international order still inevitably assumes the subject-at-the-centre as an anthropocentric legal assemblage serving neo-colonial capitalist governance strategies. Accordingly, when we speak of legal anthropocentrism in international law, we speak of a complex, intransigent nexus of convergent materio-semiotic structures serving well-rehearsed practises of privilege and oppression fundamentally dependent on (racist) patriarchy, coloniality and capitalism combined.

These patterns, if anything, intensify under contemporary neoliberalism and the rise of US imperialism: the global dominance of the TNC remains and, if anything, has gained greater potency as a mechanism for ensuring the priority of capitalist rights of property and patterns of appropriation privileged by highly uneven structural dynamics between global North and South. Legal anthropocentrism – and the power it mediates – also explains the inevitability with which business corporations have expanded 'corporate legal humanity' (Grear, 2010) through the accretion of rights originally reserved for European male property owners and then granted (incompletely) to formerly excluded human groups (Mayer, 1990). Such is the inexorability of this dynamic that even international human rights law stands colonised by TNCs and their structural priority (Baxi, 2006; Emberland, 2006; Grear, 2010; Harding, Kohl, and Salmon, 2008). Neoliberalism's 'new global constitutionalism' (Gill, 1995; Gill, 2002; Gill and Cutler, 2014) amounts to a *de facto* global constitution for corporate capital (Gill, 1995; Gill, 2002; Schneiderman, 2000), and TNC dominance, the defining characteristic of contemporary globalisation (Shamir, 2005, p. 92), means that TNCs benefit extensively from a neoliberal regime of trade-related, market-friendly human rights (Baxi, 2006) while *still* taking full advantage of an extensive power to evade accountability.

The underlying onto-epistemology of legal anthropocentrism, if anything, has intensified its destructiveness, and:

> old dichotomies [continue to] shape the operations of the new more complex systems of domination. That is to say, they mediate those systems, reflecting the uneven geography that is one of globalization's most widely remarked features. Indeed, [as] Hardt and Negri observe ... 'the geographical and racial lines of oppression and exploitation that were established during the era of colonialism and imperialism have in many respects not declined but instead increased exponentially'.
>
> *(Marks, 2003, p. 464)*

Neoliberalism, capitalism's contemporary *imperium*, operates a form of monetary authoritarianism operationalised by punitive austerity, made especially dangerous by neoliberalism's 'obliteration of the ceiling on inequality' (Moyn, 2015, p. 149) in the service of transnational capitalist power.[18] Meanwhile, the precarity generated by such systemic injustices foment multiple fear-based dynamics of 'othering', increasingly reflected in 'tough immigration and asylum policies [capitulating to] … a potent mixture of social and economic dislocation, physical risks, racism and xenophobia' (Gill, 1998, p. 6). These forces and developments simply deepen the sense in which the relentless prioritisation of law's 'subject-at-the-centre' persists and continues to threaten the future survival of multiple living beings and systems in the risk-laden Anthropocene.

What then, might emerge when legal anthropocentrism faces the planetary – when the forces and closures of juridical coloniality are brought into view *alongside* the multiplier effects of lively, agentic self-organizing planetary forces that have been there all along? And with what implications for emergent law and governance responses to the Anthropocene?

Facing the Planetary

The Anthropocene[19] is said to have collapsed the boundary between human and natural histories (Chakrabarty, 2009), yet remains marked by the histories and continuing practises of neo-colonial, eco-violent, consumptive capitalism (Malm and Hornborg, 2014; Hornborg, 2019). Since international law is now said to operate on the plane of a 'global' that complicates traditional Westphalian geo-territorial assumptions (Bethlehem, 2014), and given that international law is itself significantly shaped by the eco-colonial dynamics implicated by critical histories of the Anthropocene (Malm and Hornborg, 2014; Grear, 2015), it seems apt to ask how a new imaginary for international law could be shaped by 'facing the planetary' – especially in the light of the emergence of 'Earth system law and governance' as an international development attempting to respond to the Anthropocene predicament.

The phrase 'facing the planetary' comes from the title of a book by William Connolly (2017), who challenges the assumed centrality of human social forces by inviting his readers to face up to the role of 'the planetary' in the co-production of the Anthropocene. Connolly introduces 'the planetary' as an agentic series of 'temporal force fields'

> such as climate patterns, drought zones, the ocean conveyor system, species evolution, glacier flows, and hurricanes that exhibit self-organizing capacities to varying degrees and that impinge upon each other and human life in numerous ways. … Planetary forces have been marked by gradual change periodically punctuated themselves by rather rapid changes. To face the planetary today is to encounter these processes in their multiple intersections and periods of volatility.
>
> *(Connolly, 2017, p. 4)*

Facing the planetary, as I read it, dislodges two core assumptions of much Anthropocene discourse. First, facing the planetary emphatically dislodges any assumption concerning the *sole* centrality of human social forces in the production of the Anthropocene. Dislodging this assumption also necessarily *complicates* the operatively assumed centrality of human social forces (in this case, international law and governance strategies) as *responses to* the Anthropocene. Secondly, facing the planetary dislodges the assumption of the Holocene as a smooth, pre-existing temporality, disrupted by industrial capitalism and other 'human' social forces alone. The displacement of these two assumptions challenges what Connolly calls his key critical target, 'sociocentrism': 'the propensity to explain social processes by reference to other social processes alone' (Connolly, 2017, p. 15). Displacement of these assumptions also implies a planet-sized critique of legal anthropocentrism and of its related patterns of privilege and exclusion: Connolly notes that sociocentrism is 'often bound to notions of human exceptionalism and nature as a deposit of resources to use and master; in it humans are treated as the only entitled agents in the world' (Connolly, 2017, p. 16). Sociocentrism, as much as the underlying anthropocentrism it manifests, is uneven in highly familiar ways – not least because human exceptionalism and 'mastery' are so *predictably* patterned and 'linked to specific visions of freedom and belonging to the world' (Connolly, 2017, p. 16). These patterns suggest that the notion of 'human' agency in much Anthropocene discourse is misleading, for the notion of the 'human' here conceals an underlying set of intra-species injustices and closures reflecting the dominance of the subject-at-the-centre. Relatedly, the framing of human 'agency' in the Anthropocene narrative as that which disrupts, *on its own*, a pre-existing smooth temporality occludes the role of the planetary itself in the production of the Anthropocene.

Connolly's central argument (2017) urges attention to the planetary as an assemblage of self-organizing, lively and unpredictable forces that punctuate geological time: 'bumpy' planetary temporalities that *pre-existed* industrial capitalist processes and impacts. Importantly, Connolly argues that bringing industrial capitalism *and* the planetary into the frame *together* exposes the pivotal fact that 'the two forces together make the contemporary condition more volatile and dangerous than it would be if only the first were involved' (Connolly, 2017, p. 4). This expansion of attention raises, therefore, important considerations of justice, because any failure to see the true volatilities and dangers of the contemporary condition (by over-focusing on 'human' 'agency' and thus adopting parameters of response fundamentally conditioned by anthropocentric agentic assumptions) runs the risk of producing further injustice precisely by preventing adequate attentiveness and sufficiently nuanced responsiveness to the depth of the dangers presented by the Anthropocene. The failure to appreciate the Anthropocene as a convergence between planetary and social forces, in other words, leaves relatively untouched the very agentic assumptions underlying the predictable, uneven – and deepening – patterns of ongoing international injustice and eco-violence of colonial capitalist legal anthropocentrism. In this light, facing the planetary is legible as a normative requirement for addressing the injustices of the Anthropocene's highly unequal (and familiar) distributions of risk and vulnerability (Malm and Hornborg, 2014).

So, what might this shift of framing mean for the emergent international juridical imaginary of Earth system law and governance? What insights might facing the planetary produce for this important new intervention?

The 'Earth system' governance paradigm (ESG) was pioneered by Frank Biermann as a response to the Anthropocene in around 2005. Biermann also served as the founder and first Chair of the Earth System Governance project, a transdisciplinary research network of sustainability scholars working on the ESG paradigm. According to the project's 2010 strategy paper, ESG is defined as:

> the interrelated and increasingly integrated system of formal and informal rules, rule-making systems and actor networks at all levels of human society (from local to global) that are set up to steer societies towards preventing, mitigating and adapting to global and local environmental change and, in particular, earth system transformation within the normative context of sustainable development.
>
> *(Bierman et al., 2010, p. 203)*

Space does not permit a full analysis of ESG assumptions, but at least two seem problematic in the light of facing the planetary placed alongside a critique of legal anthropocentrism: ESG's normative reliance on the sustainable development paradigm and its assumption of the centrality of organised human agency in the form of complex, 'increasingly integrated' anthropocentric governance structures.

Sustainable development has, as is well known, been criticised for its problematic continuities with colonizing neoliberal ideology and for its overwhelming anthropocentrism alike (Adelman, 2018). It has also been widely criticised for its continuities with extractive neo-colonial economic development priorities and for – especially in practise – its structural privileging of economic growth over environmental protection.[20]

It is perhaps for these reasons that Louis Kotzé, a well-known international environmental law scholar who supports ESG aspirations, distances himself from Biermann *et al*.'s commitment to sustainable development (Biermann *et al*., 2010) when he offers his own proposals for a new juridical imaginary for international law (Kotzé, 2014). Kotzé deploys word-for-word the ESG Project's definition of Earth system governance, but, significantly, omits the final words: 'within the normative context of sustainable development' (Biermann *et al*., 2010, p. 203). His work is legible as a more critically informed attempt to, in his words, 'introduce the Anthropocene into the legal domain as a new discursive context that could hopefully assist in the appropriate future development of global environmental law and governance' (Kotzé, 2014, p. 125).

For Kotzé (2014, p. 156), the Anthropocene triggers both the potential and the need for a 'new, ground-breaking juridical imaginary' – one marked by 'a powerful new emphasis on reimagined forms of human environmental responsibility and responsiveness'. While Kotzé's analysis (2014) is focused primarily on environmental law and governance, it has far-reaching implications for international law generally, and particularly for the role/position of the state[21] because, as he puts it, the Anthropocene demands that international law should move 'beyond the state' to 'become global' (Kotzé, 2014, p. 144). This means that the state should not 'go it alone' but should now '[perceive] itself as a subject that stands in relations of responsibility and dependency with other significant actors on the geopolitical stage' (Kotzé, 2014, p. 145). Underpinning this notion of the global (complete – as Kotzé emphasises (2014) and as Biermann *et al*. also make clear (2010)[22] – with institutional interpenetrations, complexity-responsiveness and multi-level functionality) is an overarching normative appeal to care for 'the global nature of a holistic environment' (Biermann *et al*., 2010, p. 146).

Kotzé's more recent work (2018b) develops a proposal for 'three new-millennial analytical paradigms' deepening, nuancing and extending his engagement with ESG. In particular, in this more recent work, Kotzé positions his proposals against an overt critique of the 'anthropocentrism and neoliberalism' of the Sustainable Development Goals (SDGs), a critique then refracted 'through three contemporary analytical lenses of (i) the Anthropocene, (ii) the planetary boundaries theory, and (iii) the Earth system governance theory'. These lenses, 'respectively seek':

(i)   to emphasise the primacy of *Anthropos* and the extent of the human impact on the environment, while offering an ecological alternative to the prevailing destructive anthropocentrism that permeates virtually all governance institutions, norms and processes, and which diminishes global Earth system integrity;

(ii)   to offer a new and much clearer quantified vision of the limitations of the planet (expressed as boundaries) to sustain all human and non-human life in the wake of increasing anthropogenic pressures; and

(iii)   to emphasise the critical need for a normative realignment in the face of planetary shifts and to offer a new perspective on an integrated, but fragile, Earth system that should be governed by norms, institutions and processes that are better aligned with and able to respond to the integrated Earth system.

(Kotzé, 2018b, p. 43)

What, then, might 'facing the planetary' mean for such commitments?

First, facing the planetary raises important questions for framings such as the 'global nature of a holistic environment' (Kotzé, 2014, p. 146); 'the *primacy of Anthropos*' (Kotzé, 2018b, p. 43);[23] and the 'integrated, but fragile Earth system that should be *governed by* norms, institutions and processes' (Kotzé, 2018b, p. 43)[24] aligned to it. Connolly's planetary is not a 'holism', nor does it seem to align with a 'global … holistic environment' (Kotzé, 2014, p. 146).[25] While the semiotic implications of the latter phrase potentially re-install subject–object relations implicit in the dominant juridical concept of 'the environment' (as that which 'environs/surrounds' the subject-at-the-centre) (Grear, 2020a, pp. xii-xiv), 'holism' (particularly in the semiotic company of the 'global' and 'the environment') runs the related risk of producing, no matter how obliquely, a somewhat *reified* 'earth system' and a kind of (homeo)*stasis* as the backdrop for law and governance aspirations.

This nuance of stasis re-emerges in Kotzé's more recent work (2018b) when he notes that:

> the term Anthropocene suggests that the Earth is rapidly moving into a critically unstable state, with the Earth system gradually becoming less predictable, non-stationary and less harmonious as a result of the global human imprint on the biosphere.
>
> *(Kotzé, 2018b, p. 50)*

Kotzé is not here expressly drawing on his earlier reference to 'a holistic environment', but the language of the *non-stationary* invokes a temporally precedent 'stationary', while 'less harmonious' implies – in conjunction with the implicit 'stationary' – a pre-existing harmony of the kind implied by various holisms. Furthermore, the notion of 'the Earth system' when read alongside these semiotic companions, is also problematised by Connolly's notion of the planetary, which presents an altogether less 'organised' assemblage of processes, forces, ruptures and punctuations.

The central challenge presented to ESG aspirations by Connolly's call to face the planetary is, I think, the imperative to resist juridical anthropocentrism. Not only would such a rejection mean *more consistently* troubling the onto-epistemic objectifications smuggled into the notion of 'the environment' (no matter how 'holistic' in form), but it would also mean the disavowal of inferences framing the Earth as a quasi-stationary system of relatively organised 'harmony' before the advent of the Anthropocene. The framing of the planetary also problematises any juridical aspiration centred on the claim that the 'Earth system … should be *governed by* norms, institutions and processes' (Kotzé, 2018b, p. 43)[26] – and even, perhaps, the very idea that 'the Earth system' is amenable to this kind of imagined governance agency at all. Facing the planetary exposes the risk, in other words, that ESG's onto-epistemic framing merely reduces the planetary to an Earth-sized governance *object*.

Analogous potential critical concerns arise in relation to the 'the global', so central to ESG aspirations and analysis, and the third semiotic component of Kotzé's phrase 'the global nature of a holistic environment' (Kotzé, 2014, p. 146).

I think that Kotzé (2012) is correct to characterise the current move from the 'international' (the juridical order based on Eurocentric Westphalian assumptions) to 'the global' imagined as a set of nested law and governance levels.[27] Space does not allow a full discussion here, but even if the global is (rightly) to be understood as a multi-layered and institutionally complex 'space',[28] rather than as a 'level', it seems essential to avoid any construction of an ontologically 'innocent' global space. While Kotzé is alive to the colonial oppressions marking the global juridical order – and is highly critical of them – could it be that ESG global governance aspiration re-installs, albeit complexly and incompletely, a spatiality consistent with a neo-colonial hegemony of sorts – one concomitant with its production of the 'Earth system' as an object?

Stanley has argued that liberal conceptions of justice articulate and are articulated by an ontology of space that figures space as *prior to* justice projects (and which accordingly locks marginalised communities into 'fixed spatial configurations and correlations') (Stanley, 2009, p. 1000). It seems possible that the global space of the ESG imaginary might likewise obscure structural injustices produced by spatialisation itself. Stanley argues that whenever space is constructed (consciously or unconsciously) as a depoliticised ontological referent, it lends itself – all over again – to the 'liberal master subject whose unmarked and invisible body navigates through the world with all privileges intact' (Stanley, 2009, p. 1000). Space as an innocent ontological referent runs the risk of disguising a juridical operation of power deploying facially neutral constructs that continue complexly to facilitate the hierarchizing legacies of legal anthropocentrism's subject-at-the-centre. Indeed, in ESG discourse, this risk seems underlined by the conception of agency deployed. Biermann *et al.* construct 'agency' in a way that centres on 'authoritative agency', defined as 'the ability to prescribe behaviour and obtain the consent of the governed' (Biermann *et al.*, 2010, p. 204). Agency here is thus not only the implied preserve of a subset of humans, but there is a nuance of an unproblematised agentic 'power over' that draws upon older tropes of liberal law and governance agency, assumes its underlying onto-epistemic structures and echoes its deployment of 'innocent' binaries. This construction of agency co-constructs numerous 'governed' non-agents – not only the billions of humans making up 'the [putatively consenting] governed', but also the 'governed' Earth system.

In the ESG formulation of the agentic, including that of Kotzé, there appears to be more than a hint of a Promethean juridical aspiration that aligns without significant critical tension with the agentic mastery of law's subject-at-the-centre. Facing the planetary, by contrast, suggests the disruption of existing juridical constructions of the agentic. Placed alongside a critical reading of legal anthropocentrism and critiques of 'innocent spatial ontology', facing the planetary seems to suggest the need to foreground fractious politics of contestable spatialities and temporalities – contestations necessary for resisting the injustices of traditional juridical 'neutrality' and which radically trouble constructions of an unproblematised agentic 'centre'. Indeed, one of the key lessons to be drawn from critiques of capitalist coloniality in international law is the very way in which a Eurocentric ontology of space deployed 'difference-making' as a central mechanism of oppression, all the while privileging legal anthropocentrism's subject-at-the-centre. It is important to recall in this connection that the 'global', as Donna Haraway puts it (2014), has always been *highly specific* in its development. Given the intrinsically colonial shaping of 'the international',[29] and given neo-colonial intensifications visible in the 'global' as a highly uneven space of intensifying biopolitical neoliberal eco-governance (De Lucia, 2019), it is all the more important to notice *how*, and in *what ways*, spatialities and temporalities operate more complex, subtle patterns of legal anthropocentrism and modes of biopolitical closure.

In the search for a new international juridical imaginary, facing the planetary means embracing not just multi-levelled institutional, quasi-institutional and systemically embedded elements of a complex governance 'global',[30] but also embracing the Earth as lively, unruly and *agentically* active in distributed and not always predictable ways. It means thinking in terms of, and working amidst, multiply entangled, contradictory and uneasy plural globalities emerging in (and as) flows, vectors of affectivity and affect, transits of risk and contested dynamics of organisation and disorganisation. Facing the planetary also arguably means being more onto-epistemically consistent: embracing complexity all the way down and up and accepting the agentic wildness of 'the planetary' (in Connolly's sense) as being intrinsic to Earth's complex materialisations of its own generativity and intransigence. For ESG, this would imply the need to resist its current temptation to bring complexity theory into the service of a constructed legal and governance object whose ontology is shaped by persisting assumptions of juridical human agency.

Facing the planetary suggests the need to reimagine/reinvent the international legal system itself as an assemblage of epistemically limited, critically reflexive, messy practises of the kinds of 'ongoingness' imagined by Haraway in *Staying with the Trouble* (Haraway, 2016). These kinds of ongoingness call for new trouble-facing partnerships, human and non-human; for radical epistemic and agentic humility in the face of irreducible tensions; for a willingness to 'stay with' tensions; to embrace messy incompletions and extensive contingency as an intrinsic aspect of legal methodology – as an epistemological and ethical commitment fundamental to law and governance aspirations. This is not so much a matter of weaving a unified or integrated system of normative rules into a new systemic law and governance whole of some kind based on the idea of an 'Earth system' as a governance object, so much as it is a call to a juridical coming-to-terms-with the inherently *ungovernable* energies of the planetary. This kind of coming-to-terms-with augurs a radical displacement of old agentic assumptions, an overt turning away from Promethean juridical temptations and their complexly reconstituted forms of anthropocentric subject–object relations.

The proposal – in short – emerging from this brief reflection is that ESG imaginaries should seek to come to terms with the fundamental *ungovernability of their centrally imagined governance object*. Such a proposal also arguably troubles the emergent, complex governance 'subject-(at-the-centre)' constituted by, and simultaneously *as*, ESG itself.

Concluding Thoughts

Facing the planetary implies the need to relinquish familiar juridical binaries and their related reflexes of illimitable agentic desire. Legal anthropocentrism cannot deliver the much needed, non-oppressive forms of distributed responsibilisation now so urgently required.

Kotzé's work positions itself more critically in relation to the subject-at-the-centre than the ESG project appears to: he is overtly critical of sustainable development, of neoliberalism and of anthropocentrism. Even so, his global governance proposals do not (yet) completely succeed in escaping the traction of anthropocentrism. Facing the planetary fundamentally problematises the earth-sized governance object constructed by ESG aspirations, including those of Kotzé himself, as well as some of the assumptions accompanying it. Facing the planetary suggests that a radically decentralised legal imaginary of the normative agentic is yet required.[31]

Haraway (2015, p. 160) has argued that 'It matters which stories tell stories, which concepts think concepts'. International law is a Eurocentric story told by the subject-at-the-centre of legal anthropocentrism. It is a story populated by constructs and concepts such as nation states in the image of the storyteller, capitalist corporations also bearing its image – a world of 'actors' (including the state and non-state actors of the ESG imaginary) set against the background construct of a 'nature' flattened to 'environment'. Legal anthropocentrism has long conditioned the parameters of this story's 'world' as part of a juridical narratology serving the plunder of 'nature', the imposition of European civilisational priority and the institutional juridical protection of a capitalistic 'progress' predominantly serving the global North. And just as facing the co-agentic roles of the planetary and industrial capitalism in the origins of the Anthropocene allows for more realistic appreciation of both danger and injustice, so a critical reading of legal anthropocentrism, brought into conversation with the implications of facing the planetary, more thoroughly problematises the subject-at-the-centre.

Facing the planetary demands a very different starting point for the forging of any new international legal imaginary – one that eschews the subject-at-the-centre – even in – and perhaps *especially* in – emergent Anthropocene-facing law and governance proposals.

Notes

1 ECOSOC (2003, para. 20) defines a TNC as 'an economic entity operating in more than one country or a cluster of economic entities operating in two or more countries – whatever their legal form, whether in their home country or country of activity, and whether taken individually or collectively'. For a discussion of the complexities of holding TNCs to binding legal obligations, see Ruggie (2017) and Blumberg (2001).

2 Here, the 'lively' implicates the teeming agentic energies of the more-than-human and other-than-human, energies communicated by 'new materialist' (or perhaps renewing materialist) works, such as Jane Bennett's *Vibrant Matter:* the lively 'capacity of things – edibles, commodities, storms, metals – not only to impede or block the will and designs of humans, but also to act as quasi agents of forces with trajectories, propensities, or tendencies of their own' (Bennett, 2009, p. viii).

3 The concept of 'facing the planetary' deployed in this chapter draws on Connolly (2017).

4 This point is central to Halewood's exposure (1996) of liberal law's underlying onto-epistemic assumptions.

5 A fact revealed by numerous critiques of the notion of 'humanity' generally, and in law, specifically, but see Nedelsky (1990) and Davies and Naffine (2001).

6 Naffine argues that in the final analysis, and despite tensions in its production, law's notion of personhood 'fairly *systematically* helps to support a quite particular interpretation of the person, and one which has an intimate connection with its companion concept, property' (Naffine, 2001, p. 56).

7 The 'panopticon': a point from which one viewer can observe all without being seen. The analogy draws on Bentham's design for the perfect prison in which the guard occupies a central observatory tower with visual access to all cells, and all prisoners, without being visible himself.

8 Green's analysis (1998, p. 252) notes the overlap between the characteristics of the landowner and the citizen, precisely because the idealized citizen of the liberal polity was traditionally *the same person* as the landowner, 'the man with a fixed stake in his country'. This mystification between land and 'man' is very ancient: 'through all Western civilisation, from the seventh century to the fourteenth, the personal equation was largely merged with the territorial. One and all, master and man, lord and tenant, were "tied to the soil"' (Jarrett, 1935, p. 18).

9 This is an argument central to MacKinnon (1989).

10 For a useful extended discussion, see Grabham *et al.* (2009).

11 The fact that the 'permanence of land and its tangible reality were a guarantee of [its] owner's commitment' was based on the idea that his 'wealth was physically a part of the nation … entirely of a piece with the security and well-being of the nation' (Klein, 1995, p. 223), cited by Green (1998, p. 252). This also ties in with Green's argument (1998, p. 252) that the 'landowner-citizen' construct reveals its intimate continuities in central but frequently subterranean ways in the contemporary juridical order.

12 The centrality of the rationalistic masculine subject tilts all fields of law, constructing 'others' who are complexly excluded by its politics of disembodiment – even in international human rights law (Beveridge and Mullally, 1995).

13 It should be noted, however, that postcolonial states that did achieve a degree of sovereignty over their own natural resources tended to be dominated by postcolonial elites that reinforced in various ways, the anthropocentric resource-appropriative impulses underlying international law. Such entrenched arcs of quasi- and neo-colonial state (and private) power have intensified with the advent of neoliberalism, with post-Fordist elites gaining highly uneven degrees of privilege in some formerly colonized nations. For a fascinating discussion of such tensions in contemporary Africa, see Diop (2012).

14 The fundamental structures of the global North-favouring international juridical order remain decisively influential, even to the complex emergence of South–South Sino-capitalist neo-imperialism in Africa. For a discussion of the nuances of Chinese engagement in Africa and the role of European colonialism and, for example, neoliberal WTO rules, see Lumumba-Kasongo (2011).

15 'The corporation is quintessentially disembodied and also the ultimate personification of market imperatives, presenting a form of subjectivity that perfectly fits, expresses, facilitates and perhaps even intensifies the priorities of capitalism itself' (Grear, 2010, p. 97).

16 Federman's account (2003, p. 181) of the juridical emergence of the capitalist corporation as 'the new American man, the bodily expression of male power' makes entirely clear the anti-black racist origins of this privilege.

17 As Stone (1972) has noted, legal rights were originally attributed to only a very narrow class of human beings, rarely to 'others' beyond family or tribal kinship networks, and seldom to women and children.

18 A central theme in Moyn (2015); see also Moyn (2018).
19 'The Anthropocene is a period of two hundred to four hundred years (depending on who is count-ing) during which a series of capitalist, communist, technological, militarist, scientific and Christian practices became major geological forces that helped to reshape some of these [planetary] nonhuman forces' (Connolly, 2017, p. 4).
20 Arguments well encapsulated by Adelman (2018).
21 Kotzé explains that earth system governance is concerned with the human impact on planetary sys-tems, both in technological terms and in terms of 'societal steering' – a form of influence especially suited to law (Kotzé, 2018a, p. 154).
22 The 'global' in this multilevel sense is deployed throughout Biermann *et al.* (2010).
23 Emphasis on 'primacy' added.
24 Emphasis added.
25 Kotzé (2014, p. 146) appeals to the 'global nature of a holistic environment', a phrase implying envi-ronmental holism, as well as a conception of a 'global' that represents its spatiality for the juridical imagination.
26 Emphasis added.
27 'The global context is not a level but rather a space which is simultaneously national, regional and international. It consists of specific individual countries (national entities), collections of countries in specific regions (regional entities), and the entire international community of states (the international entity). Global governance is thus the sum of governance performed by the national, regional and international entities. It interlinks these phenomena and intertwines the political, social, economic and legal processes of which they consist. Governance in the global arena thus consists of all governance arrangements that exist on all of these levels; and the collection of state and non-state rules, or laws, that these actors would use to change human behaviour to such an extent that it would be conducive to their collective interests' (Kotzé, 2012, p. 301).
28 'The global context is not a level but rather a space' (Kotzé, 2012, p. 301).
29 Chakrabarty (2007, p. 7) argues that the ideology of 'progress' in the 19th century 'made modernity or capitalism look not simply global but rather … something that became global *over* time, by originating in one place (Europe) and then spreading outside it'.
30 See Kotzé's formulation of global governance (2012, p. 301).
31 For suggestive work along these lines from broadly Western legal theorists critical of the dominant juridical paradigm, see, for example, Matthews (2019); Grear (2020b); Davies (2017); Philippopoulos-Mihalopoulos (2019).

References

Adelman, S. (2015) 'Epistemologies of Mastery', in Grear, A. and Kotzé, L. (eds.) *Research Handbook on Human Rights and the Environment*. Cheltenham: Edward Elgar, pp. 9–27.
Adelman, S. (2018) 'The Sustainable Development Goals, Anthropocentrism and Neoliberalism', in Kotzé, L. and D French, D. (eds) *Sustainable Development Goals: Law, Theory and Implementation*. Cheltenham: Edward Elgar, pp. 15–28.
Anghie, A. (2005) *Imperialism, Sovereignty and the Making of International Law*. Cambridge: Cambridge University Press.
Baxi, U. (2006) *The Future of Human Rights*. Oxford: Oxford University Press.
Bennett, J. (2009) *Vibrant Matter: A Political Ecology of Things*. Durham: Duke University Press.
Bethlehem, D. (2014) 'The End of Geography: The Changing Nature of the International System and the Challenge to International Law', *European Journal of International Law*, 25(1), pp. 9–14. DOI: 10.1093/ejil/chu003
Beveridge, F. and Mullally, S. (1995) 'International Human Rights and Body Politics', in Bridgeman, J. and Millns, S. (eds.) *Law and Body Politics: Regulating the Female Body*. Aldershot: Dartmouth, pp. 240–272.
Biermann, F., Betsill, M.M., Camargo Vieira, S., Gupta, J., Kanie, N., Lebel, L., Liverman, D., Schroeder, H., Siebenhüner, B., Yanda, P.Z. and Zondervan, R. (2010) 'Navigating the Anthropocene: The Earth System Governance Project Strategy Paper', *Current Opinion in Environmental Sustainability*, 2, pp. 202–208. DOI: 10.1016/j.cosust.2010.04.005
Blanco, E. and Grear, A. (2019) 'Personhood, Jurisdiction and Injustice: Law, Colonialities and the Global Order', *Journal of Human Rights and the Environment*, 10(1), pp. 86–117. DOI: 10.4337/jhre.2019.01.05

Blumberg, P. (2001) 'Accountability of Multinational Corporations: The Barriers Presented by Concepts of the Corporate Juridical Entity', *Hastings International and Comparative Law Review*, 24, pp. 297–320.

Chakrabarty, D. (2007) *Provincializing Europe*. Princeton: Princeton University Press.

Chakrabarty, D. (2009) 'The Climate of History: Four Theses', *Critical Inquiry*, 35, pp. 197–222. DOI: 10.1086/596640

Code, L. (2006) *Ecological Thinking: The Politics of Epistemic Location*. Oxford: Oxford University Press.

Connolly, W. (2017) *Facing the Planetary: Entangled Humanism and the Politics of Swarming*. Durham: Duke University Press.

Corson, C. and MacDonald, K.I. (2012) 'Enclosing the Global Commons: The Convention on Biological Diversity and Green Grabbing', *The Journal of Peasant Studies*, 39(2), pp. 263–83. DOI: 10.1080/03066150.2012.664138

Dangerman, J. and Schellnhuber, H.J. (2013) 'Energy Systems Transformation', *Proceedings of the National Academy of Sciences of the United States of America*, 110(7), E549–E558. DOI: 10.1073/pnas.1219791110

Davies, M. (2017) *Law Unlimited: Materialism, Pluralism and Legal Theory*. Abingdon: Routledge.

Davies, M. and Naffine, N. (eds.) (2001) *Are Persons Property? Legal Debates about Property and Personality*. Aldershot: Ashgate.

De Lucia, V. (2019) *The 'Ecosystem Approach' in International Environmental Law: Geneaology and Biopolitics*. London: Routledge.

Diop, S. (2012) 'African Elites and their Postcolonial Legacy: Cultural, Political and Economic Discontent – By Way of Literature', *Africa Development*, XXXVII(4), pp. 221–235.

ECOSOC, 'Norms on the Responsibilities of Transnational Corporations and other Business Enterprises with regard to Human Rights' (26 August 2003), UN doc E/CN/.4/Sub.2/2003/12/Rev.2.

Elshtain, J.B. (1991) 'Sovereign God, Sovereign State, Sovereign Self', *Notre Dame Law Review*, 66(5), pp. 1355–1378.

Emberland, M. (2006) *The Human Rights of Companies: Exploring the Structure of ECHR Protection*. Oxford: Oxford University Press.

Esmeir, S. (2012) *Juridical Humanity: A Colonial History*. Stanford: Stanford University Press.

Federman, C. (2003) 'Constructing Kinds of Persons in 1886: Corporate and Criminal', *Law and Critique*, 14, pp. 167–189. DOI: 0.1023/A:1024703803435

Gill, S. (1995) 'Globalisation, Market Civilisation, and Disciplinary Neoliberalism', *Millennium Journal of International Studies*, 24, pp. 399–423. DOI: 10.1177/03058298950240030801

Gill, S. (1998) 'European Governance and New Constitutionalism: Economic and Monetary Union and Alternatives to Disciplinary Neoliberalism in Europe', *New Political Economy*, 3(1), pp. 5–26. DOI: 10.1080/13563469808406330

Gill, S. (2002) 'Constitutionalizing Inequality and the Clash of Globalizations', *International Studies Review*, 4, pp. 47–65. DOI: 10.1111/1521-9488.00254

Gill, S. and Cutler, A.C. (eds.) (2014) *New Constitutionalism and World Order*. Cambridge: Cambridge University Press.

Grabham, E., Cooper, D., Krishnadas, J. and Herman, D. (eds.) (2009) *Intersectionality and Beyond: Law, Power and the Politics of Location*. Abingdon: Routledge.

Grear, A. (2010) *Redirecting Human Rights: Facing the Challenge of Corporate Legal Humanity*. Basingstoke: Palgrave MacMillan.

Grear, A. (2013) 'Law's Entities: Complexity, Plasticity and Justice', *Jurisprudence*, 4(1), pp. 76–101. DOI: 10.5235/20403313.4.1.76

Grear, A. (2015) 'Deconstructing Anthropos: A Critical Legal Reflection on "Anthropocentric" Law and Anthropocene "Humanity"', *Law and Critique*, 26(3), pp. 225–249. DOI: 10.1007/s10978-015-9161-0

Grear, A. (2020a) *Environmental Justice*. Cheltenham: Edward Elgar Publishing.

Grear, A. (2020b) 'Legal Imaginaries and the Anthropocene: "Of" and "For"', *Law and Critique*, 31(3), pp. 351–366. DOI: 10.1007/s10978-020-09275-7

Green, K. (1998) 'Citizens and Squatters: Under the Surfaces of Land Law', in Bright, S. and Dewar, J., (eds.) *Land Law: Themes and Perspectives*. Oxford: Oxford University Press, pp. 229–256.

Grotius, H.(1625) *De Jure Belli ac Pacis*. Translated by Kelsey, F.W. (1925) [Online]. Available at: https://lonang.com/library/reference/grotius-law-war-and-peace/ (Accessed: 6 September 2022).

Halewood, P. (1996) 'Law's Bodies: Disembodiment and the Structure of Liberal Property Rights', *Iowa Law Review*, 81, pp. 1331–1393.

Haraway, D.J. (2014) 'Anthropocene, Capitalocene, Chthulucene: Staying with the Trouble' [Recorded lecture] University of California, Santa Cruz. 5 September. Available at: https://vimeo.com/97663518 (Accessed: 6 September 2022).

Haraway, D.J. (2015) 'Anthropocene, Capitalocene, Plantationocene, Chthulucene: Making Kin', *Environmental Humanities*, 6, pp. 159–165. DOI: 10.1215/22011919-3615934

Haraway, D.J. (2016) *Staying with the Trouble: Making Kin in the Chthulucene*. Durham: Duke University Press.

Harding, C., Kohl, U. and Salmon, N. (eds.) (2008) *Human Rights in the Market Place: The Exploitation of Rights Protection by Economic Actors*. Aldershot: Ashgate.

Hornborg, A. (2019) 'Colonialism in the Anthropocene: The Political Ecology of the Money-Energy-Technology Complex', *Journal of Human Rights and the Environment*, 1, pp. 7–21. DOI: 10.4337/jhre.2019.01.01

Ibrahim, D.M. (2007) 'A Return to Descartes: Property, Profit, and the Corporate Ownership of Animals', *Law and Contemporary Problems*, 80, pp. 89–115.

Jarrett, B. (1935) *Mediaeval Socialism*. London: Burns Oates and Washbourne.

Jung, H.Y. (2007) 'Merleau-Ponty's Transversal Geophilosophy and Sinic Aesthetics of Nature', in Cataldi, S.L. and Hamrick, W.S. (eds.) *Merleau Ponty and Environmental Philosophy: Dwelling on the Landscapes of Thought*. New York: State University of New York Press, pp. 235–257.

Kanngieser, A. and Beuret, N. (2017) 'Refusing the World: Silence, Commoning and the Anthropocene', *The South Atlantic Quarterly*, 116(2), pp. 362–380. DOI: 10.1215/00382876-3829456

Kapur, R. (2006) 'Human Rights in the 21st Century: Take a Walk on the Dark Side', *Sydney Law Review*, 28, pp. 665–687. DOI: 10.4324/9780203814031

Klein, L.E. (1995) 'Property and Politeness in the Early Eighteenth Century Whig Moralists: The Case of the Spectator', in Brewer, J. and Staves, S. (eds.) *Early Modern Conceptions of Property*. London: Routledge, pp. 221–233.

Knop, K. (1993) 'Re/statements: Feminism and State Sovereignty in International Law', *Transnational Law and Contemporary Problems*, 3, pp. 293–344.

Koch, M. (2012) *Capitalism and Climate Change: Theoretical Discussion, Historical Development and Policy Responses*. London: Palgrave Macmillan.

Koskenniemi, M. (2005) *From Apology to Utopia: The Structure of International Legal Argument*. Cambridge: Cambridge University Press.

Koskenniemi, M. (2016) 'Expanding Histories of International Law', *American Journal of Legal History*, 56, pp. 104–112. DOI: 10.1093/ajlh/njv011

Kotzé, L. (2012) *Global Environmental Governance: Law and Regulation for the 21st Century*. Cheltenham: Edward Elgar.

Kotzé, L. (2014) 'Rethinking Global Environmental Law and Governance in the Anthropocene', *Journal of Energy and Natural Resources Law*, 32(2), pp. 121–156. DOI: 10.1080/02646811.2014.11435355

Kotzé, L. (2018a) 'Reflections on Environmental Law Scholarship and Methodology in the Anthropocene', in Pedersen, O.W. (ed.) *Perspectives on Environmental Law Scholarship: Essays on Purpose, Shape and Direction*. Cambridge: Cambridge University Press, pp. 140–161.

Kotzé, L. (2018b) 'The Sustainable Development Goals: An Existential Critique alongside Three New-Millenial Analytical Paradigms', in Kotzé, L. and French, D. (eds.), *Sustainable Development Goals: Law, Theory and Implementation*. Cheltenham: Edward Elgar Publishing, pp. 41–65.

Lahey, K. and Salter, S.W. (1985) 'Corporate Law in Legal Theory and Legal Scholarship: From Classicism to Feminism', *Osgoode Hall Law Journal*, 23(4), pp. 543–572.

Lewis, S. and Maslin, M. (2015) 'Defining the Anthropocene', *Nature*, 519(7542) pp. 171–180. DOI: 10.1038/nature14258

Lumumba-Kasongo, T. (2011) 'China-Africa Relations: A Neo-Imperialism or a Neo-Colonialism? A Reflection', *African and Asian Studies*, 10, pp. 234–266. DOI: 10.1163/156921011X587040

MacKinnon, C.A. (1989) *Towards a Feminist Theory of the State*. Cambridge: Harvard University Press.

Malm, A. and Hornborg, A. (2014) 'The Geology of Mankind? A Critique of the Anthropocene Narrative', *The Anthropocene Review*, 1(1), pp. 62–69. DOI: 10.1177/2053019613516291

Marks, S. (2003) 'Empire's Law', *Indiana Journal of Global Legal Studies*, 10, pp. 449–466.

Matthews, D. (2019) 'Law and Aesthetics in the Anthropocene: From the Rights of Nature to the Aesthesis of Obligations', *Law, Culture and the Humanities*, pp. 1–21. DOI: 10.1177/1743872119871830

Matua, M. (2001) 'Savages, Victims and Saviours: The Metaphor of Human Rights', *Harvard International Law Journal*, 42(1), pp. 201–246.

Mayer, C.J. (1990) 'Personalising the Impersonal: Corporations and the Bill of Rights', *Hastings Law Journal*, 41, pp. 577–663.

McLean, J. (2004) 'The Transnational Corporation in History: Lessons for Today?', *Indiana Law Journal*, 79, pp. 363–377.

McMichael, P. (2012) 'The Land Grab and Corporate Food Regime Restructuring', *The Journal of Peasant Studies*, 39(3–4), pp. 681–701. DOI: 10.1080/03066150.2012.661369

Moyn, S. (2015) 'A Powerless Companion: Human Rights in the Age of Neoliberalism', *Law and Contemporary Problems*, 77, pp. 147–169.

Moyn, S. (2018) *Not Enough: Human Rights in an Unequal World*. Cambridge: Belknap Press.

Naffine, N. (2001) 'The Nature of Legal Personality', in Davies, M. and Naffine, N. (eds.) *Are Persons Property? Legal Debates about Property and Personality*. Aldershot: Ashgate, pp. 51–73.

Naffine, N. (2003) 'Who Are Law's Persons? From Cheshire Cats to Responsible Persons', *Modern Law Review*, 66(3), pp. 346–367. DOI: 10.1111/1468-2230.6603002

Nedelsky, J. (1990) 'Law, Boundaries and the Bounded Self', *Representations*, 20, pp. 162–189. DOI: 10.10 93/acprof:oso/9780195147964.003.0003

Neocleous, M. (2003) 'Staging Power: Marx, Hobbes and the Personification of Capital', Law and Critique, 14, pp. 147–165. DOI: 10.1023/A:1024753201618

Nibert, D. (2002) *Animal Rights, Human Rights: Entanglements of Oppression and Liberation*. Oxford: Rowman and Littlefield.

Nijman, J.E. (2007) 'Paul Ricoeur and International Law. Beyond 'The End of the Subject': Towards a Reconceptualization of International Legal Personality', *Leiden Journal of International Law*, 20, pp. 25–64.

Nixon, R. (2011) *Slow Violence and the Environmentalism of the Poor*. Cambridge: Harvard University Press.

Otto, D. (2005) 'Disconcerting "Masculinities": Reinventing the Gendered Subject(s) of International Human Rights Law', in Buss, D. and Manji, A. (eds.) *International Law: Modern Feminist Approaches*. Oxford: Hart Publishing, pp. 105–129.

Pateman, C. (1989) *The Disorder of Women: Democracy, Feminism and Political Theory*. Stanford: Stanford University Press.

Philippopoulos-Mihalopoulos, A. (2011) "… the Sound of a Breaking String': Critical Environmental Law and Ontological Vulnerability', *Journal of Human Rights and the Environment*, 2(1), pp. 5–22. DOI: 10.4337/jhre.2011.01.01

Philippopoulos-Mihalopoulos, A. (ed.) (2019) *Routledge Handbook of Law and Theory*. Abingdon: Routledge.

Plumwood, V. (1993) *Feminism and the Mastery of Nature*. London: Routledge.

Pottage, A. (2019) 'Holocene Jurisprudence', *Journal of Human Rights and the Environment*, 10(2), pp. 153–175. DOI: 10.4337/jhre.2019.02.01

Quijano, A. (2007) 'Coloniality and Modernity/Rationality', *Cultural Studies*, 21, pp. 168–178. DOI: 10.1080/09502380601164353

Richardson, J. (2000) 'A Refrain: Feminist Metaphysics and Law', in Richardson, J. and Sandland, R. (eds.) *Feminist Perspectives on Law and Theory*. London: Cavendish, pp. 119–134.

Ruggie, J.G. (2017) 'Multinationals as Global Institution: Power, Authority and Relative Autonomy', *Regulation and Governance*, 12(3), pp. 317–333. DOI: 10.1111/rego.12154

Schneiderman, D. (2000) 'Constitutional Approaches to Privatization: An Inquiry into the Magnitude of Neo-Liberal Constitutionalism', *Law and Contemporary Problems*, 63, pp. 83–109.

Shamir, R. (2005) 'Corporate Social Responsibility: A Case of Hegemony and Counter-Hegemony', in Santos, B. de S. and Rodrigues-Garavito, C.A. (eds.) *Law and Globalisation from Below: Towards a Cosmopolitan Legality*. Cambridge: Cambridge University Press, pp. 92–117.

Simons, P. (2012) 'International Law's Invisible Hand and the Future of Corporate Accountability for Violations of Human Rights', *Journal of Human Rights and the Environment*, 3(1), pp. 5–43. DOI: 10.4337/jhre.2012.01.01

Stanley, A. (2009) 'Just Space or Spatial Difference? Difference, Discourse and Environmental Justice', *Local Environment: The International Journal of Justice and Sustainability*, 14(10), pp. 999–1014. DOI: 10.1080/13549830903277417

Stone, C. (1972) 'Should Trees Have Standing? Towards Legal Rights for Natural Objects', *Southern California Law Review*, 45, pp. 450–501.

Wendt, A. (2004) 'The State as a Person in International Theory', *Review of International Studies*, 30, pp. 289–316.

Wood, E.M. (2005) *Empire of Capital*. London: Verso.

9

TOWARDS AN ECOFEMINIST CRITIQUE OF INTERNATIONAL LAW?

Karen Morrow

Introduction

It is now apparent that the Anthropocene signals a paradigm shift in the human–environment nexus, and that this in turn requires a dramatic epistemic shift in how we view what it is to be human in the world. The Anthropocene, as popularly conceived, is global in reach yet highly differentiated in its inextricably bound environmental and human impacts and origins (Morrow, 2020). The advent of the Anthropocene requires at least the dramatic reformulation (though, more likely, the reinvention) of key planetary level responses to the complex, intertwined, and escalating worldwide human and environmental impacts of the Anthropocene that are beginning to manifest (Lade *et al.*, 2020). In an age of global, interconnected, crises, international law will need to be at the vanguard of this change and play a fundamental role in identifying and effectively protecting the collective interests of humanity in a rapidly changing world. In its present form, however, it is clear that international law is woefully ill-equipped for the challenge (Vidas *et al.*, 2016).

This chapter takes a critical view of the operation of the international legal system, with particular emphasis on those areas in which it exhibits an environmental and sustainability focus, these being among the most urgent areas requiring attention in the new epoch that humanity is making. It is argued that, as the Anthropocene progresses, these relative latecomers and add-ons to international law and policy must become, and continue to be, absolutely core to how we think and act as humans in a changing biosphere. In this regard, the change that is required in international law, both systemic and in regard to specific areas of concern, is profound:

> the processes of convergence between Holocene and Anthropocene conditions will require response or transformation, including the development of new legal axioms, in accordance with the needs of the new situation – rather than responding by analogy or precedent based on the earlier situation, [which are] no longer valid.
>
> *(Vidas* et al.*, 2015, p. 11)*

However, international law in general, and thus by extension international environmental and sustainability law, are flawed products of dominant patriarchal social mores and are, as currently practiced, not only unable to engage effectively with inter-human inequalities and

DOI: 10.4324/9781003201120-12

environmental degradation but actually serve to exacerbate the problems we now face. It is suggested that an ecofeminist analysis of international law offers a potentially valuable lens through which to identify and respond to the limits of current practice. This chapter will interrogate what ecofeminism and its articulation of the shared exploitation to which women, other vulnerable or non-privileged groups, and the environment are subject may offer as a basis to do better. Ecofeminist thought broadly posits that current societal praxis, founded in part on belief systems and practices built on dualistic views of intra-human and human–nature relationships, is actively hampering the charting of a sustainable course for the Anthropocene. The consequences of dualistic, hierarchical 'othering' in a system that is both androcentric (and further refined within this to empower a select, favoured, few) and anthropocentric are readily apparent. Furthermore, the view that inter-human justice and the systemic environmental crises of the Anthropocene are inextricably linked is gaining greater currency (Laybourn-Langton *et al.*, 2019). The advent of the Anthropocene makes addressing the exclusions of current legal approaches through challenging dominant attitudes, and tackling their institutional embodiment in legal regimes and practices, both crucial and urgent. That this is an uncomfortable and enormously challenging task, questioning the very roots of dominant worldviews, is clear; but it is now unavoidable. In this endeavour, an ecofeminist analysis has much to offer in rethinking what it means to view ourselves relationally, both in terms of inter-human interactions and in interfacing with the environment, as an alternative to perpetuating the errors of a currently dominant worldview founded on illusory perceptions of human separation and superiority. This chapter outlines key features of ecofeminist thinking that could assist in these inextricably entwined endeavours. The first part of the chapter interrogates dualistic thinking and its ramifications. The second part of the chapter briefly introduces core aspects of ecofeminist thought, offering alternatives to current approaches: examining the shared oppression of non-privileged humans and nature; relational approaches; intersectionality, coalition, and allyship; the fusion of theoretical and activist perspectives and accommodating and valuing difference. The last part of the chapter looks briefly at the relationship between established feminist scholarship on international law and ecofeminism. The chapter concludes by considering what ecofeminism could offer to developing a new, sustainable, international law that better serves both people and the planet.

The Dominance of Dualism and Its Consequences

The still dominant, though increasingly contentious, worldview that shapes international law and international legal structures remains rooted in Western thought, with all that entails (Bordeau, 2004). For present purposes, a critical element of this epistemological schema is rooted initially in the Enlightenment and, in particular, the work of Descartes (Lahar, 1996, p. 2) which, although open to cogent and wide-ranging criticism (Arponen, 2016), enjoys a stubborn and pervasive epistemological legacy. Descartian thinking has variously and profoundly shaped modern, Western worldviews and the actions founded upon them; though for present purposes, we will focus on the ramifications of what has become known as Cartesian dualism. While dualism was rooted, albeit in less defined terms, in classical Greek philosophy, notably in the Platonic school (Gerson, 1986), it was Descartes' promotion of the mind/body dichotomy that really took hold of the nascent modern imagination, and which has served as a core tenet of both the Enlightenment and the scientific revolution that it spawned (Holmes, 2008).

The oppositional and hierarchical precepts of dualism have proven persuasive and immensely adaptable but also inherently problematic in their application, serving to privilege the opinions and interests of the few over the many and favouring the pursuit of human desires over the well-being of the natural world. For present purposes, the dualistic equation of the mind with rationality, humanity, and the masculine and its deemed superiority to the body, identified with the emotional, nature, the feminine, and the consequences thereof, are of vital importance. Though notionally derailed by Darwinism (Hirsch, 2020), and subject to diverse and cogent critique on numerous grounds, including failure to acknowledge the importance of human embodiment (notably in gender and racial terms) (Lock and Farquhar, 2007) and the propagation of species chauvinism (Miller, 1989), the legacy of dualism continues to dominate human thought, social ordering, and behaviour.

For present purposes, dualism, and the 'othering' that it necessarily entails (discussed below), serve to perpetuate inequality among humans and exploitation of the environment. This is the case even when the international community claims to be radically realigning human–environment and inter-human relationships (now and in the future) through the promotion of sustainable development. The latter is most often expressed in the famous 'three-legged stool' model popularised by the Brundtland Commission, which ostensibly casts environmental, social, and economic 'pillars' as 'equal' concerns that could be balanced against one another (Dawe and Ryan, 2002). However, the three pillars approach is fundamentally distorted. First, its presentation of human concerns, separating social and economic interests, is telling. The economy is itself a social construct and part of, not distinct from, society, and designating it as a discrete pillar simultaneously panders to its undue dominance and masks the fundamental human–environment dualism at the core of sustainable development. Indeed, over time, economic considerations, specifically income and wealth, have come to serve as a proxy for inequality more generally (Islam, 2015), propagating a damagingly reductive approach to the complexities of the social pillar.

Second, while distorting the treatment of human concerns within sustainability, the three pillars approach simultaneously amplifies human concerns *vis-a-vis* the environment in the notional sustainability balance. In essence, the three pillars formulation not only masks a fundamentally dualistic view of the human/environment relationship, it also ultimately augments the weight given to human interests over the environment. The three pillars approach, and the fact that the concept emerged at the environment-focused Rio Earth Summit, did however at least serve to re-establish environmental issues on the international agenda after they had fallen into lassitude following a brief flourishing in the early 1970s. However, the drive to obtain buy-in among States at the Earth Summit required that the concept of sustainable development be politically acceptable to them (French, 2005), and thus the emphasis of the Brundtland Report as presented at the conference, lay firmly on development. This was the case even though the report itself has engaged strongly with social priorities and even environmental limits (Borowy, 2014). Political expedience however ensured that, despite claims to the contrary, sustainable development as it emerged from Rio clearly foregrounded economic priorities throughout, and in consequence, it is viewed by many as simply repackaging and legitimising what can be viewed as, at best, a mildly tempered 'business as usual' approach (Kemper, Hall, and Ballantine, 2019). This critique is reinforced by the lack of ecological grounding for the concept of sustainable development, as popularly understood (Dawe and Ryan, 2002) and relatedly, the illusory nature of the 'balance' between the three pillars that it promoted (Banarjee, 2003).

While sustainable development as it emerged from the Rio Earth Summit was flawed, it could at least make some claim to enhance both consideration of the environment and of human inequality vis-à-vis economic drivers. However, this was not to last. The very flexibility of sustainable development, which holds such appeal , renders the concept open to manipulation to suit the current priorities of humanity. Over time this has allowed the social and, more substantially and consistently, the economic pillar to prevail over the environmental pillar (Morrow, 2012). Initiatives seeking to tackle societal inequalities such as the Millennium Development

Goals (United Nations, 2015) and now the Sustainable Development Goals (United Nations, 2020) aside, the economic pillar continues to effectively dominate over the social and the environmental.

The limited horizons and impact of the supposedly revolutionary sustainable development paradigm (United Nations, 2020) show how deeply embedded and pervasive dualism is, even, as Gruen argues, eroding 'our capacities for moral imagination and moral motivation' (Gruen, 2009, p. 26). It is certainly the case that the lengthy ascendancy of the sustainable development paradigm as currently practiced has simultaneously impoverished the human condition and the majority of humans as well as the planet and has pushed to the fringes alternative synthesis-based understandings of what it is to be human in relation to other human beings and to the environment.

Foundational to the systemic application of dualistic thinking is a bifurcation between subject and the object of inquiry. These ways of seeing and being in the world laid important foundations for the project of modernity and continue to underpin the conduct of much of human society. Dualistic thinking has been deployed across the globe in various guises, creating and perpetuating multiple inequalities across and within societies and globally through colonialism, racism, sexism, ableism, and so on. The adverse consequences of dualism, founded on a relationship between humanity and the environment characterised by a supposed human mastery over other elements of the natural world, and which the environment has managed to absorb through much of human history, are now increasingly being made manifest in damage and disruption on a planetary scale (UN Environment, 2019). Coming full circle, this has tended to visit the worst impacts on non-privileged humans, those who are least responsible for the damage in the first place (Levy and Sidel, 2005). We are, however, now reaching a point where privilege can no longer fully insulate the wealthy and powerful from the effects of environmental crises (Lade et al., 2020).

It is now well established that, human rights protections designed to level the playing field notwithstanding, humanity in general, and the most vulnerable in particular, are threatened by general environmental degradation and the infringement of planetary boundaries (UN OHCHR, 2019). It is true that the predicament of the most vulnerable is now an established consideration threading through in global legal and practical responses to environmental breakdown, as, for example, is the case for climate change action (Paris Agreement, 2015, Preamble; artt. 6, 7). However, how we regard human vulnerability is in a constant state of flux as the impacts of the Anthropocene become more evident. This is illustrated in the proliferating reach of climate change impacts, which are now accelerating and extending across the globe in ways that make it apparent that any nation can be vulnerable to them (IPCC, 2021b). Likewise, the rapid emergence and subsequent ebbs and flows of the COVID-19 pandemic provide salutary lessons on the global interconnections between vulnerability to Anthropocene impacts and the efficacy of responses to them (Lambert *et al.*, 2020). It can now rightly be said that, while still unevenly felt, to be human is to be vulnerable to environmental breakdown, and that while the Western and wealthy are in a better position to ameliorate adverse impacts, for now, even the most privileged can no longer completely avoid them (IPCC, 2021a).

It is ever more evident that both the roots and impacts of dualism are intimately entwined and provide a foundational element of the Anthropocene. We have now reached the logical conclusion of a social and environmental reductionism that threatens to transgress the planetary boundaries that secure a safe space in which humanity can flourish (Lade et al., 2020). Reductionism here describes the adoption of an artificial and value-laden, compartmentalised approach that fails to acknowledge both the intricacies of the environment as a complex system and of humanity's relationship with it. It is also the case that the fate of humanity and the environment is inextricably linked: the ability of the planet to support human life is at stake, notions of starting anew on another planet (Spacex, n.d.), or 'designing out' human dependence on our environment though techno-humanism (Harari, 2016), notwithstanding. Nor can securing viable human life-support on the planet ultimately benefit only the privileged few without inviting disaster in other ways. Thus, sustainable responses to the challenges we face entail securing ecological justice between humans in tandem with re-worked dealings with nature, recognising both our ultimate and utter dependence on it and the imperative need to live within planetary boundaries (Lade et al., 2020).

The Anthropocene represents a reckoning for humanity and our ways of being in the world. At its worst, it poses an existential threat for our species; at its best, it is, as Rebecca Solnit (2021) puts it when speaking of climate change, an archetype of the Anthropocene, it provides 'a mandate to build a better world, a cleaner, more equitable, more cooperative world – cooperative with nature as well as with each other'.

The case is a pressing one. The IPCC's Working Group 1 (on the physical science basis for the IPCC's Sixth Assessment report) (IPCC 2021a, in particular section D) makes it clear that climate change is already occurring, but that decisive action can still potentially ameliorate its impacts, though it recognises that this window of opportunity is finite, and that urgent action is required.

Ecofeminism as an Alternative to Dualism

Given that pursuing dualistic ways of being have led us to our present quandary, a reset on what is understood about how to be human in the world is not only desirable, but necessary. Ecofeminism, in recognising that inter-human justice and the human–environment nexus are inextricably bound together, is well placed to provide valuable insights on how this may be achieved. In this, ecofeminism shares common ground with post structuralist theory more generally and chimes with other critical discourses, such as post-colonial theory (Thompson, 2016, pp. 39–40). Ecofeminism's rejection of dualism also shares common ground with other approaches to holistic ways of thinking about the world, humanity, and environmental governance, such as relational analysis (Allen *et al.*, 2018) and integrationist multimodality (Long, 2015). These critical views, while adopting distinctive approaches, share a foundational drive to counter reductionism and underline the centrality of inter-human and human–environment connection.

As its name suggests, ecofeminism is rooted at the intersection of (radical) feminist theory and ecology. In keeping with its roots, it shares common ground with other areas of environmental thought, in particular, those framed around sustainability (Walsh, 2021), and with radical approaches to human-centred environmental concerns, including environmental justice (Verchick, 1996), and ecocentric thought such as Earth Law (Marek-Muller, 2020). At the same time, and in addition to its connections to a variety of broader philosophical approaches, considered above, ecofeminism exhibits strong affinity with areas of feminist inquiry, such as feminist citizenship (Sandilands, 1999), feminist epistemology (Code, 2006), and the ethics of care both in general (Held, 2005), and as applied to environmental concerns (Stensota, 2015). While connected to these and other cognate areas, ecofeminism offers a distinctive synthesis of feminist and environmental concerns that seeks to address what are viewed as the entwined oppressions of (particularly gendered) intra-human inequalities, and species chauvinism, which makes its approach particularly well suited to addressing dualism and its pathologies as outlined above.

At base, ecofeminist thought espouses a strongly relational ethic that extends consideration to both inter-human and human–environment nexus issues. In this, it directly opposes the inherently reductionist stance of dualism and the primacy it accords to the structures of patriarchy, which are not limited to gender inequalities, but also enfold racism, classism (Daly, 1994), ableism, and so on, and which are also expressed in human exceptionalism in which privileged humanity is considered as separate from and superior to the natural world (Manuel-Navarrete and Buzinde, 2010, pp. 136–138).

Ecofeminism eschews the idea of a single hegemonic narrative and is therefore more accurately viewed as comprising multiple ecofeminisms (Daly, 1994), which themselves encapsulate a broad spectrum of views. These range widely from essentialist ecofeminisms at one end, through to socialist ecofeminisms at the other (Morrow, 2013). Essentialist ecofeminisms tended to feature in the initial stages of the development of ecofeminist thought. Also described as affinity ecofeminisms, they argue that, based on biological characteristics and/or spirituality (Hughes, 1995), women exhibit an inherent affinity/identification with the environment (Mies and Shiva, 1993). The problematic and exclusionary bent of essentialist approaches soon became apparent and their profile has waned.

Current ecofeminist thought tends to fall into the broad categorisation of socialist ecofeminism. This too posits that women and the environment as inextricably linked, but views this connection as founded on women's reproductive role – which is not confined to bearing and raising children, but also pertains to the full range of unremunerated activities necessary to sustain life (Paltasingh and Lingam, 2014, pp. 46–47). Rooted in radical feminism, socialist ecofeminisms view the interplay between gender and environment as both personal and, at the same time, socially and politically underpinned, structured, and mandated in complex intersectional ways. Focussed on the need to effect societal change that addresses both inter-human injustice and an unsustainable characterisation of the human–environment nexus, socialist ecofeminisms promote undoing the dualism-based ideology and multifarious structures of patriarchy and replacing them with a relational, non-hierarchical ontology, that facilitates not just the survival but the flourishing of a truly sustainable form of human society on a healthy planet.

Ecofeminist thought at its core is founded on critiquing the application of dualism, as applied to both women and nature, viewing this as variously expressed in 'othering', whereby the subject is privileged over the object, with the latter deemed to be inferior to the former (Plumwood, 1993). This othering is strongly evident in the bifurcation of subject and object that underpins the gendered practice of the scientific method (Keller, 1995), itself strongly rooted in Cartesian thought. Here, ecofeminism echoes post-structural scholarship on the implications of binary oppositions. Cixous' ground-breaking feminist work on language and dualism and its implications (Cixous, 1976) provides a particularly succinct explanation of the relevant line of thought. Tellingly, she locates culture/nature alongside man/woman in her analysis of binaries. Cixous argues that all discourse repeats the same metaphor of dualistic opposition, each pairing constituting part of a 'universal battlefield' where '"victory" always comes down to the same thing: things get hierarchical' (Cixous, 1976, pp. 287–288). This approach is echoed in the ecofeminist understanding of the hierarchical nature of dualism and its ramifications: centrally, the tendency of the 'victor' (culture/man) to assume superiority over and consequently 'other' the loser (nature/woman), prompting, promoting, and justifying instrumentalisation, exploitation, and manipulation (Merchant, 1980).

On a more concrete level, (and in common with post-colonial scholarship, alluded to above) ecofeminist thought views dualism as having profoundly affected inter-human relationships. In ecofeminism, gender as a basis of othering takes centre stage, speaking to both male power and female exclusion (Morrow, 2017), and this is viewed as inextricably tied to attitudes and practices privileging (gendered) humanity over the natural world/environment that we inhabit (Plumwood, 1993). The foundational role that this plays in ecofeminist thought is clear as , for example, in Carolyn Merchant's ground-breaking work, *The Death of Nature* (1980), a formative text in the field that offered a new critical framing whereby to address these intermingled toxicities of the Western worldview. As Merchant (2006, p. 517) puts it:

> the notion of a 'Scientific Revolution' … is part of a larger mainstream narrative of Western culture that has propelled science, technology, and capitalism's efforts to 'master' nature – a narrative into which most Westerners have unconsciously been socialized and within which we ourselves have become actors in a storyline of upward progress.

This tale of mastery in turn fuelled:

> the dominant narratives of colonialism and imperialism that have helped to shape Western culture since the seventeenth century at the expense of nature, women, minorities, and indigenous peoples.
>
> *(Merchant, 2006, p. 517)*

In response to these issues, ecofeminist thought continues to concentrate on the twin axes of othering and exploitation of women and other non-privileged humans and the environment at the hands of the power holders, within states and globally. These are viewed as centrally implicated in multiple forms of injustice and in what must now, in light of environmental breakdown, be recognised as unsustainable ways of being in the world (Laybourn-Langton *et al.*, 2019). To elaborate, dichotomously framed 'othering' (MacQuarrie, 2012) has two particularly important, melded ramifications. First, it has served to cement the notion of humanity as divisible, which Brons suggests often involves an admixture of identification and opposition. The latter also feeds into a positive framing of resistance that (although only touched upon here for reasons of space), has a vivid presence in ecofeminist thought (Carlassare, 2000). At the same time, the hierarchical framing of othering facilitates attributing 'relative inferiority and/or radical alienness to the other or outgroup', thus laying the foundation for an entire range of unjust behaviours (Brons, 2015, p. 86). Second, othering has enabled what Val Plumwood terms 'the paradigm of mastery' to become the chief expression of the relationship between humanity and the environment, replacing/overwhelming other worldviews that saw/see humanity as part of nature (Plumwood, 1993). The self-justificatory fiction of the paradigm of mastery was always based on an illusion – we have never fully controlled or mastered the environment; and it is increasingly given the lie in the compound anthropogenic systemic threats to the earth system of the Anthropocene (Lade *et al.*, 2020).

Ecofeminism offers a possible corrective to the manifestations of atomistic thinking outlined above, pursuing instead holistic, relational thinking, founded on the recognition that humans are ultimately both truly 'all in this together' and embedded in the environment as ecological beings (Warren, 1994, p. 2). The latter contextualisation of the human condition absolutely refutes the paradigm of mastery by recognising humanity's fundamental dependence on the biosphere, eschewing the oppositional and antagonistic framing of mastery, and replacing it with an ethic of reciprocity. In this cause, and in simultaneously addressing the exclusions and closures based on hierarchical inter-human relationships, ecofeminist ethical approaches to correcting the ills of dualism seek to bring all human agency to the table. In so doing, they reject the idea that the scientific method is the only valid way of knowing (Wilson, n.d.), not least for its anatomising and atomising tendencies and myopic focus on the quantifiable. Ecofeminisms recognise both the value and the limits of the scientific method, concluding that it can offer only a partial account of what we need to regard as consequential in order to live well in the world. Thus, ecofeminisms advocate the recognition of additional dimensions to human understanding, in particular by admitting and valuing long discounted, or even denied, lived experience and traditional knowledges as significant in informing our thought and action.

Core Elements of Ecofeminisms

Focusing on (Addressing) the Connected Oppressions of a Patriarchal System

As discussed above, at their core ecofeminisms see dualistic, gendered, and now otherwise characterised inter-human oppressions that comprise what may be termed selective androcentrism, and environmental exploitation as intimately and inextricably intertwined. Karen Warren, for example, points to a:

> variety of so-called 'women-nature connections' – historical, empirical, conceptual religious, literary, political, ethical, epistemological, methodological and theoretical connections on how one treats women and the earth.
>
> *(Warren, 1994, p. 2)*

The multifaceted understanding of these connections means that addressing the harms that result requires harnessing the combined insights of feminist theory and ecology, each informing the other in order to address the whole (Warren and Cheney, 1991, p. 180). Connection can be expressed through what Code terms 'ideal cohabitation' (Code, 2006, p. 4), which she posits as the foundation for a just society to secure a viable future for humanity and the environment. The core elements of ecofeminism, identified to respond to this central understanding of othering and shared oppression, considered below, offer a number of interlinked strategies to respond to the challenges posed and suggest tools to construct better, inclusive, and relational ways of thinking and being in the world.

Countering Reductionism with a Relational Approach

Foundationally, then, ecofeminisms seek to counter othering and its atomistic, divisive, and inegalitarian ramifications for humanity and the natural world, with a relational, holistic, and just approach to inter-human and human–environment dynamics. They offer a relational, communal ethic in place of atomism and individualism, to counter the dualistic culture/nature binary, the supposed separation of humanity from nature (Matthews, 2017), and human exceptionalism (Manuel-Navarrete and Buzinde, 2010). Relational ecofeminist ethical approaches accord crucial importance to human embodiment, not only as it pertains to individuals but also applying it to communities at various spatial scales, and in all cases understanding them as embedded in concrete ecological contexts. The latter factor links to both the 'bottom-up' nature of the concept and practices of sustainable development and its reach from the global to the local, recognising the importance of incorporating situated knowledge and valuing lived experience that are hallmarks of ecofeminist epistemologies, considered below.

Intersectionality, Coalition, and Allyship

Given the central importance of embodiment and lived experience to ecofeminist thought and the importance that ecofeminism accords to allyship and coalition building, intersectionality has rapidly come to the fore in its development (Kings, 2017). The latter applies Kimberlé Crenshaw's pioneering and influential thinking (1991), in which she articulated the interlinked and mutually reinforcing application of gendered othering and race. Crenshaw's identification of the interplay of multiple forms of oppression, alongside recognising the importance of difference and agency in resistance, paved the way to a new understanding of the dynamics of power and powerlessness, which now extends its reach to a range of additional areas including Indigeneity, ageism, ableism, classism, heteronormativity and so on, and to the varied interplay between them. This type of holistic synthesis sits comfortably with ecofeminist thought and is particularly important in addressing the complexity of compound oppression. The latter is essential to do justice to the multifaceted nature of the real-world concerns that are in play where women experience the combined effects of compound disadvantage and adverse environmental impacts.

Ecofeminist thought recognises that an intersectional approach captures women's membership of multiple, cross-cutting constituencies based on complex, composite, embodied, identities (Ferguson, 2007, p. 40) that also provide a basis of reaching out to forge coalitions and alliances with other activists and areas of activism (King, 1995; Kirk, 1997). This type of approach has been deployed to favourable effect in a number of international law contexts, not least developing a coalition to press for the integration of gender concerns into the originally gender-blind sustainable development regime in advance of the Rio Earth Summit in 1992 (Morrow, 2013). Ecofeminist strategies were also evident in the genesis of the gender constituency under the 1992 United Nations Framework Convention on Climate Change (UNFCCC) (Morrow, 2013) and in the ongoing development of allyship in the constituency's operation (Morrow, 2017). The opportunities for forging alliances across ostensibly single-issue areas of activism, not based on identification or uniformity but rather on valued diversity and common cause, are manifold (Twine, 1997). The development of ecological masculinities provides a particularly interesting case in point, already exhibiting strong dialogic engagement with ecofeminist thought and holding much in common with it (Hultman and Pulé, 2019). There is similar, if as yet not fully realised, potential in regard to cross-fertilisation between ecofeminism and human and environmental concerns on the impacts of extractivism on Indigenous peoples, non-human animals, and the environment (Tomasello and Poirer, 2018) and in broader contexts (Laybourn-Langton *et al.*, 2019).

Combining Theory and Activism

In recognition of the importance of lived experience to our ways of being in the world referred to above, from the outset, and in common with other areas of inquiry, such as feminist political ecology (Rocheleau *et al.*, 1996), ecofeminist thought has, since its inception, combined theory and activism. This approach offers numerous important strands to ecofeminisms, not least according substantial weight to embodied or situated knowledge, which is deemed to be a core constituent of understanding, and correcting, gendered and other environmental injustices (Mies and Shiva, 1993; Rocheleau *et al.*, 1996). Valuing lived experience also raises critical issues in that such knowledge is acknowledged to be contingent, 'provisional, dynamic and changing' (Koggel, 2008, p. 179). Thus, it is reflexive in both character and application, being made and unmade by the iterative interactions of humanity and the environment that in turn shape societies, ecosystems, and, in the Anthropocene, ultimately, the biosphere. In the context of the interlinked human and environmental dimensions of sustainability and the need to respond to the changing and changeable dynamics of the Anthropocene, this is precisely the type of approach to knowledge building that is required. This type of amalgamation of activism and theoretical approaches is evident in the efforts of the gender constituency in relation to the UNFCCC, which applies feminist analysis to the realities of climate change in tandem with amplifying the voices of grassroots activists in global climate governance (Morrow, 2017).

Further utility for situated knowledge is evident in Code's concept of ecological thinking to promote an inclusive and transformative approach towards institutional processes and decision-making (Code, 2006), bringing hitherto excluded voices and perspectives to the table and enriching the knowledge base for dealing with injustice and environmental harms (Morrow, 2017). In a similar vein, the Mary Robinson Foundation's Climate Justice Initiative, for example, has long been instrumental in giving practical effect to fostering women's participation in the UNFCCC. It has not only campaigned for integrating gender balance into the regime but has also supported networking and delivered training to build women's capacity to effectively pursue opportunities to participate (Morrow, 2013). Relatedly, the significant role accorded to activism in ecofeminism also tells an important story in extending analysis beyond the well-worn ground of women's vulnerability, to also acknowledge their agency and ability to enrich our engagement with environmental issues (Arora-Jonsson, 2011). The gender constituency has taken pains to showcase grassroots activities and facilitate exchange of experience in the context of UNFCCC conferences (Morrow, 2013).

If activism enriches ecofeminist thought, so too do its academic aspects, which, as discussed above, draw upon many areas and disciplines to ground analysis and discussion from the outset, notably, across the arts, theology, philosophy, and as it has developed, (Plant, 1989) the social sciences – and more recently, politics and law. In the latter areas, gendered inequalities in both creating and experiencing adverse environmental impacts, in participation in law and politics generally, and environmental governance in particular, have provided fertile ground for scholarship (Sandilands, 1999; Morrow, 2013). While there can be tensions involved in the synthesis of activism and scholarship in ecofeminist thought, this also brings with it space for great creativity (Lahar, 1996). Both theoretical and activist perspectives are apt to be augmented when combined in networking and shared experience, insights, techniques, and strategies (Kirk, 1997; Brownhill and Turner, 2020). Furthermore, in fusing theory and activism, ecofeminism can also provide a fertile base for experiential learning (Ludlow, 2010), which in its turn is apt to inform both activism and theory.

Accommodating and Valuing Difference

Whilst dualism regards difference as providing a rationale for othering, imposing hierarchy, and ultimately domination, ecofeminist thought places positive value on difference and diversity as better representing diversity of experience and augmenting understanding (Ferguson, 2007). Whilst early ecofeminisms were justly criticised for their largely white, privileged, Western, roots (Merchant, 1980; Mies and Shiva, 1993; Sandilands, 1999), the response to this was not closure but rather the acknowledgement of the importance of respecting and accommodating difference and diversity. Vibrant additional strands of ecofeminisms have emerged driven, for example, by women of colour (Mies and Shiva, 1993; Rocheleau, Thomas-Slayter, and Wangari, 1996), and diversity continues to emerge as, for example, in queer ecofeminisms (Gaard, 2015). Ecofeminisms' embrace of inclusivity was also facilitated by the value attributed to activism and situated knowledge that is central to ecofeminist thought, outlined above – situated knowledge, by its very nature, being highly diverse and particularised in the first instance, while also providing a basis for finding common cause. The development of the gender constituency under the UNFCCC provides a case in point; it grew from a consciously cultivated inclusive, global, conversation, centred on developing a charter that a wide range of gender-focused groups could subscribe to (Morrow, 2013).

International Law, Feminism, and Ecofeminism

Law is the product of our society and its mores. It reflects society's aspirations and its deeply embedded structural flaws and inequalities (Morrow, 2020). It is therefore not surprising that there is a rich and multigenerational scholarship of feminist analysis of international law that brings home the detrimental implications of othering of women in the context of international governance (Charlesworth *et al.*, 1991; Heathcote, 2019). In particular, the now long-recognised need to view feminism as a work in progress, a process that seeks to not just amplify but also diversify women's voices, is now strongly apparent in international law (Heathcote, 2019, pp. 21–24). In this, there is a great deal of commonality with ecofeminist thinking on the disproportionate impacts of environmental degradation on women (Krupp, 2000) and their social ramifications (Mackenzie, 1984). The latter include women's relatively scant presence in international institutions and limited influence on international law and policy (Morrow, 2013; Morrow, 2020).

For present purposes, however, it is essential to recognise that international law embraces and even embeds dualism at its very core, with entrenched power holders enjoying the advantages of othering non-privileged humans and the environment (Grear, 2018). International environmental law is no exception in this (Kotzé *et al.*, 2020). This is the case despite groundbreaking work, not least by Charlesworth, Chinkin, and Wright (1991), who brought a compelling feminist perspective to bear on international law, igniting a lively, and latterly increasingly global, more diverse feminist scholarship (Heathcote, 2019).

Progress on the principles of gender equality in the international polis in the twentieth century notwithstanding, the necessary systemic change to fully realise this has remained elusive (Morrow, 2020). Thus, the reflexive application of feminist intersectional scholarship to international law (Davis, 2015) remains an active, necessary, and vital area of inquiry. The common ground that ecofeminism shares with feminist critiques of international law is, as one would expect given their partially shared ancestry, readily apparent. The ecological roots of ecofeminism ensure its salience to international environmental and sustainability law. This is, if anything, becoming ever more apparent, not least in regard to the intersectional analysis of the gendered impacts of Anthropocene threats such as climate change (Magnusdottir and Kronsell, 2021). It is also the case that, as argued above, the shift into the Anthropocene makes it imperative to ground international law in new ways that prioritise environmental restoration and prudence in the cause of the sustainability of human life. It is here that ecofeminism can make a significant contribution by recognising that to survive, let alone flourish, swift and profound changes in how humanity views and acts in relation to the environment and inter-human interactions, now and with future generations, must be recast in relational rather than exploitative terms. The planet will, though no doubt altered, survive the Anthropocene; ironically, human society as we currently know it will not.

Ecofeminism and a New, Sustainable, International Law for People and Planet

Gender is of course not the only form of inter-human othering but it has in many ways been foundational to scholarly analysis and critique of the 'multiple dimensions and modalities of social relations and subject formations' (McCall, 2005, p. 1771). It has already exposed a great deal about the toxicities of our current ways of thinking and being in the world. Ecofeminist scholarship timeously applies feminist thinking alongside ecological understanding to our present plight and not only argues that humanity must now be done with dualism and its closures but also provides a corrective to it. Dualism has been pursued beyond the point of any possible utility, and doing so has ushered in the Anthropocene, rendering the planetary conditions that prevailed in the Holocene, and which enabled societies to flourish, a thing of the past. International law must serve as a prime tool in our response to the evolving threats that humanity is unleashing, which must entail fashioning a new, truly sustainable paradigm for societal ordering and activity. But international law must also change in order to achieve this, opening the multiple closures of patriarchy that it features and which have wrought so much damage.

An ecofeminist approach to international law could offer a corrective to many of these ills, entailing at base a radical, relational process of (re)connection – as humans with one another and as humans with the non-human environment. In order to ensure the continuation of our species and secure planetary conditions that make that possible, it is imperative that we remedy the ills of disconnection (founded on erroneous premises as they are) of humanity from one another and from the environment. The view of humanity as a 'master species' perpetuated in and perpetuating modernity, whose flourishing and survival could to all intents and purposes be decoupled from the natural world, was always an illusion, and is now a threat. It has wrought huge damage on the environment and having first imposed adverse impacts on vulnerable classes of humanity, has now come full circle to threaten the very survival of our species as the cumulative, systemic, impacts of human enterprise are increasingly made manifest. Technical fixes cannot and will not deliver all that is needed. Our response to the existential crisis that humanity finds itself in must itself be more profound. Hubris must be replaced by humility. A truly sustainable future must be founded on a relational epistemology and ontology, which enables us to fashion new ways of being and doing in the world that give all of humanity and the environment of which we form part their due. Our future depends on it:

> And somebody has to say that we never need to grow forever. That we, like the trees, can reach our full growth, and mature, in wisdom and in time, that we can be enough of us. That there can be room for other breeds and kinds and lives.
>
> *Neil Gaiman, 'After Silence: For Rachel Carson'*

References

Allen, K.E., Quinn, Q.E., English, C. and Quinn, J.E. (2018) 'Relational Values in Agroecosystem Governance' *Current Opinion in Environmental Sustainability*, 35, pp. 108–115. DOI: 10.1016/j.cosust.2018.10.026

Arora-Jonsson, S. (2011) 'Virtue and Vulnerability: Discourses on Women, Gender and Climate Change', *Global Environmental Change*, 21(2), pp. 744–751. DOI: 10.1016/j.gloenvcha.2011.01.005

Arponen, V.P.J. (2016) 'A Critique of an Epistemic Intellectual Culture: Cartesianism, Normativism and Modern Crises', Journal for the Theory of Social Behaviour, 46(1), pp. 84–103. DOI: 10.1111/jtsb.12085

Banerjee, S.B. (2003) 'Who Sustains Whose Development? Sustainable Development and the Reinvention of Nature', *Organization Studies*, 24(1), pp. 143–180. DOI: 10.1177/0170840603024001341

Borowy, I. (2014) *Defining Sustainable Development for Our Common Future: A History of the World Commission on Environment and Development*. London: Routledge.

Bourdeau, P. (2004) 'The Man–Nature Relationship and Environmental Ethics', *Journal of Environmental Radioactivity*, 72(1–2), pp. 9–15. DOI: 10.1016/S0265-931X(03)00180-2

Brons, L. (2015) 'Othering, an Analysis', *Transcience*, 6(1), pp. 69–90. DOI: 10.17613/M6V968

Brownhill, L. and Turner, T.E. (2020) 'Ecofeminist Ways, Ecosocialist Means: Life in the Post-capitalist Future', *Capitalism Nature Socialism*, 31(1), pp. 1–14. DOI: 10.1080/10455752.2019.1710362

Carlassare, E. (2000) 'Socialist and Cultural Ecofeminism: Allies in Resistance', *Ethics and the Environment*, 5(1), pp. 89–106. DOI: 10.1016/S1085-6633(99)00025-X

Charlesworth, H., Chinkin C. and Wright, S. (1991) 'Feminist Approaches to International Law', *American Journal of International Law*, 85(4), pp. 613–645. DOI: 10.2307/2203269

Cixous H. (1976) 'Sorties'. Translated by Ann Liddle, in Lodge, D. (ed.) (1988) *Modern Criticism and Theory: A Reader*. London: Longman, pp. 287–293.

Code L. (2006) *Ecological Thinking: The Politics of Epistemic Location*. Oxford, Oxford University Press.

Crenshaw, K. (1990–1991) 'Mapping the Margins: Intersectionality, Identity Politics, and Violence Against Women of Color', *Stanford Law Review*, 43, pp. 1241–1300.

Daly, L. (1994) 'Ecofeminisms and Ethics', *The Annual of the Society of Christian Ethics*, 14, pp. 285–290.

Davis, A.N. (2015) 'Intersectionality and International Law: Recognizing Complex Identities on the Global Stage', *Harvard Human Rights Journal*, 28(1), pp. 205–242. DOI: 10.1525/sp.2007.54.1.23

Dawe, N.K. and Ryan, K.L. (2002) 'The Faulty Three-Legged-Stool Model of Sustainable Development', *Conservation Biology*, pp. 1458–1460. DOI: 10.1046/j.1523-1739.2003.02471.x

Ferguson, M.L. (2007) 'Sharing without Knowing: Collective Identity in Feminist and Democratic Theory', *Hypatia*, 22(4), pp. 30–45. DOI: 10.1111/j.1527-2001.2007.tb01318.x

French, D. (2005) *International Law and Policy of Sustainable Development*. Manchester: Manchester University Press.

Gaard, G. (2015) 'Ecofeminism and Climate Change', *Women's Studies International Forum*, 49, pp. 20–33. DOI: 10.1016/j.wsif.2015.02.004

Gerson, L.P. (1986) 'Platonic Dualism', *The Monist*, 69(3), pp. 352–369. DOI: 10.5840/monist198669318

Grear, A. (2018) 'Human Rights and New Horizons? Thoughts Toward a New Juridical Ontology', *Science, Technology, and Human Values*, 43(1), pp. 129–145. DOI: 10.1177/0162243917736140

Gruen, L. (2009) 'Attending to Nature: Empathic Engagement with the More than Human World', *Ethics and the Environment*, 14(2), pp 23–37. DOI: 10.2979/ETE.2009.14.2.23

Harari, Y.N. (2016) *Homo Deus: A Brief History of Tomorrow*. London: Vintage.

Heathcote, G. (2019) *Feminist Dialogues on International Law: Success, Tensions, Futures*. Oxford: Oxford University Press.

Held, V. (2005) *The Ethics of Care: Personal, Political, and Global*. Oxford: Oxford University Press.

Hirsh, A. (2020) 'The Human Error Darwin Inspired: How the Demotion of Homo Sapiens Led to Environmental Destruction', *Nautilus*, 30 September [Online]. Available at: http://nautil.us/issue/90 /something-green/the-human-error-darwin-inspired?utm_source=pocket-newtab-global-en-GB (Accessed: 13 August 2021).

Holmes, R. (2008) *The Age of Wonder: How the Romantic Generation Discovered the Beauty and Terror of Science*. New York: Pantheon Books.

Hughes, E.J. (1995) 'Fishwives and Other Tails: Ecofeminism and Environmental Law', *Canadian Journal of Women and the Law*, 8, pp. 502–530.

Hultman, M. and Pulé, P.M. (2019) *Ecological Masculinities: Theoretical Foundations and Practical Guidance*. London: Routledge.

Intergovernmental Panel on Climate Change (2021a) *Climate Change 2021: The Physical Science Basis*. Available at: https://www.ipcc.ch/report/ar6/wg1/downloads/report/IPCC_AR6_WGI_SPM_final .pdf (Accessed: 9 September 2022).

Intergovernmental Panel on Climate Change (2021b) *IPCC Press Release: Climate Change Widespread, Rapid, and Intensifying*. Available at: https://www.ipcc.ch/site/assets/uploads/2021/08/IPCC_WGI -AR6-Press-Release_en.pdf (Accessed 16 August 2021).

Islam, S.N. (2015) 'Inequality and Environmental Sustainability', United Nations DESA Working Paper No. 145. Available at: https://www.un.org/esa/desa/papers/2015/wp145_2015.pdf (Accessed: 17 August 2021).

Keller, E. (1995) *Reflections on Gender and Science*. New Haven: Yale University Press.

Kemper, J.A., Hall, C.M. and Ballantine, P.W. (2019) 'Marketing and Sustainability: Business as Usual or Changing Worldviews?', *Sustainability*, 11, pp. 780–797. DOI: 10.3390/su11030780

King, Y. (1995) 'Engendering a Peaceful Planet: Ecology, Economy, and Ecofeminism in Contemporary Context', *Women's Studies Quarterly*, 23(3–4), pp. 15–21.

Kings, A.E. (2017) 'Intersectionality and the Changing Face of Ecofeminism', *Ethics and the Environment*, 22(1), pp. 63–87. DOI: 10.2979/ethicsenviro.22.1.04

Kirk, G. (1997) 'Ecofeminism and Environmental Justice: Bridges across Gender, Race, and Class', *Frontiers: A Journal of Women Studies*, 18(2), pp. 2–20. DOI:10.2307/3346962

Koggel, C.M. (2008) 'Ecological Thinking and Epistemic Location: The Local and the Global', *Hypatia*, 23(1), pp. 177–186. DOI: 10.1111/j.1527-2001.2008.tb01173.x

Kotzé, L.J., Du Toit, L. and French, D. (2020) 'Friend or Foe? International Environmental Law and Its Structural Complicity in the Anthropocene's Climate Injustices', *Oñati Socio-Legal Series*, 11(1), pp. 180–206. DOI: 10.35295/osls.iisl/0000-0000-0000-1140

Krupp, S.J. (2000) 'Environmental Hazards: Assessing the Risk to Women', *Fordham Environmental Law Review*, 12(1), pp. 111–139

Lade S. J., Steffen, W., de Vries, W., Carpenter, S.R., Donges, J. F., Gerten, D., Hoff, H., Newbold, T., Richardson, K. and Rockstrom, J. (2020) 'Human Impacts on Planetary Boundaries Amplified by Earth System Interactions', *Nature Sustainability*, 3, pp. 119–128. DOI: 10.1038/s41893-019-0454-4

Lahar, S. (1996) 'Ecofeminist Theory and Grassroots Politics', in Warren, K.J. (ed.) *Ecological Feminist Philosophies*. Bloomington: Indiana University Press, pp. 1–18.

Lambert, H., Jaideep, G., Fletcher. H., Hammond, L., Lowe, N., Pelling, M., Raina, N., Shahid, T. and Shanks, K. (2020) 'COVID-19 as a Global Challenge: Towards an Inclusive and Sustainable Future', *The Lancet: Planetary Health*, 4(8), pp. e312–e314. DOI: 10.1016/S2542-5196(20)30168-6

Laybourn-Langton, L., Rankin, L. and Baxter, D. (2019) *This Is a Crisis: Facing Up to the Age of Environmental Breakdown*. London: Institute for Public Policy Research [Online]. Available at: https://www.ippr.org/research/publications/age-of-environmental-breakdown (Accessed 19 August 2021).

Levy, B. and Sidel, V. (eds.) (2005) *Social Injustice and Public Health*. Oxford: Oxford University Press.

Lock, M. and Farquhar, J. (2007) *Beyond the Body Proper: Reading the Anthropology of Material Life*. Durham: Duke University Press.

Long, A. (2015) 'Global Integrationist Multimodality: Global Environmental Governance and Fourth Generation Environmental Law', *Journal of Environmental and Sustainability Law*, 21(1), pp. 169–208.

Ludlow, J. (2010) 'Ecofeminism and Experiential Learning: Taking the Risks of Activism Seriously', *Transformations: The Journal of Inclusive Scholarship and Pedagogy*, 21(1), pp. 42–59.

Mackenzie, S. (1984) 'A Socialist Feminist Perspective on Gender and Environment', *Antipode*, 16(3), pp. 3–10. DOI: 10.1111/j.1467-8330.1984.tb00068.x

Magnusdottir, G.L. and Kronsell, A. (eds.) (2021) *Gender, Intersectionality and Climate Institutions in Industrialised States*. London: Routledge.

Manuel-Navarrete, D. and Buzinde, C.N. (2010) 'Socio-ecological Agency: From "Human Exceptionalism" to Coping with "Exceptional" Global Environmental Change' in Redclift, M.R. and Woodgate, G. (eds.) *The International Handbook of Environmental Sociology*, 2nd edn. Cheltenham: Edward Elgar, pp. 136–149.

Marek-Muller, S. (2020) *Impersonating Animals: Rhetoric, Ecofeminism, and Animal Rights Law*. East Lansing: Michigan State University Press.

Matthews, F. (2017): 'The Dilemma of Dualism' in Macgregor, S. (ed.) *Gender and Environment Handbook*. London: Routledge, pp. 54–70.

McCall, L. (2005) 'The Complexity of Intersectionality', *Signs*, 30(3), pp. 1771–1800. DOI: 10.1086/426800

Merchant, C. (1980): *The Death of Nature: Women, Ecology, and the Scientific Revolution*. New York: Harper and Row.

Merchant, C. (2006): 'The Scientific Revolution and The Death of Nature', *Isis*, 97, pp. 513–533. DOI: 10.1086/508090

Mies, M. and Shiva, V. (1993) *Ecofeminism*. London: Zed Books.

Miller, P. (1989) 'Descartes' Legacy and Deep Ecology', *Dialogue*, 28(2), pp. 183–202.

Morrow, K.L. (2012) 'Rio +20, the Green Economy and Re-orienting Sustainable Development' *Environmental Law Review*, 14(4), pp. 279–297. DOI: 10.1350/enlr.2012.14.4.166

Morrow, K.L. (2013) 'Ecofeminism and the Environment: International Law and Climate Change', in Davies, M. and Munro, V.E. (eds.) *The Ashgate Research Companion to Feminist Legal Theory*. London: Routledge, pp. 377–393.

Morrow, K.L. (2017) 'Ecofeminist Approaches to the Construction of Knowledge and Coalition Building – Offering a Way Forward for International Environmental Law and Policy", in Philippopoulos-Mihalopoulos, A. and Brooks, V. (eds.) *Research Methods in Environmental Law: A Handbook*. Cheltenham: Edward Elgar, pp. 298–315.

Morrow, K.L. (2020) 'Tackling Climate Change and Gender Justice -- Integral; Not Optional', *Oñati Socio-Legal Series*, 11(1), pp. 207–230. DOI: 10.35295/osls.iisl/0000-0000-0000-1166

Paltasingh, T. and Lingam, L. (2014) '"Production" and "Reproduction" in Feminism: Ideas, Perspectives and Concepts', *IIM Kozhikode Society & Management Review*, 3(1), pp. 45–53. DOI: 10.1177/2277975214523665

Paris Agreement (adopted 12 December 2015, entered into force 4 November 2016) FCCC/CP/2015/10/Add.1.

Plant, J. (ed.) (1989) *Healing the Wounds: The Promise of Ecofeminism*. Philadelphia: New Society Publishers.

Plumwood, V. (1993): *Feminism and the Mastery of Nature*, London: Routledge.

Rocheleau, D., Thomas-Slayter, B. and Wangari, E. (eds) (1996) *Feminist Political Ecology: Global Issues and Local Experiences*. London: Routledge.

Sandilands, C. (1999) *The Good-Natured Feminist: Ecofeminism and the Quest for Democracy*. Minneapolis: University of Minnesota Press.

Solnit, R. (2021) 'The IPCC's Latest Climate Report Is Dire. But it also Included Some Prospects for Hope', *The Guardian*, 13 August [Online]. Available at: https://www.theguardian.com/commentis-free/2021/aug/13/ipcc-latest-climate-report-hope?CMP=Share_AndroidApp_Other (Accessed 13 August 2021).

SpaceX (n.d.) 'Mars and Beyond: The Road to Making Humanity Multiplanetary' [Online]. Available at: https://www.spacex.com/human-spaceflight/mars/ (Accessed 12 November 2020).

Stensöta, H.O. (2015) 'Public Ethics of Care – A General Public Ethics', *Ethics and Social Welfare*, 9, pp. 183–200. DOI: 10.1080/17496535.2015.1005551

Thomas, B. (2016) *Narrative: The Basics*. Abingdon: Routledge.

Tomasello, S. and Poirier, N. (2018) 'The Intersectionality of Wildlife Conservation and Indigenous Rights', *Green Theory and Praxis Journal*, 11(1), pp. 18–34.

Twine, R. (1997) 'Masculinity, Nature, Ecofeminism." *ecofem.org e-Journal* [Online] Available at: http://richardtwine.com/ecofem/masc.pdf (Accessed: 8 September 2022).

United Nations (2015) *The Millennium Development Goals Report 2015*. Available at: https://www.un.org/millenniumgoals/2015_MDG_Report/pdf/MDG%202015%20rev%20(July%201).pdf (Accessed: 17 August 2021).

United Nations (2020) *The Sustainable Development Goals Report 2020*. Available at: https://unstats.un.org/sdgs/report/2020/The-Sustainable-Development-Goals-Report-2020.pdf (Accessed: 17 August 2021).

United Nations Environment Programme (2019) Global Environment Outlook – GEO-6: Healthy Planet, Healthy People. Available at: https://wedocs.unep.org/20.500.11822/27539 (Accessed: 9 September 2022).

UNOHCHR (15 July 2019) 'Human Rights Obligations Relating to the Enjoyment of a Safe, Clean, Healthy and Sustainable Environment', UN Doc A/74/161.

Verchick R.R.M. (1996). 'In a Greener Voice: Feminist Theory and Environmental Justice', *Harvard Women's Law Journal*, 19, pp. 23–88.

Vidas, D., Kristian, O., Fauchald, J. Ø. and Tvedt, M.W. (2015) 'International Law for the Anthropocene? Shifting Perspectives in Regulation of the Oceans, Environment and Genetic Resources', *Anthropocene*, 9, pp. 1–13. DOI: 10.1016/j.ancene.2015.06.003

Vidas, D., Zalasiewicz, J. and Williams, M. (2016) 'What Is the Anthropocene – And Why Is it Relevant for International Law?' *Yearbook of International Environmental Law*, 25(1), pp. 3–23. DOI: 10.1093/yiel/yvv062

Walsh, Z., Bohme, J. and Wamsler, C. (2021) 'Towards a Relational Paradigm in Sustainability Research, Practice, and Education', *Ambio*, 50, pp. 74–84. DOI: 10.1007/s13280-020-01322-y

Warren, K.J. (ed.) (1994) *Ecological Feminism*. London: Routledge.

Warren, K.J. and Cheney J. (1991) 'Ecological Feminism and Ecosystem Ecology', *Hypatia*, 6(1), pp. 179–197. DOI: 10.1111/j.1527-2001.1991.tb00216.x

Wilson, F. (n.d.) 'René Descartes: Scientific Method', *Internet Encyclopedia of Philosophy* [Online]. Available at: https://www.iep.utm.edu/desc-sci/ (Accessed: 08 April 2020).

10

INDIGENOUS KNOWLEDGE AND INTERNATIONAL (ANTHROPOCENTRIC) LAW

The Politics of Thinking from (and for) Another World

Roger Merino

Introduction

At the time of writing, Indigenous peoples are deeply affected by the expansion of Covid-19 over their territories. By the beginning of June 2020, civil society had reported hundreds of infected and dead people in Indigenous communities in Latin America. In Brazil and Peru, the countries most impacted by the pandemic, 1,346 and 2,191 people were infected and 151 and 349 were killed, respectively (Coordinadora de las Organizaciones Indígenas de la Cuenca Amazónica (COICA) and Red Eclesial Panamazónica (REPAM, 2020). The death ratio in Brazil is 12.6 per cent in Indigenous population whereas the national ratio is 6.4 per cent (Articulación de los Pueblos Indígenas de Brasil (APIB), 2020). In the United States, the Navajo Nation reached 5,479 infected and 248 casualties, becoming the most affected territory in the country (Lima, 2020).

The disproportionate impact of the current pandemic on Indigenous peoples is just another example of historical patterns of Indigenous peoples' marginalisation, embedded in national and international law. Despite many declarations in UN forums showing concerns about states' inaction to protect Indigenous communities, the international market for medical supplies concentrates on Western countries, making supplies less accessible for non-industrial countries where historically the Indigenous population has been excluded from basic public health and sanitation (United Nations Department of Economic and Social Affairs, Indigenous Peoples, 2020). Moreover, after years of neoliberal policies promoted by multilateral financial institutions, peripheral countries are incapable of producing medical supplies for all those sick – much less Indigenous peoples – with their precarious health systems (The Lancet, 2020).

If Indigenous peoples are the most affected by Covid-19, it is because they are at the bottom of states' priorities. In Brazil, the Mayor of Manaus, the most affected region in the country, said that Indigenous peoples are on the verge of a genocide because of government inaction (La Vanguardia, 2020). A similar claim has been made in Peru, where Indigenous organisations denounced that most infected and casualties from their communities are not being counted by

198

DOI: 10.4324/9781003201120-13

the state (Observatorio Petrolero, 2020). What is at stake here is not only an ongoing genocide in slow motion but also what Boaventura de Sousa calls an 'epistemicide' (2014), the 'murder of knowledge' of subordinated peoples produced by unequal exchanges among cultures. The infection is killing mainly the old and wise, people who possess and teach ancestral knowledge to future generations. What does national and international law do to protect and value this knowledge?

According to specialised literature, binding instruments and soft law show an emerging recognition of Indigenous knowledge (IK) at the international level (Meyer, 2001). This would be the case because international fora have actively engaged in recent decades with this knowledge, in organisations such as the Food and Agriculture Organization (FAO), the International Union for the Protection of New Varieties of Plants (UPOV), the United Nations Educational Scientific and Cultural Organization (UNESCO), the World Intellectual Property Organization (WIPO), and the World Trade Organization (WTO) (Drahos and Frankel, 2012).

Moreover, international instruments provide a broad notion of IK, covering areas such as biodiversity, education, health, literature, arts, science (United Nations Declaration on the Rights of Indigenous Peoples (UNDRIP) 2007) and, in general, intangible cultural heritage (Convention for the Safeguarding of the Intangible Cultural Heritage, 2003). IK is also conceived as a right that shapes education programs and services aimed at Indigenous populations (Indigenous and Tribal Peoples Convention (ILO Convention 169) 1989). In the fields of biodiversity and environment, some binding instruments focus on the need to protect and promote IK in using biological resources (Convention on Biological Diversity, 1992; International Treaty on Plant Genetic Resources for Food and Agriculture, 2001). Others highlight the necessary compensation and benefits to local populations for the use of IK (Convention to Combat Desertification, 1994; Nagoya Protocol, 2010), being crucial to the prior and informed consent in this regard (Mo'otz Kuxtal Voluntary Guidelines for Traditional Knowledge, 2016). Non-binding declarations highlight the contribution of IK to sustainable development (Rio Declaration, 1992) and international policy instruments emphasise how this knowledge must be included in the evaluation of environmental, social, and cultural impacts of projects (Akwé: Kon Voluntary Guidelines for the Conduct of Cultural, Environmental and Social Impact, 2004). International declarations (UNDRIP, 2007; American Declaration on the Rights of Indigenous Peoples, 2016) and forums (e.g., Intergovernmental Committee on Intellectual Property and Genetic Resources, Traditional Knowledge and Folklore of WIPO) highlight, in general, the protection of intellectual property rights over this knowledge and the application of free, prior, and informed consent for its use.

International law recognises the importance of IK as a human right that must be respected, as an asset to be protected under property rights rules, and as a condition to effectively implement environmental and development initiatives. However, scholarly contributions lack a critical analysis of the ways this recognition, rather than acknowledging the relevance of these knowledge systems to reshape current institutions and discourses, has meant the reaffirmation of hegemonic ideas and regulations. After a deep examination of the literature of international law and international development, and an analysis of international discourses and mechanisms that address IK, this chapter argues that international law has essentialised IK by conceiving it as a museum piece that must be 'conserved'. It also has commodified IK by protecting it as an asset under intellectual property rules and has instrumentalised IK by including it in Declarations and Conventions to justify environmental and development interventions. In the end, international law has obscured IK by treating it as a second-level system of knowledge, unable to support broader international rules to address environmental, sanitary, and food crises.

In fact, if properly considered in its political and epistemological dimensions, IK has the potential to reshape the anthropocentric basis of international law. Anthropocentrism, the idea that human beings are socially and politically at the centre of the cosmos, is the basis of international law theory and practice. This is the case even in those areas that are supposedly non- or less anthropocentric, such as environmental law and animal rights law because in their conception and application they reinscribe or even extend international law's anthropocentrism (see the Introduction to this volume). In the case of international Indigenous rights – and in particular IK – even though it is assumed that Indigenous legal frameworks do not rely on the ontological and hierarchical differentiation between humans and nature, international instruments and discourses recognise IK with the aim of either assimilating it into Western frameworks or instrumentalizing it to justify business-as-usual approaches to different branches of international law, from environment law to intellectual property law.

To challenge the anthropocentric way international law treats Indigenous rights, it is crucial to replace the logic of recognition of IK with the logic of sovereignty of IK. Whereas colonialism has been a vehicle for the anthropocentrism of colonial powers, Indigenous peoples – as sovereign peoples – have historically challenged the imposition of Western/Eurocentric legal ideas and institutions. Similarly, today, they oppose the use of IK for ornamental and discursive purposes. Instead, they reclaim the possibility to mindfully connect local, national, and international legal systems and policy responses to IK. For example, in the current pandemic scenario, Indigenous peoples are not simply victims. Whereas in the global North and the global South states have little interest in protecting Indigenous communities, much less in appealing to ancestral knowledge to address the pandemic, Indigenous nations are deploying actions to save their material integrity and culture. For instance, in the Amazon of Peru and Ecuador, Indigenous communities have taken self-isolation measures (such as blocking rivers and ports to limit the diffusion of the virus), as well as the reinforcement of their communal economy and traditional health systems, to reduce the spread of Covid-19 (Observatorio Petrolero, 2020; Zibell, 2020).

At the same time, many communities continue to struggle against the expansion of extractive activities that degrade forests, alerting us to the fact that it is exactly such actions that facilitate the triggering of global pandemics by destroying the habitat of animals that end up being trafficked in global consumption chains and thereby transmitting disease (Bloomfield, McIntosh, and Lambin, 2020). Struggles to preserve IK and defend the forests from extraction and commodification are inspiring proposals for urgent changes to current hegemonic discourses and practices to address environmental, food, and sanitary crises.

The next section addresses the meaning of IK by exploring its non-anthropocentric philosophical foundations. The third section discusses how international law has been unable to deal with the non-anthropocentric aspects of IK, in areas such as human rights law, intellectual property law, and environmental law. The fourth section explores the potential of IK to challenge the anthropocentric foundations of international law by focusing on two areas where IK has prompted alternative perspectives on human-nature relations: food sovereignty and the rights of nature.

The Epistemic Turn of Indigenous Knowledge and Philosophy

The philosophy and episteme that sustain Indigenous knowledge challenge the anthropocentrism of international law. The colonial project was also a project to normalise this anthropocentrism. Postcolonial and decolonial studies have shown how processes of colonisation obscured IK by producing a universal way of thinking about the world from the epistemic location of the colonisers (Castro-Gomez, 2005). This knowledge was functional to the epistemic necessities of capitalism, the need for measurement, quantification, and objectification to control resources and populations (Quijano, 2000). The European Enlightenment's key thinkers, such as Kant and Hegel, created myths of rational subjectivity outside spatial-temporal locations and power relations, locating European white men at the centre of human history (Spivak, 1999; Grosfoguel, 2007; Dussel, 1994).

In decolonial theory, the consolidation of this Eurocentrism as a scientific common sense is known as the 'coloniality of knowledge'. Knowledge is the basis of a pattern of national and international power that creates categories, concepts, and normative propositions that define the 'must be' for all the world's peoples (Lander, 2000). The naturalisation of this Western knowledge thus became the pattern to evaluate other societies (Hall, 1992) and detect their 'lacks' and 'underdevelopment' (Lander, 2000).

Epistemic disobedience, then, means rejecting the two alternatives offered by Western knowledge: to accept the inferiority of non-Western knowledge or to assimilate it (Mignolo, 2011). The third option is to advance a decolonial epistemology by delinking imposed epistemologies and revalorising subaltern systems of thinking (Mignolo, 2007). This means building knowledge from the experience of oppressed groups instead of just importing knowledge as if the oppressed groups were unable to conceptualise and interpret their own realities (Grosfoguel, 2009). A similar path is offered by Santos' epistemologies of the South (2004). He proposes amplifying the present by adding to the existing reality the knowledge and practices hidden or denied by Western modernity (sociology of absences); to then enlarge the present by adding alternatives and aspirations derived from revalorised knowledge and practices (sociology of emergences); and to undertake deliberative exchanges among the proponents of these knowledges and experiences to identify possible alliances (intercultural translation).

These critical approaches do not entail a fundamentalist rejection of Western knowledge. Instead of rejecting the use of Western categories, the idea is to think those categories from non-Western epistemologies or from a 'border thinking' to redefine their concepts and institutions (Mignolo and Tlostanova, 2006; Grosfoguel, 2009). For example, the meaning of territory, nation-state, or natural resources might have a new dimension if conceptualised from the epistemic position of Indigenous peoples.

Critical approaches do not deny the historical interlinkages between Western science and Indigenous knowledge either. This dichotomous classification, thus, must not lead to reinforce the modernisation view of 'traditional' and 'modern' thinking. Instead, it should alert to the necessity of uncovering their similarities, exchanges, and mutual learning (Agrawal, 1995) on the premise that they must be treated in a symmetrical way without neglecting their contextual character (Turnhout, 2018). For instance, the growth of native potato and quinoa by many communities of the Andes in Peru and Bolivia relies on ancestral knowledge and techniques for cultivation but also relies on modern supply chains to export those products abroad. These ancestral techniques are not a second-level knowledge in comparison to industrial agriculture and are not necessarily isolated from exchanges with other types of economic and agricultural knowledge. This analysis must also acknowledge that what is considered Western science and research has often been used as a means for colonial appropriation and displacement, and that Indigenous knowledge – rather than being merely 'traditional' – has entailed an active and transformative contestation of imposed knowledges in colonial and postcolonial contexts.

In Western academia, terms such as 'Indigenous knowledge', 'traditional knowledge', and 'ancestral knowledge', referring to a system of knowledge embedded in Indigenous cultures, first appeared in the early 1980s in the field of anthropology (Lanzano, 2013). IK has been conceptualised as a system derived from the survival tactics of Indigenous peoples, reflected in their collective memories, and preserved and improved through generations (Bruchac, 2014; Shepard and Izquierdo, 2003; Chakraborty and Sen, 2017).

However, Indigenous philosophies existed long before Western academia labelled them as 'traditional'. Indigenous philosophies are diverse bodies of knowledge within which Indigenous worlds were framed prior to European colonisation, and which also developed during colonial and postcolonial experiences (Smith *et al.*, 2016; Turner, 2006; Marsden, 2003; Whyte, 2016; 2017; Mead, 2003). Thus, IK must not be merely connected with the past (Poitras *et al.*, 2018), since Indigenous nations adapted, negotiated, and resisted the nation-state system of knowledge (Tuhiwai, Te Kahautu, and Haupai, 2016).

For example, in today's New Zealand, the Māori world weaves together people, land, natural resources, identity and well-being as shaped by, and bound up with, physical and spiritual connections (Watene and Merino, 2018). In the Andes of Latin America, *Buen vivir* (*Sumac kawsay* in Quechua and *Suma qamaña* in Aymara) is a concept rooted in Indigenous cosmologies and expresses the principles of reciprocity, complementarity, and relationality in human interactions and in relation to the cycles of nature (Merino, 2016). Such a view implies a profound respect for the differences among human beings and between human beings and the natural environment. In the Amazon, unlike Western dualism, cosmologies classify a scale of beings in which the differences between human beings, plants, and animals are of degrees and not of nature (Descola, 2004). In some Amazonian cultures, the people, certain animals, plants, and meteorological phenomena are all considered agents (human). Given specific circumstances (Greene, 2009), this explains why for them 'the original common condition of both humans and animals is not animality but, rather, humanity' (Viveiros de Castro, 2004, p. 465). In general, most Indigenous philosophies share the idea that human lives are not simply made possible, or enriched, by the natural environment, but rather bound up with, and inseparable from, the natural world (Watene and Merino, 2018).

Most IK is based on these philosophical fundamentals that exceed the Western dual approach to nature and society relations. Instead of trying to understand this deep thinking and to build from it new modes of interrelation between humans and the natural environment, state authorities and corporations often appropriate this knowledge for the purpose of justifying policy and economic interventions. This is the case, for example, in the implementation of extractive policies in the name of *buen vivir*, despite infringing on the territorial rights of Indigenous peoples in Latin America (Merino, 2016; Merino, 2018). A similar operation might be seen in international law where IK is used, on the one hand, to promote Indigenous rights, but on the other hand, to justify environmental or economic initiatives that undermine the sovereignty of Indigenous nations.

IK Under International (Anthropocentric) Law

Legal scholars argue that relevant human rights provisions from the International Covenant on Civil and Political Rights and the International Covenant on Economic, Social and Cultural Rights can be read in conjunction to provide a human rights basis for the protection of 'traditional knowledge' in a double dimension: a

> positive protection [that] entails the active assertion of IP [Intellectual Property] rights in protected subject matter, with a view to excluding others from making specific forms of use of the protected material; [and a] defensive protection [that] does not entail the assertion of IP rights, but rather aims at preventing third parties from claiming rights in misappropriated subject matter.
>
> *(Haugen, 2005, p. 665)*

This human right would have positive incidence on other rights, such as the right to life, health, work, and culture (Sree Sudha, 2018; Mandikonza, 2019; Sumida, 2017) and should be inserted into the discussions of international organisations that allegedly side-line the human rights dimension of IK (Razzaque and Ssenyonjo, 2007).

The deeper problem, however, is not the lack of a 'rights' discourse around IK in international organisations and national governments but the fact that it is usually conceptualised under Western categories and an intellectual property lens (Schuler, 2013). As has been noted, just as Indigenous territories were declared *terra nullius* in the colonisation process, IK has been treated as *gnaritas nullius* (nobody's knowledge) by IP systems, allowing IK to be appropriated in the public domain (Younging, 2015). Current frameworks are still European-based and neglect the historic exclusion of Indigenous peoples, accepting Indigenous cultural expressions if they were adapted to liberal individualist norms and requirements, such as property, creativity, and authorship (Anderson, 2015).

Indigenous movements, then, started to engage with IP as a strategy to defend their ancestral knowledge from external appropriation, in the context of the rise of global Indigenous political networks and the increasing international activism around IP rights from the second half of the twentieth century (Drahos and Frankel, 2012). Efforts were directed to protect folklore and art through copyright law, insignia through trademark law, and biotechnology and genetic resources through patent law (Schuler, 2013). This approach is understandable as a defensive move given that Indigenous peoples must often deal with the appropriation and biopiracy of their knowledge by multinational companies, governments, and national elites (Martens, 2014). This has led some scholars to argue that, if properly used (respecting the rights and consent of rightsholders), IP rights might be a crucial tool for the conservation and sustainable use of IK (Rajasekharan and Souravi, 2017).

Other scholars, however, have highlighted how IP law is ill-suited to apply to Indigenous cultures (Brown, 2003). First, IP protection is based on the lifetime of the person who has put an idea in tangible form, whereas Indigenous groups see aspects of their culture as being owned by the group (Kennedy and Laczniak, 2014). Second, the IP system's concept of the public domain as the last recipient of all knowledge requires time-period limitations for IP. From Indigenous perspectives, by contrast, certain aspects of IK are not intended for a specific period but are conceived as pertaining to the group in perpetuity (for example, sacred ceremonies, stories, prayers, and songs, among other things) (Younging, 2015). Thus, IP and Indigenous governance systems solve access problems in different ways. In the former, exclusivity of use offsets the limited duration of exclusivity, while in the latter case, the fact that the rights are unlimited in time offsets the sharing of these rights with others (Drahos and Frankel, 2012). Third, the inappropriate use of sacred cultural artifacts, symbols, or designs may not cause financial loss but can cause considerable offence to the relevant community (Anderson, 2015). Fourth, IP and IK derive from different sources. Commercial claims to intellectual property seek to serve financial interests. Legal protection for certain aspects of Indigenous cultures is grounded in the principle of self-determination and the need to ensure cultural survival. Then, some Indigenous claims to IP protection can be conceived as a partial form of reparations for past wrongs (Graham and McJohn, 2005). Fifth, IK is not simply about owning an asset for protection and compensation. Indigenous nations are demanding to be recognised as custodians, owners, and nurturers of knowledge that is valuable within and beyond Indigenous contexts (Anderson, 2010).

In sum, as conventional IP rights systems, such as copyright, patent, and trademark, have limitations in providing adequate recognition and protection for IK, there are proposals to establish a *sui generis* system of protection designed by Indigenous organisations themselves at international and national levels (Karanja, 2016; Younging, 2015; Anderson, 2010), or to emphasise on-the-ground negotiations based on mutual dignity (Brown, 2003), or to design an integrated model in which IP rights, land rights, and customary laws intersect, considering that Indigenous innovation is cosmologically linked to land rather than to laboratories (Drahos and Frankel, 2012).

These proposals seek to challenge Western legal traditions but often end trapped in property arrangements. For example, in Australia, the High Court recognised the spiritual dimension of 'Aboriginal native title' but has also made it clear that native title rights do not constitute a separate system of IP for the control of IK. Although native title is not an institution of the common law, its interpretive evolution takes place within the basic structure of Australia's statutory property institutions (Drahos and Frankel, 2012). In sum, current legal arrangements of IP are not sufficient because they neglect Indigenous self-determination (Eimer, 2020) and perpetuate the colonial mentality per which Indigenous peoples have no autochthonous and effective legal regimes to diffuse, transfer, exchange, and innovate knowledge (Tharakan, 2015), merely regimes that need to be translated and incorporated into the Western legal episteme. In this way, IP reaffirms its ethnocentrism but also its anthropocentrism. By being unable to incorporate IK in all its dimensions, non-anthropocentric aspects of IK are excluded from current arrangements. This operation is also reproduced in the field of international environmental law.

IK in Development and Global Sustainability

In the 1990s, cooperation agencies and international institutions started including IK in the discourses and practices of development and environmental conservation (Lanzano, 2013). In this context, some scholars sought to create bridges between IK and Western science. In the field of ecology, a pioneer work (Berkes, Gadgil, and Folke, 1993) portrayed IK as a diachronic knowledge based on a broad basis of knowledge-practices-beliefs of human behaviour in ecological systems, which could complement 'synchronic observations' on which Western science rests. Later, IK in environmental management was called 'traditional ecological knowledge', which focuses on the patterns of interrelation between culture and the ecosystems to produce environmental actions, responses, and adaptation mechanisms (Dudgeon and Berkes, 2003; Berkes, Colding, and Folke, 2000). IK in this view would be the result of broad and ancestral adaptation strategies, which are unique and context specific. Consequently, integrating Indigenous cultures as ecocentric models into resource management was conceived as a way to effectively implement strategies of adaptation to climate change (Makondo and Thomas, 2018; Gratani *et al.*, 2016). Under this premise, World Bank reports (1998; 2004) proposed an action frame based on the importance of IK for its development projects and partners and the need to learn from local communities (Mawere, 2014).

Views in favour of 'integrating' or 'including' IK within Western policy frames for development and conservation, however, have been blamed for co-opting and appropriating IK (Briggs, 2013) into an imposed Western epistemological framework (Louie *et al.*, 2017; Debnath, 2019) that reduces the agency of Indigenous communities (Datta, 2018). For example, the discursive construction of Indigenous subjects within the Convention of Biological Diversity emphasises 'traditionality' (traditional lifestyles, traditional knowledge) throughout the text, echoing notions of Indigenousness in colonial discourses, where 'traditional' has been imagined as an opposition to 'modern' and interpreted through the eyes of European colonisers (Reimerson, 2013). Another problem is the dominant conceptions of nature and environment, which are performative scientific representations, rather than neutral (Turnhout, 2018). For example, the concept of 'ecosystemic services' hides the logic by which, to save biodiversity, nature must have a monetary value, reducing socio-natural relationships to market transactions (Turnhout *et al.*, 2013). In fact, these ideas have reinforced the anthropocentric nature of international environmental law.

Consequently, the participation of Indigenous nations in international law making is often reduced to symbolism. During the negotiations of the Paris Climate Agreement, 51 Indigenous representatives were formally invited to participate to show the urgency of an international agreement. However, although the Paris Agreement explicitly mentions 'traditional knowledge' as an important factor to combat climate change, the majority of the diplomats rejected any direct reference to Indigenous rights. Instead of prior informed consent, the Agreement stipulates that Indigenous concerns should be 'taken into consideration' (Paris Agreement, 2015, art. 7.5). To legitimise their ambitions, dominant actors instrumentalise the participation of Indigenous representatives and deny their demands for self-determination as soon as this might endanger the fragile compromise between environmental goals and economic interests (Eimer and Bartels, 2019).

Even though principles of prior informed consent and benefit sharing emerge from the examination of both environmental and human rights instruments (Meyer, 2001), they end up being symbolic proclamations connected to an understanding of IK as a condition to implement top-down initiatives. What is neglected in policy instruments and international law related to IK is that Indigenous peoples respond to the impacts of climate change by claiming and realizing a right to environmental self-determination (Kapua'ala Sproat, 2016; Tsosie, 2007). The problem is that responses to climate change based on Indigenous peoples' choices often conflict with state priorities and private businesses, clashing with Western-imposed values and practices over what is considered as 'natural resources', such as land, water, or biodiversity. Indigenous claims in these contexts are most appropriately framed not simply as environmental issues, but as environmental justice issues aimed at restorative justice (Kapua'ala Sproat, 2016). IK, in this respect, is located at the base of Indigenous struggles for survival, for defending nature, and for extending our current legal and political imagination beyond the anthropocentric foundations of current international law.

IK Beyond International (Anthropocentric) Law

As a place-based form of thinking, Indigenous knowledge and innovation depend on land rights (Drahos and Frankel, 2012) but, more crucially, on the recognition of Indigenous sovereignty. This sovereignty might provide real meaning to international principles such as 'prior consent' and 'benefit sharing'. It would entail self-governance mechanisms rather than 'participatory' means to accede to – and exploit – IK. For instance, by conceiving Indigenous nations as sovereigns, they would not need the formal granting of IP rights over their own knowledge to have legal protection. Rather it would be assumed that the IK they claim belongs in perpetuity to the Indigenous collective and is used, kept, and spread in their own ancestral modes. Those who allege that this is not the case would have to prove otherwise. Moreover, this sovereignty could be a vehicle to develop non-anthropocentric laws and institutions inspired by IK.

Indigenous sovereignty has some recognition in international law under the right of self-determination. From the second half of the last century, international Indigenous movements engaged with the concept of self-determination used by African nations in the decolonisation era (Craven, 2010) and then included in the United Nations Charter and in the 1960 UN Declaration on the Granting of Independence to Colonial Countries and Peoples (Napoleon, 2005; Muehlebach, 2003). This notion was nuanced in the following years by international scholars who differentiated between external and internal self-determination – the former as a right to independence and the latter as a right to maintain some degree of autonomy under the authority of the nation-state. The demands of Indigenous nations were thus framed within this latter notion, as formalised in the Declaration on the Rights of Indigenous Peoples of 2007 (Oldham and Frank, 2008; Stavenhagen, 1992; Errico, 2007).

However, as Indigenous territorial rights challenge state power over the territory, Indigenous nations would express a kind of 'shared' or 'overlapped' sovereignty (Clavé-Mercier, 2018), in which the nation-state must accommodate Indigenous territorial claims. Others describe the contentious dynamics of decision-making resulting from these processes as 'multiple' or 'relational sovereignty' (Conversi, 2016; Wheatley, 2014) whereas from a social angle, Simpson (2014) emphasises how Indigenous political orders prevail within and apart from state governance in a form of 'nested sovereignty'. From an international law perspective, Indigenous global rights recognised in international instruments give place to a 'parallel sovereignty' able to overcome state national arrangements (Lerenzini, 2006), although Indigenous nations are still not fully recognised as participants in international law making (Khan, 2019). From a political point of view, a 'polymorphic sovereignty' is proposed, not dependent on territorial boundaries but on the recognition of self-government and political representation of Indigenous nations within the state (Clavé-Mercier, 2018).

Indigenous sovereignty is then a project in the making to reconfigure national and international power relations. By recognising this kind of sovereignty, Indigenous peoples would have more power to control and administer their territories and their knowledge, making space for enacting non-anthropocentric principles. This sovereignty would also allow developing territorially based and non-anthropocentric institutional innovations not only relevant for their own survival but with potential gains for humanity.

IK and Food Sovereignty

IK might contribute to national and international law making to address the food crisis. Under the notion of food sovereignty, Indigenous nations advance policy transformations to have more equitable and ecological food systems. In this regard, IK is not just an asset within the economy, but it might provide insights to build an alternative economic system. The international peasant organisation Vía Campesina formulated the concept of food sovereignty in 1996, inspired by the Zapatistas' anti-NAFTA uprising in Mexico. Stemming from a critique of neoliberal policies shared by social justice and environmental organisations, it highlights the unequal impact of international trade on national agricultural development and local ecologies.

Food sovereignty seeks to overcome the notion of food security, which emerged during the 1974 world food crisis as a right to not be undernourished. Neoliberal policies and technocratic conceptions of economic growth and free trade influenced the configuration of this concept as a development goal proclaimed by the FAO (2009). Internationally, it embodies a geographical imaginary in which inefficient and outdated agricultural economies in the South must become productive by adopting biotechnology developed in the North, while also promoting transnational agribusiness. Nationally, it sustains social programs for alleviating poverty, such as conditional cash transfers.

In its original formulation in the nineties, food sovereignty opposed food security by reclaiming the right of 'each nation' to establish national food policies with autonomy from the neoliberal international trade system. But soon after, the definition shifted the emphasis from national self-sufficiency to local and ecological self-sufficiency, as a right of peoples to define their own type of agriculture (as expressed in the grassroots Declaration 'Our World is Not for Sale' of 2007). Indigenous peoples also engaged with this concept to highlight that the sovereignty aspect of food sovereignty does not relate to 'state sovereignty' but to the sovereignty of Indigenous nations.

Although some disputes about the exact definition of food sovereignty (and its local, national, and international scale) still exist, most approaches seek to highlight the ecological and communal aspects. Thus, it currently expresses a political and ecological project that opposes extractive modes of agriculture characterised by the expansion of agribusiness and deforestation, patterns of land grabbing and land-use change, the intensive use of pesticides and transgenic, and water and land degradation. Instead, it prioritises small, agro-ecological, and local agriculture while also promoting territorial rights and autonomy for peasants and Indigenous peoples. The sustainable use of the land and livelihoods, in this model, is achieved with techniques used ancestrally by local communities. Rather than prioritizing productivity and extraction, it seeks the reproduction of the ecological and cultural conditions where those communities have ancestrally lived (Merino, 2020).

Food sovereignty thus challenges the Western-inspired and anthropocentric notion of the nation-state authority to exploit 'natural resources' on behalf of the 'nation', as an imagined community of individuals that shares socio-cultural features within a national territory. Food sovereignty questions this authority by appealing to 'multiple sovereignties' or 'relational sovereignty' to claim for broadening democratic decision-making within food systems and, at the same time, by reconceptualising notions of territory and 'resources' from the margins of the state (Grey and Patel, 2015; Daigle, 2019; Conversi, 2016). Thus, food sovereignty challenges the anthropocentric basis of the nation-state by limiting what is considered an unrestricted state power to exploit 'natural resources' and, in specific, agricultural commodities, or to grant their exploitation to transnational corporations.

In fact, agribusinesses acting under favourable national laws and international trade law do not care about the reproduction of the ecological conditions of the sites of extraction; rather, they put great pressure on national governments to facilitate the extraction of commodities even if this means the depletion of local livelihoods. Even if they claim to respect the 'sustainable use of natural resources' under corporate social responsibility codes, their view of nature as a resource is highly instrumental to extractivist ends. When food sovereignty is connected to IK, in contrast, this instrumentality is put aside by a highly respectful view of the natural environment and its reproductive cycles.

IK and Global Sustainability

Rather than being a concept that instrumentally might support the implementation of market-oriented environmental mechanisms, IK might be the basis of a new international environmental framework. The concept of 'rights of nature' (RoN) might be relevant in this regard.

Whereas the RoN might be understood as a political platform embedded in a complex genealogy and negotiations between Western critical thinking, environmental legal activism, and IK since the 1970s, from the end of the 2000s onwards it has been characterised by a strong involvement of Indigenous organisations in drafting constitutional and legal texts. It has been included in the Ecuadorian Constitution of 2008 and in specific Bolivian legislation (Law of the Rights of Mother Earth, 2010 (Law 071); Framework Law of the Mother Earth and Integral Development for Living Well, 2012 (Law 300)) as the basis for their environmental systems. Although subsequent extractivist policies have weakened the original purposes of the RoN, its explicit recognition has obliged judges and the administration to consider the interests of the environment in greater detail when enacting a decision.[1]

In New Zealand, India, and Colombia, instead of granting rights to the whole of nature, specific natural entities have been granted legal personhood. In New Zealand, on the basis of Maori

Whanganui Iwi's negotiated agreements with the Government under the Treaty of Waitangi, the government acknowledged in 2017 the Whanganui River as 'a legal entity with standing in its own right', to be managed by representatives from the Whanganui Iwi and the Crown (Gordon, 2018). In India, the 2012 Supreme Court decision on *Board of India v. A. Nagaraja* ruled that the Indian Constitution's Article 21 right to life could be extended to non-human animals, and in 2013, India's Ministry of Environment and Forests declared cetaceans 'non-human persons' in an effort to protect them from harm (Gordon, 2018). More recently in 2017, the Uttarakhand High Court ruled that the Indian rivers Ganga and Yamuna, and the Gangotri and Yamunotri glaciers, are 'legal persons' (Bajpai, 2020). In Colombia, in November 2016, the Constitutional Court ruled that the river Atrato is a subject of rights under the doctrine of guardianship, with one member of the claimant community and one delegate of the government representing the river and its basin.[2]

Internationally, there are important efforts to build an alternative institutionalisation for protecting nature's rights. The Global Alliance for the Rights of Nature (the Alliance), a network committed to the diffusion and enforcement of the RoN, promoted the Peoples Conference on Climate Change and the Rights of Mother Earth, hosted by Bolivia in April 2010. This meeting approved the Universal Declaration of the Rights of Mother Earth, modelled on the Universal Declaration of Human Rights. Article 2 of this Declaration recognises 'inherent rights' to Mother Earth (including the right to life, be respected, regenerate, clean air, water, among others), and the obligations of human beings to Mother Earth (Art. 3).

The Alliance has also fostered the creation of the International Rights of Nature Tribunal in December 2015, which investigates, tries, and decides cases involving alleged violations of the Universal Declaration of the Rights of Mother Earth. It was inspired by the International War Crimes Tribunal and the Permanent Peoples' Tribunal, fostered by citizens to investigate and publicise human rights violations. Although the Tribunal's decisions are not formally part of international law, its decisions could have performative significance for promoting RoN (Maloney, 2015; La Follette and Maser, 2017). With each case, the Tribunal recommends actions for reparation, mitigation, restoration, and prevention of further damages. It has also opened a discussion about the argument for the universal jurisdiction of the RoN.

The attempts to formalise the International Rights of Nature Tribunal have opened a discussion about the legal foundations of universal jurisdiction for the RoN. The aim is that the tribunal formally investigates any violation of these rights worldwide and effectively rules over compensation and preventive measures against environmental harms. For some, this universal jurisdiction would be based on the fact that it serves to protect basic conditions for the future exercise of constituent power (Olmos, 2018). Such conditions provide a baseline without which citizens could not meaningfully participate in democratic systems, including the ability to associate with others and live in a healthy environment where life can flourish. Therefore, the exercise of universal jurisdiction for the RoN should be seen as a means of protecting the right to self-determination or the ability of present and future peoples to participate in the formation and reformation of the states that constitute the international community.

Another view justifies universal jurisdiction by the personhood of nature itself. This entails enlarging conceptions of Westphalian personification beyond human communities (Youatt, 2017). If we assume that mountains, rivers, and other non-human entities have rights, then a key question is: Who are the legitimate arbiters of collective personhood on the global stage? Unlike Westphalian states, which are made up entirely of humans, these collective persons include both human and nonhuman actors that represent an alternative form of political life not entirely commensurable with the liberal foundations of modern nation-states (Youatt, 2017). By conceiving nature not as a singular object of technical governance, but rather as a new political subjectivity, the idea of RoN offers ontological challenges to global political ecology and overcomes the anthropocentric basis of international law.

By inspiring the institutionalisation of the RoN at national and international levels, IK is pushing the limits of the juridical and political imagination of global environmental arrangements. In this regard, it contributes to the imagining and the struggle for a post-extractive economy in which nature and human beings might flourish.

The Politics of Thinking from and for Another World

The lockdown enacted by most countries to address the diffusion of Covid-19 has meant a significant hit to the global economy but also an important breath for the natural environment. Global emissions diminished radically. The impact on the oil industry led to degrowth and to post-extractivist scholars dreaming of a post-petroleum economy. However, both industrialised and non-industrialised states are starting to take drastic measures to 'reactivate the economy', which implies the flexibilisation of environmental regulations. The new normal thus might mean a more terrible future under the mantra of a renewed 'economic development'.

International (anthropocentric) law, with its Western premises, seems unable to respond to this challenge. Even when it pretends to recognise IK as an eco-centred system of knowledge, this recognition is processed via the adaptation of this knowledge into Western economic and individualist frameworks. IK ends up being protected as some vestige of the past and in this respect, is also considered an asset inserted in IP systems for compensation aims. In the environmental arena, IK is instrumently included in international instruments and forums without providing any space for radical institutional transformations. This logic of recognition has actually meant the Western assimilation of IK.

A logic of sovereignty of IK, instead, would challenge the anthropocentric basis of international law. By being conceived not as a second-level knowledge to be manipulated or folklorised but as a valuable system of knowledge with intrinsic worth, able to influence national and global regulations, IK has the potential to enlarge our legal and political imagination to transform current institutions. The concept of food sovereignty, for example, by fostering Indigenous ways of managing local economies and agro-ecological innovations, challenges the large-scale and intensive extractive economy that deeply harms the natural environment and Indigenous territories. The concept of rights of nature also challenges this dominant economy and the anthropocentric discourses by which human beings are the masters of the Earth. The rights of nature are supporting the emergence of global institutions that oblige us to rethink the way we interact with the natural environment.

IK is not a unique way of thinking, obviously, but it is a political space where different *Indigenous knowledges* might advance their fundamental visions of political autonomy, sustainability, and communal economy. These fundamentals have allowed many Indigenous cultures, local communities, and nature itself to survive despite colonial and postcolonial aggressions. By being closely connected to non-anthropocentric and non-extractive modes of social life and interactions with the natural environment, IK might guide the way globally to really address, with long-term measures, the environmental, food, and sanitary global crises.

Notes

1 See, for example, in Ecuador Case 11121-2011-0010, Provincial Court of Justice of Loja, March 2011.
2 Case T-622/2016, Constitutional Court of Colombia.

References

Agrawal, A. (1995) 'Dismantling the Divide between Indigenous and Scientific Knowledge', *Development and Change*, 26, pp. 413–439. DOI: 10.1111/j.1467-7660.1995.tb00560.x

Akwé: Kon Voluntary Guidelines for the Conduct of Cultural, Environmental and Social Impact (2004).

American Declaration on the Rights of Indigenous Peoples (15 June 2016) AG/RES.2888 (XLVI-O/16).

Anderson, J. (2010) 'Indigenous/Traditional Knowledge & Intellectual Property', *Center for the Study of the Public Domain, Duke University School of Law*, Issues Paper, pp. 1–70.

Anderson, J. (2015) 'Indigenous Knowledge and Intellectual Property Rights', in *International Encyclopedia of the Social & Behavioral Sciences*, 2nd edn. Amsterdam: Elsevier, pp. 769–778.

Articulación de los Pueblos Indígenas de Brasil (2020) 'Racismo y subregistro de casos. Covid-19 y los pueblos indígenas', *Alerta APIB*, 18 April [Online]. Available at: https://apiboficial.org/2020/04/18/alerta-apib-02-covid-19-y-los-pueblos-indigenas/?lang=es (Accessed: 30 August 2022).

Bajpai, S. (2020) '"Righting" the Wrong: Rights of Rivers in India', *Mongabay*, 23 June [Online]. Available at: https://india.mongabay.com/2020/06/commentary-righting-the-wrong-rights-of-rivers-in-india/ (Accessed: 30 August 2022).

Berkes, F., Colding, J. and Folke, C. (2000) 'Rediscovery of Traditional Ecological Knowledge as Adaptive Management', *Ecological Applications*, 10(5), pp. 1251–1262. DOI: 10.1890/1051-0761(2000)010[1251:ROTEKA]2.0.CO;2

Berkes, F., Gadgil, M. and Folke, C. (1993) 'Indigenous Knowledge for Biodiversity Conservation', *Ambio*, 22(2/3), pp. 151–156.

Bloomfield, L., McIntosh T. and Lambin, E. (2020) 'Habitat Fragmentation, Livelihood Behaviors, and Contact between People and Nonhuman Primates in Africa', *Landscape Ecology*, 35, pp. 985–1000. DOI: 10.1007/s10980-020-00995-w

Briggs, J. (2013) 'Indigenous Knowledge: A False Dawn for Development Theory and Practice?', *Progress in Development Studies*, 13(3), pp. 231–243. DOI: 10.1177/1464993413486549

Brown, W. (2003) 'Intellectual Property Law: A Primer for Scientists', *Molecular Biotechnology*, 23(3), pp. 213–224. DOI: 10.1385/MB:23:3:213

Bruchac, M. (2014) 'Indigenous Knowledge and Traditional Knowledge', in Smith, C. (ed.) *Encyclopedia of Global Archaeology*. New York: Springer, pp. 3814–3824.

Castro-Gómez, S. (2005) *La Hybris del Punto Cero: Ciencia, Raza e Ilustración en la Nueva Granada (1750–1816)*. Bogotá: Editorial Pontificia Universidad Javeriana.

Chakraborty, R. and Sen, S. (2017) 'Revival Modernization and Integration of Indian Herbal Medicine in Clinical Practice: Importance Challenges and Future', *Journal of Traditional and Complimentary Medicine*, 7(2), pp. 234–244. DOI: 10.1016/j.jtcme.2016.05.006

Clavé-Mercier, V. (2018) 'Revisitar la Soberanía Indígena: Los Desafíos de una Reivindicación Excluida', *Relaciones Internacionales*, 38, pp. 99–119. DOI: 10.15366/relacionesinternacionales2018.38.005

Coordinadora de las Organizaciones Indígenas de la Cuenca Amazónica (COICA) and Red Eclesial Panamazónica (REPAM) (2020) 'Impacto del Covid-19 en los Pueblos Indígenas de la Oanamazonía', *Boletín*, 15 May [Online]. Available at: http://www.caaap.org.pe/2020/Documentos/Mapa-PPII-PANAMAZONIA-14.05.2020.pdf (Accessed: 30 August 2022).

Convention for the Safeguarding of the Intangible Cultural Heritage (adopted 17 October 2003, entered into force 20 April 2006) 2368 UNTS 3.

Convention on Biological Diversity (adopted 5 June 1992, entered into force 29 December 1993), 1760 UNTS 79 (CBD).

Convention to Combat Desertification (adopted 14 October 1994, entered into force 26 December 1996) 1954 UNTS 3.

Conversi, D. (2016) 'Sovereignty in a Changing World: From Westphalia to Food Sovereignty', *Globalizations*, 13(4), pp. 484–498. DOI: 10.1080/14747731.2016.1150570

Craven, M. (2010) 'Statehood, Self-Determination and Recognition', in Evans, M. (ed.), *International Law*. 3rd ed. Oxford: Oxford University Press, pp. 203–251.

Daigle, M. (2019) 'Tracing the Terrain of Indigenous Food Sovereignties', *The Journal of Peasant Studies*, 46(2), pp. 297–315. DOI: 10.1080/03066150.2017.1324423

Datta, R. (2018) 'Decolonizing both Researcher and Research and its Effectiveness in Indigenous Research', *Research Ethics*, 14(2), pp. 1–24. DOI: 10.1177/1747016117733296

Debnath, M. (2019) 'A Community under Siege: Exclusionary Education Policies and Indigenous Santals in the Bangladeshi Context', *Third World Quarterly*, 41(3), pp. 453–469. DOI: 10.1080/01436597.2019.1660634

Descola, P. (2004) 'Las Cosmologías Indígenas de la Amazonía', in Surrallés, A. and García Hierro, P. (eds.) *Tierra Adentro: Territorio Indígena y Percepción del Entorno*. Copenhagen: IWGIA, pp. 25–35.

Drahos, P. and Frankel, S. (2012) 'Indigenous Peoples' Innovation and Intellectual Property: The Issues', in Drahos, P. and Frankel, S. (eds.) *Indigenous Peoples' Innovation. Intellectual Property Pathways to Development*. Canberra: Australian National University, pp. 1–28.

Dudgeon, R. and Berkes, F. (2003) 'Local Understandings of the Land: Traditional Ecological Knowledge and Indigenous Knowledge', in Selin, H. (ed.) *Nature Across Cultures: Views of Nature and the Environment in Non-Western Cultures*. Manchester: Kluwer Academic Publishers, pp. 75–96.

Dussel, E. (1994) *1492: El Encubrimiento del Otro: Hacia el Origen del Mito de la Modernidad*. La Paz: Plural Editores, UMSA.

Eimer, T. (2020) 'What if the Subaltern Speaks? Traditional Knowledge Policies in Brazil and India', *Third World Quarterly*, 41(1), pp. 96–112. DOI: 10.1080/01436597.2019.1650639

Eimer, T. and Bartels, T. (2019) 'From Consent to Consultation? Indigenous Rights and the New Environmental Constitutionalism', *Environmental Politics*, 29(2), pp. 235–256. DOI: 10.1080/09644016.2019.1595884

Errico, S. (2007) 'The Draft UN Declaration on the Rights of Indigenous Peoples: An Overview', *Human Rights Law Review*, 7(4), pp. 741–755. DOI: 10.1093/hrlr/ngm023

Framework Law of the Mother Earth and Integral Development for the Living Well 2012 (Bolivia) [Law 300].

Gordon, G.J. (2018) 'Environmental Personhood', *Columbia Journal of Environmental Law*, 43(1), pp.49–91. DOI: 10.7916/cjel.v43i1.3742

Graham, L. and McJohn, S. (2005) 'Indigenous Peoples and Intellectual Property', *Washington University Journal of Law & Policy*, 19, pp. 313–337.

Gratani, M., Sutton, S., Butler, J., Bohensky, E. and Foale, S. (2016) 'Indigenous Environmental Values as Human Values', *Cogent Social Sciences*, 2, pp. 1–17. DOI: 10.1080/23311886.2016.1185811

Grey, S. and Patel, R. (2015), 'Food Sovereignty as Decolonization: Some Contributions from Indigenous Movements to Food System and Development Politics', *Agriculture and Human Values*, 32(3), pp. 431–444. DOI: 10.1007/s10460-014-9548-9

Greene, S. (2009) *Customizing Indigeneity: Paths to a Visionary Politics in Peru*. Stanford: Stanford University Press.

Grosfoguel, R. (2007) 'Descolonizando Los Universalismos Occidentales: El PluriVersalismo Transmoderno Decolonial Desde Aimé Césaire Hasta Los Zapatistas', in Castro-Gómez, S., Grosfoguel, R. (eds.) *El Giro Decolonial: Reflexiones para una Diversidad Epistémica más allá del Capitalismo Global*. Bogotá: Siglo del Hombre, Universidad Central, Instituto de Estudios Sociales Contemporáneos, Pontificia Universidad Javeriana, pp. 63–78.

Grosfoguel, R. (2009) 'A Decolonial Approach to Political-Economy: Transmodernity, Border Thinking and Global Coloniality', *Kult*, 6, pp. 10–38.

Hall, S. (1992) 'The new ethnicities', in Donald, J. and Rattansi, A. (eds.) Race, Culture and Difference. London: SAGE, pp. 252–259.

Haugen, H. (2005) 'Traditional Knowledge and Human Rights', *Journal of World Intellectual Property*, 8(5), pp. 663–677. DOI: 10.1111/j.1747-1796.2005.tb00273.x

Indigenous and Tribal Peoples Convention (adopted 27 June 1989, entered into force 5 September 1991) 1650 UNTS 385 (ILO Convention 169).

International Treaty on Plant Genetic Resources for Food and Agriculture (adopted 3 November 2001, entered into force 29 June 2004) 2400 UNTS 303.

Kapua'ala Sproat, D. (2016) 'An Indigenous People's Right to Environmental Self-Determination: Native Hawaiians and the Struggle Against Climate Change Devastation', *Stanford Environmental Law Journal*, 35(2), pp. 157–220.

Karanja, W. (2016) 'The Legitimacy of Indigenous Intellectual Property Rights' claims', *Strathmore Law Review*, 1(1), pp. 165–190.

Kennedy, A. and Laczniak, G. (2014) 'Indigenous Intellectual Property Rights: Ethical Insights for Marketers', *Australasian Marketing Journal*, 22(4), pp. 307–313. DOI: 10.1016/j.ausmj.2014.09.004

Khan, S. (2019) 'Rebalancing State and Indigenous Sovereignties in International Law: An Arctic Lens on Rrajectories for Global Governance', *Leiden Journal of International Law*, 32(4), pp. 675–693. DOI: 10.1017/S0922156519000487

La Follette, C. and Maser, C. (2017) *Sustainability and the Rights of Nature: An Introduction*. Boca Raton: CRC Press, Taylor & Francis Group.

La Vanguardia (2020) 'Alcalde de Manaos Teme "Genocidio" de los Indios en Amazonía por el Covid-19', *La Vanguardia*, 20 May [Online]. Avaiable at: https://www.lavanguardia.com/vida/20200520

/481304343605/alcalde-de-manaos-teme-genocidio-de-los-indios-en-amazonia-por-el-covid-19 .html (Accessed: 31 August 2022).

Lander, E. (ed.) (2000) *La Colonialidad del Saber: Eurocentrismo y Ciencias Sociales Perspectivas Latinoamericanas*. Buenos Aires: Clacso.

Lanzano, C. (2013) 'What Kind of Knowledge is "Indigenous Snowledge"? Critical Insights from a Case Study in Burkina Faso', *Transcience*, 4(2), pp. 3–18.

Law of the Rights of Mother Earth 2010 (Bolivia) [Law 071].

Lenzerini, F. (2006) 'Sovereignty Revisited: International Law and Parallel Sovereignty of Indigenous Peoples', *Texas International Law Journal*, 42, pp. 155–183.

Lima, L. (2020) 'Coronavirus en EE.UU.: La "Carnicería" de la Pandemia en la Nación Navajo, elLugar con Mayor Número de Casos Per Cápita del País', *BBC News Mundo*, 19 May [Online]. Available at: https://www.bbc.com/mundo/noticias-internacional-52642716 (Accessed: 30 August 2022).

Louie, D., Pratt, Y., Hanson, A. and Ottmann, J. (2017) 'Applying Indigenizing Principles of Decolonizing Methodologies in University Classrooms', *Canadian Journal of Higher Education*, 47(3), pp. 16 –33. DOI: 10.47678/cjhe.v47i3.187948

Makondo, C. and Thomas, D. (2018) 'Climate Change Adaptation: Linking Indigenous Knowledge with Western Science for Effective Adaptation', *Environmental Science & Policy*, 88, pp. 83–91. DOI: 10.1016/j.envsci.2018.06.014

Maloney, M. (2015) 'Building an Alternative Jurisprudence for the Earth: The International Rights of Nature Tribunal', *Vermont Law Review*, 41, pp. 129–142.

Mandikonza, C. (2019) 'Integrating Indigenous Knowledge Practices as Context and Concepts for the Learning of Curriculum Science: A Methodological Exploration', *Southern African Journal of Environmental Education*, 35, pp. 1–16. DOI: 10.4314/sajee.v35i1.13

Marsden, M. (2003) *The Woven Universe: Selected Writings of Rev. Māori Marsden*. Ōtaki: Estate of Rev. Māori Marsden.

Martens, P. (2014) 'Protection of Traditional Knowledge and Origin Products in Developing Countries: Matching Human Rights and IP Protection with Business Development Opportunities', Working Papers, 31, Maastricht School of Management. Available at: https://ideas.repec.org/p/msm/wpaper/2014-31.html (Accessed: 31 August 2022).

Mawere, M. (2014) 'Culture, Indigenous Knowledge and Development in Africa since Colonialism: The Silenced Narrative', in *Culture, Indigenous Knowledge and Development in Africa: Reviving Interconnections for Sustainable Development*. Mankon: Langaa Research & Publishing CIG, pp. 23–42.

Mead, H.M. (2003) *Tikanga Maori: Living by Maori Values*. Wellington: Huia.

Merino, R. (2016) 'An Alternative to "Alternative Development"?: Buen Vivir and Human Development in Andean Countries', *Oxford Development Studies*, 44(3), pp. 271–286. DOI: 10.1080/13600818.2016.1144733

Merino, R. (2018) 'Reimagining the Nation State. Indigenous Peoples and the Making of Plurinationalism in Latin America', *Leiden Journal of International Law*, 31, pp. 773–792. DOI: 10.1017/S0922156518000389

Merino, R. (2020) 'The Geopolitics of Food Security and Food Sovereignty in Latin America: Harmonizing Competing Visions or Reinforcing Extractive Agriculture?', *Geopolitics*, 27(3), pp. 898–920. DOI: 10.1080/14650045.2020.1835864

Meyer, A. (2001) 'International Environmental Law and Human Rights: Towards the Explicit Recognition of Traditional Knowledge', *RECIEL*, 10(1), pp. 37–46. DOI: 10.1111/1467-9388.00259

Mignolo W. and Tlostanova M. (2006) 'Theorizing from the Borders: Shifting to Geo-and Body-Politics of Knowledge', *European Journal of Social Theory*, 9, pp. 205–221. DOI: 10.1177/1368431006063333

Mignolo, W. (2007) 'The De-colonial Option and the Meaning of Identity in Politics', *Anales Nueva Época*, 9(10), pp. 119–156.

Mignolo, W. (2011) *The Darker Side of Western Modernity: Global Futures, Decolonial Options*. Durham: Duke University Press.

Mo'otz Kuxtal Voluntary Guidelines (2016) CDB/COP/13/25.

Muehlebach, A. (2003) 'What Self in Self-Determination? Notes from the Frontiers of Transnational Indigenous Activism', *Identities: Global Studies in Culture and Power*, 10(2), pp. 241–268. DOI: 10.1080/10702890304329

Nagoya Protocol on Access to Genetic Resources and the Fair and Equitable Sharing of Benefits Arising from their Utilization to the Convention on Biological Diversity (Adopted 29 October 2010, entered into force 12 October 2014) 3008 UNTS 3 (Nagoya Protocol).

Napoleon, V. (2005) 'Aboriginal Self Determination: Individual Self and Collective Selves', *Atlantis*, 29(2), pp. 31–46.

Observatorio Petrolero (2020) 'El Covid-19 y el Petróleo: Dos amenazas latentes en el territorio Amazónico Norperuano', *PUINAMUDT*, 12 May [Online]. Available at: https://observatoriopetrolero.org/el-covid-19-y-el-petroleo-dos-amenazas-latentes-en-el-territorio-amazonico-norperuano/ (Accessed: 31 August 2022).

Oldham, P. and Frank, M.A. (2008) '"We the Peoples…": The United Nations Declaration on the Rights of Indigenous Peoples', *Anthropology Today*, 24(2), pp. 5–9. Available at: DOI: 10.1111/j.1467-8322.2008.00569.x

Olmos, B. (2018) 'Assessing the Contribution of the Latin American Water Tribunal and Transnational Environmental Law', in Byrnes A. and Simm, G. (eds.) *Peoples' Tribunals and International Law*. Cambridge: Cambridge University Press, pp. 230–256.

Paris Agreement (adopted 12 December 2015, entered into force 4 November 2016) FCCC/CP/2015/10/ Add.1.

Poitras, Y., Louie, D., Hanson, A. and Ottmann, J. (2018) 'Indigenous Education and Decolonization', in *Oxford Research Encyclopedia of Education*. Oxford: Oxford University Press.

Quijano, A. (2000) 'Coloniality of Power, Eurocentrism, and Latin America', *Nepantla: Views from South*, 1(3), pp. 533–580.

Rajasekharan, P.E. and Souravi, K. (2017) 'Indigenous Knowledge and Intellectual Property Rights', in Sugathan, S., Pradeep, N. S. and Abdulhameed, S. (eds.) *Bioresources and Bioprocess in Biotechnology*. Singapore: Springer, pp. 125–142.

Razzaque, J. and Ssenyonjo, M. (2007) 'Protection of Traditional Knowledge and Human Rights Obligations: The Status of Discussion in International Organisations', *Netherlands Quarterly of Human Rights*, 25(3), pp. 401–436. DOI: 10.1177/016934410702500302

Reimerson, E. (2013) 'Between Nature and Culture: Exploring Space for Indigenous Agency in the Convention on Biological Diversity', *Environmental Politics*, 22(6), pp. 992–1009. DOI: 10.1080/09644016.2012.737255

Rio Declaration on Environment and Development (12 August 1992) A/CONF.151/26 (Vol. I) (Rio Declaration).

Santos, B. de S. (2004) 'A Critique of Lazy Reason: Against the Waste of Experience', in Wallerstein, I. (ed.) *The Modern World-System in the Longue Durée Boulder*. Boulder: Paradigm Publishers, pp. 157–197.

Santos, B. de S. (2014) *Epistemologies of the South: Justice against Epistemicide*. Abingdon: Routledge.

Schuler, L. (2013) 'Modern Age Protection: Protecting Indigenous Knowledge through Intellectual Property Law', *Michigan State International Law Review*, 21(3), pp. 752–777. DOI: 10.17613/xp86-fw43

Shepard Jr., G. and Izquierdo, C. (2003) 'Los Matsiguenka de Madre de Dios y del Parque Nacional del Manu', in Huertas, B. and Garcia, A. (eds.) *Los Pueblos Indígenas de Madre de Dios: Historia, Etnografía e Coyuntura*. Lima: IWGIA, pp. 111–126.

Simpson, A. (2014) *Mohawk, Interruptus: Political Life Across the Borders of Settler States*. Durham: Duke University Press.

Smith, L., Maxwell, T.K., Puke, H. and Temara, P. (2016) 'Indigenous Knowledge, Methodology and Mayhem: What Is the Role of Methodology in Producing Indigenous Insights? A Discussion from Mātauranga Māori', *Knowledge Cultures.*, 4(3), pp. 131–156.

Spivak, G. (1999) *A Critique of Postcolonial Reason*. Cambridge; London: Harvard University Press.

Sree Sudha, P. (2018) 'Interface between Traditional Knowledge (TK) and Human Rights in Realizing Right to Health and Health Care – An Indian Perspective', *Peace Human Rights Governance*, 2(3), pp. 331–345. DOI: 10.14658/pupj-phrg-2018-3-3

Stavenhagen, R. (1992) 'Challenging the Nation-State in Latin America', *Journal of International Affairs*, 45(2), pp. 421–440.

Sumida, E. (2017) 'Indigenous Rights Education (IRE): Indigenous Knowledge Systems and Transformative Human Rights in the Peruvian Andes', *International Journal of Human Rights Education*, 1(1), pp. 1–34.

Tharakan, J. (2015) 'Indigenous Knowledge Systems – A Rich Appropriate Technology Resource', *African Journal of Science, Technology, Innovation and Development*, 7(1), pp. 52–57. DOI: 10.1080/20421338.2014.987987

The Lancet (2020) 'COVID-19 in Latin America: A Humanitarian Crisis', *The Lancet*, 7 November [Online]. Available at: https://www.thelancet.com/pdfs/journals/lancet/PIIS0140-6736(20)32328-X.pdf (Accessed: 31 August 2022).

Tsosie, R. (2007) 'Indigenous People and Environmental Justice: The Impact of Climate Change', *University of Colorado Law Review*, 78, pp. 1625–1677.

Turner, D. (2006) *This is Not a Peace Pipe: Towards a Critical Indigenous Philosophy*. Toronto: University of Toronto Press.

Turnhout, E. (2018) 'The Politics of Environmental Knowledge', *Conservation and Society*, 16(3), pp. 363–371. DOI: 10.4103/cs.cs_17_35

Turnhout, E., Stuiver, M., Klostermann, J., Harms, B. and Leeuwis, C. (2013) 'New Roles of Science in Society: Different Repertoires of Knowledge Brokering', *Science and Public Policy*, 40, pp. 354–365. DOI: 10.1093/scipol/scs114

United Nations Declaration on the Rights of Indigenous Peoples (13 September 2007) UN Doc A/RES/61/295 (UNDRIP).

United Nations Department of Economic and Social Affairs, Indigenous Peoples (2020) 'COVID-19 and Indigenous Peoples' [Online]. Available at: https://www.un.org/development/desa/indigenouspeoples/covid-19.html (Accessed: 31 August 2022).

Viveiros de Castro, E. (2004) 'Exchanging Perspectives: The Transformation of Objects into Subjects in Amerindian Ontologies', *Common Knowledge*, 10(3), pp. 463–484. DOI: 10.1215/0961754X-10-3-463

Watene, K. and Merino, R. (2018) 'Indigenous Peoples: Self-Determination, Decolonization, and Indigenous Philosophies', in Drydyk, J. and Keleher, L. (eds.) *Routledge Handbook of Development Ethics*. London: Routledge, pp. 134–147.

Wheatley, S. (2014) 'Conceptualizing the Authority of the Sovereign State over Indigenous Peoples', *Leiden Journal of International Law*, 27(2), pp. 371–396. DOI: 10.1017/S092215651300037X

Whyte, K. (2016) 'Indigenous Experience, Environmental Justice and Settler Colonialism', in Bannon, B. (ed.) *Nature and Experience: Phenomenology and the Environment*. Rowman & Littlefield, pp. 157–174.

Whyte, K. (2017) 'Our Ancestors' Dystopia Now: Indigenous Conservation and the Anthropocene', in Heise, U., Christensen, J. and Niemann, M. (eds.) *Routledge Companion to the Environmental Humanities*. London: Routledge, pp. 206–215.

World Bank (1998) *Indigenous Knowledge for Development: A Framework for Action*. Washington, DC: World Bank.

World Bank (2004) *Pueblos Indígenas, Pobreza y Desarrollo Humano en América Latina: 1994–2004*. Washington, DC: World Bank.

Youatt, R. (2017) 'Personhood and the Rights of Nature: The New Subjects of Contemporary Earth Politics', *International Political Sociology*, 11, pp. 39–54. DOI: 10.1093/ips/olw032

Younging, G. (2015) 'Traditional Knowledge Exists; Intellectual Property is Invented or Created', *University of Pennsylvania Journal of International Law*, 36(4), pp. 1077–1085.

Zibell, M. (2020) 'Coronavirus en Ecuador: Así Hacen Frente al Coronavirus los Indígenas de la Amazonía (y Cómo se Preparan para una Postpandemia de Hambre, Desempleo y Trueque)', *BBC News Mundo*, 26 May [Online]. Available at: https://www.bbc.com/mundo/noticias-america-latina-52781113 (Accessed: 31 August 2022).

11
EARTH JURISPRUDENCE
Anthropocentrism and Neoliberal Rationality

Peter Burdon and Samuel Alexander

Introduction

For advocates of Earth Jurisprudence[1] and ecological law, anthropocentrism is a 'boo word' or a pejorative which carries only negative connotations. Nobody would openly identify as an anthropocentrist. The orthodox European interpretation presents anthropocentrism as a worldview that separates human beings from other species of life and regards the natural world as a resource for human exploitation. To support this interpretation, ecological lawyers construct anthropocentrism as a unitary concept that has developed over time to incorporate key ideas: the hierarchical 'great chain of being' from ancient Greece; 'human dominion' from Christian theology; and a dichotomy between subjects and objects from the scientific revolution.

Against the grain of this presentation, we offer an alternative reading. First, following the scholarship of Ben Mylius (2018), we divide anthropocentrism into three distinct dimensions – perceptual, descriptive, and normative. Perceptual anthropocentrism, we argue, is an inescapable part of the human condition. We engage the world through our human senses and interpret from a location that is rooted in time and place. Descriptive anthropocentrism is implicit in calls to label the current geological epoch the Anthropocene. We have, in fact, become the weather makers and are now able to influence the functioning of Earth systems. Thus, when advocates of Earth Jurisprudence (and ecological law, more generally) critique anthropocentrism, we will argue that they are really engaging its normative dimension and challenging the hubristic claim that human dominance is natural or right.

Our second argument is that the meaning of normative anthropocentrism cannot be gleaned from a purely historical analysis. Worldviews are informed by history, but they also grow and take on new forms in response to new ideas and material circumstances. To understand and respond to normative anthropocentrism today, we need to put it into conversation with current ideas and ways of thinking. We do not have the space to provide a full intellectual history here, so we limit our focus to neoliberalism, described as a political rationality. To unpack this, we first locate the logic of neoliberalism in the writing of Jeremey Bentham. While not himself a neoliberal thinker, Bentham was an early advocate of using behaviourism as a governance tool that could coerce desirable conduct without resorting to overt force. Neoliberalism has advanced this program by reshaping human subjectivity on the model of the firm or *homo economicus*. Neoliberalism also erodes all values that cannot be economised and seeks to spread instrumental

DOI: 10.4324/9781003201120-14

215

rationality to all aspects of daily life. In this way neoliberalism can be understood as a particularly extreme manifestation of the capitalist organisation of society, where the privatisation of resources, markets, and prices are the dominant institutions and mechanisms for answering societal questions about production and distribution of wealth.[2]

This dominant logic has important implications for how we understand and engage ideas like normative anthropocentrism. For example, we argue that ideas like human superiority and dominion are no longer something that needs to be explicitly advanced and defended. Instead, normative anthropocentrism survives as an implicit assumption or perhaps a starting point from which neoliberal rationality begins and spreads its logic in the world. Normative anthropocentrism has not disappeared, but it has been subsumed within a broader logic and that ought to change the way we understand its meaning and combat its effects. This leads directly to our next point which is about the risks of relying on legal rights as a method for combatting anthropocentrism and building an ecocentric approach to environmental protection. This is a popular argument from proponents of Earth Jurisprudence (Berry, 2006; Cullinan, 2011), and in response, we look at the role those neoliberal thinkers played in shaping rights discourse in the twentieth century and the way the anthropocentric focus of legal rights makes them especially vulnerable to co-option. Finally, we conclude by arguing that a contemporary reading of normative anthropocentrism requires that we also explore alternative languages and mechanisms for attaining deep notions of environmental justice.

While our chapter engages European interpretations of anthropocentrism, we hope that it is provocative and opens the door for further analysis in diverse locations, such as the global South and the colonial context. Each of these places will encounter alternative texts, experiences, and political modalities that cannot be captured in a single text or reduced through comparison. Together they represent the vast tapestry of ideas and juridical relations that those struggling for a healthy environment are confronting today.

Anthropocentrism

In Earth Jurisprudence, and ecological legal thinking more generally, anthropocentrism is presented as a mode of thinking that separates human beings from the environment and regards the natural world as a 'giant gasoline station, an energy source for modern technology and industry' (Heidegger, 1970, p. 50). In this description, anthropocentrism is unequivocally a negative idea and 'the deepest cause of the present devastation' (Berry, 1999, p. 4). It has also been described as an 'optical delusion' (Einstein cited in Bosselmann, 2008, p. 319) or 'thinking disorder' (*The 11th Hour: Turn Mankind's Darkest Hour into Its Finest*, 2007).

And yet, Ben Mylius (2018, p. 159) reminds us that anthropocentrism has a much more complex meaning that can be parcelled out into three intersecting ideas:

1. *Perceptual* anthropocentrism: which characterizes paradigms informed by sense-data from human sensory organs;
2. *Descriptive* anthropocentrism: which characterizes paradigms that begin from, center upon, or are ordered around *Homo sapiens* / 'the human'; and
3. *Normative anthropocentrism*: which characterizes paradigms that constrain inquiry in a way that somehow privileges Homo sapiens / 'the human' [passive normative anthropocentrism]; and which characterizes paradigms that make assumptions or assertions about the superiority of Homo sapiens, its capacities, the primacy of its values, its position in the universe, and/or make prescriptions based on these assertions and assumptions [active normative anthropocentrism].

This is a helpful intervention and allows us to describe Earth Jurisprudence as being concerned primarily with the third category – normative anthropocentrism. Perceptual anthropocentrism seems to us to be an inescapable part of the human condition. Writers since Friedrich Nietzsche have regarded it as integral to the way we encounter and interpret the world. In the context of a broader declamation about objectivity, Nietzsche (1989, p. 119) wrote:

> There is only a perspective seeing, only a perspective 'knowing'; and the more affects we allow to speak about one thing, the more eyes, different eyes, we can use to observe one thing, the more complete will our 'concept' of this thing, our 'objectivity,' be. But to eliminate the will altogether, to suspend each and every affect, supposing we were capable of this – what would that mean but to castrate the intellect?

Few today would challenge this argument. Nietzsche is saying that seeing and knowing always takes place from an embodied and situated existence. There is no such thing as a purely objective presentation of a social phenomenon, a concept, a thinker, or a practice. We always see the world from an angle or through a lens, even if that is not immediately perceptible to us. With this in mind, we may wish to affirm perceptual anthropocentrism as a necessary part of the human condition. Try as we might, we cannot 'think like a mountain' (Seed, 2007), and attempts in that vein risk anthropomorphising the environment and extending human dominion through arguments made in bad faith.

What about descriptive anthropocentrism? For most of human history, this idea would have been nonsensical or interpreted as a hubristic statement of human power. For centuries, the natural world has been thought of as an immense and inexhaustible system. For example, the 1972 Convention on the Prevention of Marine Pollution by Dumping of Wastes and Other Matter (London Convention) was built on the assumption that the ocean had an 'infinite ability to assimilate waste' (Harrison, 2017). More broadly, frameworks built around sustainability often include the idea of 'stationarity' (Benson and Craig, 2017), or the ability of natural systems to regenerate (Telesetky, Cliquet, and Akhtar-Khavari, 2015).

However, in light of recent scholarship on the Anthropocene, we claim that descriptive anthropocentrism has a foundational basis. For example, Earth systems scientists are saying that human beings have the power to influence the functioning and health of the Earth system as a whole (Steffen *et al.*, 2004, p. 7). Whatever additional conclusions one draws from that claim, it seems clear that human beings have unique power and that we are not like other large mammals. In fact, we go further and argue that the Anthropocene matches perfectly Mylius' description (2018) of a paradigm that causes our worldview to begin from and become ordered around human beings.[3] Whether the new geological era is formally recognised or not, the data that underpins calls for recognising that the Anthropocene is significant enough to be regarded as a paradigm shift (Hamilton, 2017, p. 13) or second Copernican revolution in terms of its importance (Angus, 2016, pp. 28, 32).

Normative Anthropocentrism: A Brief History[4]

Against the grain of orthodox environmental philosophy, we argue that anthropocentrism is not a stable or determinate concept. While it has often been presented as fully formed by advocates of Earth Jurisprudence, it is really an idea that has unfolded over time and taken shape from tensions that have arisen at different times and locations. In this section, we do not attempt to provide a general history of Anthropocentrism. Rather, we locate three origin points that have informed the European conception of anthropocentrism and engage modes of critique that commonly appear in the literature on Earth Jurisprudence.

The Great Chain of Being

In his classic study of western metaphysics, Arthur O. Lovejoy traces the origin of the philosophical concept of the 'great chain of being' to Plato's concept of plenitude. This concept covers a wide range of premises, but we use it here to refer to the notion that the universe is a *plenum formarum* in which the entire range of conceivable things does in fact exist (Lovejoy, 1960, p. 52). This notion was developed further by Aristotle and his notion of 'continuity' which held that all quantities (lines, surfaces, solids, etc.) must be continuous in time and space (Lovejoy, 1960, p. 55). While these ideas do not result logically in a hierarchical ordering of nature, Lovejoy (1960, p. 56) noted that 'any division of creatures with reference to some determinate attribute manifestly gave rise to a linear series of classes'. As evidence of this, one can detect in Aristotle a minute graduation of differences from species to species. For example, Aristotle (1984a, p. 922) noted:

> Nature proceeds little by little from things lifeless to animal life in such a way that it is impossible to determine the exact line of demarcation, nor on which side thereof an intermediate form should lie. Thus, next after lifeless things in the upward scale comes the plant, and of plants one will differ from another as to its amount of apparent vitality; and, in a word, the whole genus of plants, whilst it is devoid of life as compared with an animal, is endowed with life as compared with other corporeal entities.

It was Aristotle who first suggested the idea of arranging all animals in a single, graded scale according to their 'degree of perfection'. Aristotle constructed two formulations of this hierarchy. The first focused on the degree of development reached by the offspring at birth. From this analysis, he discerned eleven general grades, with humankind at the top and the zoophytes at the bottom (Aristotle, 1984a, pp. 1136–1137). Furthermore, this basic hierarchy is supplemented by the notion of graduation, which includes clear instrumental values (Lovejoy, 1960, p. 56). That is, the environment is conceived of as being an instrument for human use. The following passage from Aristotle (1984b, p. 1991) illustrates this analysis:

> Plants exist for the sake of animals, the brute beasts for the sake of man - domestic animals for his use and food, wild ones (or at any rate most of them) for food and other accessories of life, such as clothing and various tools. Since nature makes nothing purposeless or in vain, it is undeniably true that she has made all animals for the sake of man.

Aristotle's second formulation (1984b, p. 1991) organises the hierarchy with regard to the 'powers of soul'. The scale ranges from the 'nutritive' qualities of plants to the 'rational' attributes of human beings and then 'possibly another kind superior to this' (Aristotle, 1984c, p. 659). Importantly, each higher element in the hierarchy possesses all the qualities of those below and an additional differentiating element of their own. The second ranking had great influence on subsequent philosophy and natural history and was used by later intellectuals to justify the anthropocentric worldview. Indeed, Lovejoy (1960, p. viii) argues that 'Aristotle's hierarchy is one of the most potent and persistent presuppositions in Western thought'.

Dominium in Christianity

Our next site is notions of dominion that were developed under Christian theology. While acknowledging that scripture is subject to diverse interpretations, we highlight the way certain passages have been used to normalise and justify human dominance. In scripture, the most explicit reference to human dominium over nature is found in Genesis, Chapter 1. Here we are told that God made human beings in his own image and stationed them in a special position in relation to the rest of creation. Moreover, human beings are explicitly given dominium over all things. Genesis 1:28–31 states:

> Be fruitful and increase in number; fill the earth and subdue it. Rule over the fish of the sea and the birds of the air and over every living creature that moves on the ground.' Then God said, 'I give you every seed-bearing plant on the face of the whole earth and every tree that has fruit with seed in it. They will be yours for food. And to all the beasts on the earth and all the birds of the air and all the creatures that move on the ground – everything that has the breath of life in it – I give every green plant for food' and it was so.

Further insight can be gained through a reading of the fall of humankind depicted in Genesis 3:13–19. In this passage, God caught Adam and Eve eating from the forbidden tree. Upon receiving their confessions, God banished them from the garden and inflicted hardship upon them and the serpent that tricked Eve. In this story we see human beings, God, and nature as three *separate* entities in *conflict*. Zen philosopher Daisetsu T. Suzuki (2014, p. 116) comments: 'Man is against God, Nature is against God, and Man and Nature are against each other. God's own likeness (Man), God's own creation (Nature) and God himself – all three are at war'. Thomas Berry (cited in Jensen, 2004, p. 37) supports this interpretation of Genesis, arguing:

> There is nothing to indicate a love of existence or a capacity for intimacy with the natural world for its own sake. Not to use it for monetary or even spiritual purposes but to be present in it.

Outside of scripture, St Thomas Aquinas is responsible for fusing Greek philosophy and Christian theology. Drawing on Aristotle, Aquinas (1991, p. 75) gives unequivocal support to the concept of plenitude and affirms the primacy of human beings and the 'great chain of being'. Aquinas (1991, p. 789) argues that nonhuman animals are 'ordered to man's use' and have 'no independent moral standing'. In practical terms, this led directly to the Catholic Church regarding private property as a 'regrettable but unavoidable reality' (Pipes, 1999, p. 17). Such was Aquinas's legacy that, in 1329, Pope John XXII (cited in Pipes, 1999, p. 17) cited him directly in holding: 'Property (*dominium*) of man over his possessions does not differ from the property asserted by God over the universe, which He bestowed on man created in his Image'. An almost identical logic can also be noted in jurists like William Blackstone (1966, p. 2) who defined property as 'that sole and despotic dominion that one man claims and exercises over the external things of the world, in total exclusion of the right of any other individual in the universe'.

The Scientific Revolution and Dualism

The final moment we wish to highlight in this survey of anthropocentrism is the scientific revolution. This history is covered by virtually all advocates of ecological law because it gave rise to a mechanistic philosophy that described the environment as a fragmented, lifeless, machine (Ehrenfeld, 1978; Merchant, 1990).

Francis Bacon was one of the chief proponents of this view. According to John Milton (2005, p. 77) Bacon's 'dream was one of power over nature' derived through experiment and embodied in appropriate institutions and used for the amelioration of human life'. Bacon's scientific writings were driven by the desire to perfect eliminative induction and position it as a lasting process for re-establishing the mastery of nature that human beings had enjoyed in the biblical stories. Hans Jonas (1984, p. 140) supports this interpretation, noting that the intention of Bacon's epistemological method was to 'gain knowledge and power over nature and to utilize power over nature for the improvement of the human lot'. Bacon provides further content to this aim in *The New Atlantis.* Here Bacon (1990, pp. 34–35) argued that 'the purpose' of human society is to acquire 'the Knowledge of Causes, and Secret Motions of Things; and the Enlarging of the bounds of the Humane Empire, to the Effecting of all Things possible'.

Bacon's concern with mastering nature is premised on a perspective that positions the natural world as 'other' and as a mere 'object' for human use. From this perspective, only human beings are 'subjects' and in a position to conduct objective inquiry. In contrast, the natural world is other and under investigation. Within this framework, it is ontologically impossible to be both subject and object – 'something is either culture or it is nature; human or not human; the inquirer or the object of inquiry' (Graham, 2011, p. 29). Bacon argued that this belief in the centrality of human beings would eventually be carried over into the secular realm and maintained in practically all narratives of human evolution, even though God would be dispensed with in most scientific accounts of the origin of the universe and of our species.

The dichotomy that Bacon drew between human beings and the environment resulted in violent implications – particularly with regard to the extension of patriarchy in western rationality (Merchant, 1980). Commenting on the goal of his scientific method, Bacon (Farrington, 1949, p. 62) holds: 'My only earthly wish is … to stretch the deplorably narrow limits of man's dominion over the universe to their promised bounds … putting [nature] on the rack and extracting her secrets … storming her strongholds and castles'. At no time did Bacon (Farrington, 1949, p. 62) hide his agenda: 'I come in very truth leading you to nature with all her children to blind her to your service and make her your slave … the mechanical inventions of recent years do not merely exert a gentle guidance over Nature's courses, they have the power to conquer and subdue her, to shake her to her foundations'. Elsewhere he notes: 'We have no right to expect nature to come to us … Nature must be taken by the forelock, being bald behind' (Farrington, 1949, p. 129). He also warns that delay or more subtle methods 'permit one to clutch at nature, never to lay hold of her and capture her' (Farrington, 1949, p. 130).

Bacon's scientific method had a profound impact on the philosophical investigations of Rene Descartes. Descartes is often called the father of modern philosophy (Garber, 2005, p. 174) and sought to start philosophy anew by breaking with the dominant traditions of the seventeenth century. Descartes (1985, p. 113) followed Bacon's instruction 'never to accept anything as true if [one] did not have evident knowledge of its truth'. Hence the *Meditations* (2006) begins with a series of arguments intended to cast doubt upon everything formerly believed and culminating in the hypothesis of an all-deceiving evil genius, as a device to keep former beliefs returning. The rebuilding of the world begins with the discovery of Descartes' well-rehearsed *Cogito* argument (2006, p. 13) – 'I think, therefore I am'. This 'I' is a self, known only as a thinking thing and it is discoverable independently of the senses.

While Descartes provides a fascinating first position for philosophy, what concerns us most about his argument is the essential hierarchy and division of the world that his argument implied. According to Descartes, rational human beings know their own awareness to be certain and entirely distinct from the external world of material substance. The material world has less certainty and is perceptible only as an object. Thus, *res cogitans* – thinking substances, subjective experience, spirit, consciousness – was understood to be different and separate from *res extensa* – extended substance, the objective world, matter, the physical body, plants and animals, stones, and the entire physical universe (Descartes, 2006, p. 145). Only in human beings did the two realities come together and both the cognitive capacity of human reason and the objective reality and order of the natural world found their common source in God (Garber, 2005, pp. 174–175).

According to Descartes (2006, p. 137), the physical universe was entirely devoid of human qualities. Rather, as purely material objects, all physical phenomena could be comprehended as machines – much like the lifelike automata and ingenious machines, clocks, mills, and fountains being constructed and enjoyed by seventeenth century Europeans. God created the universe and defined its mechanical laws, but after that the system moved on its own. Such a substance was best understood in mechanistic terms, reductively analysed into its simplest parts, and exactly comprehended in terms of those parts, arrangements, and movements. As Descartes (1985, p. 139) argued: 'The laws of Mechanics are identical with those of Nature'.

Because human beings combine material and spiritual qualities, Descartes argued that we have 'rendered ourselves the lords and possessors of nature' (Descartes, 1985, p. 141). Moreover, he argued that animals were insensible and irrational machines that 'moved like clocks but could not feel pain' (cited in Nash, 1990, p. 18). In his sixth discourse on method, Descartes (1998, p. 34) held that 'coercing, torturing, operating upon the body of Nature … is not torture [because] Nature's body is an unfeeling, soulless mechanism'. For Descartes, the scientific method was much more than a tool for the attainment of objective truth. His insights were used to strengthen the anthropocentric paradigm and solidify the logic of human mastery over the environment (Graham, 2011, p. 31).

Normative Anthropocentrism and *Homo Economicus*

As indicated in the previous section, scholarship on the European conception of normative anthropocentrism tends to present it as a phenomenon that originated in Ancient Greece and was developed under Christianity and the scientific revolution (Cullinan, 2011; Berry, 2006; Burdon, 2015a).[5] These three 'moments' in world history are regarded as central to the development of anthropocentrism as a cultural idea. This analysis carries (at least) two implicit assumptions: first that anthropocentrism is a determinate concept with identifiable boundaries. Second that we can know all that we need about anthropocentrism by looking historically.

Against these assumptions, we suggest that normative anthropocentrism is an indeterminate concept which continues to evolve and take new forms. Its boundaries are less fixed than fuzzy, and it will also take new forms in response to dominant ideas, social relations, changes in technology and material conditions. Because of this it is very difficult to say anything concrete about anthropocentrism, except perhaps that it represents a tendency in human thought/action to exalt human dominion and affirm the subordinate position of the environment. Far from being a universal concept, one would also expect normative anthropocentrism to be strongest in Western capitalist societies and to have unique geographical interpretations and accents.

For these reasons our engagement with the contemporary form of normative anthropocentrism is necessarily partial and contingent. We are not seeking to hold out a new universal that can shed light on the root cause of environmental destruction. Our aim is more modest – we seek to connect normative anthropocentrism with neoliberal rationality and see how they fit together. We have chosen neoliberal rationality over other ideas such as nationalism and socialism because political scientists have identified it as one of the dominant logics of our time

(Brown, 2015). Focusing on neoliberal rationality also enables us to pick up the historical sketch we plotted in the last section and trace the origin of neoliberalism to the utilitarian philosophy and the behaviourism of Jeremy Bentham. Thus, while partial and incomplete, this analysis will provide further detail to the cultural history of normative anthropocentrism and further demonstrate the complexity of our subject.

Neoliberalism: Radical Individualism

Neoliberal political philosophy brands itself as a new and updated version of liberalism. This is useful from a rhetorical perspective and allows neoliberals to claim heritage back to thinkers like John Stuart Mill. However, branding aside, we contend that neoliberalism has more in common with Jeremy Bentham, and his advocacy of utilitarianism and behaviourism as tools for governance. Because of this relationship, we need to say something about Bentham before we can fully appreciate neoliberalism as a dominant rationality.

One of Bentham's most important contributions to this discussion is the way he called into question the increasingly popular idea that law and governance should be designed in a way to maximise individual freedom. This idea caught fire in the nineteenth century and sharpened the focus of normative anthropocentrism from human beings to the individual. Bentham's writing goes against this trend, and he was one of the early advocates for animal rights (Nash, 1990, pp. 23–24). More broadly, Bentham sought to shift the focus of politics and law away from individual liberty and toward behaviourism and shaping conduct through coercive technologies. Implicit in his corpus is the idea that human beings can be coerced or manipulated toward 'proper conduct' (Bentham, 1969, p. 96). Most famously, his writing on the Panopticon is directed toward figuring out how to use basic human psychology to manipulate and control populations without having to use overt power or force (Bentham, 2010).

Unlike liberal thinkers, Bentham was not wedded to ideas such as liberty and equality. Nor did he necessarily advocate human superiority over nonhuman animals. These things might contribute to overall happiness, but they were not sacred values. Neither were rights (Waldron, 2015) or democratic ideals related to public participation (Schofield, 2006). So, Bentham is not a liberal in the orthodox meaning of that term and he gives us a very different take on anthropocentrism from what we have encountered so far. Bentham probably affirms human superiority over the environment, but the key point is that there are no values or principles that Bentham holds absolutely – other than the principle of utility. Everything is up for grabs, so long as it can be used to promote overall happiness (Bentham, 1969, pp. 85–86). This is the key fact that makes Bentham's work so relevant for comprehending neoliberalism – both approaches to politics are willing to jettison values in the service of a larger instrumental goal – utility or economic growth. And by putting utilitarianism at the heart of his political philosophy Bentham set loose the creature that has been birthed by capitalism – the self-interested subject (Bentham, 1969, p. 96). This in turn gives rise to new dimensions of anthropocentrism which we discuss further below.

The final point about Bentham that is relevant to understanding neoliberalism as a political rationality concerns his approach to governance. As noted above, Bentham is trying to develop an approach to governance whereby people can be controlled without recourse to overt force. To do this, he needs to develop programs for manipulation and control to get us to engage in what he thinks is proper conduct. Bentham (2002, p. 321) is explicit about this:

> The greatest enemies of public peace are the selfish and the hostile passions: necessary as they are, the one to the very existence of each individual, the other to his security

> ... Society is held together only by the sacrifice that men can be *induced* to make of the gratification's they demand: to obtain these sacrifices is the great difficulty, the great task of government.

This has an important implication for Bentham's utility principle when scaled up to the level of governance. Here it is easier to spot Bentham's biases against the desires of the masses and the measures he puts in place to prevent the utility principle being used to promote hedonism (Bentham, 1969, pp. 96–97). Rather than passively accepting the desires of citizens, Bentham thinks that political leaders must play an active role in shaping them. And this can only be done by making the cost of gratifying antisocial desires very painful (Bentham, 1969, p. 98). Bentham's theory of governance seeks to raise the principle of utility to the collective and find inducements that coerce people to behave in the 'right' way (Bentham, 1969, p. 97). This is the genius of Bentham's work. He grasps that inducements to good behaviour do not lie simply in rules or systems or rewards/punishments. Rather, getting us to be proper subjects in a mass society is achieved through an organisation of space, and systems of discipline that penetrate to the very depths of the human beings and remakes us and our desires. In other words, Bentham's politics seeks to shape us at the very core of the self.

Neoliberalism has advanced this project much further than Bentham could have imagined. Here we are, thinking about neoliberalism not just as a set of economic policies (Mirowski and Plehwe, 2015), or as a political project (Harvey, 2006), but as a set of ideas or dominant logic that is pervasive in western societies. Wendy Brown (2015, p. 17) has given the fullest expression of this understanding of neoliberalism, describing it as a 'peculiar form of reasons that configures all aspects of existence in economic terms'.[6] There is a lot to unpack here. In describing neoliberalism as a 'form of reason' Brown is talking about the way dominant ideas (or logics) in society influence not only the way we think but the kinds of ideas that are 'thinkable'. This is easiest to grasp if we return to anthropocentrism and think about the ways ideas like dominion cloaked human superiority with a veneer of naturalness and inevitability. Something similar is true today from within the logic of neoliberalism. However, rather than explicitly affirming human superiority, neoliberalism has naturalised the right of business to unfettered market activity and affirmed Margaret Thatcher's dictum that 'there is no alternative' to market capitalism. The power of this logic is best captured in the oft-quoted phrase: 'It is easier to imagine the end of the world than the end of capitalism (Fisher, 2009, p. 1)'.

A second key idea in Brown's description is economisation and the metrication of daily life. The process of economisation is widespread and encompasses activities that occur both within and outside of the economy. Brown (2015, p. 31) writes: 'we may (and neoliberalism interpellates us as subjects who do) think and act like contemporary market subjects where monetary wealth generation is not the immediate issue, for example, in approaching one's education, health, fitness, family life or neighborhood'. Even the most intimate and personal parts of oneself can be coerced into this logic. For example, Brown (2015, p. 31) notes: 'one might approach one's dating life in the mode of an entrepreneur or investor ... [m]any upscale online dating companies define their clientele and offerings in these terms, identifying the importance of maximizing return on investment of affect, not only time and money'.[7]

Like the behaviourism promoted by Bentham, neoliberal rationality does not work by force or overt power. Its effectiveness lies in the way we are conscripted into self-discipline and model our own thoughts and behaviour after the model of the firm. Thus, Brown (2015, p. 33) argues: 'its project is to self-invest in ways that enhance its value or to attract investors through constant attention to its actual or figurative credit rating, and to do this across every sphere of existence'. This is an apt description of the hustle economy and the wealth generation capacity of YouTube celebrities and social media influencers. But, as Brown suggests, the reach of neoliberal rationality goes beyond these new economic actors. Even older professionals such as university academics are encouraged to comport to the model of the firm by becoming an entrepreneur of their research, competing for research money and 'networking' at conferences.

Neoliberal rationality has perfected the techniques of self-discipline first envisioned in Bentham's Panopticon. It would also seem to be morally neutral – what matters is the economisation of daily life. Brown (2006, p. 692) has advanced this reading, arguing that neoliberalism is 'expressly amoral at the level of both ends and means'. However, just as we saw with Bentham, this claim does not stack up when we move up and think about neoliberalism as a politics or tool for governance. Brown (2019, pp. 15–16) has acknowledged this in more recent work and now describes neoliberalism as a 'market-and-morals project with nihilism, fatalism, and wounded white male supremacy'. Melinda Cooper (2017) has pioneered this inquiry and unpacks a range of policy domains where patriarchal family norms have become embedded in neoliberalism. While not discussed by Cooper, an example of this in environmental politics can be seen in the way individuals are made to feel responsible for preventing climate change through consumer choices and the ethic of personal responsibility. From this perspective, the fate of the Paris Climate Treaty lies in changing light bulbs and anything resembling collective action against big polluters 'has become a target of the elite' (Lukacs, 2017).

Alongside this literature, Jessica Whyte (2019, p. 8) has presented the most convincing account of the moral foundations of neoliberalism. For Whyte, what 'distinguished the neoliberals of the twentieth century from their nineteenth-century precursors [was not] a narrow understanding of the human as *homo economicus*, but the belief that a functioning competitive market required an adequate moral and legal foundation'. The key protection described by Whyte are human rights. And in an analysis that ought to trouble advocates of environmental human rights, Whyte describes the role that neoliberal thinkers such as Friedrich Hayek played in positioning human rights as a protection for the market and private sphere (family and church) from social democratic movements. We return to this point in the next section.

What kind of normative anthropocentrism are we left with at the end of this sketch of neoliberal rationality? We have come some distance from logics that explicitly promoted values associated with human superiority and dualism. One reasonable response might be that anthropocentrism is no longer a relevant subject of critique for proponents of Earth Jurisprudence as a result of the domination of neoliberal modes of thinking that have moved on from this specific opposition between humans and nature. Its influence has been supplanted by other, more pervasive, logics. According to this view, the problem is not that human beings view themselves as separate and superior to nature. The problem is that our subjectivities have been so thoroughly reshaped that we can only see or value things in an economised or instrumental way. Even the tools that we wield to protect the environment – such as human rights – turn out to have been shaped by the architects of neoliberalism. This is, today, the true nature of anthropocentrism.

Alternatively, one might respond that anthropocentrism does not need to be explicitly stated and defended because it operates as an implicit assumption in neoliberal politics. It is an unstated starting position from which neoliberal rationality takes hold and progresses its moral and economising agenda. In the next section, we adopt the latter position and share some ideas about how advocates of Earth Jurisprudence can think about anthropocentrism now and in the future.

Engaging Anthropocentrism Today

Neoliberal rationality does not expressly advocate or defend normative anthropocentrism. But its silence on this worldview should not be interpreted as neutrality. Instead, we argue that anthropocentrism is an implicit assumption and starting position that neoliberal rationality takes for granted. Under neoliberalism, normative claims about the supremacy of human beings are not challenged. Nor is the idea that the environment is a thing that exists for human exploitation. In fact, under neoliberalism it is harder to see the environment as an entity with its own unique value. Ideas like nature having intrinsic value cannot be economised and so under neoliberal rationality they do not exist. The environment can only be thought about in economised terms and we participate in this process when we use language such as 'natural resources', 'ecosystem services' or 'natural capital'. In these examples, normative anthropocentrism is not expressed as an explicit worldview. It is an implicit assumption that works in the service of neoliberal rationality.

Simple examples of this can be found in international law which Klaus Bosselmann (2008, p. 13) has characterised as perpetuating an 'anthropocentric, resource-oriented and non-integrative approach'. The same thing could be said about World Trade Organisation (WTO) agreements and even documents purporting to protect the environment. The Sustainable Development Goals (SDGs), for example, are silent about the right of human beings to use, engage and exploit the environment for human development. That is the starting presupposition upon which the document is based. Moreover, the SDGs have been criticised for using the language of 'sustainable development' to expand markets and perpetuate neoliberal orthodoxy in the developing world (Adelman, 2017). In support of this argument, Ariel Salleh (2016, p. 953) argues that the framework seeks to realise sustainable development 'by growing gross domestic product (GDP), increasing market liberalization and free trade, as well as according more power to the World Trade Organisation'. While Heloise Weber (2017, p. 399) argues that the SDGs privilege 'commercial interests over commitments to provide universal entitlements to address fundamental life-sustaining needs'.

Beyond these examples in international law, neoliberal rationality has become a dominant logic, it has also impacted the way environmental campaigners think about their work. The most obvious example of this can be noted in campaigns for environment protection which couch their demands in an instrumental language that promotes economic growth. For example, in Australia, campaigners for the Great Barrier Reef don't talk about the intrinsic value or the importance of the ecosystem for marine life. They are incentivised to talk in terms of tourism jobs or the loss that a dead Reef will bring to the Queensland economy (Smee, 2018).[8] Of course these are valid concerns, but they are not the only or even the most important thing that will be lost if the reef collapses entirely. Wendy Brown offers another example in her reading of Barack Obama's 2013 State of the Union Address. Brown (2015, p. 25) writes:

> [E]very progressive value – from decreasing domestic violence to slowing climate change – Obama represented not merely as reconcilable with economic growth, but as driving it. Clean energy would keep us competitive – as long as countries like China keep going all-in on clean energy, so must we.

If we look further at Obama's text it is impossible to miss how environmental concerns have been economised into matters for the market, growth opportunities or as part of a broader agenda for national competitiveness. After declaring that climate change must be tackled for the sake of our children, Obama (2013) notes:

> … the good news is we can make meaningful progress on this issue while driving strong economic growth. I urge this Congress to get together, pursue a bipartisan, market-based solution to climate change … in the meantime, the natural gas boom has led to cleaner

power and greater energy independence. We need to encourage that. And that's why my administration will keep cutting red tape and speeding up new oil and gas permits.

Obama presents a master class in the logic of reductionism and his reasoning will be familiar to anyone that has spent time advocating for environmental protection. And while he does not explicitly advocate normative anthropocentrism, his indifference works to validate the status quo. As noted above, neutrality sounds prejudice free. But if normative anthropocentrism is an implicit logic in Western society, then to feign blindness to that is to affirm its role in promoting environmental exploitation.[9]

A less obvious issue connected to the intersection between normative anthropocentrism and neoliberal rationality is the proliferation of legal rights as a tool for environmental protection. Here we are not just talking about the proliferation of environmental human rights but also attempts to refashion rights as a collective power and claims about the rights of nature which have become virtually synonymous with Earth Jurisprudence (Cullinan, 2011; Berry, 2006).[10] Each of these examples differs in terms of their content and realisation in positive law. However, they are all marked by their point of origin as a tool for the protection of individual human interests (Burdon, 2015b) and should be read as part of the broader history of legal rights in the twenty-first century.

A comprehensive history of environmental rights has not yet been written and we are unable to provide one here. However, we return to a point that was raised toward the end of section three about how legal rights proliferated in the second part of the twentieth century as a tool of neoliberalism.[11] In support of this claim, Samuel Moyn (2010, p. 19) has argued that the linearity of rights after the second world war has been overstated and that it was not until the mid-1970s that rights talk spread and human rights emerged to displace other justice claims. Prior to this, social struggles were animated by more radical demands, evidenced in anticolonial struggle and the 1960s student movement (Moyn, 2010, p. 2). It was only after these more radical alternatives were defeated (in part by human rights NGOs[12]) that the language of rights gained ascendency. In fact, Moyn argues that it was precisely because human rights did not require a commitment to political and social upheaval that they emerged as the safest strategy for the ideologically disenchanted.

Seen in this light, the expansion of legal rights for ecological goals is not a radical demand. At best, they reflect what Robert MacFarlane (2019) calls our 'Anthropocene moment' – 'At once hopeful and desperate, it is a late-hour attempt to prevent a slow-motion ecocide'. But as a politics, the growth of environmental rights is predicated upon the abandonment of alternative political formations and the vision of the future those alternatives contained. Rights arguments are not about substantive or transformative change. They are not about displacing growth economics or democratising power in a way that empowers communities or builds resilience. Rather, they reflect a minimalist alternative and seek to mitigate environmental damage from firmly within the coordinates of free market capitalism (Moyn, 2010, p. 121). In a similar vein, foundational documents for environmental human rights, such as the Universal Declaration of Human Rights (1948), have also been used to promote an egoistic form of market-based individualism in developing countries (Cheah, 2007). This protection is important if that is all that is available. However, protections should not be confused with justice or with a resolution of the underlying tensions or ideological power structures such as normative anthropocentrism.

Conclusion

In this chapter, we have argued that ecological lawyers need to adopt a more nuanced understanding of anthropocentrism and affirm its perceptual and descriptive dimensions. As part of this argument, we highlighted that the Anthropocene interpellates human beings as geological agents with the unique ability to alter the Earth system. Coming to terms with this may require that we become more anthropocentric in our thinking rather than less, in the sense of developing a better sense of the history of humans' liberal governance of themselves and how it has sustained, complexified and made less apparent an ongoing sense of the disposability of the nonhuman. However, in doing so we must also guard against hubris and the arrogance that has underpinned so much of western thought and is manifest in more recent expressions like Eco modernism or techno-optimism. As ecological lawyers, our critical attention ought to be focused on normative anthropocentrism and similar positions that affirm the subordination of the environment not just to humans or humanity but specifically to human needs and greed.

With these distinctions in place, we then argued that accounts of normative anthropocentrism tend to focus on select historical moments associated with Christian theology and the scientific revolution. In this construction, normative anthropocentrism is presented as a unitary theory that can be understood in historical terms, at the expense of a closer reading of how anthropocentrism has mutated in late modernity. Against this reading, we argued that anthropocentrism is an indeterminate concept that is subject to change and flux in response to new ideas. If normative anthropocentrism is to mean anything today, ecological lawyers must engage with contemporary politics and the dominant logics of the market. This is an immense task and not something that can be accomplished in a brief chapter. However, to advance our argument, we engaged neoliberal rationality specifically and sought to understand the way human subjects are shaped on the model of *homo economicus*. Neoliberal rationality does not advance or defend normative anthropocentrism as a worldview directly. Rather, human dominion is a presumption or unspoken presupposition through which the environment is economised and reduced to purely instrumental value.

In the final part of the chapter, we sought to tease out the implications of our argument for Earth Jurisprudence and other approaches to ecological law. We began by looking at how neoliberal rationality is perpetuated by environmentalists when we think and talk in the language of natural resources or ecosystem services. We also provided examples of how campaigns for environmental protection are framed in terms of jobs and economic growth. Finally, we critically engaged with the way legal rights have become the dominant language through which environmental lawyers frame their justice claims. As part of this engagement, we pointed to the role that neoliberal thinkers played in developing human rights and how rights came to crowd out and displace more radical demands for justice and power sharing.

Our intention in providing this analysis is not to discourage our colleagues and comrades who are struggling with limited tools against the overweening power of the market. Rather we contend that we will only be able to effectively respond to normative anthropocentrism if we can perceive it clearly and understand the way it has changed (and will continue to change) over time. As part of this we also contend that ecological lawyers need to avail themselves of a wide range of languages and frameworks for articulating justice claims, rather than relying on rights and other tools fashioned under neoliberalism that may end up reproducing the very thing they seek to stand up against. Here we are reminded of Audre Lorde's (1984, p. 112) timeless injunction:

> the master's tools will never dismantle the master's house. They may allow us temporarily to beat him at his own game, but they will never enable us to bring about genuine change.

Notes

1 Earth Jurisprudence is a relatively recent legal movement which originated in the work of Thomas Berry (2006). Its fundamental goal is to move the principle of ecological integrity from the periphery

of law and legal institutions and interpretations to its foundation. For a history, see Cullinan (2011). In the authors' opinion, many of the ideas in Earth Jurisprudence are contained in environmental philosophy and ecological law. See for example Bosselmann (1999). That is why we have combined the ideas in this chapter.

2 We recognise of course that neoliberal capitalism is only one form of societal organisation, and that other forms have arisen and could arise that institutionalise different forms of property relations and market structures that might not be considered 'neoliberal'. The legal realists, critical legal, and 'social relations' theorists have done particularly astute analytical work on how 'property' and 'the market' (and thus capitalism) are concepts with many conceptions (e.g., Singer, 2000). For present purposes, however, we choose to avoid that body of critical theory in order to focus on the nexus of neoliberal rationality and anthropocentrism.

3 In saying this we acknowledge that the Anthropocene is the product of a relatively small number of humans. For debates about the appropriateness of the term Anthropocene, see Kunkel (2017).

4 This section is a condensed summary from Burdon (2015a). It also contains new material and analysis.

5 This is true of ecological thinking in law more generally. See for example Capra and Mattei (2018).

6 As discussed below, Brown (2019, pp. 10–11, 86, 102) has reconfigured this argument in more recent work.

7 Radin (2001) captures a similar idea in her notion of 'metaphorical markets'.

8 For a broader analysis of the way instrumental rationality has impacted the environment movement in Australia, see Hamilton (2011).

9 See further Marx (1978, p. 33).

10 Although Burdon (2015a) has argued that there is nothing inherent in Earth Jurisprudence that demands this interpretation.

11 The critical literature on human rights is vast. See for example Douzinas (2000).

12 See Whyte (2012; 2019, pp. 156–159).

References

Adelman, S. (2017) 'The Sustainable Development Goals, Anthropocentrism and Neoliberalism', in French, D. and Kotze, L. (eds.) *Sustainable Development Goals: Law, Theory and Implementation*. Cheltenham: Edward Elgar Publishing, pp. 15–40.

Angus, I. (2016) *Facing the Anthropocene: Fossil Capitalism and the Crisis of the Earth System*. New York: Monthly Review Press.

Aquinas, T. (1991) *Summa Contra Gentiles*. Notre Dame: University of Notre Dame Press.

Aristotle (1984a) 'History of Animals', in Barnes, J. (ed.) *The Complete Works of Aristotle*. Princeton: Princeton University Press.

Aristotle (1984b) 'Politics', in Barnes, J. (ed.) *The Complete Works of Aristotle*. Princeton: Princeton University Press.

Aristotle (1984c) 'On the Soul', in Barnes, J. (ed.) *The Complete Works of Aristotle* Princeton: Princeton University Press.

Bacon, F. (1990) *The New Atlantis*. Cambridge: Cambridge University Press.

Benson, M.H. and Craig, R.K. (2017) *The End of Sustainability: Resilience and the Future of Environmental Governance in the Anthropocene*. Lawrence: University Press of Kansas.

Bentham, J. (1969) *A Bentham Reader*. New York: Pegasus.

Bentham, J. (2002) 'Nonsense upon Stilts' in Rosen, F. and Schofield, P. (eds.) *Rights, Representation, and Reform: Nonsense Upon Stilts and Other Writings on the French Revolution*. Oxford: Oxford University Press.

Bentham, J. (2010) *The Panopticon Writings*. London: Verso.

Berry, T. (1999) *The Great Work: Our Way into the Future*. New York: Three Rivers Press.

Berry, T. (2006) 'Legal Conditions for Earth Survival' in Tucker, M.-E. (ed.) *Evening Thoughts: Reflections on Earth as Sacred Community*. Sierra Club Books, pp. 107–113.

Blackstone, W. (1966) *The Commentaries on the Laws of England*. Oxford: Oxford University Press.

Bosselmann, K. (1999) *When Two Worlds Collide: Society and Ecology*. Auckland: RSVP.

Bosselmann, K. (2008) *The Principle of Sustainability: Transforming Law and Governance* Aldershot: Ashgate.

Brown, W. (2006) 'American Nightmare: Neoliberalism, Neoconservatism, and De-Democratization', *Political Theory*, 34(6), pp. 690–714. DOI: 10.1177/009059170629301

Brown, W. (2015) *Undoing the Demos: Neoliberalism's Stealth Revolution*. New York: Zone Books.

Brown, W. (2019) *In the Ruins of Neoliberalism: The Rise of Antidemocratic Politics in the West*. New York: Columbia University Press.

Burdon, P. (2015a) *Earth Jurisprudence: Private Property and the Environment*. London: Routledge.

Burdon, P. (2015b) 'Environmental Human Rights: A Constructive Critique', in Grear, A. and Kotze, L. (eds.) *Research Handbook on Human Rights and the Environment*. Edward Elgar Publishing, pp.61–78.

Capra, F. and Mattei, U. (2018) *The Ecology of Law: Toward a Legal System in Tune with Nature and Community*. Oakland: Berrett-Koehler.

Cheah, P. (2007) *Inhuman Conditions: On Cosmopolitanism and Human Rights*. Cambridge: Harvard University Press.

Cooper, M. (2017) *Family Values: Between Neoliberalism and the New Social Conservatism*. New York: Zone Books.

Cullinan, C. (2011) *Wild Law: A Manifesto for Earth Justice*. White River Junction: Green Books.

Descartes, R. (1985) 'Meditations', in John Cottingham (ed.) *The Philosophical Writings of Descartes*. Cambridge: Cambridge University Press.

Descartes, R. (1998) *Discourse on the Method for Conducting One's Reason Well and for Seeking Truth in the Sciences*. Indianapolis: Hackett Publishing Company.

Descartes, R. (2006) *Meditations, Objections, and Replies*. Indianapolis: Hackett Publishing Company.

Douzinas, C. (2000) *The End of Human Rights: Critical Legal Thought and the Turn of the Century*. Oxford: Hart Publishing.

Ehrenfeld, D. (1978) *The Arrogance of Humanism*. Oxford: Oxford University Press.

Farrington, B. (1949) *Francis Bacon: Philosopher of Industrial Science*. New York: Schumann.

Fisher, M. (2009) *Capitalist Realism: Is There No Alternative?*. Winchester: Zero Books.

Garber, D. (2005) 'Rene Descartes', in Craig, E. (ed.) *The Shorter Routledge Encyclopaedia of Philosophy*. London: Routledge.

Graham, N. (2011) *Lawscape: Property, Environment, Law*. London: Routledge.

Hamilton, C. (2011) 'Hamilton: A New Brand of Environmental Radicalism', *Crikey*, 22 February [Online]. Available at: https://www.crikey.com.au/2011/02/22/hamilton-we-need-a-new-brand-of-environmental-radicalism/ (Accessed: 22 September 2022).

Hamilton, C. (2017) *Defiant Earth: The Fate of Humans in the Anthropocene*. Cambridge: Polity.

Harrison, J. (2017) *Saving the Oceans Through Law*. Oxford: Oxford University Press.

Harvey, D. (2006) *A Brief History of Neoliberalism*. Oxford: Oxford University Press.

Heidigger, M. (1970) *Discourse on Thinking*. New York: Harper Perennial.

Jensen, D. (2004) *Listening to the Land: Conversations about Nature, Culture and Eros*. White River Junction: Chelsea Green Publishing.

Jonas, H. (1984) *The Imperative of Responsibility: In Search of an Ethics for the Technological Age*. Chicago: University of Chicago Press.

Kunkel, B. (2017) 'The Capitalocene', *London Review of Books*, 39(5) [Online]. Available at: https://www.lrb.co.uk/the-paper/v39/n05/benjamin-kunkel/the-capitalocene (Accessed: 22 September 2022).

Lorde, A. (1984) *Sister Outsider: Essays and Speeches by Audre Lorde*. Trumansburg: Crossing Press.

Lovejoy, A.O. (1960) *The Great Chain of Being*. New York: Harper & Row.

Lukacs, M. (2017) 'Neoliberalism has Conned us into Fighting Climate Change as Individuals', *The Guardian*, 18 July [Online]. Available at: https://www.theguardian.com/environment/true-north/2017/jul/17/neoliberalism-has-conned-us-into-fighting-climate-change-as-individuals (Accessed: 22 September 2022).

Macfarlane, R. (2019) 'Should This Tree Have the Same Rights as You?', *The Guardian*, 2 November [Online]. Available at: https://www.theguardian.com/books/2019/nov/02/trees-have-rights-too-robert-macfarlane-on-the-new-laws-of-nature? (Accessed: 8 March 2023).

Marx, K. (1978) 'On the Jewish Question', in Tucker, R.C. (ed.) *The Marx-Engels Reader*. New York: W.W.Norton & Company.

Merchant, C. (1980) *The Death of Nature: Women, Ecology and the Scientific Revolution*. San Francisco: Harper and Row.

Milton, J. (2005) 'Francis Bacon', in Craig, E. (ed.) *The Shorter Routledge Encyclopaedia of Philosophy*. London: Routledge.

Mirowski, P. and Plehwe, D. (ed.) *The Road from Mont Pèlerin: The Making of the Neoliberal Thought Collective*. Cambridge: Harvard University Press.

Moyn, S. (2010) *The Last Utopia: Human Rights in History*. Cambridge: Belknap Press.

Mylius, B. (2018) 'Three Types of Anthropocentrism', *Environmental Philosophy*, 15(2), pp. 159–194. DOI: 10.5840/envirophil20184564

Nash, R. (1990) *The Rights of Nature*. Madison: University of Wisconsin Press.

Nietzsche, F. (1989) *On the Genealogy of Morals*. New York: Vintage Books.

Obama, B. (2013) 'Remarks by the President in the State of the Union Address', The White House, February 12. Available at: https://obamawhitehouse.archives.gov/the-press-office/2013/02/12/remarks-president-state-union-address (Accessed: 22 September 2022).

Pipes, R. (1999) *Property and Freedom*. New York: Alfred Knopf.

Radin, M.J. (2001) *Contested Commodities*. Cambridge: Harvard University Press.

Salleh, A. (2016) 'Climate, Water, and Livelihood Skills: A Post-Development Reading of the SDGs', *Globalizations*, 13(6), pp. 952–959. DOI: 10.1080/14747731.2016.1173375

Schofield, P. (2006) *Utility and Democracy: The Political Thought of Jeremy Bentham*. Oxford: Oxford University Press.

Seed, J. (2007) *Thinking Like a Mountain: Towards a Council of All Beings*. Gabriola Island: New Catalyst Books.

Singer, J.W. (2000) *Entitlement: The Paradoxes of Property*. New Haven: Yale University Press.

Smee, B. (2018) 'Domestic Tourism to Great Barrier Reef Falls in Wake of Coral Bleaching', *The Guardian*, 7 June [Online]. Available at: https://www.theguardian.com/environment/2018/jun/08/domestic-tourism-to-great-barrier-reef-falls-in-wake-of-coral-bleaching (Accessed: 22 September 2022).

Steffen, W., Sanderson, A., Tyson, P.D., Jäger, J., Matson, P., Moore, B., Oldfield, F., Richardson, K., Schellnhuber, H.J., Turner, B.L. and Wasson, R.J. (2004) *Global Change and the Earth System: A Planet Under Pressure*. Berlin: Springer.

Suzuki, D.T. (2014) 'The Role of Nature in Zen Buddhism' in *Selected Works of D.T. Suzuki, Volume I: Zen*. Berkley: University of California Press, pp. 113–135.

Telesetsky, A., Cliquet, A. and Akhtar-Khavari, A. (2015) *Ecological Restoration and International Environmental Law*. New York: Routledge.

The 11th Hour: Turn Mankind's Darkest Hour into Its Finest (2007) directed by Conners, N. and Conners, L. [Film] Burbank: Warner Independent Pictures.

Waldron, J. (2015) *Nonsense upon Stilts: Bentham, Burke and Marx on the Rights of Man*. New York: Routledge.

Weber, H. (2017) 'Politics of Leaving No One Behind": Contesting the 2030 Sustainable Development Goals Agenda', *Globalizations*, 14(3), pp. 399–414. DOI: 10.1080/14747731.2016.1275404

Whyte, J. (2012) 'Intervene, I Said', *Overland*, 207 [Online]. Available at: https://www.overland.org.au/previous-issues/issue-207/feature-jessica-whyte/ (Accessed: 22 September 2022).

Whyte, J. (2019) *The Morals of the Market: Human Rights and the Rise of Neoliberalism*. New York: Verso.

12
GLOBAL ANIMAL LAW, PAIN, AND DEATH

An International Law for the Dominion

Alejandro Lorite Escorihuela

Introduction: Global Animal Law and Anthropocentrism

This contribution discusses the general project of a 'global animal protection regime' in international law (O'Sullivan, Otter, and Ross, 2012) and, more specifically, the notion of 'global animal law' (Peters, 2016a). Global animal law (GAL) – a 'field of research', a 'discipline', a 'legal field of its own', and ultimately 'a manifestation and a driver of the broader ongoing "animal turn" in the social sciences and humanities' (Peters, 2016a) – is a scholarly and policy making project that seeks to identify and respond to regulatory problems concerning animal welfare in the international legal system. By addressing the problematic absence of a globally coherent legal framework organised around the principle and norms of animal welfare, GAL scholars and advocates announce a shift away from anthropocentric concerns with animals as goods (Peters, 2017a), and possibly towards more zoocentric, or rather sentiocentric concerns with animals as sentient beings (Blattner, 2016).

What anthropocentrism is and what it does are issues understood in a variety of ways in the humanities and social sciences (Hayward, 1997; Aaltola, 2010; Youatt, 2014; Garner, 2017). As a project of legal reform, GAL adopts a specific perspective on anthropocentrism as a problem, and on what moving away from anthropocentrism through law could mean. In the erudite voice of its intellectual leader Anne Peters (see generally the landmark *opus* by Peters, 2021), GAL appears as a pragmatic and keenly strategic reformist project, which cannot and should not shy away from instrumentalising anthropocentric motivations and tropes in order to achieve the higher objective of improving animal welfare (Peters, 2021, pp. 548, 557). Anthropocentrism here designates principally the centrality of human interests in law and the possible limitations of such anthropocentric normative frameworks for the purpose of advancing animal welfare. This disposition towards anthropocentrism derives from the nature of the project, which seeks prosaically to do as much as is possible to improve the lot of animals through the deployment of legal argument and institutions (Peters, 2021, p. 536).

Global animal law is a deeply realist project, both in the sense of legal realism (as expressed in the recurring notion that law is a tool box in the hands of lawyers for the pursuit of desired social ends (Peters, 2021, p. 30)) and in the sense of political realism (as is conveyed by the idea that ultimately 'the power of law as a mode of governance is limited' (Peters, 2021, p. 60)). From

DOI: 10.4324/9781003201120-15

there the task of GAL is both descriptive and normative (Peters, 2017a, p. 254): showing first where the law stands and directing it then towards where it should be with the help of the guiding principle of animal welfare. The GAL project assimilates moreover realistically that 'all human law is anthropocentric in a most banal sense, made by humans who observe and interpret the world from their standpoint, and for humans' (Peters, 2021, p. 535). As such, GAL is an appeal for legal battle on a compromised terrain where law has essentially been 'bad for animals' and a cover for what Dinesh Wadiwel coined the 'war against animals' (Peters, 2021, p. 59, citing Wadiwel, 2015).

The answer is realistic and pragmatic. Peters puts it again very insightfully. Quoting Catharine McKinnon, she announces:

> Indeed, '[c]entral dilemmas in the use of law by humans' to improve the plight of animals persist. However, these dilemmas do not condemn legal scholars and practitioners to complacency or resignation. Legal analysis can contribute to identifying animal suffering and can make normative proposals for legal reform. International law is not only a 'hollow hope' but can be made a force for good for animals.
>
> *(Peters, 2021, pp. 60–61)*

The present contribution starts with this helpful positioning of legal analysis and advocacy proposed by Anne Peters, in line with previous work on the nature and role of legal scholarship (Peters, 2017b): legal work for a social objective like animal welfare starts with what amounts to a political commitment and decision on the part of jurists. Along those lines, I propose a critical engagement with the project of GAL, based on the shared insight that law is critical to both oppression and liberation, and that critical engagement with the law is therefore a necessary part of the struggle:

> Opting out of the legal arena in the struggle for animal protection would be a bad idea, not the least because purely extra-legal activism would neglect the ongoing importance of the law in seemingly unregulated spheres of life. Although legal reform is 'never radically transformative' it still seems a *conditio sine qua non* for combatting injustice.
>
> *(Peters, 2021, p. 596)*

The specific object of attention in this dialogue with GAL's critical and programmatic endeavour consists of the nature of the harm that GAL seeks to address, as well as the relationship between the anthropocentrism of law and the nature of harm to be mitigated or reduced through the deployment of law.

To supplement the pioneering work of GAL scholarship under Anne Peter's leadership, I approach the question of harm done to animals with an understanding of anthropocentrism that differs from the view that animates, very generally, GAL. In GAL, the harm is signalled by suffering, elicited by acts of violence and negligence, and this suffering is fought as an injustice on account of its lack of rational justification. I will develop an account of the harm done to animals as constituted instead by objectification, a harm of a structural nature that informs the human–animal relation before discrete injustices occur. Such an approach accompanies, in productive tension and dialogue, the realistic reformist agenda of GAL and supplements it with the ideal and idealism of abolition; it sees in particular the solution not in Liberal justice through law, but in liberation from totalitarian control by human reason as embodied in law. My own individual argumentative trajectory creates, however, some distance with GAL, in terms of the conception of the role of law in shaping harm through anthropocentrism, now conceived as ontological domination. To borrow Maneesha Decka's elegant formulations, by its own nature '[l]aw is an anthropocentric terrain' or, more simply, 'law is an anthropocentric institution' (Deckha, 2013, pp. 784, 813). When engaging 'the animal turn' (Peters, 2016a, p. 15; 2021, pp. 38, 522), GAL as a legal field and legal discipline engages pragmatically the reality of policy and institutions within the domain of law by adopting the forms of law. I will focus on the role of the terrain itself as the space within which that turn is being taken – law as a form, a structure, and a language within which weights, worth, interest, and justice are being defined and shuffled.

Specifically, legal debate about the appropriate legal tools for advancing 'animal welfare', or even 'animal rights', presumes that law receives animals as raw data or material from outside of law and responds to harm that is being done to them out there in the world. I seek to draw attention instead to the idea that law makes animals, instead of receiving them and observing their existence; law is, among other social institutions, constitutive, as a form and a field, of the harm that is done to them, because it makes and unmakes them as objects that are both discrete and dependent on our anthropocentric world. Beyond the acts of violence targeted by anti-cruelty statutes, the focus here is on the notion that animals are victims of ontological violence, beautifully described by Tara Kennedy, as

> denying the inexhaustibility of meaning, by believing that we can conceptually grasp, in a comprehensive way, the things we encounter or, worse yet, by forcing a self-serving and calculative interpretation onto things, regardless of the way in which they show themselves to be.
>
> *(Kennedy, 2016, p. 467)*

Animals are simply made into extensions, instruments of human social processes regardless of what they may have been autonomously (Shapiro, 1990, p. 32; Plumwood, 1993, p. 52).

When we then focus on the imagined welfare of animals as legal and scientific constructions and deconstructions within global, social, and economic processes, the quest for animal welfare embraces the whole architecture of global regulation. While global law does not necessarily speak everywhere of animals, the regulatory silences of entire sectors of legal regulation contribute to the summoning and enlisting of animal beings into the networks of exploitation entangling both animals and humans (Nibert, 2002; 2013; 2017; Torres, 2007).

Following those terms, the pages below propose a critical elaboration of the cause of abolition within the debates about the contents and direction of global animal law, based on the central proposition that, as Anne Peters put it bluntly, law is indeed 'bad for animals'.

My argument unfolds as follows. First, I present global animal law's orientation as the protection of 'animal welfare'. I suggest that 'welfarism', the orientation of law towards the improvement of welfare, makes global animal law's response to anthropocentrism a version of anti-speciesism and means specifically that global animal law identifies the animal question for the purpose of regulation as a problem of discrimination. Second, I suggest that this reading of human–animal relations further implies that, as Martha Nussbaum has suggested, we can simply and progressively replace nature with justice, by extending a moral theory constructed along the lines of Liberal principles to animals (Nussbaum, 2009, p. 390). I propose that such a construction eschews the fact that animals, as animals, are not found but made by political theory and, concretely, animals are made on the terrain of law, which distributes the ontological features and political statuses of animals as pests, pets, wildlife, fish stock, offal, or real-life guinea pigs. As such, an alternative reading of the situation will be that, following Robert Cover (1985), the deployment of law is not an antecedent or a cure to violence, but violence itself. From that perspective, I propose that GAL's work of legal reform situates itself in a Liberal framing of harm as irrational or arbitrary prejudice, whereas fundamental violence on animals, which constitutes them as 'nonhuman animals' is the product of Liberal reason. To supplement GAL's fights in the trenches of the legal architecture for the sake of animal life and well-being, I suggest then that we focus on animals as victims of structural, ontological violence. We should conceptualise animal oppression as the political act of objectification, within a framing of human–animal relations that takes seriously the notion that animals are outsiders to the Liberal polity. I propose to call that normative framing of human–animal relations 'Dominionism' and describe it as Liberalism as speculatively experienced by animals being made into objects, that is, totalitarian rule characterised by erasure of autonomous worth and representation. On that basis, I conclude that highlighting the Dominion defended by international law leads to thoughts of an international law of animal liberation, and wonder about its relationship with GAL's reformist project.

If anthropocentrism is to serve as a political or ethical referent in the development of international law for animals, added sensitivity to the specifically oppressive relation of animals to law means that the focus of critique in the development of GAL must also shift towards the violent manufacturing of animals as objects of appropriation. In that light, anthropocentrism is not an ethical problem, and its critique is not aimed at speciesism, which implies Liberal rule. Anthropocentrism is a political problem, and its critique is concerned with Dominionism, which vindicates totalitarian rule.

Globalism

Global animal law is an academic, legal, and public policy project grounded in two observations on animals and law at the beginning of the twenty-first century. On the one hand, animal welfare has become a global concern and, on the other hand, the law regarding animal welfare is globally fragmented both among national jurisdictions and levels of governance (Peters, 2016a). Whereas the history of international law is littered with animal references – from seals and antelopes to turtles, blue-fin tuna, and whales – it has not paid sufficient attention to those animals' welfare (Harrop, 2013; White, 2013; Favre, 2017). International law, the mission of which is 'to regulate the relations between … co-existing independent communities or with a view to the achievement of common aims' (*Case of the S.S. Lotus (France v. Turkey)* [1927], s. 18), nonetheless displays, in Peters' diagnosis, a kind of attention to animals that is fragmented along 'anthropocentric' lines, while in addition 'virtually all aspects of (commodified) human–animal interactions possess a transboundary dimension' (Peters, 2017a, p. 254). 'International law … pays attention to collective goods, mainly for anthropocentric reasons. In contrast, the welfare of individual animals or potential rights of some animals are not yet specifically addressed by international law' (Peters, 2017a, p. 255). A lack of coherence and direction in the legal consideration given animals across the various social processes addressed by international law, such as trade, the environment, or intellectual property, is therefore what prompts 'global animal law' into existence, both as a global legal framework in line with the 'postnational constellation' and, by way of the lack of coherence it addresses, a law for globalising or globalised issues related to human–animal relations, as well as a 'research program' (Peters, 2017a, pp. 253–254).

Global animal law is therefore many things: a scholarly object of study, a set of advocacy practices, a professional project within academia, and a legal field, to mention a few. As in 'the idea of global administrative law' (Morison *et al.*, 2011, p. 2), also referred to as a 'project' or a 'field' (Kingsbury, Krisch, and Stewart, 2005) and most often explicitly distinguished from international law as inter-governmental law (Dyzenhaus, 2005), GAL can refer to the assemblage of disaggregated parts of formally heterogeneous regulation into an object of research marked, as has been said of global administrative law, by an 'ever-looming quest for a conceptual unity' (Möllers, 2015, p. 472) or, to use Kingsbury's insightful formulation, to 'tie them together apart from unity of legal sources' (Kingsbury, 2009, p. 24). As is the case for global administrative law also, a normative drive is provided to the field or project by the appeal to framing principles that will justify a shift of attention for lawyers and policy makers away from received ways of looking at the world.

In global administrative law, concern about 'accountability' (Davis and Corder, 2009) or 'legitimacy' (Dyzenhaus, 2005) prompts renewed attention to what Kingsbury, Krisch, and Stewart have called a 'global administrative space', a space where administrative law lenses may reveal what traditional international legal structures kept concealed about the true reality and extent of regulatory power at a planetary level (Kingsbury, Krisch, and Stewart, 2005, pp. 18–19). The shift in objects, from fossilised inter-sovereign relations to a heterogeneous transnational administrative process, allows one to imagine functionally or substantively diverse regulatory institutions and rules (international and national, legislative or regulatory, centralised and decentralised) under a common disciplinary analytic grid informed by the principles of administrative law (participation, transparency, reasoned decision, proportionality, reasonableness) (Cassese, 2005, p. 692) that make coherent sense of that object. For global animal law, the principle or idea that is suggested as a similarly federating ground, this time for coherent policy-making organisation concerning the transversal question of human–animal relations, is animal welfare (Peters, 2020, pp. 2–5).

Welfarism

Building Global Animal Law on the Idea of Animal Welfare

In international law generally, legal commentary is central, as reflected in the role assigned to it by the Statute of the International Court of Justice (ICJ), and commentary has been a matter of peculiar commitment for scholars because of the peculiar history of international law as a legal field (Koskenniemi, 2001; 2017). Given that international law practice and materials are scattered, 'the clarifying role of international legal scholars is crucial', in the sense that, as Peters sums up again nicely, 'academics perform a task of verbalising and ordering, which is needed for grasping an international norm and making it operational in the first place' (Peters, 2017b, p. 144). Because global animal law is a constructive or reconstructive project that is programmatically advanced by its proponents, the role and responsibility of legal scholarship in it is particularly patent as a 'subsidiary means for the determination of the law', as the ICJ Statute puts it. In the case of a project like global administrative law or global animal law, which seeks to reorganise the received structure of legal order or process (including for purposes of normative change), the role of the scholar's agency may be more obvious, as is their personal sensibility and political responsibility (Peters, 2015). The task of scholars engaged in the project of global animal law is thus to 'erect a conceptual basis and contribute to the practical development of the field by furnishing appropriate legal arguments and concepts', in such a way that

> this research would examine the practical necessity, the moral justification, and the political prospects and strategies for consolidating and strengthening the corpus of international animal welfare norms. The research programme would thus be descriptive-analytical-conceptual, and have an impetus of legal policy.
>
> *(Peters, 2016a, p. 20)*

For global animal protection through law, the organising principle provided by scholarship is predominantly, as indicated above, that of 'animal welfare', a notion used to connect fragments of animal regulation across normative levels (Favre, 2011; 2017; Sykes, 2016; Brels, 2017). In Peters' words again:

> Although the corpus of domestic, international, and local law; of state-made and privately generated norms; of hard and soft law relating to the treatment and welfare of animals is thin, it has reached a critical mass which justifies representing it as a cross-cutting matter or even as a legal field of its own, under the overarching heading of global animal law.
>
> *(Peters, 2016a, p. 21)*

Even though the segments of (international) environmental law that concern animal life directly are organised around principles of conservation and sustainable development that are arguably in tension with animal welfare (Futhazar, 2020, p. 99), it has been suggested that we can integrate them into a GAL perspective under the larger heading of 'animal protection' (Sykes, 2016). Given the methodological rift between (individual) animal welfare and (species) conservation (Favre, 2020), the scholarly outlook in designing the notion of animal protection leads unsurprisingly to a choice in favour of 'animal welfare', operationalised as an appendix to the international trade legal architecture (Sykes, 2016, p. 78) in the vein of what we already have essentially with the World Organization for Animal Health and its *Terrestrial Code* (as discussed below). In that sense, animal welfare is substantively still the guiding notion of global animal protection as a project of legal synthesis, if not codification.

Thomas Kelch, a pioneering discussant of the idea of global animal law, provides another scholarly way of 'erecting the conceptual basis' for global animal law, whether in theory or in practice, and another way of reaching out to animals with international law. Dissatisfied with the dominance of Singer-style utilitarianism as a motivating ethics for animal regulation (Kelch, 2017, p. 306), Kelch suggests mobilising other ethical approaches, such as a Feminist ethics of care (Kelch, 2016). Here also the project is concerned with the transborder reality of the animal condition and starts particularly with the treatment of animals as objects of globalised trade. Global animal advocacy, as it reacts to a globalising animal condition most concretely manifested in trade, responds (in ways that approximate Peters' own description of GAL) to the necessity of a 'common language' and overall 'theories, strategies, principles, legislation, and other campaigns that resonate across the globe' and traverse 'cultural divides that may seem impenetrable' (Kelch, 2016, p. 83). Given the nature of the problem, Kelch proposes for global advocacy a 'foundational notion ... to ground universal principles for animal advocacy', which he calls 'caring' and which refers to 'the suite of feelings and cognitions that an emotionally sound human experiences in response to focusing attention on the suffering of others' (Kelch, 2016, p. 85). Although Kelch suggests that the project of universal animal advocacy is rooted in an 'abolitionist' perspective, and although Kelch himself has no illusions about the limits of welfare-oriented reforms in animal regulation (Kelch, 2012, p. 388), nevertheless the derivation of 'animal rights' from his sophisticated care-based methodology is predominantly informed by an overarching concern for animal suffering, that is, for their specific capacity for suffering and our concomitant duty to take care of their specific well-being:

> [I]ntensive focus on the characteristics of and ways in which a creature can suffer may also provide a mechanism for deriving a set of principles relating to animals that can be expressed in terms of a set of legal rights. In this case, the rights identified would

be dependent on the nature of the animal involved. For instance, the rights we would derive from this analysis relating to a human animal would presumably be far different from those one might derive for a mollusc.

(Kelch, 2016, p. 110)

The care ethic advocated here is premised in all cases on the object of care being endowed with sentience, so that one can build an emotional if not cognitive bridge with the experience of pain or joy that we want to alleviate or promote. If the philosophical basis is singularly sophisticated, the orientation of the approach remains in line with welfarism, in this case what leading abolitionist (and anti-welfarist) legal academic Gary Francione terms 'new welfarism'. Francione succinctly maps welfarism in these terms:

[A]lthough nonhumans suffer, and their suffering should not be discounted solely on the basis of species (although species may be relevant to our qualitative assessment of animal suffering), animals do not have the sort of self-awareness that gives them an interest in their lives. Killing animals is not the problem; animals do not care that we use them; they care only about how we use them.

(Francione and Garner, 2010, p. 70)

Welfare and Objectification

Animals as Objects with the Property of Sentience

If one understands GAL as both an intellectual/professional program and a normative project invested in bringing structure and coherence to a disjointed corpus of rules concerning animals from a unifying perspective that 'guards against the fragmentation of international law' (Peters, 2017a, p. 254), two conclusions can be drawn. First, it is appropriate to identify welfarism, in a broad sense, as the ethical sensibility that animates the project of GAL, insofar as the rise of GAL responds to the factual 'globalisation of the issue of animal welfare' (Peters, 2016a, p. 17) and the corresponding lack of 'a coherent and "thick" body of law' (Peters, 2016a, p. 15). Second, the invocation of animal welfare as the federating force of GAL is arguably a response to a particular vision of anthropocentrism as a cultural or ideological force in law. It should be noted that welfarism, seen here embodied in a legal and institutional project, is not necessarily inimical to animal rights, which (beyond known ideological disputes) can be seen pragmatically, in the realistic fashion described in the introduction, as legal tools for the objective of animal welfare, considered as an interest held by animals and protected by the legal instrument of rights (Peters, 2016b, p. 43; 2021, p. 497). In this construction of rights, which relies on parallels with the pragmatic use of human rights (Peters, 2016b; 2021, p. 510), anthropocentrism is ideally displaced by a rights (or welfare) discourse that considers the autonomous interests of (non-human) animals as the foundation of legal protection measures. Welfare speaks of underlying interests, and the law's deployment is justified by the protection of those autonomous interests, ideally not defined in anthropocentric fashion.

Welfarism is thus one way of dealing ethically with an object like the animal, and all major strands of so-called animal ethics in contemporary practice and debate share such a perspective of the animal as an object. The underlying structure of the political (or moral) argument rests on the question of whether there may be factual or other bases for considering the interests of that object as relevant (does it feel? does it have a sense of self? does it have culture? can it own property?) The assignment of relevance leads to the inclusion of those interests in the cost-benefit analysis of policy making. Coekelbergh and Gunkel (2014, p. 721) have called this a 'properties approach', which they present thusly: 'First there are entities with their inherent properties, and then moral and social relations can be established based on the factual presence or absence of a particular property'. Animal ethics is dominated by the idea of enlarging the scope of moral consideration from humans outward, based on the assignment of such a property; in that sense, the object-properties approach for animals is not significantly different from the notion that human rights are founded in one way or another in the property of 'human dignity', that is, a property that justifies the claims related to the protection of individual's interests in the political community (Luban, 2015; Waldron, 2017, p. 287).

If we reframe the project of GAL from that angle, it appears constructed as a response to an object, that is, an animal, and the properties that are the object of attention under the heading of welfare. This ethical attention leads for some to the recognition of rights or personhood or selfhood. Information on those properties and their underlying object should be provided by the natural sciences on the question of animal sentience (Kelch, 2012, p. 388; Peters, 2016a, p. 22). In other words, for the purpose of regulation and institutional action, animals are approached as objects received from the outside, presumably from the natural world, and to which the law reacts appropriately – most of the information-gathering about the object being delegated to natural sciences. In that framework, the adoption of a welfarist framing for organising global regulation means that the appropriate response is ethical, that is, guided by a sense of right and wrong, based on a scientific background and within a legal frame. This sense of right and wrong is derived from a notion of harm done to animals as objects of undue violence, that is, as recipients of a particular kind of injustice; violence only becomes an injustice if the properties of the object do not justify the violence. The argument is that the relevant properties of animals make a certain treatment tantamount to injustice (Peters, 2016a, pp. 22–23).

Anthropocentrism Beyond Speciesism

As Peters points out, the 'legal protection of animals, their status, and their potential rights ultimately depend on human attitudes towards animals', and such attitudes are in turn 'influenced by habits, religion, the wealth of a society, its state of industrialization, and other cultural factors' (Peters, 2016a, p. 22). From that perspective, '[r]esearch into global animal law must … be particularly sensitive to problems of Eurocentrism, of legal imperialism, and of a North–South divide' (Peters, 2016a, p. 22). The push for 'global animal welfare standards' responds to the pragmatic and strategic imperatives of avoiding the race to the bottom in industry practices and to the 'core idea of global justice' as we adapt it to non-humans, that is, the idea that the only 'morally relevant feature' is sentience and not place of birth or nationality (Horta, 2013, cited in Peters, 2016a, p. 22). Global protection of animals is therefore based on a proposed general human attitude towards animals deriving from a sense of 'global justice' and implemented in the proposition that irrelevant distinctions among individuals, which we generally call discrimination, are problematic even in human relations with so-called non-human animals.

Technically, the invocation of welfare and ultimately sentience in order to organise the global field of animal regulation serves the purpose of circumscribing harm as a justification for legal intervention. Protection of welfare is due because there is sentience. Whether animal welfare is enshrined directly in obligations or in subjective rights leading to the imposition of obligations concerning animal welfare, the objective, from a legal viewpoint, is a pragmatic and relative rebalancing of interests between humans and animals, which is ultimately what is implied by animal ethics (Aaltola, 2005). In this case, seeing welfare as generally related to interests based in sentience allows for a consideration of the interests of both animal species and individual animals, as well as the opposability of the often fundamental interests of animals to the sometimes trivial interests of humans (Peters, 2016b, p. 49). In other words, global animal law, whether directed towards the promotion of subjective rights or not, questions the exclusively anthropocentric orientation of law, understood specifically as the irrationally systematic preference of human interests over animal interests. The importance of rights can then be seen in the pragmatic effect of rebalancing cost–cost calculations in decisions that affect animal welfare: 'animal rights would foreclose sacrificing the animals' lives for trivial grounds because their rights could be curtailed only on good grounds' (Peters, 2021, p. 398).

Although anthropocentrism is a complicated and multi-faceted philosophical issue (Hayward, 1997; Kopnina *et al.*, 2018), whether or not it is being used methodologically correctly is not as relevant here as is the question of which understanding of anthropocentrism is invoked by global animal regulation. In environmental studies, anthropocentrism in general signals excessively narrow concerns for exclusively human needs and human goals. Biocentrism, ecocentrism, or zoocentrism serve then to suggest alternative perspectives. In international environmental law, certain legal instruments have been labelled anthropocentric because they privilege the interests, or indeed the centrality, of humanity in their regulatory outlook, in contradistinction to other legal initiatives, which approach the protection of the environment as 'an objective justified in its own terms' (Sands, 2003, p. 293). Biocentrism, as an alternative to anthropocentrism, may refer to a moral equivalence of interests, associated with the notion that the privileging of homo sapiens' value is 'irrational and arbitrary' because it is based on the genetic makeup of one species among many, taken on its own independently of the network of interdependent relations among living beings. 'We might just as well refer to any other genetic makeup as a ground of superior value. Clearly we are confronted with a wholly arbitrary claim that can only be explained as an irrational bias in our favor' (Taylor, 1992, p. 117). In short, the recurring problem that is referred to under the heading of anthropocentrism in the context of policymaking is irrationality.

As far as animals are concerned, the understanding of anthropocentrism as irrational privilege accorded to humans over animals is close to the idea of 'speciesism', if we consider speciesism as the use of species-belonging for the (illegitimate) determination of treatment or worth. Richard Ryder, who coined the term, defined speciesism as '[t]he widespread discrimination that is practised by man against the other species' and proposed further that 'speciesism and racism are both forms of prejudice that are based upon appearances' (Ryder, 2017, p. 76). Peter Singer's definition of speciesism, because of its more obvious grounding in utilitarianism, hones in on the issue of relative importance of interests as it relates to the rationality of preference given to humans: 'a prejudice or attitude of bias in favour of the interests of members of one's own species and against those of members of other species' (Singer, 2002). As such, speciesism is a 'philosophically indefensible prejudice (analogous to racial prejudice) against animals' (Callicott, 1992, p. 41). Speciesism makes anthropocentrism analogous to racism, but only if racism is (narrowly and erroneously) understood to mean a bias and irrational preference, that is, an error that could be dispelled by the use of reason. That is for instance the position of reference outlined in UNESCO's *Declaration on Race and Racial Prejudice* of 27 November 1978. Seen from a perspective that draws a parallel between racism and speciesism as instances of irrationality, the call in global animal law for a heavy reliance on science is meaningful: science will ground legislation in the objective and relative weighing and balancing of interests. In the project of edifying a global architecture for animal protection, we find that the privileging of rights as a tool for the defence of human interests is speciesist in that exact manner:

> if, for example, no justification is given for attributing moral or legal rights … only to humans, if these rights are said to flow from the intrinsic dignity that comes with being human, then such a statement would be nothing but pure speciesism.
>
> *(Peters, 2016a, p. 28)*

The harm of anthropocentrism in law is thereby connected to irrational preference, and the advancement of welfare is legitimated by the scientific facts of sentience. Again, technically speaking, the principle of animal welfare draws attention to the animal–object and its property of sentience, which distinguishes animals from insentient things and brings animals closer to humans in Jeremy Bentham's sense of 'Can they suffer?' (Bentham, 1789, p. 311n). Because animals are sentient like us, being an 'animal' (or so suggests the sexism-speciesism parallel) is no less irrational, no less 'absurd' as a justification for unequal treatment, than when we invoke being a woman to justify unequal treatment among humans (Peters, 2016a, p. 33). The import of speciesism for law is thus that 'animals are one further vulnerable group that should come into the purview of the law, after women, ethnic minorities, children, or the disabled and that the members of this group (animals) require legal protection through anti-discrimination laws' (Peters, 2016a, p. 32).

The argument about the irrationality of species-based discrimination is extremely strong within a Liberal political and legal framework that privileges the eradication of arbitrary rule as a matter of principle. I would suggest that it does not exhaust the question of the proper definition of harm done to animals, and by (pragmatic) necessity sidesteps the problem of law itself as harm. From the perspective of structural violence (as opposed to punctual injustice), the focus on irrational discrimination with regard to the object–property of sentience bypasses the fact that animals, before we even start discussing their worth and interests in the law, are naturalised social constructs and not transparent holders of autonomous interests. Implicit or explicit reference to discrimination in the case of animals (most notably through certain readings that compare speciesism to racism and sexism) (Singer, 2002, p. 35; Peters, 2016b, p. 31) can conceal (but also choose strategically to ignore) the specifically anthropocentric harm done to animals that consists precisely in their objectification as 'animals'. Debating the ethics of animal use may then conceal the political nature of the subjugation of animals to our world and our laws, within which that debate is taking place, and which seeks avowedly or not to replace nature with justice (Nussbaum, 2009, p. 390), after having constructed both justice and nature. The Liberal framing of injustice and the question of discrimination or preferential treatment erases how anthropocentrism constructs animals as layered objects in the dominion of nature: a general category of non-humans, particular species, segments of functional and cultural assignations, individual beings, elements of property, or legal subjects of rights.

Liberalism

If global animal law is rooted in international law as a project, and whether it departs from international law's formalist normative commitments or not, it is part of international governance by law as an international projection of political liberalism (Morgenthau, 1965, p. 97). Martti Koskenniemi suggests that the operation of international law as a liberal project manifests itself in the notion that problems of a political nature can be solved by means understood as apolitical (Koskenniemi, 1990, p. 7). The turn to an ethical basis for the construction of global animal law considers the animal question as a question of justice and reason, not power and force. Animals are, however, not treated in an unjust manner, in the sense of arbitrary and irrational treatment amounting to discrimination. Animals are rationally constructed as objects of appropriation.

Species Inequality

The idea of discrimination relates to the idea of human dignity, that is, an inherent quality or property of all members of the human family (Universal Declaration of Human Rights 1948, Preamble, para. 1). It implies that there exist specificities within the common (biological) generality of being human. From the perspective of fundamental rights designed to protect human dignity, reliance on such individual or group specificities to make distinctions for the purpose of governmental action may be legitimate or illegitimate depending on whether the distinction constitutes a denial of the equal dignity of all humans (e.g., *Law v Canada (Minister of Employment and Immigration)* [2001]). The injustice-as-unfairness of discrimination derives from the pre-social equality among members of the human family and the equal worth of human beings as against the rest of world. As Jeremy Waldron puts it, conceptions of human dignity 'attribute a high and distinctive status to humans, a status that is supposed to contrast with the moral considerability of non-human animals', so that '[h]uman dignity presupposes an equality of worth or standing among humans, but it adds to that an additional stronger thesis … about distinctive human worth' (Waldron, 2017, pp. 3–4). Human dignity, the property of humans, is therefore strongly associated with 'distinctive equality': we are all equally different from animals (Waldron, 2017, p. 31).

The Liberal social contract as a political institution may be seen as originating in the distinctiveness of humans (because of language or reason) and as a way of protecting humans against the brutality or uncertainties of natural life. Animals are outside of the bond, because they are incapable of entering any kind of contract for lack of language and reasoning (Hobbes, 1994, p. 85) Animals are not only outside, but they represent the outside of the socially contracted polity among humans. Social bond-breakers are typically equated with predators or 'other wild beast, or noxious brute with whom mankind can have neither society nor security' (Locke, 1947, p. 126). In a social contractarian frame, harm as discrimination is a construction that makes sense as part of a social theory of justice. When it is extended to animals, one has to take into consideration the polar opposition of sociality and animality that grounds the social theory of justice. This means that animals are constructed and naturalised as outsiders, in order to ground the categories of justice that we are now seeking to apply to them. Making animals the object of justice is a continuation of human domination over animals, because they are first cast as outsiders and then (selectively) brought back into the fold of society by the framing of their needs by science and their worth by ethics.

In order to bring animals under the consideration of justice, Liberal political and moral theory have engaged in moral extension or extensionism (Nash, 1989; Plumwood, 1993, p. 159) to reach out towards animals (Nussbaum, 2009, p. 346). Given that justice can be done to animals only within the bounds of a polity, the scope of justice is widened towards the horizon of 'a gradual replacement of nature with justice', as Martha Nussbaum has suggested (Nussbaum, 2009, p. 390) or the assignation of shades of political membership or political status to varieties of animals, based essentially on their proximity to human societies and the legibility of their way of being in relation to human civilisation (Donaldson and Kymlicka, 2013).

People advocating for the extension of moral consideration to animals do not necessarily share a strictly contractualist view as far as the details of contractualist Liberalism go. But what they have inherited is the ontological feature of that spatial extension of justice: there is simply an inside and an outside of the sphere of moral, political consideration or citizenship, and this outside is constituted by individual objects with properties. That threshold may be crossed by way of 'selfhood' (Donaldson and Kymlicka, 2013) or specific 'capabilities' (Nussbaum, 2009) but will ultimately rely on something like sentience as determined by science (Donaldson and

Kymlicka, 2013, p. 31). GAL's commitment to the standard of justice-as-rationality in fighting speciesism on the basis of the common property of sentience follows the same normative logic: 'the studies on global animal law should be informed by the findings of natural science on the sentience of animals' (Peters, 2017a, p. 255).

Despite fluctuations in Liberal rhetoric, we decide whether animals can join us at the justice or democracy table, based quite obviously on our own account of them. As Peters acknowledges, 'animals did not ask us to apply our laws to them, nor did we ask them', and they 'do not have the ability to challenge us on our own terms' (Peters, 2021, pp. 535–536). The constitutive power of our account of animals is based on the notion of an objective, factual, culture-free science that vindicates political choices. At the same time, as legislation concerned with animals will show everywhere, the definition of 'animal' is still a matter of political expediency and caprice, while science is mobilised selectively depending on the contextual need for its support. The science of deciding what an animal is, or what kind of animal is before you, is not free of culture or politics. A cursory glance at the expansive literature on the 'animal question' reveals that 'the animal' is a slippery, negative, border-setting category inseparable from the relational or differential ontology of human beings, which transpires clearly in the expression 'nonhuman animal' (Agamben, 2004; Goodale, 2015). The ways in which animals are represented in human language, for instance, have been the object of numerous studies detailing the layers of local meanings attached to animals in general (being treated 'like an animal') or distinct species (being called a weasel, a pig, a bitch), as well as the complicated mechanisms that both reaffirm and conceal their subjugation (carcasses and not corpses, hide and not skin, 'service animals' and not animal servants, culling and not exterminating, and so on) (Stibbe, 2001). Language is only one cultural space where the construction of animals occurs. Law is obviously another one. 'Animals' are social and cultural constructs associated with a particular socio-political organisation as manifest in the relation between the presence or visibility of specific animals, on the one hand, and their place in local language and culture, on the other (Leach, 1964; Halverson, 1976; Tapper, 2016, p. 54).

Animals do not come to law or ethics as objects whose interests and welfare can be easily disentangled from our own. They come already constructed by culture and science with a whole set of attachments determined by time, space, and force. At a superficial level, 'animal welfare' is a set of parameters developed by humans to assess the well-being of animals in contexts that are already contexts of oppression. The welfare of animals is the welfare of 'animals in cages', 'animals in labs', and 'animals in slaughterhouses', as the OIE's *Terrestrial Code* expresses it neatly by dedicating a significant part of its Animal Welfare section to slaughterhouses (World Organization for Animal Health, 2019). Animals that are received by law are already constructed, dependent, functional appendixes of humans. Before we can imagine that we are discriminating against animals, treating them in ways that are justified or not by the fact that they are animals, anthropocentric oppression has taken place as a matter of rational ontological violence in the form of the forceful construction of animal species and individuals in relation to us, visible most immediately in the steep gradation of concern that we have for their welfare.

Dehumanisation relies on this imperially constructed and moving 'frontier of justice', a frontier constituted by the imperial gesture of subjecting all realities, outwards from the centre of power, to our local normative framing of reality. Peter Singer changed his opinion on whether we should boil lobsters to death, based on the evolving account of lobster sentience given to him by science (Carbone, 2004, p. 57). If we were talking about humans, allowing 'worth' to determine the degree or kind of acceptable violence would make for a profoundly unpalatable moral theory. That is, however, what happens to animals when we extend a moral theory that is designed for humans and is based on the 'distinctive equality' obtaining among humans. We maintain a general category of (non-human) 'animals' that encompasses a spectrum of opera-

Global Animal Law, Pain, and Death

tional differences, which renders the idea of 'animal' itself meaningless in context. Categories will moreover shift and unworthy ones can suddenly be dumped into a sub-animal category, and the worthy ones may even be extracted from the animal category (McNeil, 2008). Martha Nussbaum, for instance, ranks the worth of shrimp compared to cows within the framework of her ethical theory, and deems the killing of shrimp a harm 'less grave': 'The infliction of pain on a sentient being is a particularly grave harm. The termination of many and varied functionings is also a grave harm. Neither of these seems to be present in the case of the shrimp' (Nussbaum, 2009, pp. 386–387). An animal ethics concerned with animal welfare (or flourishing, in this case) is more specifically concerned with the gesture of humans harming animals and its legitimacy: shrimp are merely an extension of the human behaviour of eating shrimp. The focus is again and obviously the welfare of animals as part of oppressive human assemblages, regardless of the impression of disinterested concern conveyed by scientific jargon of functionings, or capabilities, or innate behaviour (*Terrestrial Animal Health Code* 2019, Art. 7.1.1). Donaldson and Kymlicka (2013) display also a measure of conceptual blindness, despite the sophistication of their approach, to the layered political structure of animals, which are always the Others from outside, subject to the unilateral and capricious judgment of humans through the implementation of arbitrary human analogies:

> [T]he types of animals that are most cruelly abused are precisely those whose consciousness is least in doubt. We domesticate species like dogs and horses precisely because of their ability to interact with us. We experiment on species like monkeys and rats precisely because they share similar responses to deprivation, fear, or rewards.
>
> *(Donaldson and Kymlicka, 2013, p. 31)*

This form of benevolence is a plausible political position, just as imperialism, colonialism, enslavement, and genocide are plausible faces of political liberalism when the polity is facing outward towards the 'savages' (Mutua, 2001; Pitts, 2006; Mills, 2014, pp. 41–62). Extensionism suggests that 'animal rights flow from the logic of human rights' (Donaldson and Kymlicka, 2013, p. 45), while it legitimates also destruction and other forms of control of animal life, whenever extensionism fails to encompass an individual or species (cats being always a vivid example of boundary-making species, especially when a government simply declares war on them) (Lynn, 2015). The political and moral outlook that advocates more justice for animals based on irrational distinction or irrational dismissal of continuity is legitimated by the inner logic of political Liberalism, as can be seen most explicitly in Donaldson and Kymlicka (2013, pp. 13–14), just as the logic of human emancipation and human rights is by necessity the model for moral consideration of animals in global animal law (Peters, 2016b, p. 42; 2020).

Now, concern for animality is the archetypal form of Liberalism's impossibility of tolerating Otherness as other, since the animal defined as *non-human* animal is always constructed with relation to human self-image. The deployment of the categories of Western political self-conception to animals (reason, interests, moral standing) is thus not a question of anthropocentrism as favouring human interests, but an anthropomorphic construction of animal interests as pseudo-human interests premised on the very primordial notion that we have the right to name and define them. These interests are tied to the properties of the animal object as individual and then added to our own self-serving and self-constituting, interest-balancing construction of the world; the extension of moral concern is an extension of our specific rationalisation of the world into sets of objects related to our absolute domination of the universe. Part of domination consists in denying and concealing the process of constitution of the dominated other as a mere extension of the dominant self (Plumwood, 1993, p. 48). The ethical operation of moral extension relies technically on the tacit 'idealized social ontology' of modern Liberalism (Mills, 2005, p. 168), which 'will abstract away from relations of structural domination, exploitation, coercion, and oppression, which in reality, of course, will profoundly shape the ontology of those same individuals, locating them in superior and inferior positions in social hierarchies of various kinds' (Mills, 2005, p. 168). In the abstract separation between humans and animals, moreover, it is clear that, as far as anthropocentrism is concerned, the *anthropos* is itself a historically situated ideological construction that forcibly displaced alternative images of humans and their relations to other interests (Chakrabarty, 2014; 2016; Grear, 2015).

Law is an important feature for the implementation, enforcement, and entrenchment of the social imaginary through which this preliminary objectifying process is rationalised. Because of the Liberal polity's reliance on law for the legitimation of violence, in law we find not the rescuer of animals but the engine of colonial expansion, in the sense of reshaping dominated local worlds and representations (both human and non-human) into extensions of a dominant worldview. From the standpoint of the dominated, imperial justice is capricious mercy, and thus law in its world-erasing power creates harm.

Legal Violence

Saying that '[l]egal interpretation takes place in the field of pain and death' is meant, according to Robert Cover (1985, p. 1601), to point to a dual relationship of legal words to violence. On the one hand, 'legal interpretive acts signal and occasion the imposition of violence upon others'. On the other hand, they 'constitute justifications for violence which has already occurred or which is about to occur' (Cover, 1985, p. 1601). He encourages us to remember that in all cases 'a legal world is built only to the extent that there are commitments that place bodies on the line' (Cover, 1985, p. 1605). Because 'law is the projection of an imagined future upon reality' (Cover, 1985, p. 1604) then 'pain and death destroy the world that "interpretation" calls up' (Cover, 1985, p. 1602).

> [I]t is precisely th[e] embedding of an understanding of political text in institutional modes of action that distinguishes legal interpretation from the interpretation of litera-ture, from political philosophy, and from constitutional criticism. Legal interpretation is either played out on the field of pain and death or it is something less (or more) than law.
>
> *(Cover, 1985, pp. 1605–1606)*

Judicial interpretation constitutes a singular moment in the dealing of pain and death that Cover is describing. But he insists that legal interpretation is what everyone does when invoking the law as against the world, whether in legislating or protesting (Cover, 1985, p. 1618). In all cases, the puzzling question of how the legal word ultimately translates into violent deed – if only considered as the erasure of the alternative worldview enshrined in an opposing reading of the law – leads to the notion that 'legal interpretation must reflexively consider its own social organization'. The gesture of interpretation itself should take into account its situation as in a moving machinery of force and destruction.

Playing with the shape and meaning of rules is made possible by the tension between a plausible Dworkinian imperative of legal coherence, on the one hand, and 'the need to generate effective action in a violent context', on the other hand. Cover calls it 'bonded interpretation' (Cover, 1985, p. 1617). In opposition to all other varieties of interpretation, legal interpretation rests on 'social cooperation', and it is a mistake (or an ideological choice) to see legal interpre-tation 'as a mental activity of a person rather than as the violent activity of an organisation of people' (Cover, 1985, p. 1628). In Cover's unmatched description, one imagines the movement of the law as it organises the world of those on which it is imposed. As such, it is a starting point for an alternative view of animal law, one that sees international law on animals as organising a world of violence rather than repelling the violence that originates from outside of law.

The clarification, elaboration, and systematisation of standards of animal welfare, just like passive attention to their operation, are constitutive of participation in the violent enterprise described by Cover. As far as the project of GAL is concerned, personal moral responsibility is not so much the issue here (especially given the realistic outlook displayed in Anne Peter's programmatic engagement), as is the notion that the cost–benefit calculation of legal work has to include the structural effect of entrenching domination by the mere gesture of speaking the law. Considering legal work as such requires extracting oneself from the internal logic that rationalises the organisation of regulation as a matter of unbiased considerations of animal well-being and health. Legal work, such as that of elaborating the legal architecture of GAL, will be seen as part of a political theory of empire, focused not on the fair treatment of the citizens of the metropolis, but on the imposition of meaning on the imperial border, where worlds are erased and realities transformed by force. The political theory or social imaginary that sees political Liberalism from the outside, as a conquering frame that fabricates the animal Other, has to highlight an alternative continuum of human–animal relations between the 'animal' and 'human' abstractions, along the spectrum that takes us from the destruction of 'pests' to medical experiments on animal bodies to the neutering of pets to the consumption of animal products and beyond. That continuum is war (Wadiwel, 2009), within the inherited natural law tradition absorbed by Hobbes, Locke, Rousseau, Pufendorf, Grotius, and others: animals as such, and humans as such, are in a condition of war that was never interrupted by compact. Simply, through our law animals are made into subjects, objects, or whatever else we want to make them, based on our reason, our values, and our force, that is, on our dominion. Animals are extracted from 'nature' as objects of consideration and control by humans, based on idealised constructions of animality that are attached to evolving and imperially favoured social imaginaries. This is not the only way of considering animals or nature, but it is a way of considering them that is in line with political Liberalism and the ways political Liberalism has associated itself with imperialism.

Dominionism

Illiberal Domination

Seeing the law as the colonially violent displacement of worlds, including alternative worlds of human–animal relations experienced by colonised peoples, recharacterises Liberalism not as the solution for animals, but as the source of their global situation. Practices in our relations to animals, from actual slaughterhouses to medical experiments, entertainment, or the use of service animals, are given meaning in the social imaginary and associated interpretive design in which they are lived and experienced by humans. Cultural anthropology tells us that there are other ways of engaging with the world, or 'worlding', in which humans do not stand in ontological opposition to the domain of animals or spirits, and in which subjective agency and representations of other beings make for alternative dynamics of appropriation and control (Ingold, 2002; 2011, pp. 112–131; Kohn, 2007; Descola, 2015, pp. 23–65). Those alternative worlds have been nearly annihilated by imperialism, but their unlikely survival confirms the fact that the West's 'modern naturalism' described by Descola (2017, p. 119) is a relative and contingent framing of reality, imposed by force on 'nature'. From there I suggest that we consider a political framing of animal domination that starts from, first, the epistemic violence characterised by the erasure of animal autonomy (including as a result of the historical destruction of alternative ways for humans of objectivising the world) and, second, the ontological violence of animal objectification. I will call Dominionism the extension of Liberalism seen from an outside perspective – the speculated position of animals or the actually suppressed position of alternative epistemes concerning the natural world. Dominionism supplements here the notion of speciesism, which limits narrowly the nature of harm to an 'injustice' by locating it in society, as we would do with racist, ableist or transphobic discrimination. The harm at the heart of Dominionism lies however in the ambivalent expulsion and socialisation of animals as objects.

The term Dominionism invokes the image of the divine grant of authority to humans over the Earth as described in Judeo-Christian Scripture (Bible, Genesis 1:28). In terms of (political) theology and political debate, the term Dominionism is associated with currents within the Christian evangelical movement in the United States and worldwide (Maltby, 2013). It signals the theocratic project of a Christian polity, deriving in general terms its authority directly from the biblical grant or 'dominion mandate' (McVicar, 2015) and implemented through the replacement of existing civil legal architecture with Old Testament biblical law (Goldberg, 2007, p. 13). Historically, the Dominionist framing provides the patriarchal basis for the kind of absolutist monarchical claims against which John Locke builds up social-contract liberalism, as an alternative foundation for the legitimacy of political power over humans (the critique of Dominion-based arguments for monarchy being the subject of John Locke's *First Treatise*) (Locke, 1997). In more contemporary terms, the Dominionist reference lays the ground for the assertion of power over the so-called natural environment, as helpfully encapsulated in Ann Coulter's words: 'God gave us the Earth … We have dominion over the plants, the animals, the seas … God said: "Earth is yours. Take it. Rape it. It's yours"' (cited in Maltby, 2008, pp. 119–120).

The idea of Dominionism is mobilised here to reveal the 'animal question' as a political question – a question concerning the distribution and circulation of power in human–animal relations. If GAL is associated with a project of Liberal justice for animals, the critical import of the frame of Dominionism is that it foregrounds 'global animal law' as a political project in the same way that international law is a Liberal political project. The frame of Dominionism thus politicises the Liberal globalist perspective and its appeal to science from a speculated animal viewpoint (Baratay, 2010; 2012a; 2012b). Alternative ontologies described by cultural anthropology render the reality of Dominionism socially and historically contingent, that is, political, from a viewpoint where suddenly reality is revealed as being made of past decisions that can hypothetically be challenged (Derrida, 1990, p. 159).

Dominionism is a formalisation of the political framing of human–animal relations that considers 'animals' as constructs of political Liberalism. It signals that harm of an animal is not an anomaly to be corrected within Liberal politics, but rather the product of totalitarian rule from a viewpoint beyond the ethics and politics of political Liberalism. As an alternative to the Liberal description of harm and justice, the Dominionist label calls for a description of ontological harm as the heart of totalitarian rule over animals.

All Nazis

Literary imagination is what we have to compensate for the silence of animals. Here I have decided to take seriously the most notorious quotation in the animal liberation movement, drawn from Isaac Baschevis Singer's short story *The Letter Writer* (1983a). The quote is as significant for what I see as the depth of its insight into totalitarian rule, as it is for the fact that it has been blatantly misused and superficially treated within the so-called animal rights movement. In the famous passage, a character named Herman wakes up from a long illness to wonder anxiously about the whereabouts of the wild mouse with which he had coexisted in his apartment

Global Animal Law, Pain, and Death

and which he had taken the habit of feeding. Assuming the death of the mouse, whom he had baptised with the biblically significant name of Huldah, his thoughts speak 'a eulogy':

> What do they know – all these scholars, all these philosophers, all the leaders of the world – about such as you? They have convinced themselves that man, the worst transgressor of all the species, is the crown of creation. All other creatures were created merely to provide him with food, pelts, to be tormented, exterminated. In relation to them, all people are Nazis; for the animals it is an eternal Treblinka. And yet man demands compassion from heaven.
>
> *(Singer, 1983a, p. 271)*

Herman then claps his hand to his mouth and says: 'I mustn't live! I mustn't! I can no longer be a part of it! God in heaven – take me away!' (Singer, 1983b, p. 272) Intolerably, all humans are Nazis to animals, even their most solicitous caretakers, such as Herman himself.

I choose to take the formal homology proposed by Singer as seriously as he chose his words carefully, here and in other similar references to the Holocaust made elsewhere in his literary work. This passage has been invoked as part of (largely unhelpful) conversations on the comparison of pains and destructions (Patterson, 2002; Davis, 2005; Sztybel, 2006; Kim, 2011; Peters, 2016b, pp. 35–36). I personally object to the reading of the quote as positing comparability of contexts, which arrogantly presumes that we have access to the truth of the facts that we are supposedly comparing, as well as the truth of the individuals concerned. The juxtaposition that is proposed by Singer here and elsewhere is not a qualitative comparison of objects, but a formal or structural homology (Jameson, 2015, pp. 43–44) between relations: Nazi to camp inmate on the one hand, and humans to animals on the other. In both cases, the encounter between dominant and dominated is made possible by a complicated network of cultural, social, political, and economic institutions and processes that lead to it (Wachsmann, 2015). In both cases the space of encounter is an elaborate material, cultural, and especially legal construction, and both parties to the encounter are historically and socially constructed.

Jews, Roma, sexual minorities, and others in Nazi death camps are nothing of what they were or may have been in their own terms, and everything that they are made to be in the Nazi concentration and extermination structure by way of propaganda and law; the Jews of Treblinka and elsewhere in the Nazi universe were Jews in the way Nazis made them to be Jews, which correspondingly made Nazis what they were. Victims of totalitarian rule are constructed objects of violence and mercy: subjects of emotion and agency only in the terms allowed by the space of encounter with humans/Nazis. Singer says that all animals are objects of totalitarian control, in the sense that their destiny and welfare are determined, but most of all defined, by others. I take Singer's juxtaposition of animals and Nazi victims as speech about the Shoah, and in particular its unfathomability, signalled by way of reference to the inaccessible complaint, and the inaccessible world of animals. It is impossible to compare, because we precisely do not have access to what would be there to compare; what we do here is suggest that formal structures of oppression echo one another. The routine oppression of animals helps us sense the enormity of the Holocaust by the evocation of structural correspondence between relations.

Only if it is read as an aberration of the liberal polity can the totalitarian space possibly be approached as an expression of irrationality, prejudice, or discrimination. As Carl Schmitt warned, the Nazi legal order is not to be measured by the principles of the constitutional system that it replaced (Schmitt, 2001, p. 11). It is other. The so-called Nuremberg Laws announce a legal system structured by the worldview of racialism, and deployed to explicitly purge society of blood-corrupting forces (Schleunes, 2005). That is not discrimination. The standard of reason that serves to measure the legitimacy of appealing to 'race' here does not connect justification of state action to human dignity, but to the racial health of the German people. The internal logic of the machinery of persecution and death is itself impeccable in a racialist, as opposed to humanist, world. As Himmler put it, '[a]ntisemitism is exactly the same as delousing. Getting rid of lice is not a question of ideology. It is a matter of cleanliness' (Office of United States Chief of Counsel for the Prosecution of Axis Criminality, 1946). There is an internal 'reason' and logic to the Nazi persecution, from the T-4 program to the Shoah, and it is encapsulated in the so-called *Blood Law* of 1935: 'purity of German blood is the prerequisite for the continued existence of the German Volk' (cited in Whitman, 2017, p. 45). Liberalism may see anti-Semitism as obvious prejudice, bias, or error, that is, as irrationality. Being anti-lice is, however, obviously rational. Himmler is translating Nazi logic by focusing on the way it is experienced within the Nazi worldview, not as it is seen from an outside perspective: Nazis *relate* to Jews like humans in general *relate* to vermin. The engulfing enormity of the claim only appears in the structural homology.

'The Jew' of the Nazis is a fabrication, first of propaganda and racialist theory and then of laws, which designate anyone who falls within that constructed category to the blood-cleansing deployment of state power. Saying that we are all Nazis to animals means that we relate to animals in a constructed space where their life and destiny are made to mean nothing outside of our own life and destiny as 'members of the human family', as the Universal Declaration of Human Rights puts it. Theodor Adorno famously said that 'the Jew' of Fascism is like 'the animal' of Kantian idealism: 'Animals play for the idealist system virtually the same role as the Jews for fascism. To revile man as an animal – that is genuine idealism' (Adorno, 2015, p. 156). Animals, as animals, are nothing except what they are to us. Anthropocentrism is not a matter of discrimination, but a matter of ontological violence, much like the ethnocentrism that animates colonial domination on the borders of the Liberal empire (Macdonald, 2010, p. 63). Dominionism is the ideological framing of animal oppression considered as totalitarian control by human reason.

Political Slaughterhouse

Dominionism as the outer envelope of Liberalism naturally finds in the liminal social institution of the slaughterhouse the central political institution of human–animal relations (Lorite Escorihuela, 2011). Nature and animals are submitted collectively to the dominion of man by God and are then, according to contractualist folklore, transformed into private property, for instance through the human labour of transforming nature. As a political institution, the slaughterhouse (actual slaughterhouse or laboratory, or kill shelter, or factory dairy farm) is a legally constructed space where the encounter between humans and nature is forever reinvented, and the ontological subjection of animals is implemented in the form of objectification of individuals into bundles of useful parts. The slaughterhouse manufactures a specific product by the operation of brute ontological force: the animal object of appropriation sacrifices their life, health, and labour by being subjugated into a condition of exploitation as material resource.

The slaughterhouse is the space that allows the fabrication of the animal as what Cary Wolfe and Jonathan Elmer insightfully termed 'subjects to sacrifice' (Wolfe and Elmer, 1995), a space where entities are routinely appropriated into a system of meaning that reshapes them as extensions of the imperial masters. Seeing the slaughterhouse as the horizon of the Liberal space, where the boundary of nature and brute force is reproduced, is significant because it ties the sacrifice of animals to normalising the possibility of sacrificing humans:

> [T]he fundamental sacrifice of nonhuman animals (in what we eat, what we wear, the testing of the products we buy) … must continue to be legitimized if the ideological work of marking human Others as animals for the purposes of their objectification and sacrifice is to be effective.
>
> *(Wolfe and Elmer, 1995, p. 146)*

Considering Dominionism as a facet, or the flipside, of political Liberalism makes it fall in line with Isaac Bashevis Singer's totalising description of animal fate at the hands of humans. More generally, it displays the political tradition centred on freedom and emancipation of humans as structurally tied to the oppression of animals. In turn animal oppression as a congenital trait of political Liberalism provides a rationalisation of Liberal imperialism, which from Locke onward has been based on some version or another of a weaponised projection of the state of nature onto 'savages' (Henderson, 2000) and variously presented degradations on the continuum of subhumanity to animality.

Dominionism and imperialism have thus unsurprisingly intertwined conceptual, if not historical, trajectories that should be excavated. The privileged 'human perspective' of anthropomorphism is as idealised as manufactured non-human animality, and the structural oppression concealed on the side of animals (i.e., the history of subjugating animals into what they are now) is equally present on the side of humans (the *anthropos* in question is the remainder of conquest, subordination, patriarchy, enslavement, and so on) (Mills, 2005; 2014; Grear, 2015). This relational idealisation of human–animal relations conceals, most importantly, the interconnections that exist between the suppressed realities of oppression on the side of animals and more familiar facts of domination on the side of humans. The relationship between oppressive institutions for animals and processes of exploitation of humans has been explored extensively, and various ways of understanding their correlation have been suggested (Nibert, 2002; 2013; Torres, 2007; Eisnitz, 2009; Adams, 2010; Pachirat, 2011). In all cases, it is not that speciesism is like sexism or racism. Rather, animal oppression is entangled with human exploitation, so that, whether it is patriarchy or the exploitation of labour, human liberation and animal liberation converge.

The entanglement of human exploitation and animal oppression must be seen in light of the logic of the above-mentioned notion of 'distinctive equality' among humans, which leads to a radical differentiation between animal oppression and discrimination. Animal oppression is the foundation of domination within the Liberal polity, because animals are made to remain outside (and so are anti-social criminals). The reason why Liberalism lives with the possibility of treating humans as unworthy others is that Liberalism is Nazism to the fabricated animal outsiders, who make the outside of the polity and the outside of justice possible by their fabricated existence. Liberalism does not have to be fair (or unfair) to animals, but it has to be fair to humans; fairness towards 'animals' will always be incoherent and capricious mercy. (Nazism does not have to be fair (or unfair) to Jews, as opposed to Aryans.) From that perspective, considering 'global animal law' as part of a Liberal political project of global governance suggests a sharper difference between animal welfare and animal liberation. From the latter perspective, GAL as a project of legal reform poses naturally the problem of legitimation, whereby the reinforcement of legal and political structuresof domination by the addition of better and fairer standards of treatment will render liberation, or emancipation from those very structures, more difficult. Shifting the focus to structural violence, and the danger of perfecting totalitarianism to the point of inescapability, leads to the conclusion that international law has to be mostly undone, not redone. In activist terms, the abolitionist perspective elicited by the evocation of totalitarianism seeks to supplement GAL with a measure of idealism to contain the dangers of cynicism carried by GAL's realist disposition.

In programmatic terms, abolitionism demands that international law be scrutinised for its capricious construction of animals into discrete functional objects in all areas of regulation, including those that do not explicitly address animals or seem commonsensically irrelevant to animal welfare. That includes therefore a critical examination of the contribution of international law to the social construction of animals through silence and neglect, because the spaces where international law does not speak of animals, as if they were not touched by all areas of human governance, are the spaces where they have been objectified out of existence.

International Dominion Law

'[L]egal thought is constructive of social realities rather than merely reflective of them' (Geertz, 2008, p. 232). When one considers the global regulation of animals as an implementation of the global Dominion and therefore not only as a colonial project, but really the metaphorical model of all colonialisms, all imperialisms, and all master–slave domination systems, then two imperative questions of global animal regulation appear, which are not immediately apparent in current GAL literature: How are animals legally made? And where are they not made?

Naming Animal Objects

First, we should pay attention to the Dominionist naming of animals in law, as the gatekeeping ontological operation that leads to the slaughterhouse. The complexity of the Dominionist operation lies in the triadic ontology of individual animals, who are (1) individual beings (who have sentience), (2) members of species (who have a collective social or asocial function), and (3) members of the animal category (which is essentially a negative group, that of non-human animals). The complex interplay between those three ontological assignments is crucial to the normalisation of animal treatment by law.

The celebrated reformed Civil Codes that admit the special nature of animals among legal beings are an important source of inspiration for the GAL project of bringing beneficial coherence to global animal governance (Peters, 2017a, p. 252). They also reveal the complexity of animals' imposed ontology. The French Civil Code says: '*Les animaux sont des êtres vivants doués de sensibilité. Sous réserve des lois qui les protègent, les animaux sont soumis au régime des biens*' (French Civil Code, artt. 514–515). The Civil Code of Québec more ambiguously says that:

> *Les animaux ne sont pas des biens. Ils sont des êtres doués de sensibilité et ils ont des impératifs biologiques. Outre les dispositions des lois particulières qui les protègent, les dispositions du présent code et de toute autre loi relative aux biens leur sont néanmoins applicables.*
>
> (*Civil Code of Québec, article 898.1*)

The dizzying ambivalence of not being property, and yet being subject to laws that treat them as property, points to animals being doubly withdrawn from society by the power of humans. The Civil Codes point in dramatic fashion to a central issue for a project on global animal law: what is an animal? How do we circumscribe the object of that corpus of rules?

Before the animal objects of the Civil Codes are (legally) *treated as property*, and while they are simultaneously *declared not to be property*, they will have to be (factually) named as animals by law based on human criteria. So, who are they? Animals are recognised collectively as existing outside of law as 'living', 'sentient', and 'with biological imperatives', only to then be subjected to law. The collective being of 'animals' (and in the generic 'global animal law' heading) is provided by the association of animals most immediately with 'nature' (biology, the environment, etc). In

Global Animal Law, Pain, and Death

the abstract, the use of the same term of 'animals' for termites and whales relies on the distinction between us and them, society and nature, subject and object (all they have in common among themselves is that they are not us, as a reverse image of Waldron's 'distinctive equality'). More concretely, the general reference to animals is used in law for the purpose of summoning concrete animals into systems of exploitation, and the law then makes and unmakes categories according to the logic of the exploitative systems that it serves.

Fish have a probable scientific referent, albeit possibly fuzzy at the edges, for zoologists. For Canadian federal law, fish can mean 'parts of fish, shellfish, crustaceans, marine animals and any parts of shellfish, crustaceans or marine animals, and the eggs, sperm, spawn, larvae, spat and juvenile stages of fish, shellfish, crustaceans and marine animals' (Fisheries Act, R.S.C., 1985, c. F-14, s. 2(1)). The legal definition departs from zoology to focus on functional human concerns, from where the definition is teleologically fashioned. Another favourite, and more ornate example, of this process of selective interpellation of animals comes from US federal legislation, in the definition of 'animals' by the *Animal Welfare Act* itself:

> The term 'animal' means any live or dead dog, cat, monkey (nonhuman primate mammal), guinea pig, hamster, rabbit, or such other warm blooded animal, as the Secretary may determine is being used, or is intended for use, for research, testing, experimentation, or exhibition purposes, or as a pet; but such term excludes (1) birds, rats of the genus *Rattus*, and mice of the genus *Mus*, bred for use in research, (2) horses not used for research purposes, and (3) other farm animals, such as, but not limited to livestock or poultry, used or intended for use as food or fiber, or livestock or poultry used or intended for use for improving animal nutrition, breeding, management, or production efficiency, or for improving the quality of food or fiber. With respect to a dog, the term means all dogs including those used for hunting, security, or breeding purposes.
>
> *(Animal Welfare Act, 7 U.S.C. 2132(g))*

Regulation here again pretends to be receiving the facts of nature mediated by science, while it simultaneously defines animals as a matter of arbitrary normativity. Objectification, the defining of a reality as an object for a subject, is the elementary individuating effect of law when deployed on animal life. From objectification the law can then pursue the work of formal commodification. In that sense, international commercial relations are an ideal place to observe the encounter of animals and law on the international plane. For instance, the harmonised system of classification of goods for the purpose of customs (International Convention on the Harmonized Commodity Description and Coding System, of 14 June 1983) and the international classification of goods and services for the purpose of trademarks (Nice Agreement Concerning the International Classification of Goods and Services to Which Trademarks are Applied, of 15 June 1957) categorise animals and animal parts among other commodities, regardless of whether they are sentient, alive, or dead. For its part, the World Organization for Animal Health (OIE), as administrator of the standards of sanitary conditions for trade in animals and animal parts, has developed its regulatory practice on animal welfare fully within the framework of international trade, where animals are present only as tradable goods and resources. As mentioned above, the OIE's *Terrestrial Health Code* includes a chapter on animal welfare, one-third of which is about organising slaughterhouses and properly killing animals. As the Terrestrial Code says it, in a surprising episode of reflexivity: '[a]nimal welfare means the physical and mental state of an animal in relation to the conditions in which it lives and dies' (World Organization for Animal Health, 2019, art. 7.1.1.). In the context of the OIE, animals are defined by the organisation's functional outlook: all of its activity, including its concern over animal welfare, is geared explicitly towards minimising disruptions to trade. More specifically, the OIE manages the standards of animal welfare from the perspective of facilitating the implementation of the so-called 'SPS' treaty on sanitary measures having an impact on trade (World Organization for Animal Health, 2011, p. v). The OIE adopted moreover what it called a 'commodity approach' to animal health (a dimension of which is animal welfare) and defined a commodity as 'live animals, products of animal origin, animal genetic material, biological products and pathological material' (World Organization for Animal Health, 2011, p. x). In this constructed legal context of interpellation, therefore, animal health is the health of animal commodities and, beyond that, the health of the trading system. That explains that directives on the welfare of animals (or animal commodities) may include the disposition on the proper killing of animals. Because animal welfare is the welfare of a commodity in a context of exchange, whether it is alive or dead is not immediately significant.

Animal Invisibility

Acknowledgement of Dominionism as the underbelly of Liberal international law leads to a second line of inquiry for abolitionists among GAL advocates: what are the animals and species that are the subject of mercy in the overall Dominion, and where are the other animals that are indirectly affected by the operation of international law? The international trade system considers animals as 'resources' as shown in landmark cases that otherwise gave early hope of environmental consciousness at the WTO (*Import Prohibition of Certain Shrimp and Shrimp Products* [1998], § 134). And so does the Law of the Sea speak of 'living resources' (UNCLOS, 1982, Part VII, s. 2). The problem is not that the animal is being treated as a resource, but that we relate to the world, animate and inanimate, as categories of resources, within which animals fall depending on the context (like fish in the context of national economies). Because we relate to animals as objects of attention, discovery, and control, to then make them objects of consumption and trade, we do not necessarily notice the ramifications of our myriad regulated actions on 'animals' that we wish to subject to our caring disposition. A global social imaginary within international law allows animals to be encountered in some places while not in others. Reality is legally constructed in each of the legal regimes that govern those respective places, so that animals ideologically vanish as objects of plausible ethical attention.

For instance, Anne Peters rightly suggests that global animal law 'needs to take into account and must be linked to international economic law, international environmental law, international human rights law, and the law of development' (Peters, 2016a, p. 22). Those subfields can be united across the process of fragmentation by the perspective of 'animal welfare' (respectively, trade in animal goods, conservation and biodiversity, regulation of use and production of service animals, and regulation of farming in sustainable agriculture). But the perspective can still remain that of the welfare of animal-objects unrelated to the systems of exploitation that have brought them to us as such. We do not necessarily ask: Why do we need to have animals in sustainable farming and development? Why trade in animal commodities? Why should we use service animals and not employ humans? In the narrow terms of examining how law deals with animals in places where they have been made visible, it is difficult to ascertain the relation of animals and their welfare to human labour conditions, or the use of military force by states, or for that matter the history of colonialism, or the march of capitalism, which all shape the overall structure of the spaces in which animals have been made visible or invisible. Because of the way labour or war or gender relations are regulated independently of 'animals', the 'welfare' of countless animals will be affected without them being visible to regulation because they have been legally constructed out of visibility (for instance, the standards of right and wrong in

Global Animal Law, Pain, and Death

humanitarian law seem to consider most animal life as unspoken collateral damage, and the law of the sea takes most of animal life as a more or less valuable accessory of our common or shared human property of the oceans).

International law constructs the facts of Australia's lawsuit against Japan over whaling as a question of treaty law and interpretation, as opposed to a case of triangular *jus ad bellum* involving Japan, Australia, and cetaceans (*Whaling in the Antarctic (Australia v Japan, New Zealand intervening)* [2014]). Hunting whales is illegal because in context it may exceed a quota or violate a moratorium; it is not illegal as an act of war against a foreign species, subject to the rules of collective self-defence that may be invoked by Australia. International law does not consider the question of beef production and trade as an issue of human dignity attached to the inhumane and degrading treatment constituted by the relationship enforced between animals and humans in slaughterhouses (*EC Measures Concerning Meat and Meat Products (Hormones)* [1997]); it is a 'process and production' question relating to the improper processing and producing of cow into tradable commodities across international lines. In this light, international legal regimes display, with regard to animals, maybe the most transparent example of what Martti Koskenniemi has called structural bias, which orients rules and legal speech according to deep-seated ways of constructing the relevant part of the world (Koskenniemi, 2009, p. 9). Across regimes, those biases coalesce as expressions of Dominionist framing. The question left to answer is what to do, if not reform.

Just like Philip Alston once suggested that having the WTO interpret human rights standards was the worst possible idea for human rights (Alston, 2002), one could imagine that international law may be the worst place to discuss animal welfare, as it simultaneously organises the global exploitation of animals. The critique of global animal law, in the most elementary sense of delimiting its extent and purpose by highlighting its political, power-based, and world-forming nature, leads in two contradictory directions. One direction concerns humans, and aims at the dismantling of the relationship at the heart of the Dominion because that relationship is the product of systems of capitalist exploitation and extraction that in all cases affect humans (Nibert, 2002). That direction leads to focusing away from animal objects and into the contents of international law that sustain the production of the Dominion on a global scale. The other direction concerns the animals themselves and the fact that 'global animal law' makes the complicated political choice of possibly participating in their construction as objects of control and mercy, because the alternative may be, from a Liberal justice perspective, much worse. The path of normative proliferation poses the additional ethical complication of adding an expansion of discourse about animals, the frenzy of animal studies, materialised into fodder of various types for the machinery of academic labour exploitation. The path of critique leads therefore in my irrelevant opinion to silence, and a call towards the unabandoned ideal of leaving animals in peace and far away from law.

Conclusion

The response to the suggestion of a critical pause in global animal law may take the shape of 'the perfect is the enemy of the good' and that we should not stop the obvious progress that is called for in the treatment of animals across the planet. Animal welfare is certainly not a superfluous or misguided pursuit, and GAL is an extraordinarily ambitious and promising project, with undoubtedly concrete consequences for actual animals. Ultimately, the increase in welfare standards and the expansion of protection for animals, which implies also suspending or eliminating harmful practices altogether, are worthwhile and desirable as such. The insistence here on the fact that global animal law is part of larger political processes that shape the world and instrumentalise animal life, serves the critical purpose of assigning possible limits to such a project; it keeps alive the possibility that one way of remaining in the legal field, as Anne Peters rightly suggests that we do, is to keep alive the ideal of an international law of animal liberation, much like we had an international law of decolonisation.

Reform and Abolition

The most important point of invoking Dominionism as the dark side, or animal-side, of Liberalism is that moral purity should be discarded as an objective. In practical terms, the invocation of Dominionism leads to a practical path of resistance involving considerable suffering and death for animals also (just like GAL acknowledges the implications of forever imperfect legal reform) if we envisage the clear conclusion that abolition of all forms of exploitation will ultimately impose the ideal of complete separation, at least until we can develop relations with animals on another civilisational premise than subject–object control. In intellectual terms, and responding here to the call upon the responsibility of scholars for the edification of a more humane international governance structure for animals, the invocation of Dominion serves to echo Hermann, from Isaac B. Singer's story: what do they know of you, the scholars?

Dominionism, understood as a political program of ontological anthropocentrism, which fashions imperially the world into a violent subject–object relationship, is what we can imagine to be animal regulation from the perspective of (anthropomorphised) animals. In other words, we are all Dominionists, in the sense of being all Nazis to animals. Here dispelling the grotesque misunderstanding of Singer as someone who is diminishing the significance of the Holocaust allows for a clarification of what it means to respond to the Dominionist condition of animals. If the life of mice is an eternal Treblinka, the answer is not more kindness or mercy from Nazis. The answer is liberating the inmates, shutting down the camp, and defeating Nazism. Seeing violence as a string of violent acts on individual persons is certainly possible, but the real problem remains that there is a structure or system that allow for acts of violence to be numerous, widespread, organised, meaningful, and normal. The operation of the camp is an ontological operation that aims at the creation of the contemporary *homo sacer* and the *Muselmann* evoked by Agamben (2004); it does not exist in a void but stands rather in a wide-ranging system of relations and institutions in all spheres of life that have normalised over time, culturally, socially, economically, legally, and otherwise, the possibility of widespread destruction of those abject entities fabricated by the camps.

Insisting on the construction of animal–objects, as opposed to the condition of individual animals caught in the webs of exploitation, highlights that the real focus of global animal law, like any other law, is the relationships among humans. The hopelessly ambiguous construction of animals in contemporary civil codes as living beings endowed with sentience that can be subject to the law of property, has dubious effects on animals themselves in practice, but definite theoretical importance in entrenching the legitimacy of the legal structure of the enslaved person within the Liberal legal system. The abstract animal, in parallel to the enslaved person, is a hybrid being thrown by law astride of the status of person and the properties of a thing (Patterson, 1985, pp. 22–23). Global animal law in that sense must be seen as rules about humans and about clarifying and organising the relations among human beings via animals; and the notion of animals as appropriable depending on contexts keeps open the legal possibility of depersoning, that is, de-humanising. In that vein, GAL's embrace of an unstable or ambivalent definition of animals adds legitimation to the fact that within the Liberal legal system there are entities that are due consideration and others that are not. Further, by considering that the object of enquiry for animal welfare is the animal-object in the cage, rather than the system of value extraction and exploitation that produces scientific and medical research for the benefit of the few and the misery of many, GAL may contribute in its own way to the entrenchment of a global system of exploitation of humans by humans. In all of this, the call to abolition in opposition to proliferation is concerned with the added legitimacy of coherence and system delivered by GAL to the Dominion.

The function of evoking Dominionism and the legitimacy of abolition is to keep GAL aware, as a possible instrument of Dominion, of the risk of being ultimately a fully coherent and overarching articulation of modern naturalism and a formalisation of our relationship to nature as a pool of resources (which as such does not exclude being kind or respectful to nature and animals). As a disciplinary project in the socio-political context of the Dominion, GAL can be seen (from the other end of imperial rule) as a human self-serving call to the proliferation and flourishing of more science, more data, more law, more debate, and more exponential growth of spoken and written word on the backs of animals. As a response to the call upon scholars to help erect global animal law, resistance to Dominionism is a call back, a call not for further legal entanglement, but a call to the ideal of provisional separation, de-domestication, and retreat.

Steps Towards International Law of Animal Liberation

Responding from an abolitionist perspective to the stated necessity of taking care of animals through (international) law means implementing a program aiming at the silence of the law on animals other than humans, as well as a retreat from the frontier of justice to leave animals to their own existences until humans can define themselves, and relate to one another, independently of a Dominionist construction of animals. In oversimple terms, which will need layers of operational detail, resistance to Dominionism means abolition, that is, the abolition of the current relation of domination and its replacement with something else. Even though we have traces of that 'something else', the change that is envisaged here concerns processes that have been running for tens of thousands of years. The change will have to be slow. Concomitantly to the critique of a 'global animal law' that does not cover all (international) law, because all (international) law affects animals, the idealised silence about animals means (1) stating the (at least provisional) illegitimacy of human rules on animals and corresponding invalidity of all such rules; (2) aspiring to the transversal undoing of the Dominionist aspects or ramifications of all regulations on non-human animals (that is, implementing the notion that controlling rule over animal destinies is a priori illegitimate); and (3) devising transitional rules to enforce separationism based on the three principles of (a) undoing the harm and minimally protecting other earthlings against the enduring effects of Dominionist rule, (b) thinking of such rules as rules of engagement and not rules of justice, and (c) framing (provisional separation) within a concern for the survival of the planet under conditions of climate catastrophe.

This silence on the direct treatment of animals should be accompanied by legal work that does not refine the legal treatment of animals but undoes the existence of such treatment in the first place. The OIE, as an organisation that deals in the health of animals as traded commodities, would be severely crippled by the abolitionist drive to revamp international law, not around the principle of animal welfare, but around the objective of liberation and the principle of 'do no harm'. Because animal welfare is the welfare of beings caught in unquestioned networks of oppression that involve and entangle humans with them, global animal law reformism, which starts with a descriptive account of all legal norms regarding explicit animal use and treatment, can be supplemented by the parallel account of 'international dominion law': an account of all of international law from the perspective of even indirect harm done to animals by the operation of all of its rules, through destruction of habitat or resources, threats to life and actual killing, and so forth. This follows from the notion that explicit rules on animals (for instance on animal welfare) are surrounded in the legal system by countless silences that amount, from the perspective of animal life, to deliberate ignorance of human impact. That representation of international law will be like the negative of international law; that is, international law seen from the speculative viewpoint of animals or of harm done to them, where international labour law or the law of the sea or the law of state responsibility can be reconstructed from combined perspectives on individual animal welfare, as well as systemic species and ecosystem conservation. As such international dominion law can be the centrepiece of a critical rereading of international law, purged of the Dominionist fantasy about subjects of power and objects of control, and towards a global framing of the process of emancipation of beings, animal, and other, as interdependent nodes of relations.

References

Aaltola, E. (2005) 'Animal Ethi'cs and Interest Conflicts', *Ethics & the Environment*, 10 (1), pp. 19–48. DOI: 10.1353/een.2005.0011

Aaltola, E. (2010) 'The Anthropocentric Paradigm and the Possibility of Animal Ethics', *Ethics & the Environment*, 15 (1), pp. 27–50. DOI: 10.2979/ete.2010.15.1.27

Adams, C.J. (2010) *The Sexual Politics of Meat (20th Anniversary Edition): A Feminist-Vegetarian Critical Theory.* New York: Bloomsbury Publishing.

Adorno, T.W. (2015) *Beethoven: The Philosophy of Music.* Cambridge: Polity Press.

Agamben, G. (2004) *The Open: Man and Animal.* Stanford: Stanford University Press.

Alston, P. (2002) 'Resisting the Merger and Acquisition of Human Rights by Trade Law: A Reply to Petersmann', *European Journal of International Law*, 13 (4), pp. 815–844. DOI: 10.1093/ejil/13.4.815

Baratay, É. (2010) 'Les socio-anthropo-logues et les animaux', *Societes*, 108 (2), pp. 9–18. DOI: 10.3917/soc.108.0009

Baratay, É. (2012a) 'La promotion de l'animal sensible. Une révolution dans la Révolution', *Revue historique*, 661 (1), pp. 131–153. DOI: 10.3917/rhis.121.0131

Baratay, É. (2012b) *Le Point de vue animal. Une autre version de l'histoire: Une autre version de l'histoire.* Paris: Le Seuil.

Bentham, J. (1789) *An Introduction to the Principles of Morals and Legislation.* Oxford: Clarendon Press.

Blattner, C. (2016) 'An Assessment of Recent Trade Law Developments from an Animal Law Perspective: Trade Law as the Sheep in Wolf's Clothing?', *Animal Law Review*, 22 (2), pp. 277–310.

Brels, S. (2017) 'A Global Approach to Animal Protection', *Journal of International Wildlife Law & Policy*, 20 (1), pp. 105–123. DOI: 10.1080/13880292.2017.1309866

Callicott, J.B. (1992) 'Moral Considerability and Extraterrestrial Life', in Hargrove, E.C. (ed.) *The Animal Rights/Environmental Ethics Debate: The Environmental Perspective.* Albany: State University of New York Press, pp. 137–150.

Carbone, L. (2004) *What Animals Want: Expertise and Advocacy in Laboratory Animal Welfare Policy.* Oxford: Oxford University Press.

The Case of the S.S. Lotus (France v. Turkey) (Judgment) [1927] PCIJ Series A No 10.

Cassese, S. (2005) 'Administrative Law Without the State? The Challenge of Global Regulation', *New York University Journal of International Law and Politics*, 37 (4), 663–694.

Chakrabarty, D. (2014) 'Climate and Capital: On Conjoined Histories', *Critical Inquiry*, 41 (1), pp. 1–23. DOI: 10.1086/678154

Chakrabarty, D. (2016) 'Whose Anthropocene? A Response', *RCC Perspectives*, (2), pp. 101–114. DOI: 10.5282/rcc/7452

Coeckelbergh, M. and Gunkel, D. (2014) 'Facing Animals: A Relational, Other-Oriented Approach to Moral Standing', *Journal of Agricultural & Environmental Ethics*, 27 (5), pp. 715–733. DOI: 10.1007/s10806-013-9486-3

Cover, R.M. (1985) 'Violence and the Word', *Yale Law Journal*, 95 (8), pp. 1601–1630. DOI: 10.2307/796468

Davis, D. and Corder, H. (2009) 'Globalisation, National Democratic Institutions and the Impact of Global Regulatory Governance on Developing Countries: Definitional Issues in Global Administrative Law : Part I', *Acta Juridica*, 2009 (1), pp. 68–89.

Davis, K. (2005) *The Holocaust and the Henmaid's Tale: A Case for Comparing Atrocities*. New York: Lantern Books.

Deckha, M. (2013) 'Initiating a Non-Anthropocentric Jurisprudence: The Rule of Law and Animal Vulnerability Under a Property Paradigm', *Alberta Law Review*, 50 (4), 783–814. DOI: 10.29173/alr76

Derrida, J. (1990) *Du droit à la philosophie*. Paris: Galilée.

Descola, P. (2015) *Par-delà nature et culture*. Paris: Folio.

Descola, P. (2017) 'Les animaux et l'histoire, par-delà nature et culture', *Revue d'histoire du XIXe siècle. Société d'histoire de la révolution de 1848 et des révolutions du XIXe siècle*, 54, pp. 113–131. DOI: 10.4000/rh19.5191

Donaldson, S. and Kymlicka, W. (2013) *Zoopolis: A Political Theory of Animal Rights*. Oxford: Oxford University Press.

Dyzenhaus, D. (2005) 'The Rule of (Administrative) Law in International Law', *Law and Contemporary Problems*, 68 (3/4), pp. 127–166.

Eisnitz, G.A. (2009) *Slaughterhouse: The Shocking Story of Greed, Neglect, And Inhumane Treatment Inside the U.S. Meat Industry*. Amherst: Prometheus Books.

Favre, B. (2020) 'Is There a Need for a New, an Ecological, Understanding of Legal Animal Rights?', *Journal of Human Rights and the Environment*, 11 (2), pp. 297–319. DOI: 10.4337/jhre.2020.02.07

Favre, D. (2011) 'An International Treaty for Animal Welfare Symposium Article', *Animal Law*, 18 (2), pp. 237–280.

Favre, D. (2017) 'When Will Concern for Animal Welfare Become Part of International Law?', *Animal Law*, 2017 (4) pp. 1+.

Francione, G.L. and Garner, R. (2010) *The Animal Rights Debate: Abolition or Regulation?* New York: Columbia University Press.

Futhazar, G. (2020) 'Biodiversity, Species Protection, and Animal Welfare Under International Law', in Peters, A. (ed.) *Studies in Global Animal Law*. Berlin: Springer, pp. 95–108.

Garner, R. (2017) 'Animals and Democratic Theory: Beyond an Anthropocentric Account', *Contemporary Political Theory*, 16 (4), pp. 459–477. DOI: 10.1057/s41296-016-0072-0

Geertz, C. (2008) *Local Knowledge: Further Essays in Interpretive Anthropology*. New York: Basic Books.

Goldberg, M. (2007) *Kingdom Coming: The Rise of Christian Nationalism*. New York: W.W. Norton.

Goodale, G. (2015) *The Rhetorical Invention of Man: A History of Distinguishing Humans from Other Animals*. Lanham: Lexington Books.

Grear, A. (2015) 'Deconstructing Anthropos: A Critical Legal Reflection on 'Anthropocentric' Law and Anthropocene 'Humanity'', *Law and Critique*, 26 (3), 225–249. DOI: 10.1007/s10978-015-9161-0

Halverson, J. (1976) 'Animal Categories and Terms of Abuse', *Man*, 11 (4), pp. 505–516. DOI: 10.2307/2800435

Harrop, S.R. (2013) 'Wild Animal Welfare in International Law: The Present Position and the Scope for Development', *Global Policy*, 4 (4), pp. 381–390.

Hayward, T. (1997) 'Anthropocentrism: A Misunderstood Problem', *Environmental Values*, 6 (1), pp. 49–63. DOI: 10.3197/096327197776679185.

Henderson, J.S. (2000) 'The Context of the State of Nature', in Battiste, M. (ed.) *Reclaiming Indigenous Voice and Vision*. Vancouver: UBC Press, pp. 11–31.

Hobbes, T. (1994) *Leviathan: With Selected Variants from the Latin Edition of 1668*. Indianapolis: Hackett Publishing.

Horta, O. (2013) 'Expanding Global Justice: The Case for the International Protection of Animals', *Global Policy*, 4 (4), pp. 371–380. DOI: 10.1111/1758-5899.12085

Ingold, T. (2002) 'From Trust to Domination: An Alternative History of Human-Animal Relations', in Manning, A. and Serpell, J. (eds.) *Animals and Human Society: Changing Perspectives*. London: Routledge, pp. 61–76.

Ingold, T. (2011) *The Perception of the Environment: Essays on Livelihood, Dwelling and Skill*. London: Routledge.

Jameson, F. (2015) *The Political Unconscious: Narrative as a Socially Symbolic Act*. Ithaca: Cornell University Press.

Kelch, T.G. (2012) 'A Short History of (Mostly) Western Animal Law: Part II', *Animal Law*, 19 (2), pp. 347–390.

Kelch, T.G. (2016) 'Towards Universal Principles for Global Animal Advocacy', *Transnational Environmental Law*, 5 (1), pp. 81–111. DOI: 10.1017/S2047102515000308

Kelch, T.G. (2017) *Globalization and Animal Law: Comparative Law, International Law and International Trade*. Alphen aan den Rijn: Wolters Kluwer.

Kennedy, T. (2016) 'The Ethics of Treating Animals as Resources: A Post-Heideggerian Approach', *Frontiers of Philosophy in China*, 11 (3), pp. 463–482. DOI: 10.3868/s030-005-016-0033-9

Kim, C.J. (2011) 'Moral Extensionism or Racist Exploitation? The Use of Holocaust and Slavery Analogies in the Animal Liberation Movement', *New Political Science*, 33 (3), pp. 311–333. DOI: 10.1080/07393148.2011.592021

Kingsbury, B. (2009) 'The Concept of "'Law'" in Global Administrative Law', *European Journal of International Law*, 20 (1), pp. 23–57. DOI: 10.1093/ejil/chp005

Kingsbury, B., Krisch, N. and Stewart, R.B. (2005) 'The Emergence of Global Administrative Law', *Law and Contemporary Problems*, 68 (3/4), pp. 15–61.

Kohn, E. (2007) 'How Dogs Dream: Amazonian Natures and the Politics of Transspecies Engagement', *American Ethnologist*, 34 (1), pp. 3–24. DOI: 10.1525/ae.2007.34.1.3

Kopnina, H., Washington, H., Taylor, B. and J Piccolo, J. (2018) 'Anthropocentrism: More than Just a Misunderstood Problem', *Journal of Agricultural and Environmental Ethics*, 31 (1), pp. 109–127. DOI: 10.1007/s10806-018-9711-1

Koskenniemi, M. (1990) 'The Politics of International Law', *European Journal of International Law*, 1 (1), pp. 4–32. DOI: 10.1093/oxfordjournals.ejil.a035781

Koskenniemi, M. (2001) *The Gentle Civilizer of Nations: The Rise and Fall of International Law 1870–1960*. Cambridge: Cambridge University Press.

Koskenniemi, M. (2009) 'The Politics of International Law – 20 Years Later', *European Journal of International Law*, 20 (1), pp. 7–19. DOI: 10.1093/ejil/chp076

Koskenniemi, M. (2017) 'Between Commitment and Cynicism: Outline for a Theory of International Law as Practice', in d'Aspremont, J., Gazzini, T., Nollkaemper, A. and Werner, W. (eds.) *International Law as a Profession*. Cambridge: Cambridge University Press, pp. 38–66.

Law v. Canada (Minister of Employment and Immigration) [2001] 1 SCR 497.

Leach, E. (1964) 'Anthropological Aspects of Language: Animal Categories and Verbal Abuse' in Lenneberg, E.H. (ed.) *New Directions in the Study of Language*. Cambridge: MIT Press, pp. 23–63.

Locke, J. (1947) *Two Treatises of Government: With a Supplement, Patriarcha, by Robert Filmer*. New York: Hafner.

Locke, J. (1997) 'First Tract on Government' in Goldie, M. (ed.) *Locke: Political Essays*. Cambridge: Cambridge University Press.

Lorite Escorihuela, A. (2011) 'A Global Slaughterhouse', *Helsinki Review of Global Governance*, 2 (2), pp. 25–30.

Luban, D. (2015) 'Human Dignity and Human Rights Pragmatism' in Cruft, R., Liao, S.M. and Renzo, M. (eds.) *Philosophical Foundations of Human Rights*. Oxford: Oxford University Press, pp. 263–272.

Lynn, W.S. (2015) 'Australia's War on Feral Cats: Shaky Science, Missing Ethics', *The Conversation*, 7 October [Online]. Available at: http://theconversation.com/australias-war-on-feral-cats-shaky-science-missing-ethics-47444 (Accessed: 15 Aug 2021).

Macdonald, G. (2010) 'Colonizing Processes, the Reach of the State and Ontological Violence: Historicizing Aboriginal Australian Experience', *Anthropologica*, 52 (1), pp. 49–66.

Maltby, P. (2008) 'Fundamentalist Dominion, Postmodern Ecology', *Ethics & the Environment*, 13 (2), pp. 119–141.

Maltby, P. (2013) *Christian Fundamentalism and the Culture of Disenchantment*. Charlottesville: University of Virginia Press.

McNeil, D.G. (2008) 'When Human Rights Extend to Nonhumans', *The New York Times*, 13 July [Online]. Available at: https://www.nytimes.com/2008/07/13/weekinreview/13mcneil.html (Accessed 5 October 2022).

McVicar, M.J. (2015) *Christian Reconstruction: R. J. Rushdoony and American Religious Conservatism*. Chapel Hill: University of North Carolina Press.

Mills, C.W. (2005) '"Ideal Theory" as Ideology'. *Hypatia*, 20 (3), pp. 165–184. DOI: 10.1111/j.1527-2001.2005.tb00493.x

Mills, C.W. (2014) *The Racial Contract*. Ithaca: Cornell University Press.

Möllers, C. (2015) 'Ten Years of Global Administrative Law', *International Journal of Constitutional Law*, 13 (2), pp. 469–472.

Morgenthau, H.J. (1965) *Scientific Man Versus Power Politics*. Chicago: University of Chicago Press.

Morison, J., Zwart, T., Anthony, G. and Auby, J.-B. (2011) 'Values in Global Administrative Law: Introduction to the Collection' in Auby, J.-B., Anthony, G., Zwart, T. and Morison, J. (eds.) *Values in Global Administrative Law*. Oxford: Hart Publishing, pp. 1–16.

Mutua, M. (2001) 'Savages, Victims, and Saviors: The Metaphor of Human Rights' *Harvard International Law Journal*, 42 (1), pp. 201–246.

Nash, R.F. (1989) *The Rights of Nature: A History of Environmental Ethics*. Madison: University of Wisconsin Press.

Nibert, D. (2002) *Animal Rights/Human Rights: Entanglements of Oppression and Liberation*. Lanham: Rowman & Littlefield Publishers.

Nibert, D. (2013) *Animal Oppression and Human Violence: Domesecration, Capitalism, and Global Conflict*. New York: Columbia University Press.

Nibert, D. (ed.) (2017) *Animal Oppression and Capitalism*. Santa Barbara: Praeger.

Nussbaum, M.C. (2009) *Frontiers of Justice: Disability, Nationality, Species Membership*. Cambridge: Harvard University Press.

Office of United States Chief of Counsel for the Prosecution of Axis Criminality (1946) *Nazi Conspiracy and Aggression*.

O'Sullivan, S., Otter, C. and Ross, S. (2012) 'Laying the Foundations for an International Animal Protection Regime', *Journal of Animal Ethics*, 2 (1), pp. 53–72. DOI: 10.5406/janimalethics.2.1.0053

Pachirat, T. (2011) *Every Twelve Seconds: Industrialized Slaughter and the Politics of Sight*. New Haven: Yale University Press.

Patterson, C. (2002) *Eternal Treblinka*. New York: Lantern Books.

Patterson, O. (1985) *Slavery and Social Death: A Comparative Study*. Cambridge: Harvard University Press.

Peters, A. (2015) 'Animal Law – A Paradigm Change', in Peters, A., Stucki, S. and Boscardin, L. (eds.) *Animal Law: Reform or Revolution?* Zürich: Schulthess Juristische Medien AG, pp. 15–31.

Peters, A. (2016a) 'Global Animal Law: What it Is and Why We Need it', *Transnational Environmental Law*, 5 (1), pp. 9–23. DOI: 10.1017/S2047102516000066

Peters, A. (2016b) 'Liberté, Égalité, Animalité: Human–Animal Comparisons in Law', *Transnational Environmental Law*, 5 (1), pp. 25–53. DOI: 10.1017/S204710251500031X

Peters, A. (2017a) 'Introduction to Symposium on Global Animal Law (Part I): Animals Matter in International Law and International Law Matters for Animals', *AJIL Unbound*, 111, pp. 252–256. DOI: 10.1017/aju.2017.70

Peters, A. (2017b) 'International Legal Scholarship Under Challenge', in d'Aspremont, J., Gazzini, T., Nollkaemper, A. and Werner, W. (eds.) *International Law as a Profession*. Cambridge: Cambridge University Press, pp. 117–159.

Peters, A. (2020) 'Introduction', A. Peters (ed.) *Studies in Global Animal Law*. Berlin: Springer Open, pp. 1–13.

Peters, A. (2021) *Animals in International Law*. Leiden: Brill Nijhoff.

Pitts, J. (2006) *A Turn to Empire: The Rise of Imperial Liberalism in Britain and France*. 1 edition. Princeton: Princeton University Press.

Plumwood, V. (1993) *Feminism and the Mastery of Nature*. London: Routledge.

Ryder, R.D. (2017) *Speciesism, Painism and Happiness: A Morality for the Twenty-First Century*. Exeter: Andrews UK Limited.

Sands, P. (2003) *Principles of International Environmental Law*. Cambridge: Cambridge University Press.

Schleunes, K. (2005) 'Nuremberg Laws 1935', in Levy, R.S., Bell, D.P., Donahue, W.C., Madigan, K., Morse, J., Shevitz, A.H. and Stillman, N.A. (eds.) *Antisemitism: A Historical Encyclopedia of Prejudice and Persecution*. Santa Barbara: ABC-CLIO, pp. 515–516.

Schmitt, C. (2001) *State, Movement, People: The Triadic Structure of Political Unity [1933]*. Corvallis: Plutarch Press.

Shapiro, K.J. (1990) 'The Human Science Study of Nonhuman Animals', *Phenomenology + Pedagogy*, 8, pp. 27–42. DOI: 10.29173/pandp15121

Singer, I.B. (1983a) 'The Letter Writer', in *The Collected Stories of Isaac Bashevis Singer*. New York: Farrar, Straus and Giroux, pp. 250–276.

Singer, I.B. (1983b) *The Collected Stories of Isaac Bashevis Singer*. New York: Farrar, Straus and Giroux.

Singer, P. (2002) *Animal Liberation*. New York: Ecco.

Stibbe, A. (2001) 'Language, Power and the Social Construction of Animals', *Society & Animals*, 9 (2), pp. 145–161. DOI: 10.1163/156853001753639251

Sykes, K. (2016) 'Globalization and the Animal Turn: How International Trade Law Contributes to Global Norms of Animal Protection', *Transnational Environmental Law*, 5 (1), pp. 55–79. DOI: 10.1017/S2047102516000054

Sztybel, D. (2006) 'Can the Treatment of Animals Be Compared to the Holocaust?', *Ethics and the Environment*, 11 (1), pp. 97–132.

Tapper, R. (2016) 'Animality, Humanity, Morality, Society', in Ingold, T. (ed.) *What is an Animal?* London: Routledge, pp. 47–62.

Taylor, P.W. (1992) 'The Ethics of Respect for Nature', in Hargrove, E.C. (ed.) *The Animal Rights/ Environmental Ethics Debate: The Environmental Perspective*. Albany: State University of New York Press, pp. 95–120.

Torres, B. (2007) *Making A Killing: The Political Economy of Animal Rights*. Oakland: AK Press.

United Nations Convention on the Law of the Sea (UNCLOS) (adopted 10 December 1982, entered into force 1 November 1994) 1833 UNTS 397.

Wachsmann, N. (2015) *KL: A History of the Nazi Concentration Camps*. New York: Farrar, Straus and Giroux.

Wadiwel, D. (2015) *The War Against Animals*. Leiden: Brill.

Wadiwel, D.J. (2009) 'The War Against Animals: Domination, Law and Sovereignty', *Griffith Law Review*, 18 (2), pp. 283–297. DOI: 10.1080/10383441.2009.10854642

Waldron, J. (2017) *One Another's Equals: The Basis of Human Equality*. Cambridge: Belknap Press.

Whaling in the Antarctic (Australia v Japan, New Zealand intervening) (Judgment) [2014] ICJ Rep 226.

White, S. (2013) 'Into the Void: International Law and the Protection of Animal Welfare', *Global Policy*, 4 (4), pp. 391–398. DOI: 10.1111/1758-5899.12076

Whitman, J.Q. (2017) *Hitler's American Model: The United States and the Making of Nazi Race Law*. Princeton: Princeton University Press.

Wolfe, C. and Elmer, J. (1995) 'Subject to Sacrifice: Ideology, Psychoanalysis, and the Discourse of Species in Jonathan Demme's Silence of the Lambs', *Boundary 2*, 22 (3), pp. 141–170. DOI: 10.2307/303726

World Organization for Animal Health (2011) *Terrestrial Animal Health Code*. 20th ed. Paris: OIE.

World Organization for Animal Health (2019) *Terrestrial Animal Health Code*. 28th ed. Paris: OIE.

WTO Appellate Body Report, United States – Import Prohibition of Certain Shrimp and Shrimp Products, (12 October 1998) WT/DS58/AB/R

Youatt, R. (2014) 'Interspecies Relations, International Relations: Rethinking Anthropocentric Politics', *Millennium*, 43 (1), pp. 207–223. DOI: 10.1177/0305829814536946

SECTION 3

Imagining a Non-Anthropocentric International Law

13

WHAT WOULD A POST-ANTHROPOCENTRIC LEGAL SYSTEM LOOK LIKE?

Ugo Mattei and Michael W. Monterossi

Introduction[1]

International law, in the perspective of the dramatic challenges that the 'Anthropocene' (Purdy, 2015) is launching on the reproduction of natural life on planet Earth, is a Janus-like institution. It can be seen, at the very same time, as part of the problem or part of the solution. It is part of the problem because, since its very foundation in the Grotius brief in favour of the East India Company which claimed the sea as a global common up for corporate grabs (Mattei and Quarta, 2018), it has worked as one of the most powerful engines of global diffusion of mechanistic ideas of extraction, aimed at capital accumulation at the expense of Nature, of the people 'without history' (Wolf, 2010), and of the commons. It is part of the solution because it has attempted, at least in theory, to tame the worst excesses of sovereign discretion and national selfishness and imperialism, through the development of some principles to be respected as international customary law or as *jus cogens*, even by the strongest political actors.

In practice, as we know, international law has not delivered even the most moderate of its promises. Since the beginning of the present millennium, a western-centric and imperialist vision of 'international human rights' has legitimised endless racist deployments of the might of the global North at the expense of the global South (see Mattei and Nader, 2008). International law under this now dominant vision is not only confirmed as anthropocentric (*human* rights) and speciesist (animals and nature are severely disrespected) but has also been left orphaned of its foundational idea, that of national sovereignty and formal (post-colonial) equality among states (Mamlyuk and Mattei, 2011). In the recent past, international law has offered subjectivity to dominant organised global actors, such as transnational global corporations or selected powerful non-governmental organisations (NGOs), thus producing a faked politicisation of the international debate outside of genuine post-capitalist evolution (Anghie, 2005; Koskenniemi, 2011; Mattei and Russi, 2012).

In spite of this evolution, it is still possible to attempt, in the interest of the commons, of non-human life, and of future generations, a *non-*, or even, *post-*anthropocentric reading of some emerging principles of both international and domestic law in an attempt to make it become slightly more part of the solution of current ecological predicaments.

In this chapter, *post-*anthropocentrism will be understood as a move away from the short-term protection of individual humans towards more collective and perennial forms of pro-

DOI: 10.4324/9781003201120-17

263

tection that would be beneficial to non-human entities, even though the legal form of said protection might take a *prima facie* anthropocentric form (e.g., the protection of future generations of humans). Indeed, by placing individual, living persons at the centre, the legal system tends to reduce its radius of action to the restricted temporal space of a human life span, thus losing sight of the economic and social effects that the undertaken policies and the legitimised actions have on entire ecosystems over time.

In this perspective, the extension of the traditional temporal regime of law to include those who are not-yet-born may paradoxically decentre the focus on 'current humans' in ways that create breathing space for more complex and comprehensive visions of international law which depart from a strict anthropocentric paradigm.

In the following pages, we shall outline how international law is evolving at present and point out how, through litigation in domestic courts, such evolution can produce the seeds of a global, ecological legal transformation.

International Law and Future Generations: Potentialities and Pitfalls

Future Generations on the Horizon of International Law

International law has paved the way for the production of a massive set of instruments and principles on which today's environmental and climate change regulatory frameworks rely. It has proved to be an exceptional laboratory in which to construe and refine an innovative conceptual framework aimed at remodelling the purely mechanistic modern world view, driven by the capitalist-based modes of production (Capra and Mattei, 2015). The concept of sustainable development, elaborated by the Brundtland Commission in 1987 and consecrated in the Rio Declaration of 1992, permitted the recognition of the intergenerational question within the legal community, inspiring both supranational and national environmental policies. The concept of precaution, introduced during the conferences on the Protection of the North Sea in 1984 and 1987, launched the 'prudential' logic within the legal system (De Sadeleer, 2002): a perspective to be pursued both at legislative and administrative levels, in order to avoid or reduce to the greatest possible degree the materialisation of uncertain scientific risks, especially when their detrimental effects on the environment and future generations are destined to manifest themselves over the medium or long term.

Both concepts have represented guiding criteria for the pursuit of a cultural and normative reprogramming of the legal system, under the aegis of a more general principle of intergenerational equity, primarily developed within the academic circles of international law (Brown Weiss, 1989; Partridge, 1981). The same expression 'future generations' constantly recurs in the wording of international treaties and declarations – from the EU Charter of Fundamental Rights to the Treaty on European Union and the United Nations Framework Convention on Climate Change (UNFCCC) – to affirm the shared purpose of stretching the boundaries of the legal-temporal regime.

These principles have also been used to promote greater horizontal integration among disciplines at the legislative level. This pathway has recently been marked, within the European Union, by the objective of instilling the logic of sustainability into the massive discipline regarding the protection of markets and consumers in a 'circular economy'.[2] The intention is to create a system of production and distribution aimed at reducing the use of non-renewable resources and moderating the consumption of goods, while employing their full capacity to produce utilities, by reusing and recycling them. Such legislation, however, will mark a step forward only to the extent that consumers will be endowed with suitable and effective remedies, enabling them to assert their sensitivity towards environmental issues in the sphere of their choices and behaviours as market actors.

Nevertheless, international law still lacks the degree of enforceability necessary to guarantee the achievement of the objectives pursued. Due to the vagueness and non-binding nature of obligations in the field of international environmental law, its effectiveness clashes with the sovereignty of the parties of the treaties, whose discretionary powers set a barrier between what political circles agreed upon at the international level and the protection and enforcement of interests of present and future citizens (Abate, 2020, pp. 3 ff). Furthermore, the policies elaborated at both European and national levels in light of those principles have often leaned towards a cost–benefit analysis which has overshadowed an evaluation of costs deferred to future generations, thus frustrating their ultimate purpose. A case in point is the European Parliament's Directive on Environmental Liability 2004/35/Ce, which, in line with the Polluter-Pays-Principle, mainly foresees the *ex-post* allocation of the costs of pollution to its perpetrator, excluding that damage for which a single polluter is not identifiable as well as that damage whose negative effects will not be manifested over a short range of time.

In fact, it is precisely in facing the climate change crisis that the limits of traditional international environmental governance have become more apparent than ever. The UNFCCC, adopted in 1992, has served as the basis for fostering international cooperation among states to prevent dangerous anthropogenic interference in the Earth's climate system. In their annual Conference of the Parties (COP), the contracting states recognised the urgency of the threat posed by global warming and elaborated specific agreements regarding the reduction path to be followed at national levels so as to limit the increase of temperatures within certain thresholds.[3] Relying on the scientific findings of the Intergovernmental Panel on Climate Change (IPCC), the 2015 COP held in Paris set the objective of keeping a global temperature rise in this century to below 2° C above the pre-industrial levels, while pursuing efforts to limit it even further to 1.5° C (Art. 3, Paris Agreement). This agreement was signed, among others, by the EU member states and the EU itself, whose legislation has established a mechanism for unpacking the greenhouse gases (GHG) reduction targets for individual decades (40% reduction by 2030, with an overall goal of 80–95% by 2050, compared to pre-industrial levels).[4] In the context of the *European Green Deal*, an even more ambitious reduction of 55% by 2030 was proposed by the EU Commission in the first half of 2021.

Despite these commitments, no effective enforcement mechanisms exist, should the Parties not comply with the objectives set by international agreements. Thus, a misalignment has been created between such objectives and national policy, which tends to give greater importance to short-term interests, which are more remunerative both in economic and political terms. The consequent postponement of mitigation and adaptation measures increases the overburdening of young and future generations. The excess of GHG due to the failure of climate policies may well lead to irreversible damage for future generations and, at any rate, reduces their ability to mitigate the effects over the long term. This is indeed the crux of the problem concerning climate change governance: it should not only be about achieving the final goal of a climate-neutral economy but also about how this path is to be pursued and how its costs are to be distributed among generations.

These frictions have recently come to light directly in courtrooms, where judges are called upon to decide an ever-increasing number of cases regarding the risks of global warming (Spitzer and Burtscher, 2017). Early climate change litigation mostly developed within the United States. However, since 2015 lawsuits have been filed in most European countries,[5] and against the EU Parliament and Council, with a trend in constant growth. Most of these are still pending, and significant court decisions are expected to be handed down in the near future.

The climate cases are usually initiated by private subjects, be they individual citizens or legal entities, such as groups and associations seeking protection for the environment. They aim to hold private actors (mostly corporations) or national governments accountable for directly contributing to hazardous climate change or for not adopting the measures necessary to reduce rising global temperatures. No matter whether the outcome of this multiplicity of lawsuits has been (or will be) successful, they have brought to the forefront a brand-new demand for social safety in consideration of long-term needs , which can no longer be ignored either by public authorities or legal academics.

The Enforceability of International Law Agreements: Lessons from Climate Change Litigation

The success of this wave of climate cases, especially those filed against states, seems to depend in large part on the willingness of courts to reconstruct the range of states' obligations towards their citizens, on the basis of a complex and systematic vision of the multilevel legal system. Given that the Paris Agreement is only (partially) binding for contracting states while not requiring compulsory reduction of greenhouse gas emissions, the main issue revolves around the legal value that the international commitments may have in interpreting those national or international laws which directly recognise justiciable rights for individuals and/or groups.

To be sure, the answer may vary according to the state under consideration. Some states may count on existing pieces of legislation and/or previous case law that facilitate the process of incorporation of international climate change law into the weft of the single, national legal system. An example can be found in the decision of first instance in the ground-breaking case of *Urgenda v. the State of the Netherlands*. The outcome of the decision is widely known. In brief, the District Court recognised the violation of the State's duty of care towards Urgenda and Dutch citizens, finding that the programmed reduction for the short term (of 20% by 2020) was insufficient for preventing the risk of climate change in an appropriate manner, since it did not take into account the fact that the costs of mitigation and adaptation are to be distributed reasonably between the current and future generations.[6] Hence, the Court ordered the limitation of the joint volume of Dutch annual emissions to within a certain limit.[7]

What is worth noting is the argumentative path followed by the District Court in order to reach its decision. Judges relied on an open standard of Dutch tort law, known as '*cellar hatch standard*' (Stein and Castermans, 2017),[8] which obliges the State to avoid situations where an already existing risk for its citizens increases, when the costs of the precautionary measures are inferior to the probability and total amount of the expected damage. The Court held that such a clause must be interpreted in accordance with the objectives and obligations imposed by climate change policy at both international and European levels; and by the right to life and the right to respect for private and family life, foreseen by Articles 2 and 8 of the European Convention on Human Rights (ECHR). According to Dutch case law, international law obligations have to be taken into account when interpreting national law open standards and concepts[9]: otherwise, it would mean that the State is indirectly violating international norms. This is what happened with regard to the climate policy pursued by the Dutch government, which disregarded – in the Court's opinion – both the Netherlands' international law agreements and the intergenerational projection that characterises its guiding principles.[10]

As can be noted, this approach relies on the specificity of Dutch Private Law, whose clauses have served to clarify the State's obligations towards its citizens, thus limiting its sphere of discretionary power in combating hazardous climate change. Similar paths appear to have been followed in other countries, both at administrative and constitutional levels.

In the French case *Commune de Grande-Synthe v. France*,[11] the Conseil d'Etat held admissible the claim by the City Municipality alleging the government's failure to take appropriate measures to reduce severe climate change-related risks endangering its territory. In its preliminary decision, the highest French administrative court stated that although the provisions of the UNFCCC and the Paris agreement are without direct effect, they must nevertheless be taken into account when interpreting domestic law, especially regarding French climate policy. The Court ruled in favour of the Municipality and ordered the State to 'take all measures necessary' to reduce GHG emissions so as to further limit hazardous climate change.[12] Similarly, in the *Notre Affaire à Tous et al. v. France*,[13] the plaintiffs alleged the violation of a right to a 'preserved climate system', which would stem from a systematic interpretation of national and international law to which France is subjected. The French Administrative Court has recognised that the State's inaction has caused ecological damage linked to climate change and awarded plaintiffs the symbolic sum of one euro for moral prejudice. The Administrative Court ordered the State to take immediate and concrete actions to comply with its duties and to repair the damage caused by their lack of intervention.[14]

In other countries, the effectiveness of international law-based commitments may be pursued by relying on those specific national, constitutional norms which foresee the protection of human rights and/or the environment. This is the route followed in the German case *Neubauer et al. v. Germany*.

The Federal Constitutional Court held that the German Federal Climate Protection Act violates the fundamental rights of the plaintiffs as guaranteed by the German Basic Law (*Grundgesetz*) due to its failure to provide sufficient emission cuts by 2030. The Court relied on international law obligations and principles to affirm that Article 20a of the Basic Law obliges the legislature not only to protect the climate and achieve climate neutrality but also to spread environmental burdens among different generations in an equitable manner. In this perspective, the Court stated that

> the Basic Law imposes an obligation 'to safeguard fundamental freedom over time and to spread the opportunities associated with freedom proportionately across generations'. Therefore, fundamental rights – as intertemporal guarantees of freedom – afford protection against policies which unilaterally offload greenhouse gas reduction burdens onto the future.[15]

The various legal systems on which these cases rely and the heterogeneity of the courts before which they have been brought does not allow for deepening the legal analysis. What is worth dwelling on is yet another, similar interpretation, which develops along the intersection between the international commitments regarding climate change and the application of the ECHR. As is well known, the Convention does not directly enshrine any general right to a healthy environment.[16] However, the European Court of Human Rights (ECtHR) has been called upon to decide a conspicuous number of cases centred on the violation of Convention rights as a consequence of environmental harm or exposure to environmental risks. Most environmentally related human-rights lawsuits have been filed on the basis of Articles 2 and 8 of the ECHR. In several cases, the ECtHR has recognised that the positive obligation for contracting states to protect such rights also finds application in connection to risks of environmental disasters, in as much as they pose a direct threat to the persons involved or affect people's private lives.[17]

The protection of these rights, together with the right not to be discriminated against (Art. 14 of ECHR), has recently been invoked with specific reference to climate change hazards. In the case of *Duarte Agostinho et al. v. Portugal et al.*,[18] six Portuguese children and young adults, supported by non-profit organisations, filed a suit against 34 ECHR contracting parties for failing to reduce the global rise in temperature, in accordance with the commitments undertaken at the international level. In brief, they argue that their right to life is threatened by the increased risks, linked to climate change, of environmental disasters such as forest fires, while affirming that their right to private life is violated by being forced to spend more time indoors due to heat waves.

At first glance, the ECHR approach to global warming-related cases appears particularly promising. Firstly, the application of the Convention unravels the matrix of sources of law needed to link the obligations found in international agreements and the norms recognising rights for individuals and/or groups. As highlighted by the ECtHR, the interpretation of the ECHR must take into account 'any relevant rules of international law applicable in the relations between the parties', according to the provision of Article 31, 3, lett. c, of the 1969 UN Convention on the Law of Treaties.[19] In light of this, it can be affirmed that the content and extent of the positive obligation of a contracting state towards its citizens, with reference to climate change, should be defined in accordance with the reduction path obligations that the party involved agreed upon at the international level.

Furthermore, Strasbourg case law in environmental matters seems able to properly face the peculiarities characterising harmful events connected to dangerous climate change. Indeed, the protection offered under Articles 2 and 8 of the ECHR is generally broader – both in space and time – than that actionable on the basis of national law, especially in those legal systems which do not expressly contemplate the protection of future generations and/or of the environment at the constitutional level.

On the one hand, the enforcement of Articles 2 and 8 of the ECHR does not necessarily require the identification of a specific endangered individual, the protection being extendable to the entire population falling under a state's jurisdiction.[20] This is also the case for environmental hazards, with respect to which the ECtHR has recognised the violation of the rights of residents for entire municipalities.[21]

On the other hand, the ECtHR has developed a concept of *danger* which encompasses those threats that are destined to materialise over the long term. As a rule, this Court guarantees protection when damage has occurred, thus considering the complainant an actual victim. However, in several cases, the ECtHR has ascertained a violation of Article 2 of the Convention by the defendant state, for failing to take practical measures to ensure the effective protection of those whose lives were endangered by an environmental disaster, notwithstanding the availability of practical information about such potential risks.[22] Hence, the ECtHR does not require – for the violation of the right to be ascertained – a short, temporal link between the risk in the case and the damaging event which represents its materialisation: as long as the state has (or could have had) assessed the potential risks of an activity (be it public or private), it has a duty to take adequate measures capable of protecting the right to life and the right for private and family life, even if those risks may materialise over the long term. In this regard, the extent of the obligations of public authorities mostly depends on the harmfulness of the dangerous activities and the foreseeability of the risks to life: the proof of which, in relation to climate change, can generally be found in the scientific knowledge elaborated by the IPCC.

The Precautionary Principle

This line of interpretation is further strengthened by the application of the precautionary principle, which represents a fundamental tool in the designing of a legal system capable of moving beyond the actual, strict anthropocentric view. The principle was initially intended to guide the hand of legislators in defining policies, to measure the correctness of the choices at the discretion of the public and to steer the activity carried out by the public administration. During recent decades, however, the principle has expanded its range of action by breaching the halls of the courts in order to furnish judges with an instrument by which they would be able to assess, on a case-by-case basis, whether a certain claim deserves protection, despite the impossibility of relying on complete scientific knowledge. In case law, its application is capable of broadening the spectrum of protection under two different, albeit connected, profiles. The first regards the core content of the precautionary principle: that is, the lack of full scientific certainty of the risk of serious or irreversible damage for which precautionary measures may be required.[23] As its definition suggests, the scientific uncertainty of a certain risk must not be used as a reason for postponing cost-effective measures which would prevent its materialisation. Such logic is particularly functional for a legal system intended to embrace future generations' interests, given that, as a rule, the risks for which there is scientific uncertainty are those whose effects are destined to manifest themselves over the long term. As clarified by the EU Commission, the precautionary principle is called upon to operate precisely when the cause–effect relationships are more difficult to prove scientifically: that is, in 'situations in which the adverse effects do not emerge until long after exposure'.[24]

The precautionary principle may also acquire relevance under another profile. As is well-known, the effects of climate change, when attributable to the hand of humans, are the result of the integration of acts or omissions on the part of multiple actors. Indeed, GHG emissions acquire their destructive force as soon as they integrate with polluting particles coming from other parts of the globe. From a legal perspective, this may impede tracing the causes of a certain risk or damaging event back to a certain action: the causal contribution of a subject, taken individually, is never what determines the production of the risk or damage that a legal action intends to remedy or eliminate. In fact, many climate change suits fail precisely on this point: the defendants succeed in sustaining that the global climate problem or the single damaging or risky event does not depend solely on their actions or emissions.

Case law may rely on the precautionary principle to supersede these technical obstacles. The duty not to postpone cost-effective measures to prevent serious environmental and health damage implies that each actor has to do 'their part' in order to tackle dangerous climate change *in time*, even when this contribution would not have avoided *per se* the risk that threatens the violation of the rights at stake. In other terms, the precautionary principle may justify the attribution of (partial) liability to a state for its failure to sufficiently reduce global warming-related risks at the local level (or to private actors for their contribution to risk creation), despite there being no absolute certainty as to whether or not the imposed reductions would be sufficient to mitigate climate change and protect citizens.[25]

Considering the wide international consensus on the precautionary principle and its ability to balance conflicting interests also in light of long-term potential costs and harm, a post-anthropocentric legal system should foster the full disclosure of its potentialities. In this sense, a wider, horizontal application of the precautionary principle in courts, in accordance with the principle of proportionality, may be pursued, not only when positive obligations of states are called into play but also when the potential risks of dangerous private activities are to be scrutinised. After all, the ecological shift which is being fostered at the political and institutional levels cannot be fully realised as long as private actors' actions and choices remain anchored to the philosophical and methodological premises of the modern age.

Towards a Post-Anthropocentric Legal System

The Intertemporal Justiciability of (Non-Human) Rights and Interests

To be sure, a more thorough integration across the multilevel legal system, as put forth in the previous section, would not be sufficient to promote the justiciability of the interests of future generations. Another factor hinders the process towards a post-anthropocentric legal regime. Indeed, the traditional logic and enforcement of human rights requires the identification of the victims of a damaging or risky event: something which may be quite impossible since most environmentally related risks, especially when associated with hazardous climate change, materialise over the long run. Therefore, even when scientific knowledge today attests, to a certain degree, to future environmental and health risks – thus becoming *foreseeable* by public authorities and/or private corporations – the lack of the 'existence' and 'judicial' presence of eventual victims impedes the ascription of liability to the responsible parties. While those who are living today are only partially affected by long-term mutations in the ecosystem, those who will indeed be most affected are not able to protect themselves by influencing today's legal actors' actions and decisions, as they do not yet exist.

The gridlock is a direct consequence of the 'modern' view of legal subjectivity which has long been characterised by a binding premise: the ontological distinction between, on one hand, the human actor with his/her body, soul, conscience, will, intent, and ability to speak; on the other, the 'thing' – that which cannot act, which does not possess will but that is simply caused. This conceptual grid came about as a consequence of a long legal and philosophical evolutionary process that has led to moulding the 'reality' of the legal subject onto the natural 'reality' of the human being. This theoretical framework was not altered by the institution of the non-human legal person – i.e., the legal entity, whose deployment was essential to pursue the emergent ideals of bourgeois society and favour the transformation of common good into capital (Capra and Mattei, 2015; Mattei and Quarta, 2018). To justify its presence, legal doctrine has elaborated different theories – either based on the fictitious nature of the legal entity or on its human-like organic structure – to confirm the centrality of the human being as the only *natural* subject of law. Legal theory has thus contributed to determining the expulsion from the political discourse of everything that is not human (be it Nature, animals, or future generations): the set of non-humans represents the objective background within which the *Anthropos* – whether it be the physical or legal person – can exercise its sovereign power (Grear, 2015).

These premises have recently been challenged by the scientific community, especially in connection with the necessity to deal with two fundamental social transformations: the ecological transition and the ongoing technological revolution (Teubner, 2006). The solicitations produced by these changes have inspired numerous works of research, especially in the fields of philosophical anthropology (Descola, 2005; Viveiros De Castro, 2014; Povinelli, 2016) and sociology (Latour, 2006; 2017) aimed at highlighting the contingent nature of the traditional *subject–human* and *object–non-human* dualism.

Recently, the topic of legal subjectivity related to non-human entities has started to stir up the legal community (Míguez Núñez, 2019). In the perspective of an ecology of law, this legal concept is more and more often linked to non-human or not-yet-human entities, in view of ensuring effective remedial structures for long-term protection, able to transcend the temporal regime of existing human beings.

Hereafter, the paper will be focused on three models of protection, inspired by a re-reading of legal subjectivity and of the linked procedural concept of *locus standi*, in an ecological and post-anthropocentric key. The first one has a more structural nature, so to say, and refers to the institution of Nature as a new legal subject. The second and third have a procedural connotation and refer to the promotion of lawsuits by groups of individuals belonging to the same generation, asserting the violation of their rights over time, and by a legal organisation acting as Attorney for future generations.

The Eco-legal Subject of Law

In recent years, there has been a notable tendency to accept the possibility of recognising legal subjectivity for some natural resources, as a means of strengthening their protection. During the 1970s, this idea started to be explored within academic circles (Stone, 1972), making an appearance in the dissenting opinion of judges particularly sensitive to the ecological problems which were emerging in that period.[26] Nowadays, however, the debate is fuelled by an increasing number of cases of recognition (through judicial decision or legislative act) of legal subjectivity for natural resources. This is the case, for example, of a park[27] and a river[28] in New Zealand and the Amazon rainforest ecosystem in Colombia.[29] In brief, this model aims at creating a *new legal framework*, through the institution of a legal person, to which an impersonal and non-human interest – that is referable to 'Nature' – is attributed, and which functions as the pivoting point for the construction of a specific legal statute for the management, use, and protection of the relative resources. To effectively operate within the legal circuit, the new subject is endowed with rights and powers assumed and exercised by a Board – in general, composed by members of the public authorities and of the communities that directly benefit from those resources – which acts in the name and on behalf of the same subject, for which it would be its mere spokesperson.

Putting aside the ideological and cosmological premises which support such cases, their analysis highlights some common elements, which are of great interest from both a theoretical and technical legal point of view. In particular, two sets of effects determined by the new legal subject deserve closer scrutiny.

Firstly, the new actor is legitimised to directly exercise its own interests in courts,[30] whenever the quality of its well-being is endangered, even over the long-term. Hence, the institution of the subject allows for a dis-intermediation from those interests (whether private or public but, in any case, assigned to an entity other than the web of natural resources to be protected), the prejudice of which traditionally constitutes the formal premise for activating the protection of environmental goods. Courts will only need to verify whether there has been a violation of the interests, referable to the brand-new actor, to maintain its integrity and the quality of its ecosystem, thus allowing for a de-individualised legal regime tailored for that exact purpose.

Beyond the capacity to institute legal proceedings, the new framework has a significant bearing on the legal regime governing the use and management of the network of resources included in it. The institution of the subject favours the construction of a form of government of the resources according to which the decisional power transcends formal ownership over resources (be it public or private) and is attributed – on the basis of specifically identified criteria – to the members of the board or to the representatives called on to manage the subject. This leads to the subtraction of the now-personified resources from the idiosyncrasies of their owner or rather from the risk of *inertia* on the part of those entitled to enforce their protection. It also promotes a participatory and polycentric form of governance, in which the representatives of public authorities are summoned to cooperate with the representatives of local communities which draw sustenance and enjoyment from the natural resources. This is what is foreseen, for example, with the subjects instituted in New Zealand. The governance of the River must be based on shared decisions between the community and the Crown, which are bound to the principles – established directly by the *Settlement Act* – meant to protect and preserve the resources over the long term.

As can be observed, the institution of the subject, transcending symbolic or purely ethical instances, assumes a highly functional value. It constitutes a legal device useful for guaranteeing effective and concrete protection beyond human beings' specific and individualistic needs, while reorganising a plurality of opposing interests, which can no longer be fashioned according to a purely proprietary paradigm.

Towards a Post-Anthropocentric Approach to Human Rights: (Young) Generations in Courts

Similar aims sustain those actions in which a group of children or young adults are designated as actors in a legal procedure. Here, it is the concept of generation that comes to the forefront as a theoretical tool for sanctioning the violation of interests of subjects otherwise lacking legal protection. Before pointing out, from a technical point of view, the main features of this second model of long-term protection, it seems useful to spend a few words on the concept of generation.

As noted above, this concept does not have an actual legal meaning, and it has been used in declarations and treaties without a precise definition. However, some insights can be drawn from the philosophical and sociological studies which have intercepted, under different profiles, the concept of generation. From their analysis, it emerges how the expression, more than denoting a specific meaning, constitutes a semantic expedient which makes it possible to intertwine under one heading two different profiles.

Declined in the plural, the term – *generations* – evokes an objective and conventional sense of temporality, produced by the passage of time, although described in an alternative way: rather than observing the history of humankind from the outside in a flow of years which structures the time of the world, such passage is represented from the inside, broken down into a succession of lives.[31] This leads the observer to acquire proximity with the future, as it will be the time of one's descendants (Resta, 1997).

In its singular form, the term – *generation* – connotes an amorphous mass of individuals, whose plane of consistency is composed by the 'common horizons of problems and answers' which they have to face (Ortega y Gasset, 1986). As has been pointed out, the glue that unites and binds one generation is not rooted in a natural bond (family) or in a conscious act of will (association); nor is it a social formation in the sense of groups constituted for specific purpose (Mannheim, 1952). It is, on the contrary, hetero-determined in the sense that the generation's identity emerges as a reflection of the sharing of certain issues; as the possibility of taking part in common historical events (Abrams, 1983); and from the 'unity of means' or resources which a specific period puts at the disposition of its generations. According to this second key – which dismisses a quantitative perspective of time in favour of a qualitative approach – the concept gives back to each generation not only *contemporaneity* but also its own *identity*, declined in a trans-individual key.

Considered from a legal angle, the identification of a *generation* may be structured around the qualification of the interests which pertain to certain common problems (Gruber, 2011), such as those tied to the depletion or degradation of natural resources, to the risks created by genetic modification, or to the accumulation of public debt, and so on. This is indeed what can be observed in the context of climate change-related lawsuits, within which the concept of generation appears to be acquiring legal significance.

The examples are numerous, on both sides of the Atlantic Ocean. The *Juliana et al. v. the United States of America et al.* case[32] was filed by a group of children, supported by environmental associations and by a scientist, acting as 'guardian for future generations'. According to the plaintiffs, the policies adopted by the government have increased the risks connected to climate change, whose catastrophic effects will mostly be manifested when their generation will be grown-up. They thus asserted that the government violated the youngest generation's constitutional rights to life, liberty, and property as well as its duty to protect public trust resources. The Court recognised that the children have adequate standing, in that the alleged harm to their personal, economic, and aesthetic interests is 'concrete and particularized' and causally linked to the defendant's conduct.

Similarly, the recent *Duarte Agostinho* case[33] brought before the ECtHR was initiated by six children and young adults, claiming that their rights had been violated by numerous states due to a failure to take appropriate measures to reduce climate change. In the case in point, the 'common horizon of problems' has been identified by the applicants, in particular, in the statistical increase in severe heat waves that the scientific community expects, with high confidence, to occur over the long term in some regions, such as Portugal. The applicants' generation as well as that of their future children and families, according to the complainants, will be affected by the increases in mean and extreme high temperatures, both in terms of higher risk of related mortality and of their ability to enjoy their private lives outdoors. Due to this effect, the applicants claim to suffer – already today, as a young generation – the prejudice arising from anxiety about the negative effects which climate change may have on them and 'the families they hope to have in future'. To date, the case has been fast-tracked by the ECtHR, and a response to the alleged violations from the defendants has been required by the Court.

Again, the concept of generation, as a centre of imputation of rights and interests, also recurs, albeit with a different twist, in the Swiss *KlimaSeniorinnen* case.[34] The claim was filed by a group of elderly people against the Swiss government. They assert that the Swiss government's failure to cope with climate change has put their health in danger, due to an increase in the likelihood of greater intensity and longer duration of heat waves, leading to a higher risk of premature death. In this way, the legal action aims at indirectly fostering an increased protection for future generations, who, at present, are not recognised as legal actors under Swiss Law (Bähr et al., 2018). It can be said that the association pursues a protection in the interests of (future) generations of elderly women, one of the most vulnerable groups facing the danger linked to climate change.

In all of these cases, the *individualistic* and *here-and-now* logic which traditionally connotes fundamental constitutional rights or human rights approaches is abandoned, to endorse an intertemporal projection of their justiciability (Düwell and Bos, 2016). The complainants do not seek protection as individuals, but as part of either today's youngest or oldest living generation. The construction of new centres of subjective imputation, of a 'generational' and thus 'superindividual' character, can constitute a technical device for rendering legally relevant those actions able to harm the ecosystem and (future) people's health over the long term, regardless of the individual who will actually suffer injury. In other words, they serve the purpose of expanding the temporal horizon of judicial evaluation and effectively enforcing human rights even in an intergenerational perspective. The way has been paved for a post-anthropocentric use of human rights.

An Italian Primer

Many of the abovementioned issues, from precaution to justiciability, to the *locus standi* of future generations, have touched ground in a recent Italian case related to the dangers of electronic pollution for growing children. This strategic case was initiated as a matter of urgency by a group of young children, their parents, and an original legal structure, a Cooperative Stock Company of mutual assistance between present and future generations. This innovative organisation was set up in February 2019 expressly providing in its bylaws its function as an Attorney for future generations in the Civil Tribunal of Turin. The action was brought to remove a 25-meter-tall structure built right in the middle of a primary school campus in the small village of Frossasco in Valnoce, a pre-alpine area within the Metropolitan City of Turin.

The defendants were two local municipalities, the Metropolitan City of Turin and the Municipality of Frossasco, and a private consortium within which a number of important big tech companies as well as some local actors, such as banks, universities, and regional governments, joined forces to create and manage sites for the exchange of Internet traffic and develop innovation projects based on the use of broadband Internet.

The plaintiffs claimed that the *principle of precaution* should be applied because significant and reputable scholarly literature shows that there is no certainty about the long-term effect on the health of developing organisms exposed for long hours to electromagnetism. The large structure positioned on the ground could host a large number of antennas of any kind with an almost certain increase in the level of electromagnetism in the school. The remedy requested was, therefore, the removal of the structure off campus to a better-suited available location and the establishment, either judicially or extra-judicially, of a *servitude in favour of future generations* capable of protecting the school campus as a sort of reservation for the unimpaired growth of young humans. Such a servitude (or similar institution, such as a trust) would be the only legal structure capable of constitutionally resisting the predictable changes in regulation limiting the power of local political institutions to protect human health against electronic pollution. Big tech industry and other special interests are already obtaining, through the capture of the political process, advantages at the expense of the health of future generations, something that in fact occurred in July 2020 after the complaint was filed. Indeed, especially in light of major investments in 5G technology, boosted by the Covid-19 induced migration of social life onto the web frontier, the lowering of standards of security and the process of deregulation has already started.

The defendants have counter-argued, thus far successfully in Court, that the formal respect of regulatory standards contained in an administrative regulation by the one antenna currently installed precludes the possibility of applying the principle of precaution, as it has already been considered by the administrative regulator. The *here and now* logic of the Italian Court frustrates the principle of precaution, since it considers it just a standard policy (that regulators should respect) rather than an actual legal principle. The most important point of the case has been, however, that a cooperative company with litigation in favour of future generations in its bylaws has been recognised as having standing, as an advocate of future generations, thus providing a first positive legal recognition of this eco-law institution.

The notion that an attorney for future generations should be present in any instance of litigation whose impact is likely to affect the long-term ecological equilibrium is an important contribution. It should be reproduced institutionally on a global scale to produce incremental transformation from the bottom up.

Conclusions: Above, Below, and Within Nature

As has been highlighted, international law has provided a vague vocabulary but no real substance to a global legal order capable of protecting the long-term interests of natural reproduction of life on the planet. Vagueness has been its structural character, denounced now for years by critical scholars, such as Koskenniemi. National legal systems are trying to fill the vacuum by deploying theoretical and practical legal solutions in order to produce principles of a very much

Toward a Post-Anthropocentric Legal System

needed ecological legal order. Such legal innovations spread around the world through well documented processes of imitation, borrowing, and diffusion that occurs from the bottom up.

Such extremely diverse and erratic evolution, which proceeds through victories and setbacks, is nevertheless the only one that can occur somewhat shielded from corporate takeover and capture the dominant legal phenomenon that has created global lawlessness and the continuous rush towards extinction. Dominant international law has developed with a disregard for nature, sharing the mechanistic vision that, from modernity onwards, has transformed the law into a mechanism of extraction. Post-anthropocentric international law should certainly get rid of the idea that humans (and their institutions) are above Nature. However, this wrong idea should not be substituted with the equally wrong (though opposite) one that humans should be located below Nature, since this opens the door to extremely dangerous visions. A dialectic between a global network of national legal systems providing and sharing ecological legal innovation, and a model of international law that looks hopelessly controlled by corporate extraction, should be able to provide a global synthesis that considers humans *within Nature*. This seems to be something worth struggling for theoretically and practically.

Notes

1 The chapter is the fruit of the exchange of ideas and of the collaboration between the two authors. However, Sections 1, 8 and 9 are to be attributed to Ugo Mattei; Sections 2 to 7 to Michael W. Monterossi, whose research was conducted as part of the project on "*Future Generations in Swiss and European Private Law*" (2019–2022), financed by the Swiss National Science Foundation.

2 Communication from the Commission, *A new Circular Economy Action Plan. For a cleaner and more competitive Europe*, COM (2020) 98 final, Brussels, 11 March 2020.

3 The threshold limit of a 2° C rise in temperatures was revised in 2018, when a lower increase of 1.5° C was identified as the limit to be respected in order to avoid higher risks due to climate change. See the Summary for Policymakers, of the IPCC Special Report *Global warming of 1.5°C* (2018).

4 As defined by the *2030 Climate & energy framework*, Brussels, 24 October 2014, available at https://www .consilium.europa.eu/uedocs/cms_data/docs/pressdata/en/ec/145397.pdf

5 The reference is, in particular, to the following cases: *Urgenda Foundation v. The State of the Netherlands,* [2015], which was based on liability for tortious acts (article 6:162, Book 6 of Dutch Civil Code) and violation of Articles 2 and 8 of the ECHR; *VZW Klimaatzaak v. Kingdom of Belgium & Others* on the rule concerning fault-based liability (art. 1382 of Belgian Civil Code); *Lliuya v RWE AG* [2015] Az. 2 O 285/15, based on the proprietary action for removal and injunction (§ 1004 of German Civil Code); *Swiss Senior Women for Climate Protection v. Swiss Federal Government* [2016], based on the alleged violation, by the Swiss government, of fundamental individual rights (art. 2 and 8 of ECHR) of a number of elderly people, grouped in and represented by the association KlimaSeniorinnen Schweiz; *Greenpeace Nordic Association et al. v. Ministry of Petroleum and Energy of Norway* [2018], against the opening of a new oil-drilling platform in the Arctic, based on the violation of the § 112 of the Norwegian Constitution which imposed the state to protect the rights of future generations to a safe and healthy environment; *Commune de Grande-Synthe v. France*, [2020], no. 427301, based on the violation of French Climate Policy; *Notre Affaire à Tous and Others v. France* [2021], against the French Government for insufficient action against climate change; *Carvalho and Others v. The European Parliament and the Council* [2019] T-330/18, based on non-contractual liability pursuant to articles 268 and 340 TFEU.

6 District Court, *Urgenda* (note 3), par. 4.76.

7 The District Court decision was confirmed by the Hague Court of Appeals [2018] and, subsequently, by the Supreme Court of The Netherlands, Civil-law Division [2019], case no. 19/00135, although in these judgements the courts based their decisions on the human rights regime set by the ECHR. Differently from the District Court, these courts have considered that, in Dutch national law, article 2 and article 8 can be applied also to protect a legal entity, such as the Urgenda Foundation, despite this not being admissible according to the ECtHR case law. See on this point L. Burgers, T. Staal, *Climate Action as Positive Human Rights Obligation: The Appeals Judgment in Urgenda v The Netherlands*, in *Netherlands Yearbook of International Law*, 2018, pp. 223 ff.

8 The Court held that potential risks linked to climate change can be assimilated to those deriving from the activities for which the standard has been previously applied and that, therefore, it is necessary to ascertain whether the state has adopted measures which, according to a reasonableness canon, are necessary to reduce the risk of hazardous climate change.

9 The Court took into consideration (in particular, the UNFCC and related agreements) protocols: the 'non harm' principle, the EU climate policy, and the principles upon which these legislations rely, among which the Court expressly recalls the principle of intergenerational fairness, the principle of sustainability, and the precautionary principle. The Court affirmed that the principle according to which national law standards cannot be applied, in a way which would imply the violation of international law obligations on the part of the state, constitutes a general principle of law.

10 More recently, the Hague District Court reached similar conclusions in relation to the contribution to climate change produced by a private company. In the *Milieudefensie et al. v. Royal Dutch Shell plc.*, (26 May 2021, C/09/571932 / HA ZA 19-379), the Court recognised the responsibility of Royal Dutch Shell in causing climate change and ordered it to reduce their aggregate annual volume of emissions.

11 See note 3.

12 Conseil D'Etat, 1 July 2021, case no. 427301.

13 See note 3.

14 Tribunal Administratif de Paris, 14 October 2021, case no. 1904967, 1904968, 1904972, 1904976/4-1.

15 Bundesverfassungsgericht, Mar. 24, 2021, Case No. BvR 2656/18/1, BvR 78/20/1, BvR 96/20/1, BvR 288/20.

16 ECtHR, *Kyrtatos v. Grecia*, [2003] n. 41666/98, 52; ECtHR; *Cordella et al. v. Italy*, [2019] 54414/13 and 54264/15, 100, concerning the failure to adopt adequate measures by the Italian state in relation to the pollution generated by the ILVA industrial plant in Taranto, Italy.

17 ECtHR, *López Ostra v. Spain* [1994] 16798/90, 51, stating that 'severe environmental pollution may affect individuals' well-being and prevent them from enjoying their homes in such a way as to affect their private and family life adversely, without, however, seriously endangering their health'; more recently, ECtHR, *Cordella et al. v. Italy*, (note 10), 157.

18 ECtHR, [2020] Request no. 39371/20.

19 ECtHR, *Nada v. Switzerland* [2012] 10593/08.

20 ECtHR, *Gorovenky and Bugara V. Ukraine*, [2012], 36146/05 and 42418/05, par. 32.

21 ECtHR, *Di Sarno et al. v. Italy* [2012], no. 30765/08, para. 110 and ECtHR, *Cordella et al. v. Italy*, (note 10), par. 172, in which the victims were identified on the basis of the city municipality in which they lived and for which scientific proof of heavy pollution was given.

22 As regards the violation of article 2, see ECtHR, *Öneryildiz v. Turkey* [2004], 48939/99, 98–101, concerning the failure to prevent a methane explosion. See also ECtHR, *Budayeva et al. v. Russia* [2008], 15339/02, 147–158 concerning the materialisation of the danger of mudslides of which authorities were informed; ECtHR, *Kolyadenko et al. v. Russia* [2012], 17423/05, concerning the risks caused by flooding related to heavy rains. As regards the violation of Article 8, references can be made to cases such as EctHR, *Taşkin et al. v. Turkey* [2004] no. 46117/99, concerning the violation of Article 8 which was considered applicable despite the government arguing that the risk might have materialised only in twenty to fifty years; and ECHR, *Tătar v. Romania* [2009], no. 67021/01, 89–97 concerning potential long-term health risks deriving from heavy metal emissions in the process of extraction of a gold mine.

23 ECtHR, *Tătar v. Romania*, 111 and 120, establishing that the state had taken effective and proportionate measures as foreseen by the principle of precaution even in the absence of scientific or technical certainty.

24 *Communication from the Commission on the precautionary principle*, COM (2000) 1 final, 2 February 2000.

25 This line of reasoning is followed, for example, by the Hague District Court in the Urgenda case (note 3), par. 4.58.

26 The reference is to *Sierra Club v. Morton* case [1972] 405 U.S. 727 USSC, in which Justice William O. Douglas argued that 'inanimate objects' should have standing to sue in court.

27 *Te Urewera Act* [2014], No. 57, 27 July 2014.

28 *Te Awa Tupua (Whanganui River Claims Settlement) Act* [2017], No 7, 20 March 2017.

29 Corte Suprema de Justicia, Sala de Casación Civil, [2018], STC4360-2018.

30 This was, after all, the primary effect which C. Stone had in mind when writing his famous work *Should Trees Have Standing?* pp. 450–467.

31 As suggested by its etymology, which points to the noun genus, generis, i.e., descendants, progeny, children. Thus, it refers to the succession between father and child, between the generation of ancestors and that of descendants.

32 [2018], 6:15–cv–01517–AA.

33 See note 12.

34 See note 3.

References

Abate, R.S. (2020) *Climate Change and the Voiceless. Protecting Future Generations, Wildlife and Natural Resources.* Cambridge: Cambridge University Press.

Abrams, P. (1983) *Historical Sociology.* Ithaca: Cornell University Press.

Anghie, A. (2005) *Imperialism, Sovereignty and the Making of International Law.* Cambridge: Cambridge University Press.

Bähr, C.C., Brunner, U., Casper, K. and Lustig, S.H. (2018) 'KlimaSeniorinnen: Lessons from the Swiss senior women's case for future climate litigation', *Journal of Human Rights and the Environment* 9(2), pp. 194–221. DOI: 10.4337/jhre.2018.02.04

Brown Weiss E. (1989) *In Fairness to Future Generations: International Law, Common Patrimony, and Intergenerational Equity.* New York: Transnational Publishers.

Burgers, L. and Staal, T. (2018) 'Climate Action as Positive Human Rights Obligation: The Appeals Judgment in Urgenda v The Netherlands', *Netherlands Yearbook of International Law*, 49, pp. 223–244. DOI: 10.1007/978-94-6265-331-3_10

Capra, F. and Mattei, U. (2015) *The Ecology of Law: Toward a Legal System in Tune with Nature and Community.* New York: McGraw-Hill Education.

De Sadeleer, N. (2002) *Environmental Principles. From Political Slogans to Legal Rules.* Oxford: Oxford University Press.

Descola, P. (2005) *Par-delà nature et culture.* Paris: Gallimard.

Düwell, M. and Bos, G. (2016) 'Human rights and future people – Possibilities of Argumentation', *Journal of Human Rights*, 15 (2), pp. 231–250. DOI: 10.1080/14754835.2015.1118341

Grear, A. (2015) 'Deconstructing Anthropos: A Critical Legal Reflection on "Anthropocentric" Law and Anthropocene "Humanity", *Law Critique*, 26, pp. 225–249. DOI: 10.1007/s10978-015-9161-0

Gruber, M.-C. (2011) 'What Is It Like to Be Unborn? Our Common Fate with Future Generations', in Mathis, K. (ed.) *Efficiency, Sustainability, and Justice to Future Generations.* Dordrecht: Springer, pp. 113–137.

Koskenniemi, M. (2011) *The Politics of International Law.* Oxford: Hart Publishing.

Latour, B. (2006) *Nous n'avons jamais été modernes: Essai d'anthropologie symétrique.* Paris: La Decouverte.

Latour, B. (2017) *Facing Gaia: Eight Lectures on the New Climatic Regime.* Cambridge: Polity Press.

Mamlyuk, B. N. and Mattei, U. (2011) 'Comparative International Law', *Brooklyn Journal of International Law*, 36 (2), pp. 385–452.

Mannheim, K. (1952) 'The Problem of Generations', in *Essays on the Sociology of Knowledge.* London: Routledge and Kegan Paul, pp. 276–320.

Mattei, U. and Nader, L. (2008) *Plunder: When the Rule of Law is Illegal.* Hoboken: Wiley-Blackwell.

Mattei, U. and Quarta, A. (2018) *The Turning Point in Private Law. Ecology, Technology and the Commons.* Cheltenham: Edward Elgar Publishing.

Mattei, U. and Russi, L. (2012) 'The Evil Technology Hypothesis: A Deep Ecological Reading of International Law', *Cardozo Law Review DeNovo*, pp. 263–277.

Míguez Núñez, R. (2019) *Le avventure del soggetto. Contributo teorico-comparativo sulle nuove forme di soggettività giuridica.* Milano: Mimesis.

Ortega Y Gasset, J. (1986) *Mirabeau o el político. Contreras o el aventurero. Vives o el intellectual.* Madrid: Alianza Editorial.

Partridge, E. (ed.) (1981) *Responsibilities to Future Generations: Environmental Ethics.* Buffalo: Prometheus Books.

Povinelli, E.A. (2016) *Geontologies. A Requiem to Late Liberalism.* Durham: Duke University Press.

Purdy, J. (2015) *After Nature. A Politics for the Anthropocene.* Cambridge, MA: Harvard University Press.

Resta, E. (1997) *Le stelle e le masserizie: Paradigmi dell'osservatore.* Bari: Laterza.

Spitzer, M. and Burtscher, B. (2017) 'Liability for Climate Change: Cases, Challenges and Concepts', *Journal of European Tort Law*, 2, pp. 137–176. DOI: 10.1515/jetl-2017-0009

Stein, E. and Castermans, A.G. (2017) 'Case Comment – Urgenda v. the State of the Netherlands: The "Reflex Effect" – Climate Change, Human Rights, and the Expanding Definitions of the Duty of Care', *McGill Journal of Sustainable Development Law*, 13 (2), pp. 303–324.

Stone, C.D. (1972) 'Should Trees Have Standing? – Toward Legal Right for Natural Objects', *Southern California Law Review*, 45, pp. 450–467.

Teubner, G. (2006) 'Rights of Non-Humans? Electronic Agents and Animals as New Actors in Politics and Law', *Journal of Law and Society*, 33, pp. 497–521. DOI: 10.1111/j.1467-6478.2006.00368.x

Viveiros de Castro, E. (2014) *Cannibal Metaphysics*. Minneapolis: Univocal Publishing.

Wolf, E.R. (2010) *Europe and the People Without History*. Berkeley: University of California Press.

14

A NON-ANTHROPOCENTRIC INDIGENOUS RESEARCH METHODOLOGY

The Anishinabe Waterdrum, Residential Schools, and Settler Colonialism

Valarie G. Waboose

Introduction

The Anishinabe people of the Great Lakes region on the North American continent, also referred to as Turtle Island, have occupied this territory since time immemorial. The Anishinabe are affiliated with an alliance called the Three Fires Confederacy, which consists of the Ojibway, Odawa, and Pottawatomi nations. According to the earliest accounts, the territory occupied by the Three Fires Confederacy stretched eastward to the mouth of the St Lawrence, across southwestern Ontario into Michigan, westward to Saskatchewan and moving northward to Manitoba and Northern Ontario, crossing Minnesota, Wisconsin, and North Dakota.

I am from the Ojibway Nation of Walpole Island Indian Reserve, also known as Bkejwanong Territory (hereinafter Bkejwanong, which means, 'where the rivers divide'), located in southern Ontario. The community is located at the mouth of the St Clair River and consists of five islands: Walpole, St Anne's, Squirrel, Bassett, and Seaway. The language spoken in this community is Ojibway and Pottawatomi.

At time of contact, there were no borders separating Bkejwanong from Algonac, Michigan, so the Anishinabe travelled this vast area freely. Today, there is an international border that separates these two communities: Bkejwanong on the Canadian side of the river and Algonac on the United States side of the border. Bkejwanong is an international border crossing community that provides access to travellers, via water vessel, into the United States and back. Having occupied this territory long before the implementation of the international border, the Anishinabe of Bkejwanong believe they have rights to cross into the United States unmolested. These rights are affirmed in the St Anne's Island Treaty (McNab, 1992, p. 140) and the Jay Treaty (Government Archives 1794, Article III).

Another important characteristic of Bkejwanong is the unceded status the territory possesses, unceded meaning that the Anishinabe never signed a treaty giving the federal government this land. This status provides the community with unique rights when dealing with provincial and federal governments.

DOI: 10.4324/9781003201120-18

279

Bkejwanong is the place where I was raised and the place that I acknowledge as my homelands. It is upon these lands, within this community, and from the Indigenous Knowledge Keepers of this community and the Midewiwin Society (Great Medicine Society) that I gained knowledge of the traditions, values, and beliefs shared in this chapter. The Midewiwin Society is also known as the society that practices the ceremonies and teachings within the Original Instructions given to Anishinabe by the Creator at the beginning of time.

This chapter is not about international law as is usually understood under Western law. It is a critique of international law's anthropocentrism through the presentation of a non-anthropocentric Anishinabe worldview. As will be seen, the Anishinabe worldview has its own legal orderings and traditions, including laws applicable between Indigenous Nations and non-Indigenous Nations. Some ideas may be familiar to Western international lawyers (treaties, diplomats, reciprocity, and so on). More importantly, this worldview includes laws that encompass relations between all of nature, including human beings. In the following section, a glimpse of the values and beliefs of the Anishinabe will be shared to provide the reader with insight into their non-anthropocentric beliefs. The chapter then examines what it might mean to give space to Indigenous research about the law more generally and develop a specifically Anishinabe legal methodology.

An Anishinabe Worldview

This section is divided into four parts to provide Anishinabe teachings that directly refute anthropocentrism. The teachings to be shared illustrate the relationships that Anishinabe have with all living things within the world and serves as the rationale for the non-anthropocentric beliefs of the Anishinabe.

The Anishinabe Creation Story

The Anishinabe Creation Story has been orally transmitted by Elders and Indigenous Knowledge Keepers from generation to generation since time immemorial. The Creation Story contains many important teachings that influence the belief system of the Anishinabe people. One such teaching within the Creation Story states:

> The Creator sent his singers in the form of birds to the Earth, to carry the seeds of life to all the Four Directions. In this way life was spread across the Earth. On the Earth the Creator placed the swimming creatures of the water. He gave life to all the plant and insect world. He placed the crawling things and the four-leggeds on the land. All of these parts of life lived in harmony with each other.
>
> *(Benton-Banai, 2011, p. 3)*

Another important aspect of this worldview is the belief that Creation was completed in a specific order – plants, insects, birds, animals, and human beings – purposely, in the order of necessity. The Anishinabe pay attention to the order of placement on the Earth because it reminds us of our obligations to those who came before us, those who already governed the territories that we later came to inhabit (Starblanket and Stark, 2018, p. 193). From the beginning, animals, birds, fish, and insects have served not only humankind but the manitous, spirits, and deities that abide in the supernatural, to enable them to fulfil their earthly purposes (Johnston, 2001, p. 115). Asch, Borrows, and Tully further state:

> The Ojibway worldview is expressed through their language and through the Law of the Orders, which instructs people about the right way to live. The standards of

conduct which arise from the Law of Orders are not codified but are understood and passed on from generation to generation. Correct conduct is concerned with 'Appropriate behaviour, what is forbidden, and the responsibility ensuing from each'. The laws include relationships among human beings as well as the correct relationships with other orders: plants, animals, and the physical world. The laws are taught through 'legends' and other oral traditions.

(Asch, Borrows, and Tully, 2018, p. 50)

In the Anishinabe worldview, the Law of the Orders, man was the last form of life placed on Mother Earth (Benton-Banai, 2011, p. 3). Stories of Naakonigewin (the Art of Big Law) through sacred Creation stories remind us that as human beings we were the last to be placed on the Earth and the most dependent of all the beings in Creation (Craft, 2019, p. 104). Because man was created last, man is totally dependent upon all living things and would not survive without the other forms of beings (Craft 2019, citing Johnston 2003, p. 30).

As humans, we also depend upon a complex web of relationships to live well, to live our mino-bimaadiziwin (the good life). Through the relationships that govern our interactions between beings, we develop a sense of our normative values and legal principles. The teachings and principles outlined here provide the understanding Anishinabe hold in relation to all living things within the Universe. It is believed the Creator created and placed all living things upon the Earth before humans to ensure the survival of humankind.

The importance of the order of all life forms created and placed upon the Earth becomes crucial in this discussion of anthropocentrism. As one can begin to see by this teaching, the Anishinabe have a contrary view of mankind's superiority over other life forms placed on Mother Earth.

Original Instructions/Naakonigewin (the Art of Big Law)

Naakonigewin are the laws given by the Creator at the beginning of time. During this time, Anishinabe was instructed how to relate to the water, Earth, plants, animals, and all beings placed on Earth. The role and responsibilities of every living element placed on Earth is contained within the Original Instructions. All living things, including man, were given Original Instructions to live by. From the Original Instructions flowed the original laws of the Universe. In essence, spiritual laws are the instructions that the Creator provides to each being (human, animal, bird, and so on) that journeys through four levels before arriving on the Earth (Craft, 2019, p. 107). Today these laws are also referred to as Indigenous Legal Orders (hereinafter ILO) and/or Indigenous Legal Traditions (hereinafter ILT).

The Original Instructions include laws that help Anishinabe to live a good life. This good life includes, but is not limited to, having respect for all living things created and placed on Earth and having gratitude for everything given to us by the Creator.

All Things Are Created Equal

The Creation Story states: as the Creator sat in total darkness contemplating Creation of this world, the only sound that could be heard was that of the she-she-gwun (shaker/rattle). As this sound echoed throughout the Universe, the Creator began creating objects within the sky realm, such as Grandfather Sun, Grandmother Moon, stars, and the planets. The Creator then placed trees, plants, animals, fish, birds, insects, waterways, mountains, valleys, and islands upon Mother Earth. At this time, all that was created was bountiful and full of beauty.

During a Repatriation ceremony, an Elder shared a vision of the world at the beginning of time. This is how he explained what he saw: the elements placed upon Mother Earth had magnificent beauty, the trees stood tall, colourful plants were abundant, the streams, rivers, lakes, and oceans were crystal clear, the soil was rich in minerals, mountains were gigantic, the air was fresh, and animals, insects, fish, birds were plentiful. With closed eyes, the view described was incredibly breath-taking.

Even though the Creator filled the world with such natural and breath-takingly beautiful things, he still wanted something more. The Creator wanted a being like the image of himself (Benton-Banai, 2011, p. 4). And so, man was created. It is explained that when man was created, the Creator took four parts of Mother Earth and blew into them using a Sacred Shell (megis) (Benton-Banai, 2011, p. 2).

Pondering the creation of Earth's elements, I recall another teaching shared at a Faculty Anishinabe Law Camp that all the elements of Creation made treaty amongst themselves when they were placed on Earth. What I understood this to mean was that each plant, animal, rock, waterway, and so on placed on Mother Earth knew its role and responsibility within the great web of life. And further, the treaties between all living things sanctioned a reciprocal relationship that each owed to the others. This teaching illustrates how all living things within Creation are interdependent upon others to survive. Because of this interdependence, all were created equal, and all are of equal importance to one another. Man is no greater than an animal, a blade of grass, or a drop of water. In the eyes of the Creator, humans were created equal to plants, insects, water life, and animals. Therefore, humans are not superior beings. All livings things within the Universe are dependent upon one another for survival, and humans must never forget this. This notion of an inter-species treaty is a far cry from the reductive sense in which treaties are understood in international law as limited agreements based on the interests of irreducible sovereigns, on the basis of Eurocentric understandings of sovereignty.

Reciprocal Relationships

Anishinabe law is all about relationships (Borrows, 2010, p. 8). These relationships exist between all beings that are part of Creation, in a variety of permutations and according to a system of generalised reciprocity (Craft, 2019, p. 105). We often refer to other beings – animals, birds, fish, insects, plants, trees, and rocks – as our older brothers and sisters, acknowledging that they were placed here before us and that we have something to learn from them, while also being dependent on them for our survival (Craft, 2019, p. 108). Our ancestors saw a kinship between plants, insects, birds, animals, fish, and human beings, a kinship of dependence: humans depending on animals and birds and insects; animals depending on insects and plants; insects depending on plants; plants depending only on the Earth, sun, and rain (Craft, 2013, p. 70). Thus, we have extended relationships with not only human life forms but with other life forms. Anishinabe believe that 'all my family' includes not just our human family, but the animals, plants, birds, fish, the water, air, and the Earth, which all form part of the great interdepending web of life (Bedard, 2008, p. 96).

The following story captures the reciprocal relationship that Anishinabe have with animals and illustrates the role of foundational narratives in creating and underpinning ILOs and ILTs:

> In a time long ago, all the Waawaashkeshiwag, Moozoong, and Adikwag, the deer, the moose, and the caribou, suddenly disappeared from Kina Gchi Nishnaabeg-ogaming. Well, maybe it wasn't so suddenly. At first, nobody noticed. The relatives of the Hoof Clan had to be very patient. But, after a while, people were starting to notice some

changes. After a long discussion, where everyone spoke what was in their hearts the people decided to send their fastest runners out in the four directions to find those hoofed ones. After some negotiation, the people learned that the Hoof Clan had left their territory because the Nishnaabeg were no longer honouring them. They had been wasting their meat and not treating their bodies with the proper reverence. The Hoof Clan had withdrawn from the territory and their relationship with the Nishnaabeg. They had stopped participating. The diplomats, spiritual people, and mediators just listened. All the parties thought about what they could give up to restore the relationship. Finally, the Hoof Clan and the Nishnaabeg agreed to honour and respect the lives and beings of the Hoof Clan, in life and in death … In exchange, the Hoofed Animals would return to our territory so that Nishnaabeg people could feed themselves and their families. They agreed to give up their lives whenever the Nishnaabeg were in need.

(Simpson, 2008, pp. 9–11,
a shortened version of the story)

This story illustrates the reciprocal relationships that Anishinabe have with animals. The story shares how the deer, moose, and caribou left the Anishinabe territory because they were not being respected by humans. The Anishinabe were wasting their meat and not treating their bodies with utmost respect. In the Anishinabe culture, it is common practice that the hunter prays and offers semaa for the hunt before they venture out. The prayer and semaa are offered to the Creation. If the hunter is successful in his hunt and the animal gives of him/herself to the hunter, the hunter will feed and clothe his family and make use of its carcass in its entirely. This reciprocal relationship between Anishinabe and animals is a practice that still exists today.

Anthropocentrism

According to Crist and Kopnina, anthropocentrism is characterised firstly by a credo of exceptionalism that exalts the human in a standalone category among all life forms (Crist and Kopnina, 2014, pp. 387–396). Kopnina further states that anthropocentrism, in its original connotation in environmental ethics, is the belief that value is human-centred and that all other beings are means to human ends (Kopnina *et al.*, 2018, pp. 109–127). Youtt adds:

In the grossest moment of anthropocentrism surrounding these practices, animals are either thought of as objects or actively treated in instrumental terms. More subtly, and more frequently, they are tacitly treated as subjects for the purposes of managing their life-processes, but that subjectivity is understood to be different in kind from that of humans, and therefore not deserving of equal treatment.

(Youtt, 2014, pp. 207–223)

Subsequently, Pointon states, the dominant worldview which evolved over time was anthropocentric in that nature was only valued as a resource for humans to exploit for their own wellbeing (Pointon, 2014, p. 779).

After centuries, this view was substantially challenged in the 1960s (Pointon, 2014, p. 779). At that time, society began to see a change as the continuous growth of technological power led to at least three major contemporary crises: an environmental crisis, an ecological crisis, and lastly a crisis of life in a broader sense (Andreozzi, 2013, pp. 7–10). The environmental crisis in this statement refers to the air that we breathe, the water we drink, and the fact that the land we stand upon is being contaminated by toxic fumes and toxic chemicals released into the water and onto the land. Further, the ecological crisis arises through lack of respect for plants, animals, and those things placed on Earth to sustain humankind. As the efforts to increase economic gains continued, the planet began to show signs of change. Some of these changes to the planet are causing depletion and destruction of elements within and around the world, such as the North Pole melting, oil spills polluting Mother Earth, animals starving and being killed for monetary gain, forests being clear-cut, and waters being contaminated by industrial waste without concern for the environmental impacts. The Earth is being abused and violated to further capitalist gain. Governments around the world need to adhere to the rights and responsibilities they have to the environment. According to Andreozzi, the notion that governments are the agents of the people should be reframed, and they should be elected as stewards of resources for future generations (McIntyre-Mills, 2013, pp. 136–155). A similar view is held by Crist and Kopnina. They state:

> The nonstop advance of the human enterprise – hitherto unproblematic in only having consequences for dismissible others – is ramifying in ways that are jeopardising people and perhaps civilization as a whole: rapid climate change, water shortages, pollution and resource depletions to mention some high-priority issues for human welfare.
>
> *(Crist and Kopnina, 2014, pp. 387–396)*

As the movement to protect animals, plants, land, sacred sites, and the health of people continues to grow, more people are educating themselves about how capitalism is damaging the Earth. In turn parents, teachers, and older members of the community must start teaching children.

Recent studies involving children and adolescents' views of nature displayed some interesting findings. Several studies have found that rural children have a higher degree of 'environmental consciousness', while urban children had feelings of fear, anxiety, and disgust on field trips to the countryside (Pointon, 2014, p. 776). From this study, it appears the perception of young people in relation to nature is impacted by the place in which they reside and the educational field trips they experience in school. This information is crucial because young people will be making decisions about the Earth and industry in the next 20 years. Children need to learn the importance of the natural world. They must learn that the world will only sustain human life if humans pay attention to the effects that industry and economic development have upon the Earth.

Humankind knew it was surrounded by everything it needed to survive. However, as man evolved, the usefulness of nature's abundance changed from a means of sustenance to a means of prosperity. To this day, the mindset towards prosperity is still prevalent around the world. By contrast, in Anishinabe society every decision that is made for the community must consider the next seven generations yet unborn. This teaching empowers leadership to act responsibly by making decisions that ensure the children yet unborn will reap the benefits of the lands, resources, Indigenous Knowledge, and way of life of the Anishinabe.

As people witness the devastating impacts that capitalism has upon the Earth, many are using their voices to address their concerns, and bridges may emerge between those inscribed within a Western tradition and those with Indigenous forms of knowledge. Experts such as David Suzuki, an environmental activist and world-renowned scientist who understands and appreciates the Indigenous worldview of equality of all living things, are speaking up for Mother Earth. Indigenous Women Water Walkers are walking across areas in Canada and the United States for water, demonstrating the importance of this resource for human survival. Greta Thunberg, a youth environmentalist, and Autumn Peltier, a youth clean water advocate, are two young teenaged women who are also speaking for the planet. More people like this need to stand up and speak for the Earth because anthropocentrism is damaging Mother Earth. The next generation of young people must continue to speak for the Earth before anthropocentric views destroy it.

It is my hope that this shortened version of key Anishinabe teachings sufficiently illustrates to the reader how Anishinabe view anthropocentrism. Although this short paper will not convince corporations to stop the devastation of the planet and does not try to provide an alternative to international law, it does speak to the Anishinabe worldview and the belief that anthropocentrism is damaging the Earth. As such, it is a critique of international law and its global extension of anthropocentric destruction.

Crafting Spaces for Indigenous Research

In the last two decades, Indigenous scholars have made significant progress introducing ILT and ILO within mainstream Canada, which may show how similar knowledge could be introduced into international law. Although ILT and ILO have been around since the beginning of time, the concept has only recently gained a foothold in legal discourse within Canada. Even though the culture and traditions of many Indigenous communities have been severely impacted by colonisation, fragments remain. The fragments are being re-assembled, and the laws of the original people of Turtle Island are being resurrected. Much of this work has been undertaken by Indigenous scholars working within higher educational institutions across Canada. As Indigenous scholars gain recognition and respect from scholars around the world for researching and writing from within their inherent paradigms, scholars from other disciplines are being educated on the possibility of radically different ontologies and epistemologies. Indigenous paradigms enable Indigenous researchers to take back and assume control of their search for knowledge (Absolon, 2011, p. 28). Often this includes rethinking the values and beliefs taught to them as children and re-examining the teachings and stories for deeper meaning. For many, this relearning is an incredibly rewarding journey. My journey is no exception and is shared here to illustrate how a non-anthropocentric Anishinabe worldview was used to create a research methodology. It shows how legal orders might be reimagined through a re-exploration of one's own journey in the world and how a more plural, less anthropocentric international law might emerge by acknowledging different situationalities.

A Personal Account of My Quest for Knowledge

The search for deeper meaning of life takes one to many places, some of which may escape privileged Western paths to expertise. My personal journey is presented as an acknowledgment that knowledge has personal dimensions rooted in one's life trajectory. My personal journey conducting research started at Trent University in a PhD program in Indigenous Knowledge. The program includes material written about ILO and incorporates teachings by Elders and Indigenous Knowledge Keepers. Being immersed in study with Elders and Indigenous Knowledge Keepers allowed me to rethink and re-examine the teachings and stories of the Anishinabe.

To begin, I must state that my decision to use an Anishinabe methodology did not come automatically. The Indigenous Research Methodology used in my dissertation came after being told by my supervisor that I needed a conceptual framework for my work. My work concerned residential school survivors and the compensation processes utilised for payment from the federal government (Waboose, 2016). For this research, 24 residential school survivors were interviewed. During an interview session one interviewee stated that he provided many interviews, and each time his words were changed to suit the interviewer. Hearing this, I made a promise not to change any of the words that he nor any other interviewees shared with me, so their words were inserted into my research exactly as shared.

The federal government of Canada wanted to 'civilise' the Indigenous people and assimilate them into mainstream white society so that settlers could continue land grabbing and exploiting the richness they found here. The federal government's plan enacted within the amended *Indian Act* (1894) permitted Indigenous children to be removed from their homes and placed in Indian residential schools (IRS). IRS were funded by the federal government and administered by churches. It was believed by removing children from their home and placing them far away from the influence of their parents and grandparents, the policy would change the heathen ways of the Indigenous children (Waboose, 2021). Former Chief Justice Beverley McLachlin notes: 'The objective – I quote from Sir John A. Macdonald, our revered forefather – was to "take the Indian out of the child". And thus, solve what was referred to as the Indian problem' (Fine, 2015).

This push to assimilate Indigenous children continued throughout the IRS era from the 1800s to 1996, and this attitude still exists today. Throughout the history of IRS, many reports were made about the health of children and conditions of the schools, but the complaints were not taken seriously. For decades, children that attended IRS never spoke about the time they spent there. But this 'learned silence' by survivors changed since the 1980s. The stories shared by most survivors depicted a gloomy and horrendous existence while attending IRS. Most details included, but were not limited to, the trauma of removal from their family and community, unusual and harsh treatment upon arrival at the school, loneliness, isolation from siblings, cruel and unusual punishment, sexual abuse, spiritual abuse, and death (Waboose, 2016; 2021).

As I sat contemplating the promise that I made to the residential school survivors I interviewed (I will explain this promise as the final part of my methodology below), I remembered the Midewiwin teachings of the Seven Fire Prophecies, and the '**voice**' of the Ma-tig-wa-kik Day-way-gun (Waterdrum) spoke to me. The vision reminded me of these words, 'if the people remained strong the Waterdrum would sound its voice again'. From that point onward, the essence of the Waterdrum led me through my research. The interconnectedness between the teachings and my research began to reveal itself and correlate before my eyes. I was able to draw an analysis between the teachings and the subject matter of my dissertation. The feeling that ran through my body convinced me this was the appropriate methodology. I have been told that when feelings such as this run through a person's body, it is the spirit of the individual responding to the situation. This is one of the most treasured moments of my study.

An Anishinabe Research Methodology

An Anishinabe Research Methodology, presented here as a methodology that is deeply rooted in a sense of self and wary of Western notions of the superiority and separateness of the mind from the body and the world, contains pieces of the lived experience, teachings, Indigenous Knowledge, and Blood Memory (memory of the ancestors) of the author. Reflecting upon this day reminds me of how immersed int mainstream academia a PhD student becomes. Although I am an Anishinabe Kwe studying in an Indigenous Studies program, colonial methodologies were at the forefront of my mind. It was only when asked to think of a conceptual framework for the research that my thoughts changed, and I began the search for a framework that was connected to my cultural identity.

Upon further research, I discovered literature written by Indigenous scholars about Indigenous research methodologies. It was reassuring to learn how other authors were advocating the use of Indigenous methodologies to resonate with our unique perspectives. According to Wilson, Indigenous methodologies are now being used in the research of Indigenous scholars throughout the world. He states that:

> Within the past decade though, research and researchers have begun to change. More is being done to bring Indigenous communities into the research process, and the usefulness of the research is becoming more visible and beneficial to communities. A precursor for this change has been the growing number of Indigenous people who have excelled in academia and who focus their study on their own peoples. These new Indigenous scholars have introduced Indigenous beliefs, values and customs into the research process and this in turn has helped research to become much more culturally sensitive to Indigenous peoples.
>
> *(Wilson, 2008, p. 15)*

By using research methodologies that are culturally relevant and sensitive to Indigenous people, Indigenous researchers can use their stories as theoretical frameworks within which they can interpret other stories, teachings, and experiences. Margaret Kovach explains that:

> Conceptual frameworks make visible the way we see the world. Within research, these frameworks are either transparent (i.e., through form) or not, yet always present. The rationale for explicit representation of one's conceptual framework is that it provides insight into a researcher's beliefs about knowledge production, in general, and how those beliefs will impact the research project. The content and form of the conceptual framework itself assists in illustrating the researcher's standpoint, thus giving the reader insight into the interpretative lens that influences the research.
>
> *(Kovach, 2009, p. 41)*

Kathleen Absolon further argues that:

> The past, present and future intersect and much of our research is about searching for truth, freedom, emancipation and ultimately finding our way home. Finding our way home means searching to return to our roots and finding the dignity and humanity intended by the Creator.
>
> *(Absolon, 2011, p. 55)*

The Anishinabe conceptual framework of the Waterdrum incorporated into my research includes six elements that are required to dress the Waterdrum: the tree, the stones, the deer hide, the deer hide lacing, the water, and the wooden plug. Simpson, Kovach, and Absolon state that using our own conceptual framework helps Indigenous researchers, such as myself, better understand Western research practices. Kovach further states that:

> We carry our framework, which is not inherently good or bad, around with us and it is through this framework that we view the data.
>
> *(Kovach, 2009, p. 41)*

As previously stated, sitting and contemplating a conceptual framework for my research, I saw correlations between the Seven Fires Prophecies (Benton-Banai, 2011, p. 93) and the story of the residential school survivors. The story of the resurrection of the Waterdrum is told in the

Seven Fires Prophecy teachings. The Seventh Fire prophecy speaks to specific events that would occur because of the Sixth Fire. The seventh prophecy specifically states that:

> The seventh prophet that came to the people long ago was said to be different from the other prophets. He was young and had a strange light in his eyes. He said, 'In the time of the Seventh Fire an Osh-ki-bi-ma-di-zeeg (New People) will emerge. They will retrace their steps to find what was left by the trail. Their steps will take them to the Elders who they ask to guide them on their journey. But many of the Elders will have fallen asleep. They will awaken to this new time with nothing to offer. Some of the Elders will be silent out of fear. Some of the Elders will be silent because no one will ask anything of them. The New People will have to be careful in how they approach the Elders. The task of the New People will not be easy.
> If the New People will remain strong in their quest, the Waterdrum of the Midewiwin Lodge will again sound its '**voice**' (emphasis added).
>
> *(Benton-Banai, 2011, p. 93)*

The correlation drawn between the Seven Fires Prophecy and the subject matter of my research was the '**voice**' within the Waterdrum. In my work, the residential school survivors wanted their voices to be heard rather than the voices of the researcher/writer. I knew their '**voices**' needed to be heard, just as the voice within the Waterdrum needed to be heard in the Seven Fires Prophecy.

This story tells how a tree transformed into a Waterdrum:

> The old man knew that some essential part of the Lodge was missing. He tried to relive the time that he was with the Seven Grandfathers to see if there was some part of the instructions that he had missed. He meditated all day to try to discover the missing part of the ceremony. When the Sun was sinking in the West at the finish of the day, the old man turned from watching the sunset and looked to the East. He saw in the eastern sky what looked like a huge tree. The tree appeared to be coming closer. The old man thought, 'This must be what is missing'. As he watched the tree turned into the very same vessel that the Seven Grandfathers had in front of them when he visited their Lodge in his childhood.
>
> *(Benton-Banai, 2011, p. 68).*

The Anishinabe that follow the teachings of the Midewiwin Society use two different types of Waterdrums in the Midewiwin Lodge: the Grandfather Drum and the Little Boy Waterdrum (Benton-Banai, 2011, p. 71). The Grandfather Drum can be recognised by the hoop placed at the top of the drum, while the Little Boy Waterdrum is tied together with seven small, round stones. In this section, when referring to the Waterdrum, it is the Little Boy Waterdrum being referenced.

The Waterdrum is constructed using various elements taken from Mother Earth. The elements required to assemble a Waterdrum include a hollowed tree, a piece of deer hide, seven round stones, approximately six feet of leather lace, a wooden plug, and water. Once the various parts are gathered, they are placed on the Drumkeeper's altar (a place where sacred objects are placed) in preparation for dressing. Putting the pieces of the Waterdrum together is referred to as dressing the Little Boy.

The teaching of the first Midewiwin ceremony describes a very important aspect of the Waterdrum. In this teaching, it is stated that after the Little Boy was dressed, the Old Man pulled the plug out of the hole of the wooden drum and then held the drum up to the Creator, and he blew his breath into the drum (Benton-Banai, 2011, p. 71). This teaching illustrates how the Creator gave the Waterdrum life and how the breath of the Creator is linked to all life (Benton-Banai, 2011, p. 71). Breathing life into the Waterdrum is similar to the first breath of a new-born baby. When a new-born takes its first breath, the baby releases a loud wail. Similarly, a Waterdrum makes an unbelievably beautiful sound when it is tapped with a drumstick. The first breath represents new life and a new beginning.

Upon studious contemplation, this analogy emerged: the dressing of the Little Boy can be compared to preparing for a research project. When assembling a Waterdrum, the builder must search for a tree that will be used to construct the drum. Not just any tree will suffice. It must be a tree which can perform the duties required of a Waterdrum. The following pages will explain the various parts of the Waterdrum. Note that each of the elements that are used to construct the Waterdrum are elements taken from Creation: the hollowed base is carved from a tree, the deer hide from an animal that gave itself to the hunter, the round stones from the shores of Lake Superior, the water from a river, lake or stream; and tobacco placed in the wooden base picked from the Earth. The various parts of the Waterdrum are explained below to illustrate how I interpreted their usage within my research.

Tree: Anishinabe Conceptual Framework

The hollowed tree stump carved into the Waterdrum was the conceptual framework employed for my dissertation. This framework was borrowed from the teachings of the Waterdrum transmitted in the Three Fires Midewiwin Lodge. The drum is the heartbeat of Mother Earth. As long as the heart continues to beat, the life of the Anishinabe people will remain. The powerful and beautiful voice of the Anishinabe is the being that accompanies the sounding of the Waterdrum. The importance of the Waterdrum should never be underestimated, for it is the voice of the Anishinabe. In my work, it was the voice of the residential school survivors.

Stones: Methodological Principles

The stones tied around the Waterdrum are used to illustrate the various methodological principles required to undertake this study. To illustrate these principles, this section is divided into subsections to represent the stones used to tie the Waterdrum. In the circle of stones that fastens the deer hide to the base of the drum, no stone is more important than the other. Each of the seven stones used to tie the Waterdrum are equally important to the overall construction of the Waterdrum as are the methodological principles of a dissertation.

Stone 1: Relationality

In Anishinabe society, as previously mentioned, relationships are extremely important to the peaceful co-existence of the community. Relationships not only exist within the human realm but also within the spiritual and physical realms as well (Absolon, 2011, p. 125). According to Anishinabe teachings, all living things within the universe are interconnected therefore relationships are crucial to the survival of humankind and the universe. In my research on residential school survivors, personal relationships were utilised.

Stone 2: Outsider/Insider – Respecting Local Protocols

Another consideration requiring prior thought and consideration was how to enter each Aboriginal community without offending community members. Going into an Aboriginal community without prior contact or a contact person would be a grave mistake because community members do not think favourably of research conducted within their community by outsiders. According to Linda Smith:

> Most research methodologies assume that the researcher is an outsider able to observe without being implicated in the scene. This is related to positivism and notions of objectivity and neutrality ... Indigenous research approaches problematize the insider model in different ways because there are multiple ways of both being an insider and an outsider in indigenous contexts. The critical issue with insider research is the constant need for reflexivity.
>
> *(Smith, 1999, p. 137)*

This sort of question typically eludes international legal research methodology, including when it comes to Indigenous peoples, who have a long history of being studied without permission for the purposes of producing normative outcomes that are ultimately detrimental to them.

Stone 3: Ethical Considerations

There were many ethical considerations to consider before undertaking my research. Depending upon the type of research undertaken, the type of questions that could be asked was determined by ethical parameters. I was committed that no question should cause potential harm during the interview process. Some of the ethical considerations to be considered prior to interviewing survivors had to be examined before the research is undertaken. Every possible situation that could arise has to be examined and a response and plan considered beforehand. Again, this can be contrasted with an approach to 'victims' in international law that is often instrumental, generic, and not steeped in ethical questions about what it means to speak for others, especially for those who have less power and privilege.

Stone 4: Creating the Time and the Place

Another consideration decided beforehand was the place where the interview would be conducted and the time required. To ensure the individual felt safe and comfortable each participant states the time and place for the interview. Safety is one of the most important aspects of the interview process.

Stone 5: Reciprocity

Reciprocity in Aboriginal communities is taught when children are very young. For everything a person takes from Creation, something must be given back. The practice of offering tobacco is important to Aboriginal people in many areas of Turtle Island. Giving back can take many shapes and forms, such as gifts and information. In a research context, Kovach states:

> They say that we traditionally knew about portal, the doorway, how to get knowledge and that it was brought to the people by sharing, by community forums, by sitting in circles, by engaging in ceremony, by honouring your relationship to the spirit. When we do that, the spirit will reciprocate and we will be given what we are needed.
>
> *(Kovach, 2009, p. 154)*

The belief in reciprocity dates back to the Original Instructions that were given to man when lowered to the Earth. Reciprocity is not alien to international law, but it is not a common principle guiding mainstream international legal research. Thinking about what it might entail in that context can be enriched by non-anthropocentric research methodologies, such as that derived from the Anishinabe Waterdrum – methodologies that acknowledge reciprocity and respect as central to acquiring knowledge.

Stone 6: The Interview Questions and Process

Part of the interview process includes questions formulated before the interview takes place. In my work, one of the important parts of data gathering was the type of questions that would be asked and the process that would be taken to find the answers.

Stone 7: Cultural Protocols

When conducting research in an Aboriginal community, cultural protocols are important to consider when planning the research strategy. Learning proper protocol for various Aboriginal communities is extremely important for conducting the research in a respectful manner because not all protocols are the same in every Indigenous community. Simpson states:

> When we put our tobacco down and ask for help to solve a problem, to come up with a strategy or so that the Stone we threw ripples through the world in a positive way, we are asking the implicate order to visit our action.
>
> *(Simpson, 2011, p. 146)*

Again, attention to cultural protocols might enrich international law research by emphasising the situatedness of any legal tradition, the need to understand and respect it on its own terms, and to do the deep cultural legwork required to not simply incorporate it in one's pre-existing belief system.

Deer Hide: Positionality and Self-Care

The deer hide stretched over the wooden base of the Waterdrum represented the positionality within the research. An interview process can be emotionally draining. For this reason, it is crucial that a person takes care of themselves, mentally, physically, emotionally, and spiritually.

Deer Hide Lacing: Important Considerations

The lacing securing the deer hide over the top of the drum is seen as the considerations taken into account before conducting the research. Before the interview process started, the implications of the research had to be considered from many different angles for both researcher and those who will be interviewed. Before the interview process begins, every stone has to be turned over and thought through to ensure no preventable harm will come to the participants.

Water: People, the Participants of this Research

The water poured into the Waterdrum represented the participants that were interviewed: the life-force of the study. Choosing who will be interviewed must be considered in the sense that, will I get the information that I am searching for.

Valarie G. Waboose

The Wooden Plug: My Word, the Promise I Gave

The wooden plug that secures the water in the Waterdrum represents the promise made to the participants interviewed. While being interviewed, several spoke adamantly about not being quoted properly or their story not being told accurately in other research studies in which they had participated. These statements were deeply disturbing, and as a result, I gave them my word that they would not be misquoted. For many the process was intimidating and difficult. And for some, being interviewed yet again was unwarranted and seen as another exercise of being misunderstood or misinterpreted.

While this may be an unrealistic expectation, I gave my 'word'. Further, I sealed my interview commitment with the tobacco offered to the participant, therefore I was bound by my promise. Within Anishinabe society when a person gives their 'word' that they will do something for another person, their word is generally accepted as a given. For the Anishinabe, giving your word is a sacred covenant. This is especially true because I was speaking to Elders who understand and practice these sacred ways. So, when I gave my word to the survivors, I made a commitment to them to share the truth as they spoke it.

Conclusion

I want to thank you for taking this journey with me. I have a deeply embedded worldview that reflects the Anishinabe teachings. An understanding that views all living things within the world as being equal. In the natural order of things or Law of the Orders, man is of least importance to the continuation of this planet. This is the underlying reason Anishinabe, such as I, see all of Creation through a non-anthropocentric lens and learn from nature in formulating our laws and research methodologies. Without humankind, capitalism, or international law, Mother Earth could sustain itself. However, humankind cannot survive without the riches of this planet. It is with this understanding that this chapter was shared.

In my view, anthropocentrism is in direct conflict with the Original Instructions given to mankind by the Creator. Because of the impact that anthropocentrism is having upon the Earth, it has never been more urgent to rethink international law's relationship to nature, capitalism, and appropriation. Anthropocentrism must be re-evaluated and included in future dialogue to address the narrow and inaccurate understanding of nature and animals and their place within the world. Those that disagree with anthropocentrism must continue to unsettle and query the discourse before the world is completely depleted of its natural resources.

My personal feelings about the state of our planet largely reflects the understandings about my place in the natural order of Creation and other teachings reflected in this paper. It is my belief that possessing Indigenous Knowledge creates a responsibility. In this regard, my responsibility is to write and educate others about the Anishinabe worldview (including its legal orderings, traditions, and research methodologies) towards anthropocentrism and how this mentality is impacting the planet. It is my responsibility, but it is not mine alone. It is also the responsibility of all peoples around the world. Anishinabe knowledge is not universalistic in the way that international law claims to be (a law applicable to all peoples) but is universalistic in that its particularism speaks usefully to universal concerns that should animate international law.

There is an urgency to help the planet survive for the next seven generations. People around the world need to seriously assess the impacts that capitalism is having upon the Earth and better understand the mediating role that international law has long had in propagating and legitimising capitalism. They must accept they were placed last on Earth and relearn how to treat other beings with respect. Humankind must understand completely that they are not needed on Earth for the natural world to survive. What they must learn is that they, humankind, need the natural world to survive. If corporations are allowed to continue to destroy the natural world, they will eventually destroy themselves.

Economic gains should not be the driving force when assessing the future of this planet. There must be a balance between what is needed for human survival and what is wanted by corporations. Capital gains for corporations must not be the deciding factor; the impacts upon the natural world must be the deciding factor. Thinking about the future of the planet is the only way the next seven generations will reap the benefits of this planet.

As an Anishinabe Kwe who views the Original Instructions of the Creator as supreme law, I am suggesting that the law's relationship to anthropocentrism – from the substance of law to the way knowledge is acquired, and the manner in which research conducted – requires serious re-examination because it is endangering the planet. As part of my responsibility, I share this paper and its non-anthropocentric Indigenous research methodology, which is based on and reflects the Original Instructions given by the Creator to the Anishinabe on Turtle Island.

References

Absolon, K.M.E. (2011) *Kaandossiwin: How We Come to Know*. Halifax: Fernwood Publishing.

Andreozzi, M. (2013) 'Relationships over Entities in Relations', *Beyond Anthropocentrism*, 1(1), pp. 7–10.

Asch, M., Borrows, J. and Tully, J. (2018) *Resurgence and Reconciliation: Indigenous - Settler Relations and Earth Teachings*. Toronto: University of Toronto Press.

Benton-Banai, E. (2011) *The Mishomis Book: The Voice of the Ojibway*. 2nd ed. Minneapolis: The University of Minnesota Press.

Berard, R.E.M. (2008) 'Keepers of the Water: Nishnaabe-kwewag Speaking for the Water', in Simpson, L. (ed.) *Lighting the Eighth Fire: The Liberation, Resurgence, and Protection of Indigenous Nations*. Winnipeg: Arbeiter Ring, pp. 89–106.

Borrows, J. (2010) *Drawing Out Law: A Spirits Guide*. Toronto: University of Toronto.

Craft, A. (2016) *Breathing Life into the Stone Fort Treaty: an Anishinabe Understanding of Treaty One*. Saskatoon: Purich Publishing Limited.

Craft, A. (2019) 'Navigating Our Ongoing Sacred Legal Relationship with Nibi (Water)' in Borrows, J., Chartrand, L.N., Fitzgerald, O.E. and Schwartz, R. (eds.) *Braiding Legal Orders: Implementing the United Nations Declaration on the Rights of Indigenous Peoples*. Waterloo: Centre for International Governance Innovation, pp. 100–110.

Crist, E. and Kopnina, H. (2014) 'Unsettling Anthropocentrism', *Dialectical Anthropology*, 38(4), pp. 387–396. DOI: 10.1007/s10624-014-9362-1

Dewdney, S.H. (1975) *The Sacred Scrolls of the Southern Ojibway*. Toronto: University of Toronto Press.

Fine, S. (2015) 'Chief Justice Says Canada Attempted "Cultural Genocide" on Aboriginals', *The Globe and Mail*, 28 May [Onlne]. Available at: https://www.theglobeandmail.com/news/national/chief-justice-says-canada-attempted-cultural-genocide-on-aboriginals/article24688854/ (Accessed: 30 January 2021).

Johnston, B. (2001) *Manitous: The Spiritual World of the Ojibway*. St. Paul: Minnesota Historical Society Press.

Johnston, B.H. (2003) *Honour Earth Mother: Mino-audjaudauh Mizzu-Kummik-Quae*. Wiarton: Kegedonce Press.

Kopnina, H., Washington, H., Taylor, B. and Piccolo, J.J. (2018) 'Anthropocentrism: More than Just a Misunderstood Problem', *Journal of Agricultural and Environmental Ethics*, 31(1), pp. 109–127. DOI: 10.1007/s10806-018-9711-1

Kovach, M. (2009) *Indigenous Methodologies: Characteristics, Conversations and Contexts*. Toronto: University of Toronto Press.

McIntyre-Mills, J.J. (2013) 'Anthropocentrism and Well-being: A Way Out of the Lobster Pot?', *Systems Research and Behavioral Science*, 30(2), pp. 136–155. DOI: 10.1002/sres.2131

McNab, D. (1992) *Review Article in Treaties and an Official Use of History. Canada: Indian Treaties and Surrenders from 1680 to 1890. Volume 1, 2, 3*. Saskatoon: Fifth House Publishers.

Pointon, P. (2014) '"The City Snuffs out Nature": Young People's Conceptions of and Relationship with Nature', *Environmental Education Research*, 20(6), pp. 776–794. DOI: 10.1080/13504622.2013.833595

Simpson, L. (2008) *Lighting the Eighth Fire, The Liberation, Resurgence and Protection of Indigenous Nations*. Winnipeg: ARP Books.

Simpson, L. (ed.) (2011) *Dancing on our Turtle's Back*. Winnipeg: Arbeiter Ring Publishing.

Smith, L.T. (1999) *Decolonizing Methodologies: Research and Indigenous Peoples*. Dunedin: University of Otago Press.

Starblanket, G. and Kiiwetinepinesiik Stark, H. (2018) *Towards a Relational Paradigm – Four Points for Consideration: Knowledge, Gender, Land and Modernity in Resurgence and Reconciliation: Indigenous – Settler Relations and Earth Teachings*. Toronto: University of Toronto Press.

Waboose, V.G. (2016) *Re-Living the Residential School Experience: An Anishinabe Kwe's Examination of the Compensation Processes for Residential School Survivors*. Peterborough: Trent University.

Waboose, V.G. (2021) 'The Children Have Awakened Canada', *TWAILR: Reflections*, 12 August [Online]. Available at: https://twailr.com/the-children-have-awakened-canada/#easy-footnote-bottom-12 -4130 (Accessed: 6 October 2022).

Wilson, S. (2008) *Research Is Ceremony: Indigenous Research Methods*. Black Point, Nova Scotia: Fernwood Publishing.

Youatt, R. (2014) 'Interspecies Relations, International Relations: Rethinking Anthropocentric Politics in Millennium', *Journal of International Studies*, 43(1), pp. 207–223. DOI: 10.1177/0305829814536946

15

NON-HUMAN ANIMALS AS EPISTEMIC SUBJECTS OF INTERNATIONAL LAW

Vincent Chapaux

Introduction

It is no secret that there is a strong project in current scholarship whose aim is to conceptualise an alternative legal organisation of the human/non-human nexus. The contours of this academic project are not easily defined, as scholars involved are not always (in fact are usually not) from a legal background. The project arose in the field of philosophy before reaching the social sciences and the humanities (Deckha, 2011, p. 212). I count the work of philosophers such as Peter Singer, Tom Regan, Sue Donaldson, Will Kymlicka, and Corinne Pelluchon as being part of this project together with biologists such as Marc Bekoff and (international) legal scholars such as Anne Peters. What all these works have in common is their will to offer alternative legal propositions to the current organisation of human/non-human relations. They might not be published in legal journals or collections, but I will refer to them as the current animal law scholarship as they are all explicitly referring to the way animal law is (or should be).

At first sight, current animal law scholarship seems to be an intrinsically post-anthropocentric project. First, because its very purpose is to bring back focus on non-human animals in a legal system that is currently profoundly structured around the Anthropos. Second, because most of the authors of this field embrace the normative position that human and animal interests should be better balanced. In fact, it is hard to find a single piece of academic literature in the last decade that would be pleading for the *reinforcement* of animal exploitation, a plea that would arguably be redundant considering the current situation. In the specific field of international law, even authors who embrace a fairly descriptive approach, position themselves normatively in favour of profound legal changes in favour of the animals, such as their legal personification in order to grant them rights (Peters, 2021, p. 421).

But on close inspection, the general post-anthropocentrism of current animal legal scholarship should be nuanced because the notion of anthropocentrism accepts at least three definitions.

First, there is a perceptual anthropocentrism, i.e., a way to see the world that is informed by data received or gathered by the senses of the human body. In this most basic understanding, every human worldview is anthropocentric as it cannot be informed by anything else than the human senses. Second is a descriptive anthropocentrism, a worldview that 'in some way begins from, revolves around, or is ordered according to the species *Homo sapiens*'. In this form

DOI: 10.4324/9781003201120-19

of anthropocentrism 'objects of contemplation are defined (...) by reference to, by comparison with or in terms of their relations to the Anthropos that is the centre of the paradigm'. Third, there is a normative anthropocentrism, a worldview of a prescriptive nature, in which the centrality or the superiority of the Anthropos is either implicitly or explicitly considered normal or just (Mylius, 2018).

I argue that animal law scholarship has been mainly focused these last decades on the question of normative anthropocentrism. Numerous authors have identified the normative manifestations of the anthropocentrism of the current (international) legal system. They often have suggested alternatives to restructure this system in a post-anthropocentric fashion. But while doing so, the field has been somewhat blind to the profound *descriptive* anthropocentrism of the (international) legal system, whose main notions and structures were crafted with the Anthropos as a central epistemological figure. Consequently, and because the deconstruction of the descriptive anthropocentrism of legal systems has not occurred, the current propositions of animal legal scholarship are still profoundly structured around legal notions (rights, territory, citizenship, etc.) that have been crafted by and for the Anthropos before being 'extended' to non-human animals. In short, my point in the first part of this chapter is that the post-anthropocentric project in (international) animal law is still profoundly anthropocentric in its legal imaginary.

In the second part of the chapter, I explore what it would mean to include animals in the construction of an alternative legal imaginary. I first and foremost underline the important risks that are involved when bringing animals or 'nature' into the construction of international law, a habit traditionally reserved to natural law theorists and social Darwinists. I follow up by bringing in the philosophy of Vinciane Despret to see how animals could be considered not only as bearers of rights and duties but as epistemic subjects whose worldview could help us redefine the legal structure of both human and non-human societies.

A Post-Anthropocentric Project with an Anthropocentric Legal Imaginary

Current post-anthropocentric projects in animal law are often the heirs of the way the animal question has been structured in Anglophone philosophy at the end of the twentieth century in books such as *Animal Liberation* (Singer, 1977), *The Moral Status of Animals* (Clark, 1977), *The Case for Animal Rights* (Regan, 1983), *Moral, Reason and Animals* (Sapontzis, 1987), *Animals, Property and the Law* (Francione, 1995), and *Animals and Why They Matter* (Midgley, 1998).

These writings have in common four important traits. First, they are all involved in a form of ethical philosophy, meaning that they are dedicated to two major questions: 'What should I do?' and 'What sort of person should I be?' (Shafer-Landau, 2012, p. xi). Second, all these authors start to answer these questions with the idea that something is not ethically right with the way current societies organise the relationships between humans and animals. Third, they consider that, in current societies, humans have been given too much power over animals. Fourth, they all prescribe solutions that would change this power balance *in favour* of animals. In that sense, and even though the differences between these authors are numerous (Jeangène Vilmer, 2008), this literature is unquestionably engaged in a fight against normative anthropocentrism.

At the same time, this literature is profoundly and unapologetically anthropocentric on a descriptive level as one of its main goals is to prove that animals are, to a certain extent 'just like humans' and should be treated 'the same way'. Based on this logic of non-discrimination, it defines animals *by reference to* or *by comparison with* the Anthropos.

This is the explicit argument of Peter Singer who presents the fight for animal liberation as an extension of the fight for 'Blacks and Women' (Singer, 1977, p. 1). His categories of thinking are therefore logically structured around an idealised *human* figure, and the protection animals should receive is measured with regard to the proximity of these animals with an abstract *Anthropos*:'As long as we remember that we should give the same respect to the lives of animals as we give to the lives *of those humans at a similar mental level*, we shall not go far wrong' (Singer, 1977, p. 22).This does not mean, of course, that the interests of humans shall systematically prevail over those of the animals. Quite the opposite. In certain instances, the interests of certain animals could be equal or even superior to those of a particular human being. But the arbitration between these interests still stands to be realised by reference to, or by comparison with the abstract figure of the *Anthropos*:'The preference, in normal cases, for saving a human life over the life of an animal when a choice has to be made is a preference based on the characteristics that normal humans have' (Singer, 1977, p. 22).

The philosophy of Tom Regan is similarly constructed.The rights and duties of animals are structured around the notion of moral agents.These moral agents are not defined abstractly but explicitly construed on what he calls 'normal adult human beings'. Consequently, the distribution of rights and duties amongst animals is based on criteria presented as objectives (those of the moral agents) but which are in fact the characteristics of the human adults (self-awareness, experiential welfare, etc.) (Regan, 1983, p. 153).

This general project of 'extension' to the animals of human social, political, and legal structures has been pursued and refined in more recent writings. One of the most iconic works in that regard is Sue Donaldson and Will Kymlicka's *Zoopolis* (Donaldson and Kymlicka, 2011).The project of this book is to explore how various political and legal concepts could be 'adapted from the human to the animal context' (Donaldson and Kymlicka, 2011, p. 14). What these authors are suggesting is to extend the liberal theory of the state and international relations to our relationships with animals (Braverman, 2015). Central notions therefore include sovereignty, territories, citizenship, or migration. The general idea is to consider domesticated animals as citizens; wild animals as sovereign on separate territories; and liminal animals (those who share a territory with us without being domesticated – such as pigeons or foxes) as 'denizens', i.e., quasi-citizens with a reduced set of rights and responsibilities (Donaldson and Kymlicka, 2011, p. 214).

These academic writings are profoundly intertwined with numerous forms of legal activism centered on the extension of human legal protections to the animals such as the *Great Ape Project*, pleading for the extension of the right to life, the protection of individual freedom, and the prohibition of torture to the great apes in the name of their similarity with the human species (Cavalieri, 2015). This logic is also currently tested in various tribunals around the world. Procedurally, lawyers mainly resort to the use of the notion of *habeas corpus* and try to extend its application to non-human animals. The first known writ of habeas corpus on behalf of a non-human being was filed in Brazil in front of the 9th Salvador Criminal Court.The petitioners were seeking the release of Suíça, a chimpanzee who was detained at the Zoo Getúlio Vargas.The writ was built explicitly on the work of some of the above-mentioned authors, such as Peter Singer and Gary Francione.The main argument was that the biological proximity of humans and chimpanzees should trigger the application of the same protection to both species (Santana and Santana, 2005). From 2014 onward, the same logic was used in the United States in numerous cases, by the Nonhuman Rights Project, an association presided by Steven Wise, whose vocation was apparently triggered by reading Peter Singer's book in the 1970s (Nonhuman Rights Project, 2022). The same argument has also been used in Argentina and Colombia (Chapaux, 2015).

In short, these projects seek to get past the most anthropocentric facets of our legal systems, but they are doing so by using legal categories that have been conceptualised by and for the

Anthropos. A valid question would be to ask if notions built by one group can ever be used to build a profoundly critical approach of the said group (Lorde, 2003). In the case of anthropocentrism, it might prove very difficult, as the very process of using human categories to describe the world is a performative act of normative nature in favour of the human. To a certain extent, descriptive anthropocentrism is always intrinsically normative, although in a less active way (Mylius, 2018, p. 184). Describing *is* prescribing (Bourdieu, 1981). But in this chapter, my goal is not to *assess* these projects in any way. I have no normative claim regarding these authors and activists. My only point is to highlight that they have been focusing on the deconstruction of one specific form of normative anthropocentrism. And I think that it is conceivable that we could rethink social and legal notions not only to the benefit of the animals but to a certain extent *with* them.

Epistemic Humility: Bringing Animals back into the Construction of International Law

Inviting the animals to the table of the legal imaginary of our societies might seem appealing, but it is an unquestionably dangerous road. To avoid the pitfalls of previous legal traditions and philosophies, I relied on the work of philosopher Vinciane Despret. With her help, I explore how animals could help us rethink central notions of international law such as the concept of territory.

A Dangerous Path

To start, it is important to underline that the current separation between human and non-human in the conceptualisation of legal norms is, historically, the exception rather than the norm. For the longest time, the quest for the 'natural' law of societies has been one of the main endeavours of legal scholars and philosophers. 'Nature' and 'animal behaviour' were mobilised to establish the legitimacy of various sets of social and legal norms.

This tradition is sometimes complicated to fathom when observed from a twenty-first century perspective, as it relied on a completely different understanding of the notion of nature:

> Not very long ago, one could delight on the curiosities of the world without making any distinction between the information obtained from observing animals and that which the mores of antiquity or the customs of distant lands presented. 'Nature was one' and reigned everywhere, distributing equally among humans and nonhumans a multitude of technical skills, ways of life, and modes of reasoning. Among the educated at least, that age came to an end a few decades after Montaigne's death, when nature ceased to be a unifying arrangement of things, however disparate, and became a domain of objects that were subjects to autonomous laws.
>
> *(Descola, 2013, p. xv)*

The problem with this 'natural law' approach is that it relied entirely on the way 'nature' or 'animal behaviour' was constructed, and this construction could differ enormously from one author to another.

The Tavoillot brothers have demonstrated this point very elegantly. Pierre-Henri Tavoillot teaches philosophy at Sorbonne University. His brother François is a beekeeper in Haute Loire. Together they studied how the figure of the bee was mobilised by political philosophers

Animals as Epistemic Subjects

throughout the ages and showed that it had been used to plead in favour of almost every imaginable type of political organisation.

For Seneca the Younger, Xenophon, Pliny the Elder, and Thomas Aquinas the beehive constitutes living proof that Nature favours monarchy as an ideal form of political organisation. Political theorists who show a preference for a strict categorisation of human groups, on the other hand, see in the beehive a confirmation that aristocracy (or a form of caste system) is the most efficient way of structuring human communities. Republicans, for their part, mobilised the figure of the bee to glorify its laborious and collectivist character and deny any legitimacy to the idea of monarchy. According to them, in the beehive, the real power rests in the worker bees, who only tolerate the queen and the males to the extent that they are useful for the perpetuation of the species. The bee was also mobilised in anarchist writings. Proudhon, for example, sees in the beehive a horizontal and contractual social organisation. No vertical logic would govern it. The beehive would be an anarchic prototype as the freedom of each individual would not depend on any institution and would only be limited by other individuals. Finally, the beehive would be a feminist project, as it would be the example of an efficient matriarchal system (Tavoillot and Tavoillot, 2015).

This simple list of examples shows that the mobilisation of animal behaviour to justify the legitimacy of one particular sociolegal organisation is highly problematic as, historically, it appears to consist in the construction of a version of the natural world that would be similar to one's political project, before reimporting this reified version of 'nature' into the human political debate as an argument to legitimise its natural essence and timelessness.

Aside from this well-known tradition, other and more recent academic projects have offered to use nature and the animals to propose specific norms. Numerous projects were pursued during the nineteenth and the early twentieth century, with theories trying to borrow the legitimacy of biology, as a science, to prescribe legal and political actions such as human population control, human 'races' hierarchisation or biological determination of criminal behaviour (Lamy, 2014, p. 178). Most of these research agendas were abandoned after the Second World War, a conflict during which the link between political projects and human or animal biology was durably discredited.

But the will to use biology, in particular animal biology, to prescribe modes of legal organisation did not die entirely in that period. One of its last incarnations was the *sociobiology* project of entomologist Edward Wilson. Wilson's thought was summarised in his 1975 book *Sociobiology: a new synthesis* (reprinted in 2000 (Wilson, 2000)), in which he offers to create a general theory of social behaviour, based on the Darwinian evolutionary principles, and applicable to both humans and animals. In his theory, various highly complex human behaviours, such as watching a football game (Wilson, 2000, p. 143), bonding with friends as teenagers (Wilson, 2000, p. 703), and performing arts (Wilson, 2000, p. 704) are explained, at least in part, by the need to provide some form of evolutionary advantage to the human species. As Wilson explained later, his theory is based on how he sees the human mind:

> I believe that the human mind is constructed in a way that (...) forces it to make choices with a purely biological instrument. If the brain evolved by natural selection, even the capacity to select particular aesthetic judgments and religious beliefs must have arisen by the same mechanistic process (...). The brain exists because it promotes the survival and multiplication of the genes (...). The human mind is a device for survival and reproduction and reason is just one of its various techniques.
>
> *(Wilson, 1978, p. 2)*

Based on this Darwinist logic, Wilson proceeds in establishing what certain human social rules ought to be, from the general definition of ethical standards of justice (Wilson, 2000, p. 701) to ones structuring the organisation of human sexuality (Wilson, 1978). The whole project is based on the observation of animal behaviour.

Wilson's project triggered an enormous amount of reaction inside and outside academic circles (Lamy, 2014) and even though research in *sociobiology*, *per se*, is not currently an active field of investigation, the influence of Wilson's work on specific research agendas is still quite important. Entire journals based on Wilson's general idea are currently active, particularly in the field of psychology (evolutionary psychology) and economics (e.g., bioeconomics, Collins, Baer, and Weber, 2016).

All these projects show that one of the other dangers of bringing animals back into the construction of legal structures is to import with them some form of social reductionism; in Wilson's case, the idea that social rules are or should be constructed to provide an evolutionary advantage to the human species.

With a Little Help from Vinciane Despret

The previous pages are not pleading in favour of bringing animals back into the process of legal creativity. So far, animals seem to have been instrumentalised as a rhetorical argument to sustain already anthropocentrically defined political projects or used to oversimplify human behaviour through various forms of biological reductionism. Before letting go of the idea altogether, however, I wanted to explore how Vinciane Despret's philosophy could offer a way of involving animals in legal conceptualising without stumbling into the above-described pitfalls.

Vinciane Despret is a Belgian philosopher who has been writing for more than twenty-five years on the philosophy of ethology and on human–animal relations (see Buchanan, Chrulew, and Bussolini, 2015). She brings animals in to 'open up our imagination to other ways of thinking, to break away from certain routines (…) to make other stories possible' (Despret, 2022, p. 134) and to 'multiply worlds' (Despret, 2022, p. 29). At the heart of her philosophy lies a profound decentring, a move away from the human point of view to embrace an animal perspective. The goal is to try and think like an animal (Despret, 2015). In this regard, Despret's proposition can be seen as an extension to non-human animals of the ethnomethodology project, which aimed at understanding human actions 'from within', i.e., from the actor's point of view in order to reveal their practical meaning (Psathas, 1980, p. 4).

The second important element of her proposition is its purpose. The author explicitly refuses to instrumentalise animals to solve human problems or to deduce from the observation of nature of the animals any objective or natural law applicable to human societies. She brushes aside this tradition and embraces the point of view developed by Engels regarding social Darwinism:

> The whole Darwinian theory of the struggle for existence is simply the transference from society to animate nature of Hobbes' theory of the war of every man against every man and the bourgeois economic theory of competition, along with the Malthusian theory of population. This feat having been accomplished (…) the same theories are next transferred back again from organic nature to history and their validity as eternal laws of human society declared to have been proved. The childishness of this procedure is obvious, it is not worth wasting words over.
>
> *(Engels quoted in Priest 2017, p. 574)*

The animal convening in Despret's work is at the same time more modest and more radical than numerous political philosophy writings. More modest in that it does not suggest replacing but rather simply questioning and problematising human political and legal structures. The observation of the animal is therefore modest in that it does not carry any normative legitimacy.

But it is nonetheless radical, as it invites the observer to use animal behaviour to break usual patterns of thinking and traditional representations, which prevent the observer from imagining an alternative reality. The project involves seeing in any animal's behaviour something different than an instinct or a preference but an individual worldview, a potential episteme for an entire society. It is an epistemological rupture, designed not to put aside common knowledge in favour of a scientific one, but rather to put to the test the accumulated and unconscious worldviews of a specific individual, including in his/her scientific practice. It is an epistemological rupture as a counterpart of the resistance in a psychoanalytical sense (Fabre 2013, p. 63), the one that prevents a patient from accessing undesired truths and, consequently, from curing the neurosis (Canavêz and Miranda, 2011, p. 150). It is the possibility of revealing and going beyond the internalised and practical recognition of the social arbitrariness that Pierre Bourdieu called the doxa (Deer, 2008).

Confronting International Law with Some Random Cats' Behaviour

According to Vinciane Despret, the animal can help us 'unlearn' (déprendre) a central notion of international law and politics: the notion of 'territory', a concept whose current definition is, according to her, profoundly linked to the notion of private property. Animals are not the only beings that could be observed to that end. Other humans, such as those described by anthropologist Philippe Descola (2018), could help us reach the same goal. But animals, and birds in particular, are serious candidates as well (Despret, 2020).

The question is how? How can we jump from the rawness of the behaviour of a particular animal to a proposition that would challenge our views on specific human institutions such as the notion of territory? Despret gives us some hints about how *not* to proceed. She warns us that we should be careful about the notions we use because, more often than not, when we involve animals 'the very way the problem is formulated and presented ends up excluding those interrogated because it requires them to reply in terms set out in advance' (Despret, 2022, p. 29). Second, regarding the specific notion of territory, she emphasises that it should be looked at as a process rather than a given. Territories do not exist without territorialisation and we should therefore focus on 'acts of territorialisation' rather than on 'territories' (Despret, 2022, p. 93). Third, she emphasises that talking about 'animals', as if they were a valid category, is part of the problem (Despret, 2022, p. 91), as there is no unifying behavioural theory behind every animal act.

If we follow her advice, the question we should be asking to help the animals 'open our imagination' to a different form of international law is not 'how do animals organise their territories' but rather 'how does a particular animal pose acts of territorialisation regarding the physical space that surrounds him/her and how could this vision challenge the conception of space management currently embedded in international law'. Following this idea concretely, as a legal scholar, is more than a bit uncomfortable as it forces us to confront what may well appear to be a rather sophisticated set of legal institutions with the randomness of some animal's behaviour regarding their surroundings. I offer up my own thinking in what follows as an example of how one might be challenged by the kind of legal imaginaries suggested by the human observation of the non-human animal world.

Here is, to begin with, a list of four acts of territorialisation that can be found in Vinciane Despret's *Living as a Bird* (2022). First, some cats do not define territory as a space that would be theirs at the exclusion of every other member of their species. Rather, they organise a shared use of the space based on schedules. Second, some tricoloured blackbirds only assert dominance over a low platform of cattails, but everything above this vegetation remains free for any conspecific use: The sky belongs to everyone. Third, common coots display a very different behaviour regarding the use of the space depending on whether their surroundings are frozen or not. And finally, it seems that some red-winged blackbirds are not looking for unoccupied spaces to settle but rather for spaces that are already populated with conspecific members, the quality of the settlement being linked to the number of neighbours.

The question remains of what could be done with this to revitalise thinking about the law? How do these animals question our understanding of human physical space organisation through international law? I have no definitive answer to these questions and certainly no methodology that could help systematise the epistemological rupture that the encounter of animal behaviours with international law would supposedly create. The only thing that I can offer at this stage is what this encounter produced in me, when I confronted what I knew of international law with Despret's first example: The way some cats organise their relationship with their surroundings (relying on a shared and scheduled use of the space rather than on the idea that a territory must fall under an exclusive and permanent jurisdiction).

My first reaction was that the way physical spaces were organised among States in international law did not have much in common with the way these cats were apparently sharing theirs. After all, the very idea of the Westphalian system, of which current international law is the heir, is that every piece of emerged land is placed under the exclusive jurisdiction of one, and only one, State. On this land and other land appurtenances (territorial sea, airspace, etc.), the State exercises the completeness of its competencies to the exclusion of any other authority (Salmon, 2001, p. 1076). The right to 'use' the land at one state's discretion is conceptually limitless, as long as it does not cause injury to other States. As is well established in international law, each State merely has an 'obligation not to knowingly allow its territory to be used for acts contrary to the rights of other States' (*Corfu Channel Case (UK v Albania)* [1949], p. 22). This rule applies in a context of use of force, such as in the *Corfu Channel Case* but has been echoed in environmental disputes on the transboundary emission of polluted fumes (*Trail smelter case (United States, Canada)* [1941]), or the potential contamination of shared waters (*Pulp Mills on the River Uruguay (Argentina v. Uruguay)* [2010]). But apart from that, a state's territory can be used at will, without sharing it with other equals, as cats do.

I then realised that this assertion was a bit rash. Certain types of space management include a form agreement between more than one State as is the case with mandates, trusteeships (during the colonial period), or other and more recent forms of international administrations in Kosovo or East-Timor. However, I had to admit that these legal structures did not profoundly challenge the Westphalian narrative as their existence is precisely justified by a project of 'State building', ultimately designed to (re)create independent States able to exercise their exclusive jurisdiction on a designated piece of emerged land after the completion of a form of 'civilizing mission' or reconstruction (Wilde, 2008). I therefore concluded that physical spaces in international law were conceptualised as objects of current or future exclusive jurisdiction of a particular State and that this legal structure was very far from the behaviour of cats.

But while this assessment was mainly correct, I remembered that various physical spaces on and around the planet were legally designed precisely to escape the appropriation by one particular State. Among the main examples is Antarctica, which is under the joint governance of the twelve States currently parties to the Antarctic Treaty (*The Antarctic Treaty* 1959, art. IV). Even though technically the claims of exclusive sovereignty over Antarctica are suspended and not abandoned (Grob, 2007, p. 462), the continent has been administered jointly by the Parties for decades. Activities relating to mineral resources (other than scientific research) are prohibited (Protocol on Environmental Protection to the Antarctic Treaty 1991, art. 7). Second comes outer space: 'outer space, including the moon and other celestial bodies, is not subject to national appropriation by claim of sovereignty, by means of use or occupation, or by any other means' (Treaty on principles governing the activities of States in the exploration and use of outer space, including the moon and other celestial bodies 1967, art. II). Third come the High Seas, which are 'open to all states', including for fishing purposes (United Nations Convention on the Law of the Sea 1982 art. 87). And finally comes 'the Area', which is defined as the 'sea-bed and ocean floor and subsoil thereof, beyond the limits of national jurisdiction' (art 1, §1). According to the United Nations Convention on the Law of the Sea (UNCLOS), 'The Area and its resources are the common heritage of mankind' (art. 136). No State can claim or exercise sovereignty over any part of the Area or its resources, which are to be exploited by an international authority to the benefit of 'mankind as a whole' (art. 137).

I then thought that cats were, in a way, sharing their physical space as humans when it comes to the exploitation of 'the Area'. The main difference was that cats were actually not organising themselves to *exploit* the resources of their surroundings but were leaving the physical space almost untouched, only collecting food and depositing scent marks to inform the other cats of their passing (Lorenz, 2002, p. 32).

That is when cats' behaviour reminded me of the potential usefulness of the notion of *usufruct* to organise our relation to the living and non-living beings surrounding us (Barrow, 1990; Ost 2003, p. 306; Sambon, 2019). The notion of usufruct refers to "the right to the beneficial ownership of a thing, the proprietorship of which is in another" (Cochran, 1888, p. 264). This legal logic is typically used to allow a person to live in the house of the deceased partner who owned it. But the concept was not always used to organise ownership between humans. As Erin Drew reminds us,

> a common thread runs through English theological, moral, and literary writing on the subject of humans' relationship to the nonhuman world throughout the long eighteenth century: that humans are not the true owners of the nonhuman world, in either a spiritual or a literal sense. Rather, humans are the usufructuaries of another party, usually either God or future generations of living beings, to whom they are accountable for the long-term stability of the land.
>
> *(Drew, 2016, p. 196)*

So, here is what forcing a completely unrelated cat's behaviour into my vision of international law did to my own legal imagination. I am now wondering how we could organise our physical space management differently. I know this is conceivable as we already did it with 'the Area' (although the reality of seabed exploitation is apparently moving away from the original idea of common heritage of mankind (Feichtner, 2019)). I wonder if the notion of usufruct might help us reorganise our relation to living and non-living entities that surround us. And finally, I am wondering if, as cats, we wouldn't be better with a scheduled rotation of the use of the land, rather than allowing one State or one institution to seize its fruits. Wouldn't it be a great way to ensure an actual representation of all the States rather than to entangle sovereignty and jurisdiction in ways that, apparently, end up reproducing the domination logics of the past (Chimni, 2022)?

Conclusion

Current animal legal scholarship is simultaneously engaged in a profound disanthropocentrisation of the (international) legal system and a reinforcement of the anthropocentric idea that humans are the only beings that should be involved in the conceptualisation of complex legal notions such as 'freedom', 'territory', 'right', or 'citizenship'. The reason behind the acceptance of this descriptive anthropocentrism might simply be strategic: Contesting the normative *and* descriptive anthropocentrism of (international) law *at the same time* might just be too much (Deckha, 2011, p. 210). But it is not inevitable. In fact, the idea of bringing back some form of animal legal creativity is not unheard of (Otomo and Mussawir, 2014). Authors have recently been advocating in favour of forms of 'more-than-human' legalities (Braverman, 2015) or legal creations based on 'multispecies account of the world' (Leth-Espensen and Svensson, 2021).

The question remains as to how these integrations of non-human perspectives should be conducted. The previous pages showed how dangerous it could be to bring back the animals in the construction of what will ultimately be a human legal system, in the sense that it will not be crafted by non-human animals. But the writings of Vinciane Despret offered us a modest way out. A thin line that we could walk to humbly recognise non-human animals as potential epistemic subjects or allies.

This I what I tried to do by confronting the notion of territory in international law with random cats' behaviours. I don't know if this kind of opposition between human and animal epistemes is doing much. But I do think that it might humbly contribute to the construction of a post-anthropocentric legal system in at least three ways.

First, and ironically, it helps us to *denaturalise* international law, i.e., 'to unsettle its self-evidence as the presumed universal mode of relation to earth; to demote its status from the universal to the particular, to make it strange' (Storr, 2022). Second, it gives scholars a means to challenge the fundamental tenets of international law, which is arguably the only way international legal systems could face the current ecological crisis (Natarajan and Khoday, 2014, p. 542). Third, it has a broader performative effect when it comes to the general consideration of the animals as subjects that matter, not only as resources or bodies to be preserved, but as complex beings with specific individual worldviews.

This is a start. But it mainly helps us 'deconstruct' and only marginally helps us 'reconstruct' an alternative episteme, a task that I feel ill-equipped to undertake. As a lawyer, I have acquired technical skills aimed at understanding and applying legal norms. As a critical lawyer, I was trained in the general deconstruction of legal systems, especially in tracking the power relations they help reproduce. But when it comes to the construction of new legal ideas and the deployment of our legal imagination, I feel that my training has been profoundly unresourceful. It is a paradox as imagination and creativity are, in fact, at the core of law as a practice (Del Mar, 2020). In this sense, even though it might not seem like a priority, teaching legal scholars more methods, techniques, and legitimacy to construct alternative legal epistemes might be a necessary step in building a post-anthropocentric international law.

References

Antarctic Treaty (adopted 1 December 1959, entered into force 23 June 1961) 402 UNTS 71.

Barrow, E.G.C. (1990) 'Usufruct Rights to Trees: The Role of Ekwar in Dryland Central Turkana, Kenya', *Human Ecology*, 18(2), pp. 163–176.

Bourdieu, P. (1981) 'Décrire et prescrire', *Actes de la Recherche en Sciences Sociales*, 38 (1), pp. 69–73.

Braverman, I. (2015) 'More-than-Human Legalities: Advocating an "Animal Turn" in Law and Society', in Ewick, P. and Sarat, A. (eds.) *The Handbook of Law and Society*. Hoboken: Wiley Press, pp. 307–321.

Buchanan, B., Chrulew, M. and Bussolini, J. (2015) 'Vinciane Despret', *Angelaki*, 20 (2), pp. 1–3. DOI: 10.1080/0969725X.2015.1039819

Canavêz, F. and Miranda, H. (2011) 'Sur la résistance chez Freud et Foucault', *Recherches en psychanalyse*, 12 (2), pp. 149–157.

Cavalieri, P. (2015) 'The Meaning of the Great Ape Project', *Politics and Animals*, 1, pp. 16–34.

Chapaux, V. (2015) 'Articuler droits animaux et droits humains: Leçons des tribunaux new-yorkais, argentins et indiens', *Droits Fondamentaux*, 13, pp. 1–16.

Chimni, B.S. (2022) 'The International Law of Jurisdiction: A TWAIL Perspective', *Leiden Journal of International Law*, 35 (1), pp. 29–54.

Clark, S.R.L. (1977) *The Moral Status of Animals*. Oxford: Oxford University Press.

Cochran, W. (1888) *Students' Law Lexicon: A Dictionary of Legal Words and Phrases*. Cincinnati: R. Clark & Co.

Collins, J., Baer, B. and Weber, E.J. (2016) 'Evolutionary Biology in Economics: A Review', *Economic Record*, 92 (297), pp. 291–312.

Corfu Channel Case (UK v Albania) [1949] ICJ Rep 4.

Deckha, M. (2011) 'Critical Animal Studies and Animal Law', *Animal Law*, 18, pp. 207–236.

Deer, C. (2008) 'Doxa', in Grenfell, M. (ed.) *Pierre Bourdieu: Key Concepts*. Durham: Acumen, pp. 119–130.

Del Mar, M. (2020) *Artefacts of Legal Inquiry: The Value of Imagination in Adjudication*. Oxford: Hart Publishing.

Descola, P. (2013) *Beyond Nature and Culture*. Chicago: University of Chicago Press.

Descola, P. (2018) 'Anthropologie de la nature', *L'annuaire du Collège de France. Cours et travaux*, 2018 (116), pp. 481–497. DOI: 10.4000/annuaire-cdf.13058

Despret, V. (2015) 'Thinking Like a Rat', *Angelaki*, 20 (2), pp. 121–134.

Despret, V. (2020) 'Ce que les oiseaux nous apprennent des territoires'. Interview with Vinciane Despret. Interviewed by Julie Mouvet for *Regards Croisés*, Université de Namur, 19 February [Online]. Available at: https://www.youtube.com/watch?v=4Cimoj_6OP4 (Accessed: 10 October 2022).

Despret, V. (2022) *Living as a Bird*. Cambridge: Polity Press.

Donaldson, S. and Kymlicka, W. (2011) *Zoopolis: A Political Theory of Animal Rights*. Oxford: Oxford University Press.

Drew, E. (2016) '"Tis Prudence to Prevent th'Entire Decay": Usufruct and Environmental Thought'. *Eighteenth-Century Studies*, 49 (2), pp. 195–210.

Fabre, M. (2013) 'Rupture épistémologique et travail sur les représentations', *Recherches en éducation*, 2013 (17). DOI: 10.4000/ree.7939

Feichtner, I. (2019) 'Sharing the Riches of the Sea: The Redistributive and Fiscal Dimension of Deep Seabed Exploitation', *European Journal of International Law*, 30 (2), pp. 601–633.

Francione, G.L. (1995) *Animals, Property, and the Law*. Philadelphia: Temple University Press.

Grob, J. (2007) 'Antarctica's Frozen Territorial Claims: A Meltdown Proposal', *Boston College International and Comparative Law Review*, 30 (2), pp. 461–484.

Jeangène Vilmer, J.-B. (2008) 'Les principales positions', in *Éthique animale*. Paris: Presses Universitaires de France, pp. 71–104.

Lamy, G. (2014) 'Dépasser le débat nature-culture: Le pari de la sociobiologie d'Edward Wilson. Petite histoire d'une controverse', in Coutu, B. (ed.) *De la dualité entre nature et culture en sciences sociales*. Montréal: Editions libres du carré rouge, pp. 181–209.

Leth-Espensen, M. and Svensson, M. (2021) 'Beyond Law's Anthropocentrism. A Sociolegal Reflection on Animal Law and the More-than-human Turn', *Scandinavian Studies in Law*, 67, pp. 35–50.

Lorde, A. (2003) 'The Master's Tools Will Never Dismantle the Master's House', in Lewis, R. and Mills, S. (eds) *Feminist Postcolonial Theory: A Reader*. Edinburgh: Edinburgh University Press, pp. 25–28.

Lorenz, K. (2002) *On Agression*. London: Routledge.

Midgley, M. (1998) *Animals and Why They Matter*. Athens: University of Georgia Press.

Mylius, B. (2018) 'Three Types of Anthropocentrism', *Environmental Philosophy*, 15 (2), pp. 159–194.

Natarajan, U. and Khoday, K. (2014) 'Locating Nature: Making and Unmaking International Law', *Leiden Journal of International Law*, 27 (3), pp. 573–593.

Nonhuman Rights Project (2022) *Our Story*. Available at: https://www.nonhumanrights.org/our-story/ (Accessed: 6 April 2022).

Ost, F. (2003) *La nature hors la loi: L'écologie à l'épreuve du droit*. Paris: Éditions La Découverte.

Otomo, Y. and Mussawir, E. (2014) *Law and the Question of the Animal: A Critical Jurisprudence*. London: Routledge.

Peters, A. (2021) *Animals in International Law*. Leiden: Brill Nijhoff.

Priest, G. (2017) 'Charles Darwin's Theory of Moral Sentiments: What Darwin's Ethics Really Owes to Adam Smith', *Journal of the History of Ideas*, 78 (4), pp. 571–593.

Protocol on Environmental Protection to the Antarctic Treaty (adopted 4 October 1991, entered into force 14 January 1998) 2941 UNTS 3.

Psathas, G. (1980) 'Approaches to the Study of the World of Everyday Life', *Human Studies*, 3 (1), pp. 3–17.

Pulp Mills on the River Uruguay (Argentina v. Uruguay) (Judgment) [2010] ICJ Rep 14.

Regan, T. (1983) *The Case for Animal Rights*. Berkeley: University of California Press.

Salmon, J. (ed.) (2001) *Dictionnaire de droit International Public*. Bruxelles: Bruylant.

Sambon, J. (2019) 'L'usufruit, un modèle pour le droit d'usage du patrimoine environnemental', in Gutwirth, S. and Ost, F. (eds.) *Quel avenir pour le droit de l'environnement?* Bruxelles: Presses de l'Université Saint-Louis, pp. 173–195.

Santana, H.J. and Santana, L.R. (2005) *Petition for a Writ of Habeas Corpus 833085-3/ 2005, 9th Salvador Criminal Court, Salvador, Bahia. Brazil (translation)* [Online]. Available at: https://www.animallaw.info/case/suica-habeas-corpus (Accessed 10 October 2022).

Sapontzis, S.F. (1987) *Morals, Reason, and Animals*. Philadelphia: Temple University Press.

Shafer-Landau, R. (2012) *Ethical Theory: An Anthology*. 2nd edn. Chichester: Wiley-Blackwell.

Singer, P. (1977) *Animal Liberation: A New Ethics for Our Treatment of Animals*. New York: Avon.

Storr, C. (2022) 'Denaturalising the Concept of Territory in International Law', in Natarajan, U. and Dehm, J. (eds.) *Locating Nature: Making and Unmaking International Law*. Cambridge: Cambridge University Press, 179–199.

Tavoillot, P.-H. and Tavoillot, F. (2015) *L'Abeille (et le) Philosophe. Étonnant voyage dans la ruche des sages*. Paris: Odile Jacob.

Trail smelter case (United States, Canada) [1941] 3 Reports of International Arbitral Awards 1905.

Treaty on principles governing the activities of States in the exploration and use of outer space, including the moon and other celestial bodies (adopted 27 January 1967, entered into force 10 October 1967) 610 UNTS 205.

United Nations Law of the Sea Convention (adopted 10 December 1982, entered into force 16 November 1994) 1833 UNTS 397 (UNCLOS).

Wilde, R. (2008) *International Territorial Administration: How Trusteeship and the Civilizing Mission Never Went Away*. Oxford: Oxford University Press.

Wilson, E.O. (1978) *On Human Nature*. Cambridge: Harvard University Press.

Wilson, E.O. (2000) *Sociobiology: The New Synthesis*. 25th anniversary edn. Cambridge: Belknap Press.

16

GROUNDING ECOCIDE, HUMANITY, AND INTERNATIONAL LAW

Tim Lindgren

Introduction[1]

In March of 2019, the Australian Peoples' Tribunal (APT) for Community and Nature's Rights ran a citizens' inquiry into the health of the Barka/Darling River and Menindee Lakes. In the wake of floodplain harvesting, neoliberal water unbundling, and long-running colonial dispossessions, the Tribunal travelled rural towns and Indigenous communities to hold public hearings on the ecological and social discontinuities of the Australian Murray-Darling Basin. Attending hearings in the towns of Mildura and Wentworth as part of an ongoing effort to make sense of peoples' tribunals and international law, I listened to local interlocutors telling stories about their lived pasts and presents under forces of colonialism, statecraft, and capitalism.

Crafting a scenography for local inhabitants to speak about life and law outside the formal legal system, the Tribunal opened a space for more free travels across the spectrum of legal interpretation and enunciation. Whilst far from all of the Tribunal interlocutors, some local farmers used this opportunity to walk international law into the Murray-Darling Basin, mobilising whatever they deemed relevant to the temporal and spatial specifics of their lived experiences. As legal norms were read outside the interpretative repertoire of international legal bodies, these interlocutors reached for justice not necessarily beyond the law but above that of domestic law. That is, some Murray-Darling Basin inhabitants reached towards a universal and global realm of international law under a desire to save the situated assemblages of human and non-human life from ecocidal violence on the ground.[2]

In this chapter, I want to pause and hold on to this moment of reaching; a reaching towards the universal promises of international law in the struggle for a functional ecological life down on earth. This is a reaching that many international lawyers, including myself, have done almost intuitively at some point in their career. Whilst reflections on the Tribunal and its law are for other places than this, I want to think about this reaching by reading it in the context of the broader struggle for an international crime of ecocide under international criminal law (ICL). Because just like some of the local interlocutors that reached for international law as a response to the shifting grounds of ecocidal violence beneath their feet, many advocates for an international crime of ecocide search for a universal and global international law able to bring modern humanity back into a relationality of care and reciprocity with the soil.

DOI: 10.4324/9781003201120-20

Over the course of the past decade, international jurists, activists, and academics have been advocating for an international crime of ecocide as a fifth missing crime against peace to be included under the 1998 Rome Statute, giving the International Criminal Court (ICC) jurisdiction to prosecute environmental crimes (Higgins, 2015). Building on a longer institutional history where an international crime of ecocide has figured at various points throughout the making of ICL, ecocide advocates reach for the universal punitive promises of the ICC in an effort to address large-scale environmental destruction (Gauger *et al.*, 2013). Resting against the universal jurisdictional aspirations of the ICC, an international crime of ecocide is broadly imagined to speak an end to the culture of impunity for environmental crimes under the international legal order and bend the discipline of international law into an ecocentric enterprise (Greene, 2019).[3] In recent years, the proposal has also taken hold in contemporary debates on international law-making. States such as Vanuatu and the Maldives have called for consideration of ecocide at the ICC assembly, and international lawyers such as Philippe Sands have drafted a much debate official definition of the crime of ecocide (Minkova, 2021).

Whilst the struggle for an international crime of ecocide is increasingly diverse in aims and hopes, many advocates still look to the universal promises of the ICC under an assumption that international law's universal reach has the ability to bring modern humanity into a relationality with the soil beneath our feet. That is, that this law can span the globe in its totality and arrest ecocidal practices, and that the expansion of international law's universality thus turns the discipline into an ecocentric legal regime that fosters reciprocity and care with the land. Without foreclosing the struggle for an international crime of ecocide, I want to problematise this notion that the anthropocentrism of international law can be transcended by expanding its universal reach. I want to trouble the reach for international law under the call for an international crime of ecocide and consider what legal orders and relationalities with the soil we might render invisible unless we think carefully about how we turn to the current international legal order of things.

I thus offer a critical redescription of the call for an international crime of ecocide as a way of decelerating the speed at which we reach for more international law. Taking on slowness in thought and action as a methodological and jurisprudential practice of vigilance (Pahuja, 2013; Barr, 2016), I attempt to arrest the moment in which we reach – 'to suspend it mid-air' in order to think carefully about where we are going, and what we might lose, when we engage the desire for more international law under an international crime of ecocide. Slowing down the leap towards more and better international law, I hope to shift register: to engage a semantic stepping out of the normal conventions of international law and have us think more carefully about what legal worlds we speak into being as we turn to the universal and global promises of international law for a better ecological future.

In doing so, I do not offer a general consideration of the layers of violence bound up in ecocidal practices. Be it as a legal imaginary or a conceptual trope, ecocide has always been known by the colonised as part and parcel of a wider portfolio of colonial practices and contingencies that result in cultural and physical genocide (Fanon, 1961; Saro-Wiwa, 2000; Watson, 2018a). Explored extensively by Damien Short and Martin Crook over the past decade, the phenomenon of ecocide has been strongly theorised as a form of structural genocide, endemic to colonialism and capitalism (Crook and Short, 2014; Short, 2016; Crook, Short, and South, 2018). Whilst situated in this literature, I rather focus our attention on the call for an international crime of ecocide to reflect on the undergirding aspiration to overcome international law's anthropocentrism by investing further in its universal reach. The piece is thus part of a limited, but hopefully growing, critical conversation about the limits and possibilities of an international crime of ecocide (Minkova 2021; Cusato and Jones, 2022).

I begin by setting out the historical context and broader search for an international crime of ecocide under international law, describing how many ecocide advocates look to the punitive promises of criminal law and the universal promises of international law as a catalyst for a legal relationality of reciprocity with the soil. I then, in the second part, consider what it means to make appeals to international law under the notion that an international crime of ecocide has the ability to capture the globe in its totality from a position of universal law. Holding on to the very image of the globe itself, I detail how the leap of faith towards more of the current international legal order draws our attention away from the materiality of life on earth. Directing our attention to an abstract global realm rather that the ground beneath our feet, this leap, I suggest, renders transcendental an international law that risks enforcing, rather than addressing, the violence of ecocide. Without staying vigilant to what spatial imaginaries we engage when we make appeals for an international crime of ecocide, I thus argue, we risk grappling with a myopia; a myopia that upholds the established international legal order of things as the only law there is and render other legal orders, with other relationalities with the soil, invisible.

Rather than assuming that a more expansive and inclusive international law will facilitate a future demarcated by environmental justice and survival, I aspire to have us think carefully about what may become possible if we dare to suspend, momentarily, the reach for the current international legal order. In the final section, I hence suggest that we might do well in slowing down the speed at which we reach and instead embrace a jurisdictional listening before we rush to bring more international law into being. Embracing slowness and listening in our jurisprudential thought might allow us to encounter the jurisprudential work of Kumbumerri and Munalijahlai scholar Christine F. Black, detailing a legal life patterned into the land itself (Black, 2010). Or it might invite a reconning with the inter-species Indigenous internationalism that Michi Saagiig Nishnaabeg author Leanne Betasamosake Simpson describes in her work (Simpson, 2017). Reading these scholars as a gesture of legal meetings rather than appropriation, I hope to invite us to consider what it would mean to slow down and listen to the standing of these internationals and laws as we respond to the anthropocentrism of international law – laws that practice the legal life we might desire from an international crime of ecocide in the first place.

Looking for Law

Ecocide emerged as a legal trope in the world of international law after the American plant biologist Arthur W. Galston introduced the term at the Conference on War Crimes and the American Conscience in 1970 in Washington, DC (Gauger *et al.*, 2013). Galston was part of a larger group of American scientists that denounced the 'environmental destruction and potential human health catastrophes arising' from the United States' herbicidal warfare program Operation Ranch Hand in Vietnam (Zierler, 2011, p. 14). Condemning the portfolio of chemical agents deployed by the United States, Galston suggested that 'wilful and permanent destruction of environment' ought to be considered an international crime on par with genocide under 'the term *ecocide*' (Galston citied in Zierler, 2011, p. 18). Mobilising the spectres of the Nuremburg trials, Galston 'asked the international community … to come together' and formulate a 'proposal against ecocide' as an international crime in itself (Cusato, 2017, p. 494; Galston cited in Zierler, 2011, p. 18).

Whilst Galston advocated for ecocide as a war crime and did not consider it as part of a broader pattern of systemic environmental violence bound to modernity's pasts and presents (Zierler, 2011), the concept would take on a life of its own (Short, 2016). In 1972, ecocide emerged in the inaugural speech of the United Nations Stockholm Conference on the Human Environment given by the Swedish Prime Minister Olof Palme, who 'spoke explicitly … of the Vietnam War as an "ecocide"' (Gauger *et al.*, 2013, p. 5). In 1973, the international law and war crimes expert Richard A. Falk, who was part of a non-governmental working group for a UN Convention on Ecocidal Warfare in Stockholm in 1972, published a draft International Convention on the Crime of Ecocide (Falk, 1973). Calling into question the limited reach of the 1948 Genocide Convention, Falk argued for an international crime of ecocide as a war crime hinged on *mens rea* requisite of intent (Crook and Short, 2014).

What follows in the wake of Falk's drafting of an ecocide convention is a long institutional history where ecocide, at various points in time, came alive during the shaping of ICL. Be it the Legal Committee of the UN General Assembly or the International Law Commission, the inclusion of an international crime against the environment was widely discussed over a forty-year period (Gauger *et al.*, 2013, pp. 7–8). And whilst the concept of criminalising widespread ecological destruction in peacetime, without *mens rea* requisite of intent, travelled through the drafting of the 1998 Rome Statute, it was eventually excluded from the final draft in 1996 (Greene, 2019).

Falling outside of the scope of ICL, ecocide has yet taken on a renewed life over the past decade. In 2010, the late environmental lawyer and activist Polly Higgins introduced a proposal for an international crime of ecocide before the International Law Commission, suggesting an amendment of the Rome Statute to include ecocide as a fifth crime against peace (Higgins, 2015). Higgins imagined the international crime of ecocide as a crime of strict liability, defined broadly as 'the extensive destruction, damage to, or loss of ecosystem(s) of a given territory, whether by human agency or by other causes, to such an extent that peaceful enjoyment by the inhabitants of that territory has been severely diminished' (Higgins, 2015, p. 3). Turning to ICL and embedding the crime of ecocide under the jurisdiction of the ICC, Higgins hoped, would offer a radically different avenue of accountability by allowing the Court to reach persons of senior responsibility, ranging from state officials to CEOs of transnational corporations.[4]

In the wake of the 2010 proposal, Higgins also launched the *Eradicating Ecocide* campaign (Higgins, 2015). Over the course of the last decade, this campaign carried the legal imaginary of ecocide into social movements and justice organisations, inspiring not only a subset of local ecocide campaigns and organisations but also giving life to the concept in mock trials and international peoples' tribunals (MacCarrick and Maogoto, 2018; Crook, Short, and South, 2018). Today, the most prominent international advocacy is carried forward by Ecological Defence Integrity (EDI) and its *Stop Ecocide* campaign,[5] founded by Higgins and Jojo Mehta in 2017. As a result of this advocacy, a number of states and political figures has also come out in support of an international crime of ecocide. For example, in late 2019 Vanuatu and the Maldives called for consideration of ecocide at the ICC's assembly.[6] In 2021, an Independent Expert Panel co-chaired by Philippe Sands and Dior Fall Sow furthermore offered an official definition of ecocide, instigating a broader conversation about what it would mean to amend the Rome Statute (Heller, 2021; Minkova, 2021).

Much of this campaign for an international crime of ecocide also conceived of itself as a more comprehensive response to the failure of domestic and international environmental law to prevent extensive ecological destruction. Locating 'humanity … at an existential crossroad' (Mehta and Merz, 2015, p. 4), ecocide activists and scholars alike indeed speak to an international crime of ecocide as an opportunity for a paradigmatic and structural shift towards law of an ecocentric nature. Contesting the paralysis of soft law under international environmental law and the jurisdictional limitations of domestic procedures, an international crime of ecocide is imagined to be 'changing the principles and assumptions that underlie the current [international] legal system' and the world it ushers into being (Greene, 2019, p. 26).

Particularly, many ecocide advocates suggest that an international crime of ecocide imposes an ecological prudence in international law itself, thus speaking an end to the culture of impunity where transnational corporate actors travel free from prosecution for environmental violence. As Higgins and *Stop Ecocide* have argued, an international crime of ecocide would impose a 'duty of care' on corporate actors and government officials by forcing them to take personal responsibility for their conduct, as they otherwise find themselves liable under ICL (Higgins, 2015). That is, an international crime of ecocide is believed to offer an avenue to prosecute individuals of seniority under a universal jurisdictional reach, thus inserting a principal deterrence from ecocide that would force the very structures of corporate practice to take shape around a more ecologically sound social metabolism (What is ecocide?, 2019). As *Stop Ecocide* puts it, a

> law making ecocide a crime, bringing with it the prospect of senior executives and complicit government ministers facing criminal proceedings, will be a strong deterrent against ecocidal practices and will be *the catalyst* for finding new, sustainable ways of operating.
>
> *(What is ecocide?, 2019; emphasis added)*

An international crime of ecocide is, in other words, suggested to offer 'a legal … life insurance policy … not only for the environment but for communities, for future generations and for the wider life of our Earth' (What is ecocide?, 2019).

In this sense, an international crime of ecocide is seen as a force for a new ecological future through law; a future located in the possibility of overcoming international law's anthropocentric forces and transforming it into a force for the planet as a whole (Higgins, 2015; Greene, 2019). Under this order of things, international law, as an otherwise anthropocentric endeavour, is presented as the catalyst of societal transformation. As Martin Crook, Damien Short, and Nigel South argue, the movement for an international crime of ecocide aims to establish ecocide as an international crime 'on the grounds that the nature of the crimes currently covered by the Rome Statute … is clearly anthropocentric, reflecting a concern only with atrocities and actions committed by humans with human implications and impacts' (Crook, Short, and South, 2018, p. 304). An international 'crime of ecocide in the mandate of the ICC', the campaign suggests, 'would broaden the Court's scope to include crimes that are *ecocentric* in nature and that affect the ecosystem of the planet' (Crook, Short, and South, 2018, p. 304). As such, the desire for an international crime of ecocide is buttressed by both a critique of the procedural limitations of international law and a faith in the possibility of international law – as a universal enterprise – to usher into being a different social cognisance and ecological relationality.

Coming to international law from various locations of critique and faith in the law, many proponents of an international crime of ecocide thus suggest that the expansion of the international legal order might offer an avenue to address, or at least contest, structural and large-scale ecocidal violence. Reading international law through a perhaps more instrumental perspective, these voices thus move towards international law under various shades of hope in its global reach. As an institutional space both with the possibility of punitive measures and universal jurisdictional aspirations, advocates for an international crime of ecocide look to the ICC and international law as a legal space able to span the globe and arrest the forces of ecocidal violence that otherwise go unchecked.

Whilst not necessarily under a blind faith that locates international law 'on the side of the angels' (Pahuja, 2011, p. 1), this broader reaching towards international law for a crime of ecocide thus makes a move towards the global under a faithfulness that the punitive jurisdictional reach of the ICC may force modern humanity to actually care for and attend to the fabric of life that define its existence. And it is precisely this moment of reaching towards the universal and global promises of an international crime of ecocide that I want to pause on and think more carefully about. Without foreclosing the turn to international law as a necessity for survival, a deceleration allows us to better consider what might be at stake when we move with great speed and determination towards more universal and global international law in order to give birth to a better ecological future.

Finding Global Law

In order to slow down and think more carefully about this movement towards international law for an international crime of ecocide, we might do well in following an increased attentiveness in critical legal thought to the materiality of modern life under the geological era of the Anthropocene (Philippopoulos-Mihalopoulos, 2017). Whilst a somewhat separate intellectual tradition to that of international law, attention to this scholarship offers a way to pause and dwell on the urge for global law as a response to contemporary environmental concerns, and to consider what is at stake when we leap towards international law under a faith in its global and total reach (Birrell and Dehm, 2020).

Dominant narratives on the Anthropocene have persistently collapsed this new geological time into a 'species thinking' that universalises the responsibility and experiences of the current environmental crisis under an untouched totalising humanity, or *Anthropos*, as the modern human (Grear, 2015). Embedded in 'a larger Eurocentric development story', mainstream accounts on the Anthropocene paint male and white modern rationality, technoscientific solutions, and sustainable development on the 'human condition', obscuring differences of race, gender, and class along a global colonial divide (Baskin, 2019, p. 150; Davis and Todd, 2017). Yet, as Kathleen Birrell and Julia Dehm remind us, critical voices have also used the Anthropocene thesis as an impetus to interrogate how commonplace global responses to ecological harm renders it largely impossible to formulate a legal life that translates into a reciprocity with the ground (Birrell and Dehm, 2020).

Reading against the universalising stories of the Anthropocene as a Promethean human destiny, Bruno Latour has argued that the contemporary environmental crisis demands that the modern human shifts register away from the totalising 'global thinking' embedded in modern western thought (Latour, 2017). For long, Latour argues, western modernity has been obsessed with knowledge that encapsulates the whole and its parts under a totality; a totality that exits the confines of the particular to *see the world completely* under an extraterritorial God's-eye 'view from above' and yet 'from nowhere' (Haraway, 1988). Reading through Peter Sloterdijk's '*spherology*', Latour suggests that this strive for 'complete singularity' in 'Western philosophy, science, theology, [law] and politics lies in the fact that' western modern thought 'has infused all the virtues into the figure of a Globe' (Latour, 2017, p. 123). Taking a global view as the point of departure, this occurs 'without paying the slightest attention to the way in which that Globe might be built, tended, maintained, and inhabited' (Latour, 2017, p. 123).

Whilst the strive towards a universal and totalising position held together in the image of the globe might appear as an innocent preposition, it has dramatically flattening effects upon the situated particularities of life down on earth. Symbolic of a 'real "white man's burden"' captured through the image of the 'blue marble' visible from space, this act of defining the world in its totality from the position of the transcendent leads to a slippage of abstraction from the particular to the universal (Latour, 2017, p. 122). And in this slippage, the particular of western modernity is transported into that which we conceive of as the universal totality. Once reproduced as a transcendental totality, it is released from its earthly and situated context and given life as all-encompassing and global, allowing western modern thought to define the very contours of the universe, and also life on earth.

But in locating this globe and the strive for totality so central to modern western thought, Latour also draws our attention to something much less obvious, and earthlier. This metaphysical strive towards universal knowledge and values also rips our attention away from the particularities of life down on earth itself. As if one could actually live, feel, and breath from the outer layers of the atmosphere, this tour de force in Eurocentric reduction makes it seem like we are largely independent and above the 'vast web of systems and processes' of life on Earth (Matthews, 2019, p. 667). As Latour explains, whether

> we are dealing with … Humanity, or Nature taken as a whole, the danger is always the same: the figure of the Globe authorizes a premature leap to a higher level by *confusing the figures of connection with those of totality.*
>
> *(Latour, 2017, p. 130)*

The leap towards a global totality, in other words, unifies 'too quickly what first needs to be composed' by paying attention to the social and ecological relations, entanglements, and encounters of life on the ground (Latour, 2017, p. 138).

And it is this irreducibility of the earthly that may also force us to think more carefully about what form of law, with what spatial discourse, we demand more of when we attempt to 'transcend' the anthropocentrism of international law. Taking on Latour's work on the globe and his reading of James Lovelock's Gaia thesis, Daniel Matthews suggests that a critical reading of the Anthropocene brings into purview an '*earthly life*' (Matthews, 2019, p. 670). Referring 'not to an understanding of human life at *global* or even *planetary* scales', Matthews argues that the ecological uncertainties of the present, quite physically, demand that we attend to the irreducibility of the earthly web of life and soil beneath our feet.

Reading against what Sloterdijk calls the 'backdrop ontology' of modern thought, where the natural world simply provides the scene from which human dramatology is staged, Matthews details how the ecological discontinuities of the Anthropocene are making the certainty of that very background less stable and less total (Matthews, 2019). Whilst the scenery itself has always moved for those attentive enough, what happens under the current ecological uncertainties is that the soil itself – the backdrop – begins to move in response to modern industrious activities to such an extent that it becomes largely impossible to ignore. Or put better, as the agency 'of human and non-human forces that defines our current' conditions are changing faster than before, it also becomes increasingly impossible to understand and dominate nature through modern regimes of property, sovereignty, and other tropes of mastery so central to western thought (Adelman, 2015).

Under an extraordinary twist of 'faith', then, the shifting grounds of modern life demand that the modern human, if it's to survive, also shifts register: that he attends the soil rather than rushing to capture its entirety under an abstract universality of global knowledge and values. Rather than seeing 'the earth as something that is already unified and enclosed', the materiality of the ground 'insists on an earth that is irreducibly "down here"' – a constellation where the particular and irreducible comes first, and any global spatial imaginary comes second, and never as a totality (Matthews, 2019a, p. 674). The increasingly unstable ecological conditions of the present, in other words, unsettles the very understanding of the grounds of modern humanity. It demands that the modern human – or the *Anthropos* – directs attention to the weaved and situated patterns of connectivity and interspecies assemblages in place. Or put differently, it demands that the *Anthropos* leaves behind the universalised global thinking and begin to pattern political and *legal life* into the interconnections of life and knowledge on the ground (Matthews, 2019). That the *Anthropos* makes law and legal life that moves with the forces that shape our lived material realities.

Read in this manner, we might also begin to sense what is at stake as we reach into the universe of international law in search of new legal groundings to make our stand. As Kathleen Birrell and Julia Dehm have suggested, the spatial and temporal vernaculars available under the current international legal order seem ill-equipped to foster the situated legal relationality and sensitivity that Matthews and Latour argue to be necessary (Birrell and Dehm, 2020). Quite the contrary, the image of the globe and the extraterritorial view of defining the world from a metaphysical universe 'out there' lives a particularly authoritative life under international law.

As Usha Natarajan and Julia Dehm have detailed, the very notion of the 'environment' in international law stems from an environmentalism situated in a western history and culture. Under this particular environmentalist discourse, the modern legal subject is assumed to hold 'a conceptual posture external to Earth, allowing it sufficient distance to look back and see a single globe' in its totality (Natarajan and Dehm, 2019). As aptly captured by the first photograph of Earth from Apollo 8 on Christmas Eve in 1968, this abstract juridical humanity is 'compelled to venture beyond the world because "it is only from such an external position that the boundaries of the world can be drawn and knowledge of what exists guaranteed"' (Natarajan and Dehm, 2019). Representative of the larger claim to universality by international law as a discipline 'justified and dynamized by the continued assertion of universal values from a professed position of external objectivity', the figure of the globe offers a position from which international law can cast itself 'around everything (i.e. "the environment")' (Natarajan and Dehm, 2019). Viewing the world from the universe, modern humanity may organise and regulate the environment under mutual discourses of exploitation and stewardship, claiming that it is 'we who surround the environment, and not the other way around' (Natarajan and Dehm, 2019).

In this same vein, the image of the globe plays a particularly important role in normalising the territorial imaginary of international law as a universal order of sovereign states. As maybe the ultimate universalising trope of modern legal thought, international law follows a territorial order where the *totality* of the globe is divided into modern nation-states that enjoy sovereignty over territory – states that, after their formations, are imagined to return to the global to regulate the globe as an enclosed totality (Nesiah, 2003; Eslava and Pahuja, 2019). This 'territorial matrix' of states, as Birrell and Dehm suggest, 'has narrowed "the active boundaries of geopolitics and geo-economics", privileged specific practices of spatial authority and disciplined postcolonial imaginaries of "independence", and delimited the possible relations between law and place' (Birrell and Dehm, 2020, p. 12).

Whilst the worlds undone by the Berlin Conference, the Mandate System, the Bretton Woods System, or other projects of European world-making tell other tales, under this territorial order of things the sovereign state is indeed imagined as the only possible political entity and mode of human organisation there is (Anghie, 2004; Eslava, Fakhri, and Mesiah, 2017; Getachew, 2018). Those that make claims to political and legal life that runs along other relationalities with the soil fall outside the spectrum of legal intelligibility, painted as cultural practices stuck in a primitive past. With race at its apex of foreclosability and 'whiteness' as 'the ownership of the Earth forever and ever', those not abiding by the totality of this order of things are excluded from this universal and global community of modern and developed 'humans' of sovereign states (Du Bois, 1920, p. 18; Pahuja, 2011; Eslava and Pahuja, 2019).

This spatial discourse – a spatial discourse authorised by a universalising and totalising thinking that obscures and distracts from the earthly materiality of life on the ground – thus offers little respite for legal thought that takes ecological reciprocity as its axis of engagement (Mickelson, 2014; Birrell and Dehm, 2020). From sovereignty and property to development and progress, the normative groundings of international law are indeed situated in a view of the world where the 'environment' comes alive through extraction and domination, 'managed by special regimes' of environmental law (Natarajan and Khoday, 2014, p. 573; Storr, 2016; Porras, 2014). With the 'very origins of international law' found 'in doctrines put forward to allow private actors from the North to exploit natural resources in the South', 'the discipline plays a crucial role in transforming a unified planet into discrete sovereign territories, in converting nature into exchangeable property' and 'separating an atomized human individual from the intertwined mesh of life' (Natarajan and Dehm, 2019). Indeed, as Birrell and Dehm argue, much more 'grounded … articulations of justice', based on 'new [legal] imaginaries and representational strategies', are needed if we are to address the current forms of ecological 'violence and injustice' (Birrell and Dehm, 2020, p. 20).

Read through this story, the reach towards the universal promises of international law for an international crime of ecocide may also be better understood. Whilst ecocide advocates of different shades leap towards international law under a sense of faith that its global and universal form may bring 'humanity' back to care for the soil, the very reach itself lends support to the expansion of an international legal order that tends to foreclose the possibility of a grounded legal life. Without vigilance over what spatial order of things we engage when we demand more of international law, we indeed risk further normalising a world where sovereignty, temporal progress, capitalism, developmentalism, and other formative pillars of the international legal order figure as transcendental.

Because with the international legal order of things seen as predominately universal and given, international law takes on a much more instrumental valence. The promises and possibilities for global justice through international law comes alive primarily as an assemblage of rules and norms that function as technical instruments of global governance. Released from their European and colonial epistemic roots, these rules and norms are imagined as flexible enough to be taken out, reimagined, and then plugged back in again to make international law speak of regenerative flows rather than extractivism and accumulation administrated from above. With a less grounded, or perhaps hastened, engagement with the interknit of norms and values that lay out the architectural fabric of international law, the reach towards international law thus risks enforcing, rather than disrupting, the Promethean dream of modern legal thought. That is, the very desire to transcend international law's anthropocentric tendencies by including the ecological under its universal reach risks enforcing an international legal order to which environmental injustice might be endemic.

Indeed, as those such as Christine Schwöbel-Patel, Christopher Grevers, and Tor Krever remind us, the very discipline of ICL does not lend itself easily to contestation of structural and racialised forms of violence that fall outside the vernaculars of intelligibility of western modernity, and particularly not the layers of harms of global capitalism (Krever, 2013; Grevers, 2014; Schwöbel-Patel, 2021). As an institution committed to 'universalising concepts, such as a singular, shared "humanity"', any ungrounded or unqualified faith in the universal reach and punitive possibilities of the ICC lends the criminalisation of ecological violence to Eurocentric and neoliberal world-making under international law (DeFalco and Mégret, 2019, p. 77; Schwöbel-Patel, 2021). In rushing too hastily towards the global by appealing to the universal jurisdictional possibilities of the ICC as an institution capable of global punitive justice, we might thus lose sight of where we stand and where we are going. Without staying vigilant to what spatial imaginaries we engage, we might grapple with a myopia over the fact that the spatial order of international law renders transcendental a set of norms that enforce, rather than address, the violence of ecocide and ecocidally induced genocide.

By gazing into an assemblage of Anthropos-centric legal life, where sovereign nation-states and an untouched juridical humanity reign, we henceforth risk enforcing stories of this legal world as the only legal world there is. In hoping to overcome international law's anthropocentrism without staying vigilant to what we demand more of, we lend support to international law's spatial, temporal, and relational order of things as transcendental and universal to the 'human condition' and to law itself. In looking towards international law for an international crime of ecocide without being more specific about how and why we, as international jurists, come before this legal world, we thus also foreclose the possibility of other legal forms, or other internationals and laws, that may be better suited to achieve what an international crime of ecocide is imagined doing. Because as we slow down our reach towards the universal promises of international law, we see that an ungrounded rush to international law indeed risks further foreclosing the possibility of living and listening to the soil through law at all.

Grounding International Law

What may it mean, then, to decelerate the speed at which we travel with and towards international law; to slow down the movement for more of this international law in itself? Beyond a heightened vigilance over how we engage and live with international law, an attentiveness to its architectural fabric also invites us to think in less determined ways about the international legal order and its law. Rather than rushing to invent new legal regimes that reinforce the sovereign state as the only entity with the authority to speak law into being, such slowness invites the international jurist to think more carefully about what legal forms we aspire to in order to bring modern life down to earth. That is, it invites us to pause the notion that we can transcend international law's anthropocentrism by amending its jurisdictional limits so as to include the ecological world under its universal gaze, and to reflect on other laws with other relationships with the soil. Suspending and thus provincialising the universality of the current international legal order of things, we may arrest ourselves to be more attentive – to listen rather than rushing to speak law into being.

Following Shaunnagh Dorsett and Shaun McVeigh's attentiveness to jurisdiction beyond sovereignty and Olivia Barr's invitation to a jurisprudential method of slowness, the international jurist and ecocide advocate may hence do well in slowing down the reach for more international law and engage a jurisdictional listening as an initial point of departure (Dorsett and McVeigh, 2012; Barr, 2016). That is, a listening that arrests us in the moment which we tend to assume authority to speak law into being, and in which we assume that more of this law paves the path for new ecological futures. Rather than taking for granted that jurisdiction flows from sovereignty *post facto*, and that the international legal order of things is simply a mendable aggregation of rules, those that speak for more international law as a way of transcending international law's anthropocentrism may hold an obligation to listen to authorisations of law bound to other legal orders (Dorsett and McVeigh, 2012; McVeigh, 2019). To 'slow down the movement between the search for law and the discovery of sovereignty' and to take seriously other laws authorised differently, arresting ourselves to listen better before we rush to imagine more international law into being (Pahuja, 2013, p. 70).

To cultivate a sense of jurisprudential slowness and jurisdictional listening thus demands that the international jurist that leaps towards more international law thinks carefully about what existing legal worlds we travel past as we make this leap of faith. Considering the place I write from and the locality where we began this chapter, a jurisdictional listening imposes an obligation to engage with the forms of law already patterned into the continent of Australia; laws that far outpace the Westphalian fable that otherwise defines most legal possibilities within the realms of western modernity. Under such engagement, international jurists may come into encounter with the jurisprudential tales told by those such as Christine Black, patterning our legal existence into the soil by reading the *land as the source of the law* (Black, 2010).

Through a dialogical encounter with Indigenous jurisprudence across Australia and elsewhere, Black has articulated a *talngai-gawarima* jurisprudential structure where law itself exists in the land, and knowledge of that law comes through movement with the land under walking, sensing, feeling, and story-telling (Black, 2010, p. 4). Reading through a meeting with Buntji Senior Law Man (SLM) Bill Neidjie, Black details how the 'force or energy' of the land – or the *Djang* – forms a balance of relationships that defines the geobiophysical materialities of place. With law as literally embedded in place, this balance 'comprises legality' in itself; a legality that must be felt under 'a subjectivity that ensures the predominance of *feeling* … for the web of interconnected relationships that patterns humans into their environment' (Black, 2010, p. 184). To be a 'lawful person' is to maintain balance and to be in constant dialogical *legal* encounter (through feelings) with the materiality of the soil, recognising that 'we are patterned into the galactic *physis* that requires us to be mindful that our actions have intergalactic implications as well as neighbourly implications' (Black, 2010, p. 164).

Unlike the universal promises of international law, where law is a transient and 'place-less' human abstraction, Black's account thus hinges on a sensibility and attentiveness to law as patterned into the fabric of the land along temporal and spatial orders of relationality. Whilst it is neither my objective nor my place to define or reify legal orders that are not mine, the material implications of such jurisprudential structure give shape to a different set of legal practices than what is currently available under international law. Under international law, law is defined as a distinctly human imaginary, and authority to exist in place is mediated through the acquisition of sovereignty – an acquisition that allows for a universal and transient global law to be applied immediately across a vastly unknown landscape (Barr, 2016). But with the land as the law, the authority to exist in place is formed through practices of legal enunciation based on an affective attentiveness to the land itself.

As Black explains with reference to Māori jurisprudence, when arriving at new land 'the Māori would set up a *waananga* (house of learning), or a type of research station, to consider the lay of the land and the laws that may be perceived through the patterns found in the landscape of the place before' moving onto the land (Black, 2010, p. 153). Through such legal practice, 'a sacred contract would … be negotiated between the Māori, through reading the signs, and the Law in the landscape', in which the 'spirit' of the 'new land, or renewed environment, is … addressed as "titleholder"' (Black, 2010, pp. 152–153). The use of the land would thus remain 'conditional on keeping the sacred bond', and be 'cultivated for the preservation of the contract rather than just for material gain' (Black, 2010, p. 153).

In Māori jurisprudence, the authority to exist in a place is thus constantly renegotiated and woven through a legal practice of attentiveness and care to the mosaic of laws in place. Rather than land being seen as a static resource awaiting acquisition and appropriation under '"discovery" on behalf of a sovereign', the relationship with the land is constantly negotiated, and the land is thus 'received' instead of 'taken' (Black, 2010, p. 153). Shifting the locus of the law from an abstract universal 'humanity', law is 'posited in the Land' and authority mediated by reciprocal encounters in place.

Under such legal relationality with the soil, human rights and responsibilities also take on a more grounded form. In her encounter with Ngarinyin SLM David Mowaljarli, Black describes the legal relations of totem animals – a relation where human rights and responsibilities are patterned into the land itself. For a human with a totem animal, according to Ngarinyin law, the 'human's right is to bring into being the assigned species', whilst the 'responsibility is to keep actualizing that right to continue the species' (Black, 2010, p. 58). As Black explains, these legal relationships hence 'pattern a management system of resources – a pattern that, if applied to the whole of Australia, would require humans to see themselves as responsible for a specific food'

(Black, 2010, p. 58). Under this legal practice, human rights are not operative strictly on a universal plane. Rights and obligations are rather grounded in the ecological processes that maintain our existence, binding human practice – as a matter of law – to the land, as well as the global and cosmological forces that shape place.

The human right and responsibility is hence to act and engage with the ecological, rendering mastery itself implausible; the responsibility of continuation forecloses the possibility of mastery for capital accumulation because such practice is diametrically opposed to a thriving species and the legally bound feelings for that species. Punishment for a breach of this 'Law of Relationship' is accordingly not guided by punitive measures, nor always direct (Black, 2010, p. 184). With a breach unsettling the balance of things, 'punishment … may be "a long time coming"' and may appear elsewhere (e.g. climate change), but manifests in the loss of ecological life and the affective pain associated with such loss (Black, 2010, p. 153).

What we begin to encounter, then, when we interact with the Indigenous jurisprudential tales described by Black, is a mosaic of laws and woven legal relationships already existing in place. That is, we encounter legal practices that pattern a relationality into the soil and the cosmos, offering a different legal life to that of the international legal order, be it in regard to territorial sovereignty, criminal responsibility, or legal subjectivity. In drawing attention to Black's jurisprudential accounts here, I am not attempting to assert a romanticised dualism that envisions Indigenous legal orders as 'nature' from the viewpoint of western law. Nor am I suggesting that international jurists should 'incorporate' whatever they like into international law, as if legal knowledge and practice can simply be extracted from its place to save international law from its anthropocentric oblivion. Rather, I seek to draw attention to the presence of other legal orders, legal orders that are often rendered invisible in the leap towards a more universal international law; legal orders whose full recognition in a place such as Australia could constitute precisely the kind of practices needed to tackle the ecological uncertainties of the present.

And it is perhaps also with an attentiveness to this plurality of legal orders that we can begin to sense what it would mean to invest less in this international legal order. In *As We Have Always Done*, Leanne Betasamosake Simpson offers a powerful account on Indigenous resurgence that, amongst other things, introduces us to a Nishnaabeg internationalism (Simpson, 2017). Detailing an Indigenous internationalism that is grounded in place, attentive to cosmological encounters, and deeply relational, Betasamosake Simpson describes a legal internationalism that operates not simply between First Nations but between human and more-than-human worlds.

Weaved together with her account on the diplomatic and legal procedures between human nations, she tells a story about a treaty making process between the Nishnaabeg and the Hoof Clan – the deer that roam Nishnaabeg territory. Forged in a time when the deer left Nishnaabeg territory because of disrespect to their clan and their habitat by the Nishnaabeg, the treaty ensures mutual coexistence based on care, reciprocity, and respect between the Nishnaabeg and the Deer nation. Acknowledging the agency of the deer, the treaty includes a set of mutual obligations as cohabitants in the same territory. As Betasamosake Simpson explains, the 'Deer clan, or nation, in this story has power, agency, and influence. They have knowledge that is now shared and encoded in the ethics and practices of hunting deer for the Nishnaabeg' (Simpson, 2017, p. 61).

Maintained by ceremonial and political practices that impose obligations to honour the Deer nation and care for their habitat, the treaty thus instils an inter-species legal obligation that binds them, as a matter of law, through the land. Instead of bringing the ecological world under the vanguard of a universal law, Nishnaabeg internationalism appears to reckon with the mutual obligations situated at the heart of the international relations that we have – be it with each other or the more-than-human worlds. As Betasamosake Simpson explains, 'Indigenous internationalism isn't just between peoples' (Simpson, 2017, p. 58). Rather, it includes 'radiating relationships with plant nations, animal nations, insects, bodies of water, air, soil, and spiritual beings in addition to the Indigenous nations with whom we share parts of our territory' (Simpson, 2017, p. 58).

What seems to emerge in this story is thus a radically different understanding of international legal relations; an account that reminds us that the Westphalian international legal order and its relationship with the land are far from the only possible international law there is. That there are other internationals and laws that have practiced a legal conduct of relationality with the land long before the birth of the Westphalian system, and that they continue to do so. Perhaps, then, a reading of Betasamosake Simpson and Black's accounts on Indigenous internationalism and jurisprudence invites us to consider if the very struggle to overcome international law's anthropocentric tendencies by demanding more of this law is to approach the wrong end of the spectrum. That is, if the struggle to address international law's anthropocentrism and connect modern humanity with the land lies not so much in expanding its universal reach so that it includes the ecological but perhaps in making international law less total. Because just like the local interlocutors that reached for international law at the Australian Peoples' Tribunal in Mildura and Wentworth, we may risk overlooking, or render invisible, legal orders that already practice the legal life that we desire from an international crime of ecocide in the first place.

Unless we slow down and think about where we are going when we reach for more international law under an international crime of ecocide, we indeed risk foreclosing the possibility of other internationals and laws organised along a relationality with the soil that goes beyond the nation-state, capitalism, and modernity. In this sense, a jurisdictional listening invites international jurists and others alike to peel back enough layers of colonial and Eurocentric capture to consider what kind of legal life is possible under the current international legal order, and what law we demand more of. That is, an invitation to engage in further critical reflexivity about what humanity, and what relationship with the land, we invoke when we turn to the universal promises of international law, and what forms of international and law we need to sustain us through the future.

What this means to the struggle for an international crime of ecocide is not an endpoint. It would be problematic to foreclose the struggle for an international crime of ecocide considering that few of us have the privilege to choose the tools we use as we struggle against ecocide and for survival. However, it poses a critical invitation to slowness; an invitation to suspend the speed at which we travel towards the universal and global promises of international law, or at least an invitation to be explicit about the tension in our turn to this universality. As we address the anthropocentrism of international law and rush to find ways to survive the future with law, we would thus do well in thinking carefully about where we stand and where we are going. To think about how we, as international jurists, take responsibility for international law. And to consider how international law might be decelerated in a manner that allows us to recognise other legal orders and pattern our legal life with attentiveness to the ground (Dorsett and McVeigh, 2012). A patterning of legal life that takes the soil as the site of grounding, understanding both where we stand and our interconnectivity with the legal worlds around us.

Conclusion

Leaning against Salman Rushdie's novel *The Ground Beneath Her Feet* to think about the changing grounds of 'third world' feminisms, Vasuki Nesiah has argued that in finding "'ground on which to make our stand'" we yet also 'struggle with whether "we are mostly given that territory"' in the first place (Nesiah, 2003, p. 30). This tension between where we stand and how we think about the 'ground beneath our feet', she argues, imposes 'seismic tremors' that may dislocate the political and legal grounds completely, allowing us to take up new territories for the political projects of hope we pursue (Nesiah, 2003, p. 38). As such, 'even the fractured and shifting ground beneath our feet evinces cracks of opportunity and hope' (Nesiah, 2003, p. 38).

In this chapter, I have slowed down our reading of the struggle for an international crime of ecocide in order to think both about the 'ground beneath our feet' and the 'new territories we may take up' as we struggle for better futures through the international and law. Both materially and metaphorically, I have argued, those of us that reach for international law and its universality in order to transcend the discipline's anthropocentrism and connect modern humanity with the land may do well in thinking carefully about where we stand and where we are going. To consider what territories we render transcendental when we leap towards more of the same universalising and totalising international law. In so doing, I have suggested that we might want to act with a sense of slowness before we move, counteracting the accelerating speed at which western modernity tends to rip apart the very ecological conditions of life on the ground (Barr, 2016). As a way of interrogating the ground beneath our feet both intellectually and materially, slowness offers a way for us to listen to other laws and think differently about our place. It allows us to imagine new groundings from which we can make our stand.

In this sense, this piece should not be read as a nihilist rejection of international law, nor a demand to simply transcend law for better politics under an assumption that all law is violence. Quite the contrary: it is an invitation to think differently about the internationals and the laws we live with. Listening before speaking more international law into being offers a respite from which we might see and think differently all together; to think more carefully about the fact that there has, as Irene Watson and other Indigenous scholars make clear, always been multiple internationals and laws and that we live under legal relationalities with the soil (Watson, 2018b). Be it a cultivation of the *inter gentes* of international law as 'a law *between* peoples and individuals' that Frédéric Mégret speaks to or the decolonial horizons of Joseph Slaughter's reading of international law as an interlanguage, the pause invites us to rethink what territories we want to take up and how we might relate to those territories as we travel spaces of law (Mégret, 2016, p. 3; Slaughter, 2019). To dream about and take up other legal grounds and encounters alongside other internationals and laws as we make and remake our stand again and again.

Notes

1   I express my gratitude to the editors, Vincent Chapaux, Frédéric Mégret, and Usha Natarajan, for their unstinting generosity in preparing this chapter. Thanks to Sundhya Pahuja, Shaun McVeigh, Kathleen Birrell, and Julia Dehm for the several conversations and shared reading seminars that have shaped the ideas in this piece. I also thank Danish Sheikh, Eliana Curato, and Emily Jones for reading with encouragements, including Daniel Ricardo Quiroga-Villamarín and other participants of Lucern Doctoral Forum in Law & Humanities for thoughtful commentary.
2   This is not a commentary on all interlocutors, nor on the rich work of the APT. For example, some Barkandji people attended and detailed how the *Barka* patterns a fabric of conduct. My engagement with this was too shallow, however, to adequately engage. After the completion of this chapter, the APT published comprehensive recommendations that touched on ecocide and Indigenous law. Available at: https://tribunal.org.au/sessions/2019-barka-darling-inquiry/report-and-recommendations/.
3   The call for an international crime of ecocide is rapidly developing, and some recent developments are either only briefly commented on or not included in these reflections.
4   See Stop Ecocide International, 'Making Ecocide a Crime' [Online]. Available at: https://www.stopecocide.earth/making-ecocide-a-crime.
5   See Stop Ecocide International, 'Who we are' [Online]. Available at: https://www.stopecocide.earth/who-we-are-.

6 See Stop Ecocide International, 'Leading states' [Online]. Available at: https://www.stopecocide.earth/leading-states.

References

Adelman, S. (2015) 'Epistemologies of Mastery', in Grear, A. and Kotzé, L.J. (eds.) *Research Handbook on Human Rights and the Environment*. Cheltenham: Edward Elgar, pp. 9–27.

Anghie, A. (2004) *Imperialism, Sovereignty and the Making of International Law*. Cambridge: Cambridge University Press.

Barr, O. (2016) *A Jurisprudence of Movement: Common Law, Walking, Unsettling Place*. London: Routledge.

Baskin, J. (2019) 'Global Justice and the Anthropocene: Reproducing a Development Story', in Biermann, F. and Lövbrand, E. (eds.) *Anthropocene Encounters: New Directions in Green Political Thinking*. Cambridge: Cambridge University Press, pp. 150–168.

Birrell, K. and Dehm, J. (2020) 'International Law & the Humanities in the "Anthropocene"', in Chalmers, S. and Pahuja, S. (eds.) *Routledge Handbook of International Law and the Humanities*. London: Routledge.

Black, C.F. (2010) *The Land is the Source of the Law*. London: Routledge.

Crook, M. and Short, D. (2014) 'Marx, Lemkin and the Genocide–Ecocide Nexus', *The International Journal of Human Rights*, 18(3), pp. 298–319. DOI: 10.1080/13642987.2014.914703

Crook, M. Short, D. and South, N. (2018) 'Ecocide, Genocide, Capitalism and Colonialism: Consequences for Indigenous Peoples and Glocal Ecosystems Environments', *Theoretical Criminology*, 22(3), pp. 298–317. DOI: 10.1177/1362480618787176

Cusato, E.T. (2017) 'Beyond Symbolism', *Journal of International Criminal Justice*, 15(3), pp. 491–507. DOI: 10.1093/jicj/mqx026.

Cusato, E.T. and Jones, E. (2022) 'The Imbroglio of Ecocide: A Political Economic Critique of the International Criminalisation of Ecological Destruction' (unpublished).

Davis, H. and Todd, Z. (2017) 'On the Importance of a Date, or Decolonizing the Anthropocene', *ACME: An International Journal for Critical Geographies*, 16(4), pp. 761–780.

DeFalco, R.C. and Mégret, F. (2019) 'The Invisibility of Race at the ICC: Lessons from the US Criminal Justice System', *London Review of International Law*, 7(1), pp. 55–87. DOI: 10.1093/lril/lrz002.

Dorsett, S. and McVeigh, S. (2012) *Jurisdiction*. London: Routledge.

Du Bois, W.E.B. (1920) *Darkwater: Voices from within the Veil*. New York: Harcourt, Brace and Company.

Eslava, L. and Pahuja, S. (2019) 'The State and International Law: A Reading from the Global South', *Humanity*, 14 May [Online]. Available at: http://humanityjournal.org/issue11-1/the-state-and-international-law-a-reading-from-the-global-south/ (Accessed: 6 October 2022).

Eslava, L., Fakhri, M. and Mesiah, V. (eds.) (2017) *Bandung, Global History, and International Law: Critical Pasts and Pending Futures*. Cambridge: Cambridge University Press.

Falk, R.A. (1973) 'Environmental Warfare and Ecocide – Facts, Appraisal, and Proposals', in Thee, M. (ed.) *Bulletin of Peace Proposals 1973*. Oslo: Tromsö: Universitersforlaget, pp. 80–96.

Fanon, F. (1961) *The Wretched of the Earth*. Penguin.

Gauger, A., Rabatel-Fernel, M.P., Kulbicki, L., Short, D. and Higgins, P. (2013) *The Ecocide Project: 'Ecocide is the Missing 5th Crime Against Peace'*. London: Human Rights Consortium.

Getachew, A. (2018) *Worldmaking after Empire*. Princeton: Princeton University Press.

Grear, A. (2015) 'Deconstructing Anthropos: A Critical Legal Reflection on "Anthropocentric" Law and Anthropocene "Humanity"', *Law and Critique*, 26(3), pp. 225–249. DOI: 10.1007/s10978-015-9161-0

Greene, A. (2019) 'The Campaign to Make Ecocide an International Crime: Quixotic Quest or Moral Imperative?', *Fordham Environmental Law Review*, 30(3), pp. 1–48.

Grevers, C. (2014) 'International Criminal Law and Individualism: An African Perspective', in Schwöbel, C. (ed.) *Critical Approaches to International Criminal Law: An Introduction*. London: Routledge.

Haraway, D. (1988) 'Situated Knowledges: The Science Question in Geminism and the Privilege of Partial Perspective', *Feminist Studies*, 14(3), pp. 575–599. DOI: 10.2307/3178066

Heller, K.J. 'Skeptical Thoughts on the Proposed Crime of "Ecocide" (That Isn't)' (2021) [Online] Avaliable at: http://opiniojuris.org/2021/06/23/skeptical-thoughts-on-the-proposed-crime-of-ecocide-that-isnt/ (Accessed: 13 October 2022).

Higgins, P. (2015) *Eradicating Ecocide: Laws and Governance to Prevent the Destruction of Our Planet*. 2nd edn. London: Shepheard-Walwyn.

Krever, T. (2013) 'International Criminal Law: An Ideology Critique', *Leiden Journal of International Law*, 26(3), pp. 701–723. DOI: 10.1017/S0922156513000307

Latour, B. (2017) *Facing Gaia: Eight Lectures on the New Climatic Regime*. Cambridge: Polity.

MacCarrick, G. and Maogoto, J. (2018) 'The Significance of the International Monsanto Tribunal's Findings with Respect to the Nascent Crime of Ecocide', *Texas Environmental Law Journal*, 48(2), pp. 217–238.

Matthews, D. (2019) 'From Global to Anthropocenic Assemblages: Re-Thinking Territory, Authority and Rights in the New Climatic Regime', *The Modern Law Review*, 82(4), pp. 665–691. DOI: 10.1111/1468-2230.12426

McVeigh, S. (2019) 'Critical Approaches to Jurisdiction and International Law', in Allen, S., Costelloe, D., Fitzmaurice, M., Gragl, P. and Guntrip, E. (eds.) *The Oxford Handbook of Jurisdiction in International Law*. Oxford: Oxford University Press

Mégret, F. (2016) 'Foreword', *Inter Gentes*, 1(1), pp. 3–4.

Mehta, S. and Merz, P. (2015) 'Ecocide – A New Crime Against Peace?', *Environmental Law Review*, 17(1), p. 3–7. DOI: 10.1177/1461452914564730

Mickelson, K. (2014) 'The Maps of International Law Perceptions of Nature in the Classification of Territory', *Leiden Journal of International Law*, 27(3), pp. 621–639. DOI: 10.1017/S0922156514000235

Minkova, L. (2021) 'The Fifth International Crime: Reflections on the Definition of "Ecocide"', *Journal of Genocide Research*, 25(1), pp. 62–83. DOI: 10.1080/14623528.2021.1964688

Natarajan, U. and Dehm, J. (2019) 'Where Is the Environment? Locating Nature in International Law', *TWAILR: Reflections*, 30 August [Online]. Available at: https://twailr.com/where-is-the-environment -locating-nature-in-international-law/ (Accessed: 6 October 2022).

Natarajan, U. and Khoday, K. (2014) 'Locating Nature: Making and Unmaking International Law', *Leiden Journal of International Law*, 27(3), pp. 573–593. DOI: 10.1017/S0922156514000211.

Nesiah, V. (2003) 'The Ground Beneath Her Feet: "Third World" Feminisms', *Journal of International Women's Studies*, 4(3), pp. 30–38 .

Pahuja, S. (2011) *Decolonising International Law: Development, Economic Growth and the Politics of Universality*. Cambridge: Cambridge University Press.

Pahuja, S. (2013) 'Laws of Encounter: A Jurisdictional Account of International Law', *London Review of International Law*, 1(1), pp. 63–98. DOI: 10.1093/lril/lrt009

Philippopoulos-Mihalopoulos, A. (2017) 'Critical Environmental Law as Method in the Anthropocene', in Philippopoulos-Mihalopoulos, A. and Brooks, V. (eds.) *Research Methods in Environmental Law: A Handbook*. Cheltenham: Edward Elgar, pp. 131–158.

Porras, I. (2014) 'Appropriating Nature Commerce, Property, and the Commodification of Nature in the Law of Nations', *Leiden Journal of International Law*, 27(3), pp. 641–660. DOI: 10.1017/S0922156514000247

Saro-Wiwa, K. (2000) *Genocide in Nigeria: The Ogoni Tragedy*. London: Saros International Publishers.

Schwöbel, C. (2021) *Marketing Global Justice the Political Economy of International Criminal Law*. Cambridge: Cambridge University Press.

Short, D. (2016) *Redefining Genocide: Settler Colonialism, Social Death and Ecocide*. London: Zed Books.

Simpson, B.L. (2017) *As We Have Always Done Indigenous Freedom through Radical Resistance*. Minneapolis: University of Minnesota Press.

Slaughter, J.R. (2019) 'Pathetic Fallacies: Personification and the Unruly Subjects of International Law', *London Review of International Law*, 7(1), pp. 3–54. DOI: 10.1093/lril/lrz003

Storr, C. (2016) 'Islands and the South: Framing the Relationship between International Law and Environmental Crisis', *European Journal of International Law*, 27(2), pp. 519–540. DOI: 10.1093/ejil/chw026.

Watson, I. (2018a). 'Aboriginal Relationships to the Natural World: Colonial "protection" of Human Rights and the Environment', *Journal of Human Rights and the Environment*, 9(2), pp. 119–140. DOI: 10.4337/jhre.2018.02.01

Watson, I. (2018b) 'First Nations, Indigenous Peoples: Our Laws Have Always Been Here', in Watson, I. (ed.) *Indigenous Peoples as Subjects of International Law*. New York: Routledge, pp. 96–119.

'What Is Ecocide?' (2019) [Online]. Available at: https://ecocidelaw.com/ecocide-law-2/ (Accessed: 11 June 2020).

Zierler, D. (2011) *The Invention of Ecocide: Agent Orange, Vietnam, and the Scientists Who Changed the Way We Think about the Environment*. Athens, Georgia: University Georgia Press.

17

FORMLESS INFINITE

Law beyond the Anthropocene and the Earth System

Elena Cirkovic[1]

Introduction

The rise of the commercial space industry has potential to endanger the preservation of outer space environment(s), such as the Earth's orbits, or surfaces of celestial bodies. This chapter explores long-term sustainability issues related to increasing activities in the Earth's orbital space and its interconnectedness with the Earth System (Magalhães et al. 2016). The term 'Earth System' refers to the Earth's interacting physical, chemical, and biological processes. An orbit is a regular, repeating path that one object in space takes around another one. An object in an orbit is called a satellite. A satellite can be natural, like Earth or the moon. Many planets have moons that orbit them. A satellite can also be man-made, like the International Space Station. The International Telecommunication Union (ITU) Radio Regulations define an orbit as 'the path, relative to a specified frame of reference, described by the centre of mass of a satellite or other object in space subjected primarily to natural forces, mainly the force of gravity' (International Telecommunication Union, 2020, s. 1.184). Planets, comets, asteroids and other objects in the solar system orbit the sun. Most of the objects orbiting the sun move along or close to an imaginary flat surface called the ecliptic plane. The proposed definitions of the boundary between the Earth's atmosphere and outer space, are unclear, considering orbital and suborbital trajectories used by space vehicles. Recent investigation of the inner edge of outer space from historical, physical, and technological viewpoints proposes 80 km as a more appropriate boundary than the currently popular 100 km Von Kármán line (McDowell, 2018). Planets, comets, asteroids, and other objects in the solar system orbit the sun. Most of the objects orbiting the sun move along or close to an imaginary flat surface. This imaginary surface is called the 'ecliptic plane'.[2] 'Celestial bodies' are masses of natural matter located at or above an altitude of 100 km above the Earth's sea level. This term includes primarily planets, natural satellites, comets, stars, and asteroids. 'Space debris' is defined as 'all non-functional, artificial objects, including fragments and elements thereof, in Earth's orbit or re-entering into Earth's atmosphere. Human-made space debris dominates over the natural meteoroid environment, except around millimetre sizes'.[3]

An exponential growth in the debris population in Earth's orbit creates environmental problems and poses a potential threat to the continues space access. For instance, space technology has been instrumental for Earth System monitoring it provides information for half of the 50 Essential Climate Variables (ECVs) (Hakimdavar *et al.*, 2018). Different orbital characteristics are

DOI: 10.4324/9781003201120-21

better suited for certain applications. Specific features and technical specifications of orbits are beyond the scope of this chapter.

Adequately addressing orbital environmental challenges requires a coordinated international and transdisciplinary approach to the evolving global regulatory framework. The chapter employs the Earth System Law (Cadman, Hulbert, and Simonelli, 2022) and complex systems (CS) approaches (Folke et al., 2005; Cosens, 2017; Wilson and Massimiliano, 2023) for the rethinking of international law of outer space. In addition to these approaches there is a need to advance a new thinking for the protection of Earth System and interplanetary environments. The ongoing anthropogenic environmental degradation in outer space is unpredictable and disruptive. A new understanding of the cosmic and post-anthropocentric politico-juridical space—or the 'cosmolegal'—might provide an alternative framework. The 'cosmolegal' argument responds to the 'cosmic' need to reorient the anthropogenic self-understanding of humanity's law because humanity has been approaching the 'infinity' of space with a legal consciousnesses firmly rooted in terrestrial human values.

The chapter is divided into two broad sections: the first section elaborates on the proposed cosmolegal approach in reference to the CS and post-human approaches to the Earth System and outer space; the second section provides an analysis of international law's legal architecture for outer space, and environmental problems in the Earth's orbit.

The Cosmolegal

Anthropocentric thinking has been dependant historically on the human self-understanding of its subjective positioning, as well as cosmic imaginations. As this thinking changed and evolved over time, the human mind came to understand that Earth and its various systems, is the third planet orbiting the Sun. Earth's axis of rotation is tilted 23.4 degrees with respect to the plane of Earth's orbit around the Sun. This tilt causes our yearly cycle of seasons. During part of the year, the northern hemisphere is tilted towards the Sun, and the southern hemisphere is tilted away. With the Sun higher in the sky, solar heating is greater in the north producing summer there. Less direct solar heating produces winter in the south. Six months later, the situation is reversed. When spring and fall begin, both hemispheres receive roughly equal amounts of heat from the Sun. Earth is the only planet that has a single moon. The Moon stabilises the Earth's wobble, which has made the climate less variable over thousands of years. Earth sometimes temporarily hosts orbiting asteroids or large rocks. They are typically trapped by Earth's gravity for a few months or years before returning to an orbit around the Sun. Some asteroids will be in a long 'dance' with Earth as they both orbit the Sun. Some moons are bits of rock that were captured by a planet's gravity, but Earth's Moon is likely the result of a collision billions of years ago. When Earth was a young planet, a large chunk of rock smashed into it, displacing a portion of Earth's interior. The resulting chunks apparently clumped together and formed its Moon.

This 'Introduction 101' can be found on NASA's website with other elaborations and explanations for the broader public. As this chapter primarily employs a social science perspective, the author shall not dare venture further beyond her expertise. However, this very basic understanding of where she lives as a human being is essential in the self-understanding and an elaboration of a legal thinking that needs to remove itself from the anthropocentric self-positioning, which, arguably, has been proposed to the Western history of thought since the Copernican principle. Nevertheless, it appears necessary to continue the argument that 'man' and eventually, or hopefully, 'human' is not the centre of the universe or the Earth System.

The treatment of this subject in law has also received various iterations, and especially in the natural-positivist divide. This chapter will not revive that history, as its focus is on the current relationship between mainstream international legal practice and both the Earth System and ongoing human activities in the so-called 'outer space'.

Outer space continues to evoke human imaginaries concerning new worlds to be discovered and settled, adventure, and futurism (Cirkovic, 2022). This approach, as some have argued, stands in contrast to the 'global commons' approach to outer space and to other domains, such as the deep seas (Feichtner and Ranganathan, 2019)

Notably, scientific, technological, and futuristic imagination and innovation do not have to carry negative connotations, and their overarching objectives can differ, reflecting the drives of scientific research, human (animal) curiosity, and the expansion of knowledge, to uses of 'space for Earth' or even 'space for space' in the context of potential settlements and other infrastructure. Thus far, however, these uses have been framed in economic terms (Weinzierl and Sarang, 2021). For instance the European Space Agency (ESA) is starting to rely on titles such as 'ESA's Commercialisation Gateway' with the stated goal 'to make Europe a hub for companies with global ambitions in the space sector — and the ideas and drive to change the world for the better'.[4] ESA website identifies the new 'market trend', commercial space opportunities: the case of environmental monitoring. These uses focus primarily on commercial profit, and space becomes a vast vacuum that can be used for anything: from the disposal of nuclear waste to the exploitation of bountiful energy supplies that are available in space.

In the face of this overwhelmingly 'commercial space age', the cosmolegal asks for the dislocation of international law from its distinctly local and territorial inter-national status and 'into the cosmos'. While such a direction of travel seems like a science fiction novel or proposal, human activities in space are ongoing and intensifying. What once seemed to be an imagined future is now a growing reality. The cosmolegal addresses the 'commercial space age' with a normative proposal that seeks to extend ecological concerns to outer space and recognise Earth System's (and therefore, human) embeddedness in space.

International law's anthropocentric concern with global environments have delineated inner and outer environments, where the 'outer' environments consist of the spaces beyond the atmosphere and beneath the lithosphere. This brackets what tends to count as the human environment to the space between the surface of the Earth and the limits of our atmosphere (Olson and Messeri, 2015). No territorial claims to any space-time of outer space had been advanced prior to the 1967 Outer Space Treaty (OST) (Treaty on Principles Governing the Activities of States in the Exploration and Use of Outer Space, including the Moon and Other Celestial Bodies 1967); and, unlike the oceans, which are still framed in terms of 'drawing lines in water' (Jones, 2016), in outer space, there is no difference between the 'territorial' and the 'international', neither are there 'exclusive economic zones' nor 'protected areas'. And unlike the atmosphere, in outer space there are no exclusive national 'airspaces' and therefore no 'no-fly zones'. Brandau (2015) described the early satellites and their uses as 'Weapons practically "mapped" the void long before notions of outer space as a battleground were seriously discussed outside of fiction, and before the spatiality of outer space was acknowledged as something very unlike anything on Earth' (p. 248).

Importantly, much of what we discuss as 'space activities' is taking place in the orbit. Theoretically, there is an infinite number of orbits. However, the number of useful orbits is finite. The use of any one orbit reduces the usability of other orbits that may intersect of have close approaches to the occupied orbit (Blount, 2023). In the 1960s, the international community agreed to regulate the assignment of slots in the geostationary orbit (GEO) belt through the ITU to manage the use of the limited spectrum available. Today, any company or nation planning to launch a satellite to GEO must apply to the ITU for an orbital slot. It is beyond the scope of this analysis to engage with various domestic regulations of orbital usage. For instance, orbital assignments for low Earth orbit (LEO), medium Earth orbit (MEO), and highly eccentric orbits (HEOs) have been addressed by some domestic regulatory bodies, such as the Federal Communications Commission (FCC) in the United States. FCC manages and allocate spectrum for satellites in these orbits. It is important to mention that GEO is of particular importance because of its specifications. As ESA website explains:

> Satellites in geostationary orbit (GEO) circle Earth above the equator from west to east following Earth's rotation – taking 23 hours 56 minutes and 4 seconds – by travelling at exactly the same rate as Earth. This makes satellites in GEO appear to be 'stationary' over a fixed position. […] GEO is used by satellites that need to stay constantly above one particular place over Earth, such as telecommunication satellites. This way, an antenna on Earth can be fixed to always stay pointed towards that satellite without moving. It can also be used by weather monitoring satellites, because they can continually observe specific areas to see how weather trends emerge there. Satellites in GEO cover a large range of Earth so as few as three equally spaced satellites can provide near global coverage. This is because when a satellite is this far from Earth, it can cover large sections at once. This is akin to being able to see more of a map from a metre away compared with if you were a centimetre from it. So, to see all of Earth at once from GEO far fewer satellites are needed than at a lower altitude.
>
> *(ESA, 2020)*

These specifications mean that GEO does get treated as a 'resource' (Blount, 2023) with potential user allocation.

The approach proposed in this chapter and elsewhere (Cirkovic, 2021a; 2021b; 2022a; 2022b) does not imply that outer space resources, including the orbit, should have no human use, or that space technology only causes contamination of Earth's orbits. Rather, the question it addresses is: what type of socio-technical and environmental developments are a priority for the governmental and commercial space sectors? In the broadest sense, the environment of outer space forces humans to grapple with the questions of anthropocentrism and infinity. The proposed term 'cosmolegal' merges cosmology—a branch of astronomy that involves the scientific study of the large-scale properties of the universe as a whole—with law, to develop a different understanding of politico-juridical space, which is ordinarily Earth- and human-centric (Cirkovic, 2021a; 2022b). Cosmolegality attempts to construct an inclination in thinking towards this infinity without seeking to re-establish a totality.

In operating this displacement and shift of focus, the cosmolegal also counters the international aspect (the reduction of international law to its cartographic and terrestrial history and to the regulator of matters between states) of current global legal architectures. Incursions into outer space, combined with ongoing challenges in the Earth System, such as global warming, means that there is already a 'cosmic', and/or complex, unpredictable, and planetary scope to issues that are an inextricable part of human life. In contradistinction then, to the current state-centric assumptions of international law, the proposal for a cosmolegal approach to law making necessitates participation of a far wider range of actors, requiring more inclusive and transdisciplinary processes that do not rely primarily on state-level decision-making.

The cosmolegal proposal relies, in part, on the conceptualisation of an agency that extends beyond the self-reflexivity and intentionality of humans to include all non-humans or what/who humans have considered to be non-human (Frank, 2015; Cirkovic, 2021). One consequence of this viewpoint is that the cosmolegal does not attempt to provide a complete picture of world dynamics. Rather, it leaves a space of disjunction for future contingencies and uncertainties on Earth and in outer space, such as the unpredictable consequences of human activities there. The cosmolegal responds to the fact that reality is dynamic and unpredictable, as is the unknowable (Hamzić, 2019).

Detailed study of the meaning of legal interpretation and/or enforceability of international law is beyond the scope of this chapter. It is difficult (likely impossible) to capture all the unknowns of the cosmos in anthropocentric, anthropogenic international law, which emerges from specific forms of self-organisation of the human species. The cosmolegal thus recognises the particularity of international law and characterises it as a highly specific system that forms just one part of Complex-Adaptive Systems (CAS) and, even more broadly, as interrelating with Complex-Physical Systems (CPS).

CPS are systems that have fixed components and rules of operation but can still have unexpected behaviours that in turn can create new rules of operation and even new systems (Holland, 2014) (For instance, a climate system comprising of natural elements and operating under laws of atmospheric chemistry and physics can produce unexpected weather patterns/systems). CAS agents can learn and adapt their behaviour to the behaviour or action of other agents in the system (McPherson *et al.*, 2021; Cosens *et al.*, 2017). The behaviour of the system, as a whole, can be arbitrarily rich and complex. Both CS and the anthropogenic regime actors (in this case, international law and lawmakers) can recalibrate their practice. In such a process, first, the dominant practice becomes undermined because of societal or environmental changes (e.g., exponential growth of space debris; rapid technological innovation; more human actors in outer space; climate change), and the novel practice eventually either fills the vacuum or emerges as an entirely novel thread of thinking and practice.

This is important in addressing the question of how international law might extend beyond the orbital and towards the cosmic, which requires a different understanding and discussion of spacetime.[5] This does not imply that 'planets' or 'asteroids' would have the agency of humans—after all, agency does not require cognition (Bennet, 2010). The cosmolegal challenges the current construction of distinctions and disparate attributes of the world—including the construction of agency. Or as Grear argues,

> inspired by new imaginaries of lively, porous embodiment and of Earthly materialities expressing their own incipiencies and tendencies, the onto-epistemic frame opens expansively. Into the direct purview of questions of justice, injustice and ethics come the jostling propensities and capacities of organic and inorganic forms of liveliness.
>
> *(Grear, 2020, p. 360)*

The apparent fracturing and subject–object division in human understanding, practice, and regulation of the planetary, orbital, and interplanetary domains stems from a reductive human understanding thereof. Accordingly, the cosmolegal, instead of being the mirror of a permanently split human subjectivity, would recognise and respond to the indeterminate nature of the world 'beyond' law, embracing the actuality of a materiality that is not defined by strict binaries of subject–object and human–non-human. The cosmolegal hypothesis is that law's current presentation as performing and enacting interventions based on reasons in mental operations, or on incentives in utility calculations, or on moral decisions, is still too anthropocentric. Cosmolegality requires a different ontological approach (Burton *et al.*, 2012) and recognises the possibility of many co-present variables.

The 'agency' of the non-human could also determine its capacity to impact the human. Opening possibilities for the rethinking of the world of law making in this direction is necessary because, as noted above, new activities (such as the mining of asteroids and celestial bodies), which continue to extend the borders of current international law into the cosmos, are becoming possible—along with the concomitant risks of unintended or unexpected reactions of various other-than-human environments. Is it safe for humans to settle Mars? Will humans bring other species to extra-terrestrial settlements? What are the ethical implications of this? Again, the objective of cosmolegality is not to argue against the possibility of human species moving to other planets. Rather, it insists that human self-perception and relevant anthropogenic laws of societal organisation will have to reckon with more than just human interests. The space environment is always reacting, and this is already evident in the congestion and increased risk of activities in Earth's orbits. The question is what is overlooked when space is constructed as a passive empty receptacle of human agency and its products? The answer is partially given in Grear's engagement with Harraway's sympoiesis, as she argues 'Both cognitively and affectively the imaginary of embodiment becomes a space of holobiont relationalities—even the human body is not straightforwardly "human" in the sense that the human body is also in large part a complex, shifting and contingent community of microbiota, viruses and other tiny non-human holobionts. What might sympoietic normativities look like? First, they would not look, I suspect, like ambitious and "overarching" aspirations for Earth System law/governance in the Anthropocene' (Grear, 2020, p. 361).

The development of future-oriented space norms requires strong transdisciplinary cooperation. Even if the learning curves are steep, they are necessary for a step beyond existing socio-technical imaginaries (Jasanoff and Kim, 2015).

Plural Knowledges and Space Narratives

With respect to the current state of the art, the cosmolegal emerges from the 'law-beyond the human' and 'posthuman' approaches, as well as from CS thinking (Cirkovic, 2022). The cosmolegal research horizon builds on the hypothesis of profound discontinuity. An interrelation of various subsystems in one CS model cannot easily be addressed by conventional law and policy making. Complex models interrelates the environment, social impact, human decision-making, and technology as domains with complex interactions, among many other dynamics, known and unknown (Cirkovic, Rathnasabapathy, and Wood, forthcoming). Traditional disciplinary approaches are accustomed to addressing each 'area' individually in detail. Yet by considering certain groupings (such as the economic valuations that combine technology and societal impact), and capturing them together, it is possible to overcome important challenges that lie at the intersections of these domains. For cosmolegality, such complexity thinking and 'trans-sectional-disciplinary' approaches do not refer merely to different disciplines but also to what humans came to identify as domains of the Earth System, or outer space.

Importantly, for cosmolegality, scholarship addressing agencies of the non-human cannot ignore other ontologies, which include Indigenous knowledge—and not merely as a 'cultural' concession.

Indigenous peoples' traditional knowledge has become a reference in various reports on SDGs and the Earth System. However, researchers often re-essentialise or romanticise traditional ecological knowledge (TEK) without an acknowledgement that 'Indigenous peoples' are not a monolithic constant: for instance, terminology such as 'Indigegogy' is largely situated in North America (Indigegogy is a term coined by Cree Elders Peggy and Stan Wilson (Wilson and Wilson, 2015) to frame Indigenous knowledge, literature, and scholarship and decolonial Indigenous practice). Importantly, in the context of outer space, institutions such as NASA have adopted initiatives including NASA's Indigenous People's Capacity Building Initiative (which assists Indigenous communities' use of satellite remote sensing for natural and cultural resource management[6]), in the dominant and ongoing debates on the future of outer space law, in which Indigenous voices have not previously been included. The dominant discourse amounts to, in effect, a discourse of 'outer space colonialism': Mars, the Moon, and other planets are somehow assumed to be inherently available for the use of ('advanced') humanity.

The 'colonising' of outer space evokes the frontier-focused (colonial) history of international law and the outdated construct of the 'openness' of empty spaces or *res nullius* (Cirkovic, 2007). While colonialism as such has been outlawed in modern international law, imperial semantics persist in outer space law (Casumbal-Salazar, 2017). In response to the dangers of space coloniality, some astronomers have warned against the use of colonial frameworks in relation to other planets (Mandelbaum, 2018; Prescod-Weinstein *et al.*, 2020). More recently, aerospace engineers (Rathnasabapathy *et al.*, 2020) and space environmentalists have also been proposing and enacting decolonial transdisciplinary projects in the academic context.[7] However, while collaborations between space agencies, such as NASA, aim to include Indigenous knowledges, the marginalised status of Indigenous peoples as non-state actors in international law has ongoing implications at the level of inter-governmental decision-making.

The rhetoric of the right to space highlighted by some corners of the outer space sector reflects not only anthropocentric assumptions, but also specific values of *imperium* and *commercium*. Outer space is imagined as a limitless resource—a space frontier—and this metaphor of the frontier, with its associated images of pioneering, homesteading, claim-staking, and taming, has been persistent in the history of international law and colonialism.

However, outer space, or the cosmos, has been part of CS approaches. In Indigenous knowledges,[8] the cosmos is intimately connected to the overall belief system: the lived cosmovision. Importantly, however, this article cannot properly engage with or represent various Indigenous knowledges and sciences; that work needs to be done by Indigenous scholars and practitioners who also have familiarity with that knowledge.

More simply, anthropogenic and anthropocentric law is designed to regulate human behaviour and is built on assumptions about how human beings behave and demonstrate agency, while non-human 'objects' are not recognised as having agency. Yet, to the extent that non-human beings, things, or processes are not under total human control and act in unpredictable ways, with generative agency of their own, old assumptions about the uniqueness of human agency can no longer apply. The implications of this are more radical than the implications of approaches such as 'rights of nature', as rights discourses imply that some right has to be claimed, fought for, and granted by a human agent. Taking seriously the extent to which non-humans have agency means that the terms of reference for law have to change. Bringing this argument into contact with the multiple complexities presented by outer space and by the agentic significance of planetary and extraplanetary forces leads to the need for something like the cosmolegal approach.

The cosmolegal requires the 'undoing' of current boundaries between the natural and social sciences. As STS scholars have argued, scientists are situated in social settings that influence knowledge production concerning 'nature' (Jasanoff, 2016). A reference to physics provides just one example of the necessary dissolution of currently assumed disciplinary boundaries. Richard Feynman, for example, has argued that some physics theories are difficult for humans to think about and to accept because they do not conform to 'common sense' and reasonableness. The theory of quantum electrodynamics, for instance, describes nature as *absurd* (Feynman, 1985, p. 10), and perhaps unsurprisingly, the application of quantum mechanics beyond physics has thoroughly challenged the nature–culture divide in the social sciences. Unknowability, multiplicity, and divergence would be central to cosmolegality's methodological and epistemological commitments. The natural (or social) sciences cannot assume a God-like view (Jasanoff, 2016), whether it is cantered in a sovereign or the assumption of human nature and its 'rational economic' proclivity for innovation and/or commerce. The cosmolegal operationalises this affirmation and is aligned with increasing scholarly interest in critical approaches to anthropocentrism arising from the circumstances of climate change and environmental degradation more generally, which demands reflexivity in relationships between human and non-human. Therefore, the transdisciplinarity called for here is not an assumption that other disciplines or sciences are more 'practical'. There is already a dominant narrative of 'practical' works that 'get things done' emanating from forms of management in governmental, military, or commercial sectors. Rather, such transdisciplinarity requires a self-reflectivity and the displacement of specific interests identified as somehow being crucial to human nature: ownership, exploitation, utility.

Concerning the relationships between outer space, transdisciplinarity, and, specifically, Indigenous knowledges, cosmolegality does not assume that 'Indigenous' peoples or knowledges have no interests in commercial activities. Such assumptions have been complicit in colonial impositions, which have often sought to essentialise Indigenous knowledge, and to produce a frozen rights approach to Indigenous practices to prevent economic competition with settler communities.[9] The theme of Indigenous ontologies and ecology is incredibly significant in the current calls for and development of environmental regulation in outer space. However, Indigenous knowledge cannot be treated either as monolithic, or as a 'tool' for non-Indigenous scholars and practitioners to advance their specific arguments and/or interests. That is not why it is a reference for cosmolegality. It stands as a source of particular knowledge that is helpful in ongoing human attempts to understand the 'cosmos', as it is important to: (1) acknowledge that such knowledge predates western theoretical approaches of 'posthumanism' (Braidotti, 2013), 'new-materialisms' (e.g., Latour, 2005; Barad, 2007), and so on; (2) emphasise that this knowledge, like any other, has not been static since its inception; (3) acknowledge that these very contradictions and debates are products of colonialism.

Transdisciplinarity, here, means the decolonisation of international law, as much as it means an engagement with other disciplines and ontologies. In this sense, space is not just the new frontier, but is also a reminder of already present existential problems of humanity and ongoing studies of 'human nature'. Such questions, however, are beyond the scope of this article.

Space Environment and International Law

The atmosphere extends upwards into the lower regions of space. While carbon dioxide (CO_2) causes global warming in the lower atmosphere (below 30 km), it causes the opposite effect at higher levels within the atmosphere (global cooling) and a decrease in atmospheric density (Brown *et al.*, 2021). What does this mean in relation to global warming? The CO_2 molecules can gain energy by collisional excitation or by absorption of infrared radiation. They can also lose energy through collisions with other atmospheric molecules or emission of infrared radiation. In the lower atmosphere, the 'greenhouse effect' results from the emitted infrared radiation being quickly absorbed by other molecules due to the higher atmospheric densities. In the high atmosphere, any energy emitted is much more likely to be lost to lower altitudes or space due to the thinner atmosphere.

These density reductions have been simulated for increasing CO_2 concentrations up to an altitude of 500 km by computationally modelling the Earth's atmosphere. For reference, the International Space Station (ISS) orbits at around 400 km. Density reductions up to the year 2100 have been given for the four CO_2 concentration scenarios published by the Intergovernmental Panel on Climate Change (IPCC) (Masson-Delmotte *et al.*, 2021). A reduction in density at these altitudes means a decrease in the amount of atmospheric drag that orbiting objects experience. This increases the amount of time it takes for their orbit to decay. If the 1.5°C target is met, orbital lifetimes will be 30% longer than those in the year 2000. Even if CO_2 levels were to stabilise or even reverse, the number of orbital debris would still be expected to double. At present, more research is required for humans to understand the severity of the problem—with the impact of the sun's solar cycle also known to be an important factor in atmospheric density changes. The findings may also pose challenges for regulators and satellite operators and other companies seeking to build large constellations of thousands of satellites. The International Telecommunication Union (ITU) and national regulatory filings indicate the possibility of around 100,000 satellites in the next decade.

In addition, satellite constellations, in their number and magnitude, present a uniform threat to the 'night sky'. Some astronomers, observers, scientists, and local communities, inter alia, gained recognition as affected stakeholders (or, affected national entities)[10] as they raised the issue of radio pollution and 'light pollution' emanating from LEO and affecting their respective activities. In many instances, the workings and functioning of large-scale, government-established, run, and funded astronomical observatories have experienced errors in their astronomical observations. At the request of the United Nations Committee on the Peaceful Uses of Outer Space (COPUOS), the UN Office of Outer Space Affairs, the International Astronomical Union, and Spain organized a Conference on 'Dark and Quiet Skies for Science and Society' in 2021 which published a variety of initial findings and draft recommendations (UNOOSA, 2022). The report considered all sources of electromagnetic interference that currently hinder or jeopardize the clear view of the unpolluted night sky, including the recent issue of satellite constellations, have been taken into account. It highlighted measures that prevent or lessen their adverse effects. As the report emphasizes, recommendations have been formulated with a focus on both technical and economic feasibility. The report analysed all artificial interference that could have a negative impact on the visibility of the night sky. These interferences were logically grouped into three categories according to type. The first category referred to the effect caused by the artificial emission of visible light during the night, also known as ALAN (Artificial Light At Night). The second category referred to the impact that the very large number of communication satellites in LEO would have on astronomical observations. The third category referred to the interference that radio broadcasting, both by terrestrial and satellite sources, had on observations by radio telescopes. In the specific wording of the report, excessive, ubiquitous, and improperly directed illumination had a very negative impact on three main aspects: the ability of our citizens to view a pristine, starry sky; the efficiency of scientific observations of cosmic phenomena by amateur and professional astronomers; and the bio-environment, including human health. Most species of flora and fauna are highly sensitive to the natural cycle of light and dark, and artificial light at night can have negative effects on them. The majority of animals, including most invertebrates and over three-quarters of mammal species, are nocturnal. Additionally, about 40% of bird species migrate, and approximately 80% of them do so at night (Sjoberg and Muheim, 2016). These animals rely on the natural light of stars and the Milky Way to navigate. However, the increase in artificial light at night, driven by the rise of energy-efficient solid-state lighting such as LEDs, is causing problems for humans and wildlife alike. Satellite imagery has shown that the emission of artificial light is increasing by 2% annually, twice the rate of population growth. Based on this data, a model suggests that more than one-third of the world's population can no longer see the Milky Way due to skyglow (UNOOSA, 2022, p. 16).

The growing understanding of environmental impact of human activities in outer space is prompting a variety of activities and scholarship at global level. However, a careful analysis of recent reports on space debris, artificial light, and other environmental problems also demonstrates that the space sector—broadly defined and including governmental and non-governmental actors, academia, the civil society and some local communities—has yet to engage with other developments in international law, and more specifically environmental law and climate litigation. With respect to interdisciplinarity and inclusion of plural voices, including indigenous knowledges, space sector publications and reports still have very limited participation and indigenous knowledge is portrayed in broad, sensationalist, and even outdated terms (e.g., Milligan, 2015). The frontierist ethos of 'discovery' seeks to reinvent the wheel in outer space, and while specifics of extraterrestrial spacetime requires novel understanding, research, and law making, broader principles of international law could still apply. This is without recognition of existing limitations of the international legal ontology—which this edited volume seeks to address. The solution requires the usual calls for greater participation not only of a variety of stakeholders, but also ontologies and in the case of outer space law-moving the discipline away from its more Cold-War-security-militaristic origins (Cirkovic, forthcoming).

In the present context of space governance, the anthropogenic and anthropocentric novel, emergent, and potentially dominant practice does not necessarily have all the necessary societal, ecological, and even cosmic understanding. Nor does it consider all interests and possible outcomes. Phenomena such as space debris and/or climate change have impacts on multiple societal sectors and production practices and, therefore, can gradually diminish the feasibility of current practices and production models, including international law.

As appears to be a trajectory in lex specialis-lex generalis interactions, dynamics, developments, and relationships (ILC, 2006), in an event of significant new trends and understandings, there appears to be an initial resistance to diversification and reform, leading to internal splits within the discipline where new movements start do demand and then develop and implement new reforms, and where international law gains new interpretation and elaboration. The results of various efforts could be both, soft and hard law. Non-exhaustive list of considerations might include: rigorous and transparent impact assessments; transdisciplinary and pluriversal understanding of the environmental, social, cultural and economic risks of outer space activities; the precautionary principle, ecosystem approach, and the polluter pays principle implementation; consideration of an entire lifecycle of each activity and a transformation to a resource-efficient circular economy (including responsible terrestrial mining practices); transparency, inclusiveness and public consultation mechanisms in all stages of knowledge production (academic and civil society) as well as decision-making processes (law and policy), and, where relevant, respect for free, prior and informed consent (FPIC) of indigenous peoples. However, the objective and proposal in this chapter is a broader reorientation of thinking the outer space not as an additional ecosystem, but as an already and always intertwined aspect of the Earth System. In other words, the Earth System is already part of outer space.

Anthropocentrism of International Law

Analysis of state practice before the OST demonstrates that the international community considered the free use of space as involving its environmental protection (UNGA, 1962, paras. 21–22; UNGA, 1966, para. 47). After Project West Ford, the International Astronomical Union (IAU) passed a 1961 resolution noting 'the grave danger that some future space projects might seriously interfere with astronomical observations', emphasising 'that no group has the right to change the Earth's environment in any significant way without full international study and agreement' (International Astronomical Union, 1961). In 1963 the *Institut de Droit International* passed a resolution stating that space activities

> which may involve a risk of modifying the natural environment of the Earth, of any of the celestial bodies or in space in a manner liable to be prejudicial to the future of

Formless Infinite

scientific investigation and experiment, the well-being of human life, or the interests of another State, necessarily affect directly the interests of the whole international community (Institut de Droit International, 1963, para 12). The Brundtland Report (1987, ch. 10, para. 80) referred to debris regulation as 'clearly overdue'.

The eventual adoption of voluntary Space Debris Mitigation Guidelines of the Committee on the Peaceful Uses of Outer Space contributed to binding legal requirements some national legislations (UNOOSA, 2021). The United General Assembly Resolution 68/74 on *Recommendations on national legislation relevant to the peaceful exploration and use of outer space* (UNGA, 2013) identified the need to regulate all aspects of a space activity, i.e., (i) launch and placement of satellites, (ii) operation of such satellites in Earth's orbits, including their impacts; and (iii) de-orbit and end-of-mission life. The *Guidelines for the Long-Term Sustainability of Outer Space Activities of the Committee on the Peaceful Uses of Outer Space* (UN COPUOS, 2021), Guideline A.2, specifies that States consider regulations and policies that would minimize the impacts of human activities on Earth as well as in the outer space environment; and (encourage advisory input from affected national entities during the process of developing regulatory frameworks governing space activities. However, the concept of national entities remains restrictive in the context of anthropogenic activities which impact the complex outer space-Earth systems. For instance, astronomical societies and associations around the world have been presenting evidence regarding the impacts of large constellations on government-based, run, and operated observatories. Guideline A.3 further requires, amongst other stipulations, that in fulfilling the requirements under Article VI of the OST for continuing supervision, during all phases of a mission life cycle and assessment of all risks to the long-term sustainability of outer space activities associated with the space activities and to take steps to mitigate such risks to the extent feasible.

So far, these proposals do not move beyond the imagined dichotomy between Earth System. The impact (or risks) of operation of a satellite constellation are felt in the Earth System and/or in outer space environment, and in all phases of a mission life cycle. For instance, MIT's Space Enabled Research Group has explored the framework Environment-Vulnerability-Decision-Technology (EVDT) (Reid, Zang, and Wood, 2019). The framework captures the environment, social impact, human decision-making, and technology as four domains with complex interaction. And as argued elsewhere:

> space infrastructures are linked to Earthly infrastructures, and both are products of, among others, interlinked geopolitics, socio-scientific- technological progress, economic interests, all of which in turn, have contributed to environmental degradation. The role of dominant political, economic, security and other legal and policy decision-making processes in determining whether the use of such infrastructure supports socio environmentally constructive or destructive practices has yet to be determined.
>
> *(Cirkovic, Rathnasabapathy, and Wood, forthcoming)*

In that analysis, we argued for sustainable value creation in outer space in relation to corporate social responsibility and space sustainability.

However, this approach, which is meant for governmental and private sector decision-making stakeholders reaches the glass ceiling of what is possible within the current socio-technical and legal systems. The question that needs to be answered is: how do we continue the conversation with different sectors, but with the de-anthropocentric, and perhaps even cosmolegal approach to the issues at hand? As Amitav Ghosh wrote in *The Nutmeg's Curse* (2022) 'the climatic changes of our era are nothing other than the Earth's response to four centuries of terraforming' (p. 83). The shift in an understanding of not only Earth System agencies, but also broader intricacies of the 'cosmic' is necessary for an account of how non-human agencies respond to and shape human past, present, and future. Possible actions for sustainability can include the following: recognising complexity of change; human intervention in various complex systems (and their unintended/unknown results); clear normative objectives (and recognition of what the outcomes might be in each case). The interconnections between social and ecological systems are complex and operate at different spatial and temporal scales. They are characterized by feedback effects, a dynamic and rapidly changing environment, and nonlinear processes in orbit. All these factors contribute to uncertainties in assessing the consequences of decision-making, which can affect various stages of implementation, from designing monitoring programs to estimating the costs and impacts of management measures. Predicting the long-term and cumulative effects of such complexities and constant changes poses a challenge to knowledge production, especially in responding to issues such as orbital debris. Hence, there is a need to recognize, acquire knowledge, and propose more holistic and interdisciplinary approaches to complex problems, including space debris. However, decision-makers have yet to implement these approaches in practice (Cirkovic, Rathnasabapathy, and Wood, forthcoming).

Humans depend on outer space and the Earth System in myriad ways, yet both environments are undergoing rapid degradation due to human activities. The growing interest in ecosystem approaches to conservation in the Earth System has attempted to promote a value system which would recognise non-human agencies and their inherent significance beyond any immediate relevance for human survival.

Preliminary Conclusion

The above discussion is non-exhaustive and ongoing. The speed and uncertainty of anthropogenic degradation in the Earth System poses a challenge to the existing anthropocentric and anthropogenic institutions, including modern international law. The complex social-eco-logical-technological systems are slow to adapt or transform to these changes (Preiser et al., 2018). This is, in part, due to the formal state and international legal structures, formed as the juridical expression of various expressions of *imperium* and *commercium*, which are currently extending beyond what has been perceived as planetary boundaries and into outer space (Cirkovic, 2022). Some scholarly responses are arguing that emergent forms of adaptive governance appear to be particularly effective in managing complexity (Cosens *et al.*, 2017). However, despite the complexity of Earth System and outer space, the critical legal approaches and histories have already demonstrated the less complex drive of state and commercial interests. This latter scholarship now needs to extend beyond the 'global' and into the cosmic, or even 'cosmolegal'.

The objective of the cosmolegal is not to argue for a capacity for governments to work across jurisdictional and sectoral boundaries in times of rapid change. Rather, it argues for an epistemological change. In this sense, it is a normative and theoretical proposal.

Complex systems faced with disturbances, such as climate change and orbital debris, may shift into another state, and that shift may be difficult to reverse. Governance focused on optimisation of engineered infrastructure rather than considering the realities of the outer space 'environment' may lack the space to adapt, pushing systems closer to tipping points. The cosmolegal proposal is meant to address, in law, non-linear tipping points and uncertainty. Complex systems are inherently unpredictable, and as a result, the very concept of law would need to be

Formless Infinite

re-thought to not only recognise different types of uncertainty but also to develop new institutions to manage, navigate, and resolve such unknowns.

Among the plural interests of states and individual regimes there is no agreement as to what we would like to achieve. However, the discussions at UN COPUOS regarding space debris, while still very divided, point to a problem that can have repercussions of planetary proportions. The planetary and existential nature of climate change in the Earth System and the extension of environmental degradation into outer space stands perhaps as the ultimate challenge to the establishment of the managerial and technocratic focused regulatory regimes. International law remains state-centric, while at the same time referring to fluid and orbital spaces such as the seas, environmental pollution, and outer space. The state-focused approach leaves open the possibility for the prioritisation of both public and private interests. In contrast, the inclusion and/or recognition of the Earth System and Earth's orbit, as capable of re-acting to human action and not only as property, a resource, or a protectorate, would allow global law to respond in a more informed effort to the current consequences of global warming.

The epistemological challenges of climate change and outer space cannot rely on either descriptive understanding of legal fragmentation or the normative calls for different prioritisations within the existing regimes and asymmetries. Neither can it help with the new dominance and visions of formal global law. The understanding and reform of each regime or a system require normative and functional modifications and a recasting in this regard. Phenomena such as the space debris are happening outside of the existing (human) languages of 'social' systems, although they have the simultaneous capacity to affect all of them.

Instead, the proposed cosmolegality departs from general legal anthropocentric obsessions, by pushing the boundaries of legal thought to include the Earth System and outer space as capable of acting and reacting, even if in a non-sentient or cognitive manner, in relation to human actions, and affect the very existential aspect of humanity. Such a rethinking of the law requires the flexibility of disciplinary language and a re-definition of law, with also a clear intention to address its structural biases.

Notes

1 This paper was written with the help of the Arctic Avenue Grant, University of Helsinki.
2 'For Educators' section of NASA [Online]. Available at www.nasa.gov.
3 See European Space Agency (ESA), Frequently Asked Question [Online]. Available at: www.esa.int/Safety_Security/Space_Debris/FAQ_Frequently_asked_questions.
4 See 'From Idea Generation to Business Success' section of ESA Commercialization Gateway [Online]. Available at: https://commercialisation.esa.int/.
5 In physics, spacetime is any mathematical model which fuses the three dimensions of space and the one dimension of time into a single four-dimensional manifold. Spacetime diagrams can be used to visualise relativistic effects, such as why different observers perceive differently where and when events occur.
6 See *The Indigenous Peoples Pilot* [Online]. Available at: https://appliedsciences.nasa.gov/Indigenous-peoples-pilot.
7 See, for example, the Satellite Constellations 2 Workshop SATCON2 [Online]. Available at: https://noirlab.edu/satcon2.
8 For instance, the University of Calgary has a webpage dedicated to first nations sky lore in Canada. Available at: https://science.ucalgary.ca/rothney-observatory/community/first-nations-skylore.
9 In Canada, see Van der Peet case *R. v. Van der Peet* [1996]. Critics of the Van der Peet test also point out that the test situates Aboriginal cultural practices in the past. Critics argue that both the ruling and the test rely on the notion that Aboriginal cultures and traditions are static and unchanging and ignore the inherently dynamic, adaptive nature of culture. Legal experts and Aboriginal leaders have further criticised the court system for being ethnocentric and failing to apply the same criteria to non-Aboriginal

populations. To do so would mean that only pre-contact European practices, for example, would be considered integral to Euro-Canadian culture.

10 UN Committee on the Peaceful Uses of Outer Space, *Guidelines for the Long-Term Sustainability of Outer Space Activities*, 27 June 2018, UN Doc. A/AC.105/2018/CRP.20, Guideline A.2 recommends States to amend their national laws, regulations, and policies and asks that, in doing so, the views and concerns of nationally affected entities should be considered.

References

Barad, K. M. (2007) *Meeting the Universe Halfway: Quantum Physics and the Entanglement of Matter and Meaning.* Durham: Duke University Press.

Bennet, J. (2010) *Vibrant Matter: A Political Ecology of Things.* Durham: Duke University Press.

Blount, P.J. (2023) 'Legal Issues Related to Satellite Orbits', in Blount, P.J. and Hoffmann, M. (eds.) *Oxford Research Encyclopedia of Planetary Science.* Oxford: Oxford University Press.

Braidotti, R. (2013) *The Posthuman.* Cambridge: Polity Press.

Brandau, D. (2015) 'Demarcations in the Void: Early Satellites and the Making of Outer Space', *Historical Social Research*, 40(1), pp. 239–264.

Brown, M. K., Lewis, H. G., Kavanagh, A. J. and Cnossen, I. (2021) 'Climate Change in Space: Thermospheric Density Reductions in LEO and the Impact on Debris Environment', in Proceedings of the 8th European Conference on Space Debris.

Burton, I., Dube, O.P., Campbell-Lendrum, D., Davis, I., Klein, R.J.T., Linnerooth-Bayer, J., Sanghi, A. and Toth, F. (2012) 'Managing the Risks: International Level and Integration across Scales', in Field, C.B., Barros, V., Stocker, T.F., Qin, D., Dokken, D.J., Ebi, K.L., Mastrandrea, M.D., Mach, K.J., Plattner, G.-K., Allen, S.K., Tignor, M. and Midgley P.M. (eds.) *Managing the Risks of Extreme Events and Disasters to Advance Climate Change Adaptation.* Cambridge: Cambridge University Press, pp. 393–435.

Cadman, T., Hurlbert, M. and Simonelli, A.C. (eds.) (2022) *Earth System Law: Standing on the Precipice of the Anthropocene.* London: Routledge.

Casumbal-Salazar, I. (2017) 'A Fictive Kinship: Making "Modernity," "Ancient Hawaiians," and the Telescopes on Mauna Kea', *Native American and Indigenous Studies*, 4 (2), pp. 1–30.

Cirkovic, E. (2007) 'Self-Determination and Indigenous Peoples in International Law', *American Indian Law Review*, 31(2), pp. 375–399. DOI: 10.2307/20070792

Cirkovic, E. (2021a) 'The Earth System, Hydrosphere, and Outer Space: Cosmo-legal Approaches', *Völkerrechtsblog*, 9 February [Blog]. Available at: https://voelkerrechtsblog.org/the-Earth-system-hydro-sphere-and-outer-space-cosmo-legal-approaches/ (Accessed: 1 July 2021).

Cirkovic, E. (2021b) 'The Next Generation of International Law: Space, Ice, and the Cosmolegal Proposal', *German Law Journal*, 22(2), pp. 147–167. DOI: 10.1017/glj.2021.4

Cirkovic, E. (2022a) 'International Law beyond the Earth System: Orbital debris and interplanetary pollution', *Journal of Human Rights and the Environment*, 13(2), pp. 324–348. DOI: 10.4337/jhre.2022.02.01

Cirkovic, E. (2022b) 'Hyperbole', in Goodrich, P., Gandorfer, D. and Gebruers, C. (eds.) *Research Handbook in Law and Literature.* Cheltenham: Edward Elgar, pp. 296–313.

Cirkovic, E. (forthcoming) 'Weaponization' in Blount, P.J. and Hoffmann, M. (eds.) *Oxford Research Encyclopedia of Planetary Science.* Oxford: Oxford University Press.

Cirkovic, E., Rathnasabapathy, M. and Wood, D. (forthcoming), 'Promoting Sustainability Value in Earth's Orbit'.

Cosens, B.A., Craig, R.K., Hirsch, S.L., Arnold, C.A., Benson, M.H., DeCaro, D.A., Garmestani, A.S., Gosnell, H., Ruhl, J.B. and Schlager, E. (2017) 'The Role of Law in Adaptive Governance', *Ecology and Society*, 22(1), 30. DOI: 10.5751/ES-08731-220130.

ECOSOC (20 May 1968) 'Letter dated 20 May 1968 from the Permanent Representative of Sweden Addressed to the Secretary General of the United Nations', UN Doc E/4466/Add.1.

ESA (30 March 2020) 'Types of Orbits' [*Online*]. Available at: https://www.esa.int/Enabling_Support/Space_Transportation/Types_of_orbits (Accessed: 8 March 2023).

Feichtner, I. and Ranganathan, S. (2019) 'International Law and Economic Exploitation in the Global Commons: Introduction', *European Jornal of International Law*, 30(2), pp. 541–546. DOI: 10.1093/ejil/chz026

Feynman, R.P. (1985) *Quantum Electrodynamics.* London: Penguin.

Folke, C., Hahn, T., Olsson, P. and Norberg, J. (2005) 'Adaptive Governance of Social-Ecological Systems', *Annual Review of Environment and Resources*, 30, pp. 441–473. DOI: 10.1146/annurev.energy.30.050504.144511

Frank, R.M. (2015) 'Skylore of the Indigenous Peoples of Northern Eurasia', in Ruggles C. (ed.) *Handbook of Archaeoastronomy and Ethnoastronomy*. New York: Springer, pp. 1679–1686.

Grear, A. (2020) 'Legal Imaginaries and the Anthropocene: "Of" and "For"', *Law and Critique*, 31, pp. 351–366. DOI: 10.1007/s10978-020-09275-7

Ghosh, A. (2022) *The Nutmeg's Curse: Parables from a Planet in Crisis*. Chicago: University of Chicago Press.

Hakimdavar, R., Wood, D., Eylander, J., Peters-Lidard, C., Smith, J., Doorn, B., Green, D., Hummel, C. and Moore, T.C. (2018) 'Transboundary Water: Improving Methodologies and Developing Integrated Tools to Support Water Security' [Online]. Available at: https://ntrs.nasa.gov/api/citations/20180002176/downloads/20180002176.pdf (Accessed: 18 October 2022).

Hamzić, V. (2019) 'What's Left of the Real?', in Fassin, D. and Harcourt, B.E. (eds.) *A Time for Critique*. New York: Columbia University Press, pp. 132–150.

Holland, J.H. (2014) *Complexity: A Very Short Introduction*. Oxford: Oxford University Press.

ICAO (LC) (1956) ICAO Doc A10-WP/30, LE.1.

Indigenous Community Members of the Lhaka Honhat (Our Land) Association v. Argentina, Merits, Reparations and Costs, Inter-American Court of Human Rights Series C No 400 (6 February 2020).

ILC (2006) 'Fragmentation of International Law: Difficulties Arising from the Diversification and Expansion of International Law', UN Doc A/CN.4/L.682 and Add.1.

Institut de Droit International (1963) 'The Legal Regime of Outer Space', Available at: https://www.idi-iil.org/app/uploads/2017/06/1963_bru_02_en.pdf (Accessed: 18 October 2022).

International Astronomical Union (1961) Resolution No. 1, 6 Proc IAU 4.

Jasanoff, S. (2016) *The Ethics of Invention: Technology and Human Future*. New York: WW Norton & Company.

Jasanoff, S. and Kim, S.-H. (2015) *Dreamscapes of Modernity: Sociotechnical Imaginaries and the Fabrication of Power*. Chicago: University of Chicago Press.

Jones, H. (2016) 'Lines in the Ocean: Thinking with the Sea about Territory and International Law', *London Review of International Law*, 4(2), pp. 307–343. DOI: 10.1093/lril/lrw012

Latour, B. (2005) *Reassembling the Social: An Introduction to the Actor-Network Theory*. Oxford: Oxford University Press.

Lovell, B. (1968) 'The Pollution of Space', *Bulletin of the Atomic Scientists*, 24, pp. 42–45. DOI: 10.1080/00963402.1968.11457750

Magalhães, P., Steffen, W., Bosselmann, K., Aragão, A. and Soromenho-Marques, V. (eds.) (2016) *The Safe Operating Space Treaty: A New Approach to Managing Our Use of the Earth System*. New castle upon Tyne: Cambridge Scholars Publishing

Mandelbaum, R.F. (2018) 'Decolonizing Mars: Are We Thinking About Space Exploration All Wrong?', *Gizmodo*, 20 November [Online]. Available at: https://gizmodo.com/decolonizing-mars-are-we-thinking-about-space-explorat-1830348568 (Accessed: 18 October 2022).

Martin, J.D. (2017) 'Prestige Asymmetry in American Physics: Aspirations, Applications, and the Purloined Letter Effect', *Science in Context*, 30(4), pp. 475–506. DOI: 10.1017/S0269889717000242

Masson-Delmotte, V., Zhai, P., Pirani, A., Connors, S.L., Péan, C., Berger, S., Caud, N., Chen, Y., Goldfarb, L., Gomis, M.I., Huang, M., Leitzell, K., Lonnoy, E., Matthews, J.B.R., Maycock, T.K., Waterfield, T., Yelekçi, O., Yu, R. and Zhou, B. (eds.) *Climate Change 2021: The Physical Science Basis*. Cambridge: Cambridge University Press.

McDowell, J. (2018) 'The Edge of Space: Revisiting the Karman Line', *Acta Astronautica*, 151, pp. 668–677. DOI: 10.1016/j.actaastro.2018.07.003

McPherson, M.L., Finger, D.J.I., Houskeeper, H.F., Bell, T.W., Carr, M.H., Rogers-Bennett, L. and Kudela, R.M. (2021) 'Large-scale Shift in the Structure of a Kelp Forest Ecosystem Co-occurs with an Epizootic and Marine Heatwave', *Communications Biology*, 4, 298. DOI: 10.1038/s42003-021-01827-6

Milligan, T. (2015) *Nobody Owns the Moon: The Ethics of Space Exploitation*. Jefferson: McFarland Books.

Muheim, R., Sjöberg, S. and Pinzon-Rodriguez, A. (2016) 'Polarized Light Modulates Light-dependent Magnetic Compass Orientation in Birds', *Proceedings of the National Academy of Sciences*, 113(6), pp. 1654–1659.

Olson, V. and Messeri, L. (2015) 'Beyond the Anthropocene: Un-Earthing the Epoch', *Environment and Society*, 6, pp. 28–47. DOI: 10.3167/ares.2015.060103

Preiser, R., Biggs, R., De Vos, A. and Folke, C. (2018) 'Social-Ecological Systems as Complex Adaptive Systems: Organizing Principles for Advancing Research Methods and Approaches', *Ecology and Society*, 23(4), p. 46. DOI: 10.5751/ES-10558-230446

Prescod-Weinstein, C., Walkowic, L., Tuttle, S., Nord, B. and Neilson, H.R. (2020) 'Reframing Astronomical Research Through an Anticolonial Lens – for TMT and Beyond', *arXiv: Instrumentation and Methods for Astrophysics* [Online]. Available at: https://arxiv.org/abs/2001.00674 (Accessed: 18 October 2022).

Rand, L.R. (2016) *Orbital Decay: Space Junk and the Environmental History of Earth's Planetary Borderlands*. PhD thesis. University of Pennsylvania.

Ranganathan, S. (2016) 'Global Commons', *European Journal of International Law*, 27, pp. 693–717. DOI: 10.1093/ejil/chw037

Rathnasabapathy, M., Wood, D., Letizia, F., Lemmens, S., Jah, M., Schiller, A., Christensen, C., Potter, S., Khlystov, N., Soshkin, M., Acuff, K., Lifson, M. and R. Steindl (2020) 'Space Sustainability Rating: Designing a Composite Indicator to Incentivise Satellite Operators to Pursue Long-Term Sustainability of the Space Environment', in 71st International Astronautical Congress (IAC).

Reid, J., Zeng, C. and Wood, D. (2019) 'Combining Social, Environmental and Design Models to Support the Sustainable Development Goals', *IEEE Aerospace Conference, Big Sky, MT, USA, 2019*, pp. 1–13. DOI: 10.1109/AERO.2019.8741623

Sjöberg, S. and Muheim, R. (2016) 'A New View on an Old Debate: Type of Cue-Conflict Manipulation and Availability of Stars Can Explain the Discrepancies between Cue-Calibration Experiments with Migratory Songbirds', *Frontiers in Behavioral Neuroscience*, 10(29), pp. 1–12. DOI: 10.3389/fnbeh.2016.00029

Treaty on Principles Governing the Activities of States in the Exploration and Use of Outer Space, including the Moon and Other Celestial Bodies (adopted 27 January 1967, entered into force 10 October 1967) 610 UNTS 205 (Outer Space Treaty).

UNCOPUS (2021) *Guidelines for the Long-Term Sustainability of Outer Space Activities of the Committee on the Peaceful Uses of Outer Space*. Vienna: United Nations.

UNGA Res. 68/74 (11 December 2013), UN Doc. A/RES/68/74.

UNGA Summary record of the 1492nd meeting (17 December 1966) A/C.1/PV.1492.

UNGA Verbatim record of the 1293rd meeting (6 December 1962) A/C.1/PV.1293.

UNGA Verbatim record of the 1294th meeting (7 December 1962) A/C.1/PV.1294.

UNOOSA (2022) 'Dark and Quiet Skies II for Science and Society: Working Group Reports'. Available at: https://www.iau.org/static/science/scientific_bodies/working_groups/286/dark-quiet-skies-2-working-groups-reports.pdf (Accessed: 8 March 2023).

'US Space Needles Criticized by British' (1963) *Chicago Tribune*, 17 May, p. 6.

Weinzierl, M. and Sarang, M. (2021) 'The Commercial Space Age Is Here', *Harvard Business Review*, 12 February [Online]. Available at: https://hbr.org/2021/02/the-commercial-space-age-is-here (Accessed: 18 October 2022).

Wilson, A. and Vasile, M. (2023) 'Life Cycle Engineering of Space Systems: Preliminary Findings', *Advances in Space Research*. DOI: 10.1016/j.asr.2023.01.023

Wilson, P. and Wilson, S. (2015) 'Indigeogy' [Keynote Presentation at Chiefs of Ontario, Charting Our Own Path Forward Education Symposium, Thunder Bay, ON]. Available at: http://education.chiefs-of-ontario.org/es2015 (Accessed: 18 October 2022).

INDEX

1972 United Nations Conference on the Human Environment (Stockholm Conference) 84, 88, 105, 309
1982 United Nations Convention on the Law of the Sea (UNCLOS) 102, 303
1992 Rio Declaration on Environment and Development 89
1992 United Nations Framework Convention on Climate Change (UNFCCC) 157, 191, 264
1994 Agreement on Technical Barriers to Trade (TBT) 65
1994 Agreement on the Application of Sanitary and Phytosanitary Measures (SPS) 65

Absolon, K. 287
AI swarming drones 123, 132–134
AI-powered drones 132
Anand, A. P. 113
animal invisibility 252
animal welfare 71, 231–240, 242, 251–252
Anishinabe 279: Creation Story 280; Waterdrum 279, 286–289, 291; *see also* Indigenous knowledge
anthropocentric bias 2, 147
anthropocentrism: beyond speciesism 238; and ecocentrism 92, 97; global animal law and 231
Anthropos-centric legal life 315
Aquinas, T. 219, 299
Aristotle 39, 218
armed conflicts 122; in the Anthropocene 122; contemporary 124; conventional weapons in 129; ecology of 133; in the ILC 130; natural environment in 123; posthuman ecology and 127
article XX GATT 71
Australian Peoples' Tribunal (APT) 307

Bacon, F. 87, 220
Bentham, J. 222, 240
Berry, T. 219
Bkejwanong 279

celestial bodies 303, 323
cellar hatch standard 266
climate change 42, 157–158, 186–187, 264–270, 273; *see also* environmental degradation; Indigenous knowledge
coloniality of knowledge 201
commercial space industry 323
Commune de Grande-Synthe v. France 267
Complex-Adaptive Systems (CAS) 327
conventional drones 133
cosmolegal 324–325
COVID-19 73–74, 186, 198, 209, 274
crime against the environment 310
criminalisation of ecological violence 315
critical approaches to international law 6, 145
Critical Legal Studies (CLS) 145

deconstructing human–nature power 145
deer hide lacing 291
Deleuze-Guattarian 122, 128; *see also* war-machines
Descartes 86, 184, 220–221
Despret, V. 7, 296, 298, 300
dominant patriarchal social mores 183
dominionism 245–46; naming animal objects 250; slaughterhouse 248; totalitarian 234
dominium in Christianity 219

Earth Jurisprudence 227n1; scientific revolution and dualism 219
Earth orbit 323, 326

Index

'Earth system' governance paradigm (ESG) 173–177
Earth System law 328
ecocidal violence 307, 311
ecofeminism 156–157, 183; allyship 190; as an alternative to dualism 187; approach to international law 194; coalition 190; core elements of 190; critique of international law 183; dominance of dualism 184
eco-legal subject of law 271
environmental degradation 55, 87, 154, 193, 330
environmental ethics 36, 55, 86, 283
environmental justice 6, 145, 157–158
eternal Treblinka 247, 254
Eurocentrism and coloniality 167
European Convention on Human Rights 266
European Enlightenment 4, 201
European Green Deal 265
European Space Agency (ESA) 325
exploitation of nature 14, 23, 25, 29, 154

feminism: inclusivity 192; feminist approaches to international law 155; feminist theory 187; intersectionality 190; oppressions of a patriarchal system 190
fragmentation of international law 10, 237
free seas doctrine 104–105, 109, 113–114

Galston, A. W. 309
General Agreement on Tariffs and Trade (GATT) 27, 62
Geneva Conventions (1949) 124
Gillespie, A. 91
Great Ape Project 297
great chain of being 218
Grotian doctrine 104–105, 113

Hague Regulations (1907) 124
healthy environment 44, 48, 208, 216, 267
hierarchical precepts of dualism 184
Homo Economicus 221
human environment 84, 87, 194, 325
human legal protections to the animals 297
human rights 35; approaches to the environment 43; as an anthropocentric moment 37; the anthropocentrism of 45; The Committee on Economic, Social, and Cultural Rights (CESCR) 43; the European Court of Human Rights (ECtHR) 41, 267; as humanist pursuit 37; Inter-American Court of Human Rights (IACtHR) 42; international human rights law 40–41; natural rights vs. nature's rights 39; non-human animal 36, 40, 128, 238, 295; non-human world 38, 91, 94; in opposition to nature 38; register of rights beyond humans 47
human warfighters 127, 133–134
human–artificially intelligent (AI) 122, 134

Indigenous internationalism 309, 318–19
Indigenous knowledge 199; beyond international law 205; in development 204; global sustainability 204, 207; instrumentalised 199; second-level knowledge 201, 209; United Nations Declaration on the Rights of Indigenous Peoples (UNDRIP) 199; Western knowledge 201; *see also* NASA; sovereignty
Indigenous legal traditions 281; *see also* Indigenous knowledge
Indigenous movements 203, 205
Indigenous research methodology 285, 293
Indigenous women 156, 284
inhuman bodies 6, 161
intellectual property 65, 199–200, 202–203, 234; and Indigenous governance 203
International Astronomical Union 331–332
International Covenant on Civil and Political Rights 202
international crime of ecocide 307–312, 316, 319
International Criminal Court (ICC) 308
international criminal law 307
international dominion law 250
international environmental law 29, 40, 84, 94, 154, 310; anthropocentrism and the violence of 93; biopolitical critique 96–97
international humanitarian law 122, 124, 127; humanizing mission 124, 131
international law and future generations 264
international law for people and planet 193
international law of animal liberation 234, 254–255
International Telecommunication Union 323, 331
intra-human and human–nature relationships 184
intra-human justice 35

Kelch, T. 236
Kennedy, D. 147–148
Koskenniemi, M. 147–148, 151
Kotzé, L. 174–176

law of the sea 102; common heritage of mankind 106; conservation of marine resources 109; future of the 103, 113; imperial and colonial heritage 103–106; maximum sustainable 109–111; ocean space 103, 106, 109
legal anthropocentrism 165–167, 169–173, 177
liberalism 240
Lovejoy, A. O. 218

Maneesha Decka 233
Marxist and neomarxist approaches to international law 146, 151
Massacres of El Mozote and nearby places v. El Salvador 42
military: collateral damage 125; disciplinary system 134; human military commanders 133;

Index

human-military 124; military dog 126; military dolphin 127; mission 133; personnel 129; proportionality 124; submarines 127; tactics 129; technologies 127; US military 127; vehicles 129; *see also* AI swarming drones; international humanitarian law; war-machine

Mother Earth 51, 208, 281–282, 284–285, 288; Universal Declaration of the Rights of Mother Earth 49, 208

Mylius, B. 146, 151, 158, 215–217

Naakonigewin 281
NASA 324, 329
natural law 15–17, 21, 39, 245, 298
Nazis 246–249, 254
neoliberalism 150, 158, 171, 215, 222
New Approaches to International Law (NAIL) 145, 147
non-anthropocentric 46, 56, 156, 200, 206, 280
non-human: entities 47, 108, 127, 208, 270; exclusion of the non-human animal 36; modernist approaches to the 40; Other 128, 132; rights 47, 49, 50, 56
normative anthropocentrism 217, 221

Odawa 279
Ojibway 279
ontological anthropocentrism 86, 254
Outer Space Treaty (OST) 325

Pardo, A. 102–103, 107, 116
Part VII of UNCLOS 114
Platonic school 184
post-anthropocentric 46, 123–124, 270, 272–275, 296
posthuman critical approaches to law 146, 158
Pottawatomi 279
Precautionary Principle 268
primacy of the human 5, 39, 47

rights and duties of animals 297
Rio+20 Summit 90
Robinson, M. 192

satellites in geostationary orbit 326
Seals Panel Report 63
Selden, J. 105
Singer, I. B. 246
social Darwinism 300

sovereignty 5, 13, 205–207; abuse of 41; colonial peoples 26; doctrine of 14, 22; food sovereignty 200, 206–207, 209; globalising Locke 18; Indigenous 202, 209; moderate sovereignty 17; personal sovereignty 23; *post facto* 316; postcolonial lives of 24; property formation 14; sovereign obligation to cultivate the earth 18; standards of civilisation 23

Space Debris Mitigation Guidelines 333
space environment and international law 330
species inequality 241
stewardship 56, 88, 91–92, 97, 114, 314
structuralism and the environment 149; approaches to international law 147
sustainable development 28–29, 43, 68, 87, 149, 185; Sustainable Development Goals (SDGs) 174; United Nations Conference on Sustainable Development 90

tech-entities 132–133
territorialisation 301; *see also* Despret, V.
the sea as a legal person 111
Third World Approaches to International Law (TWAIL) 152
Three Fires Confederacy 279
trade law: animals as goods 65; appellate body 63; clinical separation of trade from non-trade 62; commodification of animals 62; commodification of nature 61; commodification of the living 61; denial of subjectivity 70; economic growth 73; instrumentality 67; ownership 70; status of animals 62
transnational corporate actors 311; *see also* international crime of ecocide
transnational corporation 165, 170–171

Urgenda v. the State of the Netherlands 266

war-machines 122, 124, 128, 132–135
Waterdrum 279, 287–289, 286; *see also* Anishinabe
welfarism 233, 235
Western and patriarchal mode of reasoning 156
Western modernity 4, 45–46, 201, 312, 316, 320
Western tradition 38; *see also* Anishinabe
Westphalian international legal order 319
women and nature 156, 188
World Intellectual Property Organization (WIPO) 199
World Trade Organization (WTO) 65, 199, 225

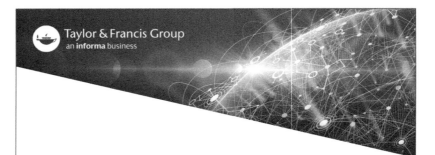